International Conference for Young Scientists, Specialists, and Postgraduates on Nuclear Reactor Physics (ICNRP-2016)

Journal of Physics: Conference Series Volume 781

Moscow, Russia
5 – 9 September 2016

Editors:

Anton Smirnov
Rynat Bahdanovich
Ekaterina Proshkina

ISBN: 978-1-5108-3684-6
ISSN: 1742-6588

Printed from e-media with permission by:

Curran Associates, Inc.
57 Morehouse Lane
Red Hook, NY 12571

Printed by Curran Associates, Inc. (2017)

For permission requests, please contact the Institute of Physics at the address below.

Institute of Physics
Dirac House, Temple Back
Bristol BS1 6BE UK

Phone: 44 1 17 929 7481
Fax: 44 1 17 920 0979

techtracking@iop.org

Additional copies of this publication are available from:

Curran Associates, Inc.
57 Morehouse Lane
Red Hook, NY 12571 USA
Phone: 845-758-0400
Fax: 845-758-2633
Email: curran@proceedings.com
Web: www.proceedings.com

Table of contents

Volume 781

International Conference for Young Scientists, Specialists, and Postgraduates on Nuclear Reactor Physics 2016 (ICNRP-2016)

5–9 September 2016, Moscow, Russian Federation

Accepted papers received: 1 December 2016
Published online: 7 February 2017

Preface

011001
OPEN ACCESS
International Conference for Young Scientists, Specialists, and Postgraduates on Nuclear Reactor Physics 2016 (ICNRP-2016)

011002
OPEN ACCESS
Peer review statement

Papers

Features of fast reactors

012001
OPEN ACCESS
Inherent Safety Characteristics of Advanced Fast Reactors

A S Bochkarev, A S. Korsun, V.S. Kharitonov and P.N. Alekseev.....1

012002
OPEN ACCESS
Radiogenic lead as coolant, reflector and moderator in advanced fast reactors

E G Kulikov.....9

012003
OPEN ACCESS
Neutronic calculation of fast reactors by the EUCLID/V1 integrated code

D A Koltashev and A A Stakhanova.....15

012004
OPEN ACCESS
The concerted calculation of the BN-600 reactor for the deterministic and stochastic codes

E V Bogdanova and A N Kuznetsov.....20

012005
OPEN ACCESS
Concerning advantages in using ^{208}Pb as such a FR coolant

G Khorasanov, E Zemskov and A Blokhin.....26

012006
OPEN ACCESS
On fundamental quality of fission chain reaction to oppose rapid runaways of nuclear reactors

G G Kulikov, A N Shmelev, V A Apse and E G Kulikov.....30

012007
OPEN ACCESS
Low-power lead-cooled fast reactor loaded with MOX-fuel

E R Sitdikov and A M Terekhova.....37

012008
OPEN ACCESS
Atomistic simulation of defects formation and structure transitions in U-Mo alloys at swift heavy ion irradiation

L N Kolotova and S V Starikov.....41

012009
OPEN ACCESS
Improved safety fast reactor with "reservoir" for delayed neutrons generating

G G Kulikov, V A Apse, A N Shmelev and E G Kulikov.....45

012010
OPEN ACCESS
Mechanistic approach for nitride fuel evolution and fission product release under irradiation

A P Dolgodvorov and V D Ozrin.....51

012011
OPEN ACCESS
Contribution of neutron-capture reactions in energy release in the fuel core of BN-600

R B Bahdanovich, V I Romanenko and G V Tikhomirov.....55

Ecological issues and prospects of closed nuclear fuel cycle

012012
OPEN ACCESS
Closed fuel cycle with increased fuel burn-up and economy applying of thorium resources

G G Kulikov and V A Apse.....62

012013
OPEN ACCESS
Radioactive waste from decommissioning of fast reactors (through the example of BN-800)

A A Rybin and O A Momot.....68

Management and non-proliferation of nuclear materials

012014
OPEN ACCESS
The remote methods for radwaste and SNF control

O Ivanov, V Stepanov, A Danilovich and V Potapov.....72

012015
OPEN ACCESS
X-ray fluorescence-based method for the quantitative determination of uranium in the aqueous solutions

S Dubrovka, S Chursin and V Verkhoturova.....80

012016
OPEN ACCESS
Nondestructive Determination of ^{235}U Enrichment of Uranium Samples in the Presence of Actinides (^{232}Th, ^{236}U, ^{237}Np)

Yu V Stogov.....85

Challenges in closed nuclear fuel cycle implementation

012017
OPEN ACCESS
The basic features of a closed fuel cycle without fast reactors

E A Bobrov, P N Alekseev and P S Teplov.....92

012018
OPEN ACCESS
Influence of uncertainties of isotopic composition of the reprocessed uranium on effectiveness of its enrichment in gas centrifuge cascades

A Yu Smirnov, A R Mustafin, V A Nevinitsa, G A Sulaberidze, A A Dudnikov and V E Gusev.....96

012019
OPEN ACCESS
The scheme for evaluation of isotopic composition of fast reactor core in closed nuclear fuel cycle

I S Saldikov, M Yu Ternovykh, P A Fomichenko and A S Gerasimov.....101

Calculation for the needs of operation and design of nuclear reactors

012020
OPEN ACCESS
On some perturbation theory singularities for reactivity effects calculations

B D Abramov.....109

012021
OPEN ACCESS
On monitoring and forecasting of graphite stack temperature in transient modes

A M Zagrebaev, N V Ovsyannikova and R N Ramazanov.....113

012022
OPEN ACCESS
Use of molybdenum as a structural material of fuel elements for improving the safety of nuclear reactors

A N Shmelev and B K Kozhahmet.....117

012023
OPEN ACCESS
Optimization of the energy complex "NPP-accumulator" in case of force majeure

G Zaluzhnaya and A Zagrebaev.....122

012024
OPEN ACCESS
Equalization of energy density in boiling water reactors (as exemplified by WB-50). Development and testing of WB -50 computational model on the basis of MCU-RR code

Yu B Chertkov, V V Disyuk, E Yu Pimenov and N V Aksenova.....127

012025
OPEN ACCESS
Validation of deterministic and Monte Carlo codes for neutronics calculation of the IRT-type research reactor

M V Shchurovskaya, V P Alferov, N I Geraskin and A I Radaev.....131

012026
OPEN ACCESS
Development of coupled neutronics/thermal–hydraulics test case for HPLWR

P Pham, I D Gamtsemlidze, R B Bahdanovich, S P Nikonov and A D Smirnov.....136

012027
OPEN ACCESS
The relationship between hydraulic resistance and heat transfer in the crossflow-line of staggered tube bundles

V N Fedoseev, A S Korsun, M I Pisarevsky and Y N Balberkina.....144

012028
OPEN ACCESS
Minimization of the energy loss of nuclear power plants in case of partial in-core monitoring system failure

A M Zagrebaev, R N Ramazanov and E A Lunegova.....148

012029
OPEN ACCESS
Generation of XS library for the reflector of VVER reactor core using Monte Carlo code Serpent

K I Usheva, S A Kuten, A A Khruschinsky and L F Babichev.....153

012030
OPEN ACCESS
Calculation evaluation of multiplying properties of LWR with thorium fuel

I V Shamanin, V M Grachev, V V Knyshev, S V Bedenko and N G Novikova.....160

012031
OPEN ACCESS
Modernization of the VVR-TS reactor core for the increasing of the radionuclides production

R V Fomin, V V Kolesov and O Yu Kochnov.....166

012032
OPEN ACCESS
Improvement of Modeling HTGR Neutron Physics by Uncertainty Analysis with the Use of Cross-Section Covariance Information

V F Boyarinov, A V Grol, P A Fomichenko and M Yu Ternovykh.....170

012033
OPEN ACCESS
Experimental analysis of liquid-metal reactor scram rod kinematic characteristics

F D Konovalenko and S I Kondrashov.....188

012034
OPEN ACCESS
Application of ATHLET/DYN3D coupled codes system for fast liquid metal cooled reactor steady state simulation

V Ivanov, A Samokhin, I Danicheva, N Khrennikov, J Bouscuet, K Velkov and I Pasichnyk.....195

012035
OPEN ACCESS
Nuclear Safety of RBMK Storage Pool under Seismic Impact

A Fedosov.....201

012036
OPEN ACCESS
Recent neutronics developments for reactor safety studies with SIMMER code at KIT

A Rineiski, M Marchetti, L Andriolo and F Gabrielli.....205

012037
OPEN ACCESS
The search of the best mode of the reserve power supply consumption during the nuclear reactor's emergency shutdown procedures in case of force majeure circumstances

A M Zagrebaev and A V Trifonenkov.....210

Computational reactor tests, codes and data bases

012038
OPEN ACCESS
Test case specifications for coupled neutronics-thermal hydraulics calculation of Gas-cooled Fast Reactor

F Osuský, R Bahdanovich, G Farkas, J Haščík and G V Tikhomirov.....215

012039
OPEN ACCESS
Non-invasive studies of multiphase flow in process equipment. Positron emission particle tracking technique

B V Balakin, T C H Adamsen, Y-F Chang, P Kosinski and A C Hoffmann.....224

012040
OPEN ACCESS
Nuclide radioactive decay data uncertainties library

D S Barabanova and G M Zherdev.....231

012041
OPEN ACCESS
Simulation of motions of the plasma in a fusion reactor for obtaining of future energy

Askhat Zhumabekov.....236

012042
OPEN ACCESS
The use of the neutronic calculation code CORNER for evaluating the protection of fast neutron reactor and CNFC equipment

M E Shekhanova.....240

012043
OPEN ACCESS
The choice of core unit cells boundaries in surface harmonics method by the test tasks solving example

A Elshin, A Pakhomov and V Riuchin.....244

012044
OPEN ACCESS
Multi-scale model for point defects behaviour in uranium mononitride

S Starikov, A Kuksin and D Smirnova.....249

012045
OPEN ACCESS
Test problem for thermal-hydraulics and neutronic coupled calculation fore ALFREAD reactor core

A Filip, G Darie, I S Saldikov, A D Smirnov and G V Tikhomirov.....254

012046
OPEN ACCESS
Preparation macroconstants to simulate the core of VVER-1000 reactor

V Y Seleznev.....261

012047
OPEN ACCESS
Specifications for a coupled neutronics thermal-hydraulics SFR test case

A Tassone, A D Smirnov and G V Tikhomirov.....268

012048
OPEN ACCESS
The optimal approach for the processes of verification and validation of NPP software and hardware complexes

S Korolev, A Tolokonsky and V Rogov.....274

012049
OPEN ACCESS
Computational simulation of thermal hydraulic processes in the model LMFBR fuel assembly

M V Bayaskhalanov, I G Merinov, A S Korsun and M N Vlasov.....280

012050
OPEN ACCESS
Test case for VVER-1000 complex modeling using MCU and ATHLET

R B Bahdanovich, E V Bogdanova, I D Gamtsemlidze, S P Nikonov and G V Tikhomirov.....284

012051
OPEN ACCESS
Non-destructive methods of control of thermo-physical properties of fuel rods

A B Kruglov, V B Kruglov, V S Kharitonov, P G Struchalin and A G Galkin.....291

012052
OPEN ACCESS
Adaptive Neural Network Algorithm for Power Control in Nuclear Power Plants

Al Masri Husam Fayiz.....295

Nuclear education & knowledge management

012053
OPEN ACCESS
Experience in implementation of «Nuclear Knowledge Management» course at the National Research Nuclear University MEPhI

N I Geraskin and A N Kosilov.....304

012054
OPEN ACCESS
Experience and advantages in implementation of educational program in network form at Department «Closed nuclear fuel cycle Technologies» of National Research Nuclear University «MEPhI»

A G Beygel', K V Kutsenko, A A Lavrukhin, E P Magomedbekov, V A Pershukov, V L Sofronov, E A Tyupina and A N Zhiganov.....309

012055
OPEN ACCESS
Computer modeling and simulators as part of university training for NPP operating personnel

M. Volman.....313

012056
OPEN ACCESS
Open web system of Virtual labs for nuclear and applied physics

I S Saldikov, V V Afanasyev, V I Petrov and M Yu Ternovykh.....317

012057
OPEN ACCESS
Effectiveness evaluation of the R&D projects in organizations financed by the budget expenses

D Yakovlev, E Yushkov, A Pryakhin and M Bogatyreova.....324

012058
OPEN ACCESS
Specifics of MS training in the area of nuclear materials safe management for new-comers in nuclear power

N I Geraskin and V B Glebov.....331

012059
OPEN ACCESS
The quantum-mechanical approach to construction of quantitative assessments of some documentary information properties (on example of nuclear knowledge)

A A Lebedev, N V Maksimov and E V Smirnova.....336

012060
OPEN ACCESS
Cognitive competence of graduates, oriented to work in the knowledge management system in the state corporation "Rosatom"

V Kireev, A Silenko and A Guseva.....342

012061
OPEN ACCESS
Modernization of physical protection educational laboratories in the National Research Nuclear University MEPhI

N I Geraskin and A A Krasnoborodko.....347

012062
OPEN ACCESS
Improvement of the course "Management of intellectual property" based on the mixed state contract in the field of scientific R&D

D Yakovlev, A Ageev, E Yushkov and M Bogatyreova.....351

012063
OPEN ACCESS
International academic program in technologies of light-water nuclear reactors. Phases of development and implementation

N I Geraskin and V B Glebov.....358

012064
OPEN ACCESS
Application of the Institution of Exclusive Rights in the Field of Science

D Yakovlev, E Yushkov, A Zanardo and M Bogatyreova.....364

012065
OPEN ACCESS
Model and technology of lingware support in tasks of the nuclear knowledge management system

A S Gavrilkina, O L Golitsyna and V A Fedorova.....368

IOP Conf. Series: Journal of Physics: Conf. Series **781** (2017) 011001 doi:10.1088/1742-6596/781/1/011001

PREFACE

We are pleased to introduce the Proceedings of the International research conference **«International Conference for young scientists, specialists and post-graduates on Nuclear Reactor Physics 2016 (ICNRP-2016)»** (5-9 September 2016, Health resort «Volga», Moscow, Russia) organized by the National Research Nuclear University MEPhI, with ROSATOM partnership.

Representatives of research organizations and universities from twelve countries (Russia, Germany, Norway, Finland, Kazakhstan, Belarus, Italy, Slovakia etc.), delivered their presentations on various topics. The major topics are features of fast reactors, calculation for the needs of operation and design of nuclear reactors, computational reactor tests, codes and databases. Over a hundred people from 37 organizations attended the conference. More than 93 papers were presented.

The received papers were reviewed according to the standards of the Journal of Physics: Conference Series and developed by the organizers' scientific criteria. This volume of the journal includes 65 papers devoted to various branches of nuclear reactor physics and technology.

During the conference, various sports activities were held, as well as a workshop on the problems of nuclear education in Russia. Most of the participants, according to the results of the survey were satisfied and expressed a desire to take part in the next conference in 2018.

The organizing committee is very grateful to the:

- Participants of the conference for their valuable contribution with the delivered presentations and interesting papers,
- Conference program committee chairman Strikhanov M.N., rector of National Research Nuclear University MEPhI,
- Program committee co-chairs: Caruso G., professor, Sapienza University of Rome, Hascik J., professor, Technical University of Bratislava, Janardhanan N.K., assistant professor, Jawaharlala Nehru University, Pershukov V.A., deputy director general, Rosatom, Tikhomirov G.V., dean of Physical-technical faculty, NRNU MEPhI, Tikhomirov V.V., professor, Belarusian State University,
- Competitiveness Program of National Research Nuclear University MEPhI for the financial support,
- Personnel of Journal of Physics for the support and communication.

Editor: Anton Smirnov
Co-editors: Rynat Bahdanovich, Ekaterina Proshkina

Peer review statement

All papers published in this volume of *Journal of Physics: Conference Series* have been peer reviewed through processes administered by the proceedings Editors. Reviews were conducted by expert referees to the professional and scientific standards expected of a proceedings journal published by IOP Publishing.

IOP Conf. Series: Journal of Physics: Conf. Series **781** (2017) 012001 doi:10.1088/1742-6596/781/1/012001

Inherent Safety Characteristics of Advanced Fast Reactors

A S Bochkarev[1,2], A S. Korsun[1], V.S. Kharitonov[1] and P.N. Alekseev[1,2]

[1]National Research Nuclear University MEPhI (Moscow Engineering Physics Institute), Moscow, Russia
[2] NRC «Kurchatov Institute», Moscow, Russia

E-mail: alseboch@mail.ru

Abstract. The study presents SFR transient performance for ULOF events initiated by pump trip and pump seizure with simultaneous failure of all shutdown systems in both cases. The most severe cases leading to the pin cladding rupture and possible sodium boiling are demonstrated. The impact of various features on SFR inherent safety performance for ULOF events was analysed. The decrease in hydraulic resistance of primary loop and increase in primary pump coast down time were investigated. Performing analysis resulted in a set of recommendations to varying parameters for the purpose of enhancing the inherent safety performance of SFR. In order to prevent the safety barrier rupture for ULOF events the set of thermal hydraulic criteria defining the ULOF transient processes dynamics and requirements to these criteria were recommended based on achieved results: primary sodium flow dip under the natural circulation asymptotic level and natural circulation rise time.

1. Introduction

Nowadays the one of the most important advanced NPP requirements is the safety requirement. NPP safety is a result of safety management, legal and engineering measures. Engineering measures employed so far include those that enhancing inherent safety performance of reactor. According to [1] «inherent safety refers to the achievement of safety through the elimination or exclusion of inherent hazards through the fundamental conceptual design choices made for the nuclear plant». It should be noted that inherent safety doesn't exclude the implementation of the active and passive safety systems in reactor design, but it rather ensures the reliance and diversity of reactor safety performance. The aim of the implementation of inherent safety principles in a reactor design is to reach such a safety level that there would be no combinations of transient initiating events that could violate the integrity of NPP safety barriers. In this case the initiating events could be both internal (any NPP system failure) and external ones (resulted from specific to NPP site impacts of natural phenomenon or human activity) including terrorism. According to [2] «a set of design extension conditions shall be derived on the basis of engineering judgement, deterministic assessments and probabilistic assessments for the purpose of further improving the safety of the nuclear power plant by enhancing the plant's capabilities to withstand, without unacceptable radiological consequences..». Thus a need arises to simulate such events and their consequences, to design additional engineer features and to provide their design-basis justification. That leads to increased timing budgets, physical resources and money supply required. The enhancing of inherent safety in advanced reactors designs makes possible to eliminate any beyond basis events on the early stage of conceptual design of reactor. Hence it would be not necessary to provide any additional safety management, to take any engineering measures with respect to beyond basis events on the stage of NPP operating.

2. ULOF transient analysis

The results of the inherent safety analysis presented in this article are obtained by using the one-dimensional DYANA code [3-6] for inherent safety analysis of fast liquid metal cooled reactors. Estimated sodium temperature and mass flow obtained from LOHS+LOF analysis via DYANA code [6] were in reasonable agreement with those obtained from PHENIX benchmark end-of-life test [7].

Under station blackout or loss of pump power supply conditions there is gradual reduction in primary sodium flow rate takes place due to pump trip. Pumps run-down time under all primary pumps trip conditions is sufficient to hold the temperature of the reactor components (fuel, cladding, primary sodium, reactor vessel, for LWR there are critical power ratio and steam-zirconium reaction margin) within acceptable limits and to remove reactor decay heat if reactor is scrammed. Besides, reactor design also includes auxiliary power supplies (batteries, diesel generators etc) providing primary pumps operating in the mode of decay heat removing (generally low flow rates about 5-6% of nominal value [8]). Natural circulation of primary sodium is developing if all power supplies failed. The most dangerous initiating event is pump seizure. Even single pump seizure may cause the excursion of acceptable temperature of fuel cladding.

ULOF transient dynamics depends not only on combination of initiating events but on the initial state conditions and parameters like initial power right before the accident, control rod position etc.
The results of ULOF transient analysis for fast sodium cooled reactor are presented in this chapter. MOX-fueled reactor 2800 MW of thermal power [9, 10] is under investigation. Seizure of all primary pumps (case A) and total loss of primary pumps power supplies (case B) are considered separately. Both accidents are under «without scram» and PDHRS system failure conditions. All secondary systems are normally operating. Initial state provides 100% nominal power.

Firstly, it is necessary to explore the conditions leading to violating of integrity of NPP safety barriers. There are 5 physical safety barriers for reactor under investigation (fuel matrix, steel cladding, reactor vessel and primary devices vessels, secondary devices vessels, containment). The safety barriers are termed through the acceptable temperatures (temperature criteria) of reactor critical components: fuel, cladding, primary sodium and reactor vessel. The values of all used temperature criteria are listed in table 1. The melting point is chosen for fuel temperature criteria because fuel melting causes increasing in gaseous fission products pressure under the cladding, besides the phase change gains the fuel swelling process. All these factors may cause the fuel cladding break. The value of cladding criteria is defined by the stress limits of fuel pin under the specific reactor conditions. The boiling temperature is set as the primary sodium criteria, because the sodium boiling can produce a high positive reactivity excursion due to the sodium void effect (if it is not negative). Finally the stress strain behavior of reactor vessel under the specific reactor conditions defines its temperature criteria.

Figures 1 and 2 illustrate the core components temperature excursion under cases A and B conditions accordingly. Case A is observed to be the most dangerous one (fig. 1) followed by cladding temperature unacceptable excursion on 8 sec. and sodium boiling onset on 15 s. of transient. Case B also initiates unacceptable cladding temperature excursion on 9 s. and sodium boiling onset on 15 s. of transient. This is very specific for ULOF transient to produce exceeding of cladding temperature criteria.

Table 1. Temperature criteria (Maximum acceptable temperatures of reactor critical components).

Component	Temperature criteria, K
Coolant	1153
Cladding	1073
Fuel	3023
Reactor vessel	1023

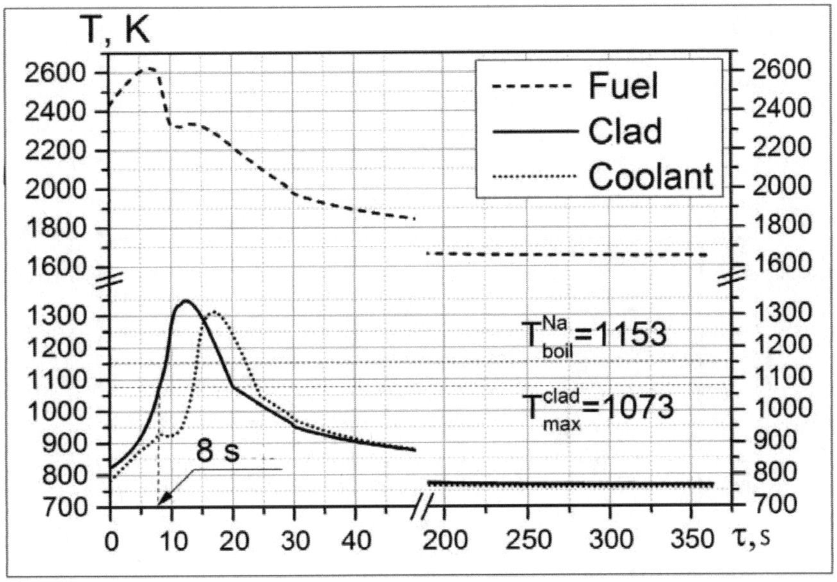

Figure 1. Peak temperatures of fuel, cladding and primary sodium under case A conditions.

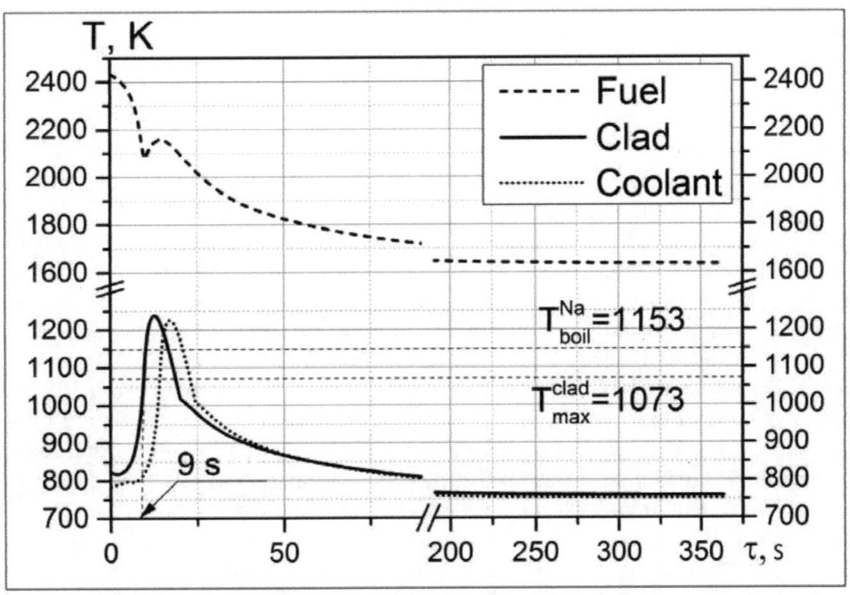

Figure 2. Peak temperatures of fuel, cladding and primary sodium under case B conditions.

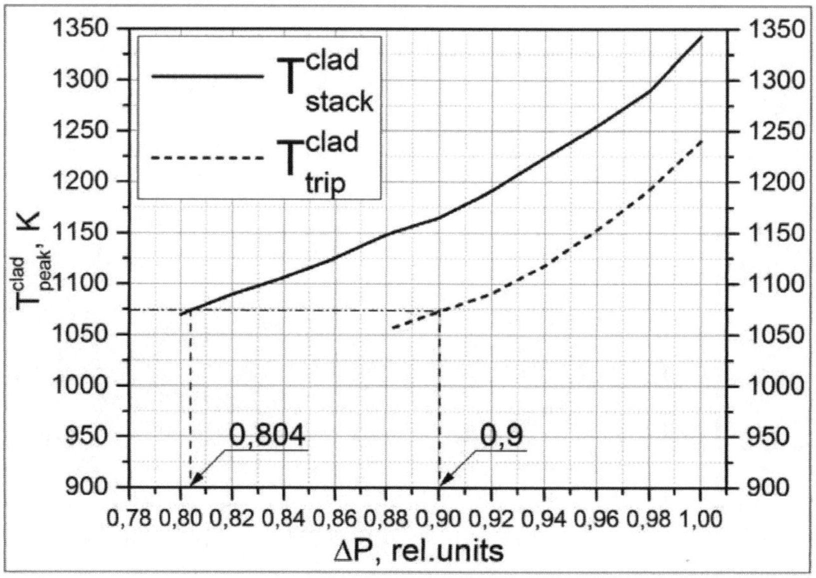

Figure 3. Peak temperature excursions under cases A and B conditions versus the pressure drop of primary circuit.

Next consider how do the values of some important reactor parameters impact on the peak temperature of the reactor components in term of the inherent safety.

The reducing of primary circuit pressure drop leads to the mitigation of ULOF transient. Temperature excursions are run down the acceptable values if primary circuit pressure drop reached $0.804\Delta P_0$ and $0.9\Delta P_0$ under cases A and B conditions accordingly (Fig. 3). To maintain the integrity of NPP safety barriers in terms of inherent safety of the reactor it requires the peak temperatures of the reactor components to stay within acceptable safety limits even under the most severe and low-probability conditions. In this work the most severe accident scenario is under the case A conditions. Fig. 4 illustrates that as the primary circuit pressure drop is reduced to $0.804\Delta P_0$ the reactor meets the requirement of safety barrier integrity under both case A and B conditions.

The safety barriers integrity conditions can be defined through the criteria characterizing the ULOF transient dynamics. The authors recommend such criteria as F_{NC} (natural circulation onset level), ΔF (coolant flow rate drop under the natural circulation onset level) and Δτ (time from coolant flow rate drop under the natural circulation onset level till natural circulation onset) (Fig. 5). Now to mitigate the ULOF transient the requirements to F_{NC}, ΔF and $\Delta\tau$ should be worked out.

The normalized primary sodium flow rate under case A conditions with reduced to $0.804\Delta P_0$ pressure drop is presented in fig 5. Referring to the fig. 5 recommended criteria values are shown in table 2. The peak cladding temperature is within acceptable limits, primary sodium boiling margin is 90° (fig. 4) if the requirements to the pressure drop and criteria are met.

The increasing of primary pump run-down time leads to the mitigation of ULOF transient under case B conditions. Temperature excursions are run down the acceptable values if run-down time reached the value of $1.8\tau_0$ (fig. 6). Only case B is considered because the increasing of primary pump run-down time under case A conditions obviously is not effective.

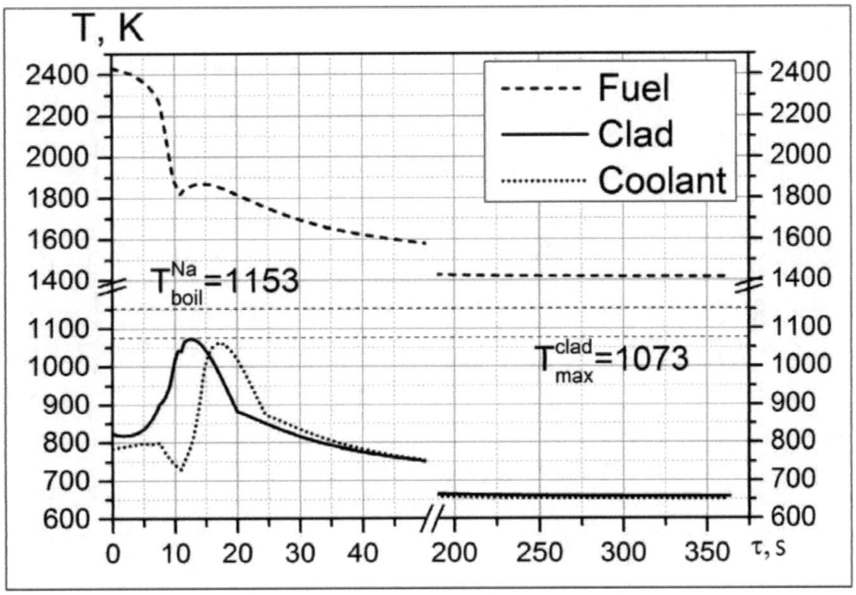

Figure 4. Peak temperatures of fuel, cladding and primary sodium under case A conditions with reduced primary circuit pressure drop.

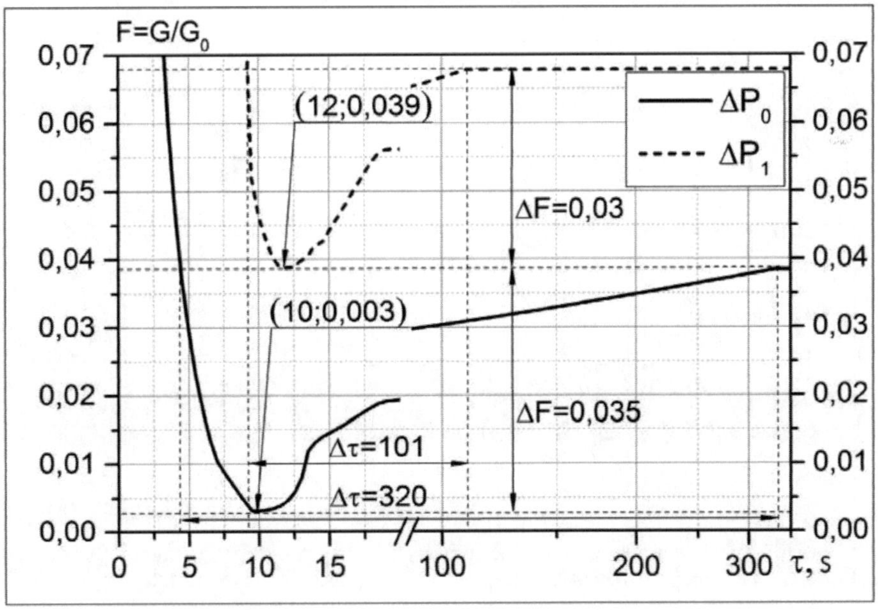

Figure 5. Primary sodium flow rate under case A conditions with reduced primary circuit pressure drop.

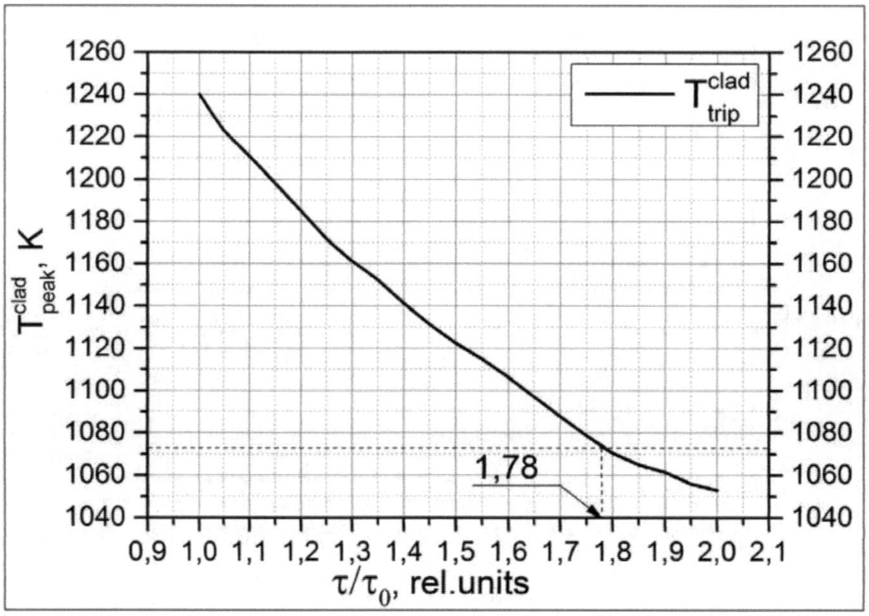

Figure 6. Peak temperature excursions under the case B conditions versus primary pump run-down time.

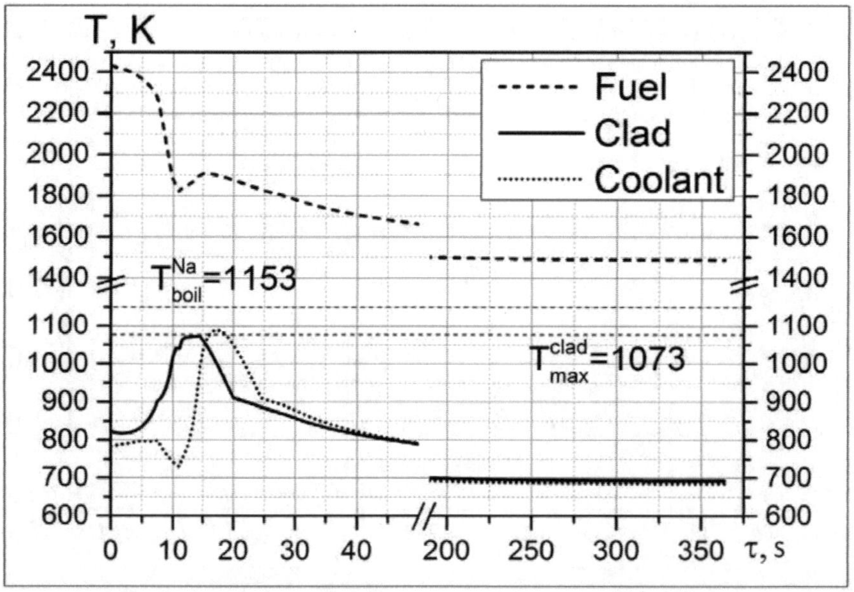

Figure 7. Peak temperatures of fuel, cladding and primary sodium under the case B conditions with increased primary pump run-down time.

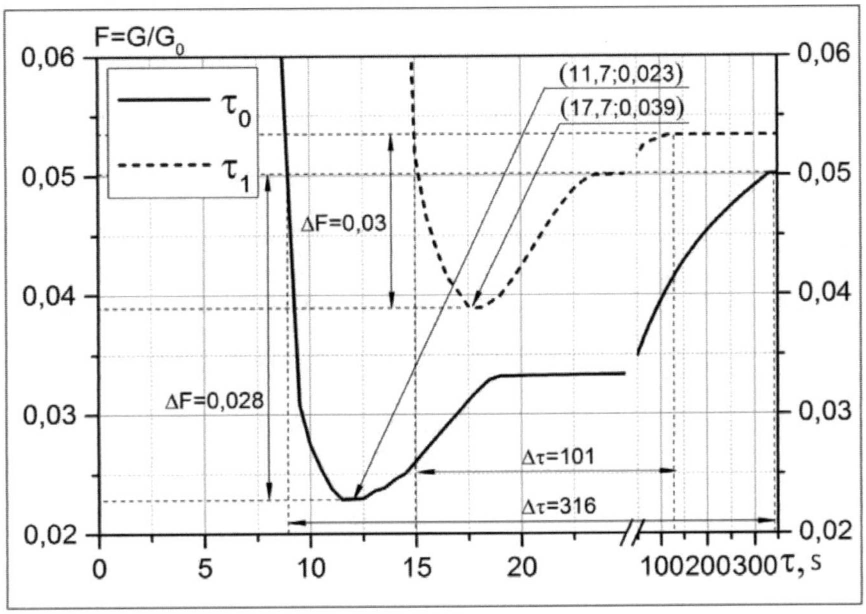

Figure 8. Primary sodium flow rate under case B conditions with increased primary pump run-down time.

Table 2. Requirements to the criteria characterizing the ULOF transient dynamics under case A conditions.

Criteria	Restriction
$\Delta\tau$	$\leq$ 101 s.
ΔF	$\leq$ 0,03 rel. units
F_{NC}	$\geq$ 0,07 rel. units

Table 3. Requirements to the criteria characterizing the ULOF transient dynamics under case B conditions.

Criteria	Restriction
$\Delta\tau$	$\leq$101 s.
ΔF	$\leq$ 0,015 rel. units
F_{NC}	$\geq$ 0,053 rel. units

The normalized primary sodium flow rate under case B conditions with encreased to $1.8\tau_0$ run-down time is presented in fig. 8. Referring to the fig. 8 recommended criteria values are shown in table 3. The peak cladding temperature is within acceptable limits, primary sodium boiling margin is 65° (fig. 7) if the requirements to the run-down time and criteria are met.

It should be noticed that integrity of NPP safety barriers is achieved under all investigated transient initiating events if the requirements to the criteria characterizing the ULOF transient dynamics are met.

3. Conclusion

ULOF transient performance analysis for pool-type MOX-fueled sodium fast reactor 2800 MW of thermal power is carried out. Two ULOF scenario cases were considered: all pumps seizure and all pumps trip. Both cases were followed by simultaneous scram and PDHRS failure. It was found out that the most severe case is all pumps seizure followed by exceeding of maximum acceptable fuel cladding temperature and possible sodium boiling.

The impact of various features on SFR inherent safety performance for ULOF events was also analyzed. The decrease in primary pressure drop and increase in primary pump run-down time were investigated. Performing analysis resulted in a set of recommendations to varying parameters in terms of enhancing the inherent safety performance of SFR under investigation. In order to exclude the safety barrier rupture for ULOF events the set of thermal hydraulic criteria characterizing the ULOF transient dynamics (fig. 5, 6) and requirements to them (tables 2, 3) were recommended based on achieved results: F_{nc} (natural circulation onset level), ΔF (coolant flow rate drop under the natural circulation onset level) and $\Delta\tau$ (time from coolant flow rate drop under the natural circulation onset level till natural circulation onset). To develop the refined requirements for the proposed criteria it is necessary to analyze SFR performance for ULOF, UTOP, ULOCA, ULOHS events and their overlaps coupled with uncertainty analysis. It is also necessary to take into account the heat removal through PDHRS even in a failure mode by heat-conductivity through the HX walls and to refine the acceptable temperatures of critical components of reactor with respect to reactor inherent safety.

References

[1] *Safety related terms for advanced nuclear plants* 1991 (Vienna: International Atomic Energy Agency) IAEA-TECDOC-626

[2] *Safety of Nuclear Power Plants: Design. Specific Safety Requirements 2.1* 2012 (Vienna: International Atomic Energy Agency)

[3] Alekseev P N and Bochkarev A S 2013 Potential of the Internal Self Protection of a Reactor. *Nuclear physics and engineering* **4** № 5 pp 1–5

[4] Bochkarev A and Alekseev P 2014 Methodology for Inherent Safety Analysis of Fast Reactors. *Technical Presentation on ICONE 22*. 31269

[5] Bochkarev A S, Alekseev P N, Korsun A S and Kharitonov V S 2015 Inherent safety analysis of fast reactors. *XIV International Conference NPP Safety and Personnel Training*. Russia, Obninsk

[6] Bochkarev A S, Alekseev P N, Korsun A S and Kharitonov V S 2016 Modeling of natural circulation for the inherent safety analysis of sodium cooled fast reactors. *Izvestiya Wysshikh Uchebnykh Zawedeniy, Yadernaya Energetika*. No. 3 pp. 129-138

[7] *Benchmark Analyses on the Natural Circulation Test Performed During the PHENIX End-of-Life Experiments* 2013 (Vienna: International Atomic Energy Agency) IAEA-TECDOC-1703 169 p

[8] Floyd E Dunn 2002 Impact of design options on natural circulation performance of the AFR-300 advanced fast reactor. *International Congress on Advanced Nuclear Power Plants.* Hollywood.

[9] *Status of Fast Reactor Research and Technology Development* 2012 (Vienna: International Atomic Energy Agency) IAEA-TECDOC-1691

[10] *Fast Reactor Database 2006 Update* 2007 (Vienna: International Atomic Energy Agency) IAEA-TECDOC-1531

Radiogenic lead as coolant, reflector and moderator in advanced fast reactors

E G Kulikov

Department of theoretical and experimental physics of nuclear reactors, National Research Nuclear University MEPhI (Moscow Engineering Physics Institute), Kashirskoe highway, 31, Moscow, Russia

E-mail: EGKulikov@mephi.ru

Abstract. Main purpose of the study is assessing reasonability for recovery, production and application of radiogenic lead as a coolant, neutron moderator and neutron reflector in advanced fast reactors. When performing the study, thermal, physical and neutron-physical properties of natural and radiogenic lead were analyzed. The following results were obtained:
1. Radiogenic lead with high content of isotope ^{208}Pb can be extracted from thorium or mixed thorium-uranium ores because ^{208}Pb is a final product of ^{232}Th natural decay chain.
2. The use of radiogenic lead with high ^{208}Pb content in advanced fast reactors and accelerator-driven systems (ADS) makes it possible to improve significantly their neutron-physical and thermal-hydraulic parameters.
3. The use of radiogenic lead with high ^{208}Pb content in advanced fast reactors as a coolant opens the possibilities for more intense fuel breeding and for application of well-known oxide fuel instead of the promising but not tested enough nitride fuel under the same safety parameters.
4. The use of radiogenic lead with high ^{208}Pb content in ADS as a coolant can upgrade substantially the level of neutron flux in the ADS blanket, which enables effective transmutation of radioactive wastes with low cross-sections of radiative neutron capture.

1. Introduction. Thorium, uranium and radiogenic lead

As it was written in the IAEA Red Book-2014 [1], the evaluated global thorium resources were 6.4 million tons in total. The largest thorium resources are placed in the following countries: India – 0.85 million tons, Brazil – 0.63 million tons, Australia and the USA – about 0.6 million tons each.

Demands of the world industry for thorium are very small, mainly because of its radioactivity. According to information from the US Mining Bureau [2], in 2009 only about 5000 tons of monazite ores were mined in India, 1200 tons in Brazil and 600 tons in Malaysia. In total, only about 1200 kg of thorium were consumed by the world industry in 2007.

The monazite sands constitute the most important source of rare-earth elements and thorium. General chemical formula for the monazites is usually written as $(Ce,Th)PO_4$ though the monazites contain, in addition to cerium, lanthanum, praseodymium, neodymium and some other rare-earth metals, as well as uranium and radiogenic lead in addition to thorium. As a rule, thorium content in the monazites covers the range from 2.5% to 12%, uranium content – from 0.1% to 0.4% content of rare-earth metals – from 55% до 68% and relatively low percentage of radiogenic lead.

The radiogenic lead in the monazites is a final material produced by consecutive radioactive decays of main thorium and uranium isotopes. As is known, ^{232}Th is the start isotope for the chain of isotopic transformations caused by α- and β-decays of intermediate chain members. Totally, decay chain of ^{232}Th numbers twelve members, products of seven α-decays and five β-decays. The decay chain ends with stable lead isotope ^{208}Pb.

Similarly, uranium isotopes in the monazites are the start isotopes for their own decay chains. The decay chain of ^{238}U numbers fourteen members (eight α-decays, six β-decays) and ends with stable lead isotope ^{206}Pb. The decay chain of ^{235}U numbers eleven members (seven α-decays, four β-decays) and ends with stable lead isotope ^{207}Pb. So, final product in the decay chains of main thorium and uranium isotopes is the radiogenic lead. However, uranium-produced lead significantly differs from thorium-produced lead because of different final isotopes. Indeed, finally ^{232}Th transforms into lead isotope ^{208}Pb, ^{238}U and ^{235}U – into other lead isotopes, ^{206}Pb and ^{207}Pb, respectively.

That is why the natural minerals can contain, in addition to the "common" lead with a certain isotopic composition, a whole family of the radiogenic leads in uranium, thorium and mixed uranium-thorium ores. Isotopic compositions of the radiogenic leads can cover a rather wide range in dependence on elemental structure and age of the ore deposits. It is evident that the radiogenic leads extracted from mixed uranium-thorium ores with dominant thorium fraction (the monazite sands) are characterized by high content of ^{208}Pb.

In the authors' point of view, just the radiogenic lead obtained as a by-product of the monazite reprocessing plants represents a special interest for advanced projects of various nuclear facilities thanks to some unique neutron-physical properties of stable lead isotope ^{208}Pb. The radiogenic lead with high content of this lead isotope is contained in the monazite sands where thorium fraction is substantially larger than uranium fraction.

Currently, the radiogenic lead can be extracted as a by-product of the existing technologies used for recovery of rare-earth metals from mixed deposits of uranium-thorium ores. Some experimental data were published in Ref. 3 on recovery of the radiogenic lead with high content of ^{208}Pb from the reprocessing tails of Brazilian monazites with application of ion-exchange and electrochemical technologies. It was demonstrated that ion-exchange refinement of the thorium reprocessing tails allowed them to extract only 33% of radiogenic lead while the more expensive electrochemical refinement was capable to extract up to 98% of radiogenic lead with content of ^{208}Pb at the level of 88%.

The table 1 presents experimental data about elemental structure of some monazites and about isotopic compositions of the radiogenic leads extracted from these monazite deposits.

Table 1. Elemental structure of monazite ores and isotopic compositions of the radiogenic leads [5-8].

Ore deposits	U / Th / Pb [weight %]	^{204}Pb/ ^{206}Pb / ^{207}Pb / ^{208}Pb [atomic %]
Guarapari, Brazil	1.3 / 59.3 / 1.5	0.005 / 6.03 / 0.46 / 93.5
Manitoba, Canada	0.3 / 15.6 / 1.5	0.010 / 10.2 / 1.86 / 87.9
Mt. Isa Mine, Australia	0.0 / 5.73 / 0.3	0.038 / 5.44 / 0.97 / 93.6
Las Vegas, USA	0.1 / 9.39 / 0.4	0.025 / 9.07 / 1.13 / 89.8
South Bug, Ukraine	0.2 / 8.72 / 0.9	0.010 / 6.04 / 0.94 / 93.0
"Common" lead	———	1.4 / 24.1 / 22.1 / 52.4

Today Russia possesses thorium ores with evaluated resources of 0.15 million tons of thorium dioxide [1]. About 85 000 tons of thorium-bearing monazite concentrate are stockpiled in Sverdlovsk region. These thorium reserves were partially mined in the Russian Federation and partially imported from Mongolia, China and Vietnam. The monazite concentrate is under reprocessing now only in order to reduce its radiation effects on the environment. If we suppose that Russian monazite concentrate contains 1.5% of the radiogenic lead, like Brazilian monazite sands, then total amount of the radiogenic lead in the stockpiled monazite concentrate is equal to about 1200 tons.

In Russia the rare-earth metals, molybdenum and titanium are extracted from uranium and thorium-bearing ore deposits at Kola Peninsula [4]. Full-scale industrial plants are in operation there for recovery of these useful materials. Besides, several technological lines are in operation there for recovery and isolation of all radioisotopes to prevent their negative radiation effects on population and the environment. So, only one technological stage should be introduced into the existing technological lines, namely extraction of the last material in uranium and thorium chains of radioactive decays, i.e. extraction of the radiogenic lead. Obviously, the radiogenic lead could be extracted with application of the well-developed chemical technology which is currently in usage for recovery of "common" lead from natural lead-bearing ores.

2. Radiogenic lead in fast reactors

In the 1980s, on initiative and under leadership of Professor Victor Orlov, the works under the project of fast lead-cooled reactor BREST have started [9]. The use of lead as a coolant offers the following evident advantages: high boiling point, weak chemical interactions with air and water, low neutron-induced activation, good neutron-reflecting properties, relatively small neutron absorption and moderation. The radiogenic lead with high ^{208}Pb content is able to strengthen additionally these positive properties. Isotope ^{208}Pb is a double-magic nuclide with closed neutron and proton shells. Thanks to this fact, micro cross-sections of neutron absorption by ^{208}Pb are substantially lower than those of other lead isotopes and a lot of light nuclides.

The closure of neutron and proton shells explains the fact that excitation levels of ^{208}Pb nuclei are significantly higher than those of other lead isotopes. Therefore the energy threshold of inelastic neutron scattering by ^{208}Pb is significantly larger than that of other lead isotopes. This fact explains weak inelastic neutron scattering by ^{208}Pb. As a result, operation of fast reactors could become safer due to the favorable reactivity effect of coolant density caused by the lower spectral component of this effect. Moreover, small micro cross-sections of neutron capture by ^{208}Pb decreases one else component of the coolant density reactivity effect, namely the component related with neutron absorption [10].

In order to verify theoretical prerequisites on potential advantages caused by the use of radiogenic lead with high ^{208}Pb content as a coolant, neutron reflector and neutron moderator, appropriate numerical studies were carried out for fast lead-cooled reactor BREST-OD-300 [11]. The computer code TIME26 [12] was used in the numerical studies. The computer code TIME26 can determine main neutron-physical parameters of fast reactors within the frames of 26-group diffusion approximation and one-dimensional geometrical models. Micro cross-sections of neutron reactions were taken from the evaluated nuclear data library ABBN-78 and processed by the auxiliary computer code ARAMACO-C1 [13] to take the self-shielding effects into account for each reactor zone.

The computations have demonstrated that the use of ^{208}Pb as a coolant of the fast reactor with mixed uranium-plutonium nitride fuel made it possible to obtain some remarkable improvements [10]. For example, if ^{208}Pb is substituted for "common" lead, then the values of effective neutron multiplication factor K_{EFF}, fuel breeding ratio in the reactor core (BRC) and the coolant density reactivity effect (CDRE) could be remained at the same level by appropriate variations of plutonium fraction in fuel, pitch of fuel lattice and height of the reactor core. The computations have shown that the former values of K_{EFF}, BRC and CDRE may be preserved by insignificant reduction of plutonium fraction (from 13.8% to 13.6%), but by substantial widening of fuel lattice (pitch of fuel lattice must be increased from 13.6 mm to 23.6 mm) and by substantial increasing the height of the reactor core (from 110 cm to 300 cm). These changes have led to the following results:

- The same coolant heating-up in the widened lattice of elongated fuel elements can be achieved with significantly lower coolant velocity (about two-fold reduction).
- Pressure drop for coolant pumping through the widened fuel lattice can be five times lower.

Project of fast reactor BREST-OD-300 presumes the use of high-density mixed uranium-plutonium nitride fuel in order to provide small variations of the reactor criticality at full-power operation, i.e. reactivity must be well below effective fraction of delayed neutrons during full time of the reactor operation. The computations have demonstrated that substitution of the radiogenic lead for "common" lead could allow us to come back to the more assimilated and widely used MOX-fuel without substantial losses in the advantages listed above.

Design of fast reactor BREST-OD-300 does not include radial and axial blankets (fuel breeding zones) to enhance safety of the reactor operation. The favorable negative feedbacks, which could appear if ^{208}Pb is used as a reactor coolant, allowed us to bring back these fuel breeding zones. The computations have demonstrated that addition of axial blanket fueled with natural uranium nitride could increase total fuel breeding ratio roughly on 0.14 with insignificant worsening of other parameters.

One other advantage of the radiogenic lead is a possibility for substantial elongation of prompt neutron lifetime. The value of prompt neutron lifetime in project of fast reactor BREST-OD-300 is equal to about 0.5 μs. If ^{208}Pb is used as a coolant and neutron reflector (50-cm thick), then prompt neutron lifetime becomes longer by a factor of 3. If this reflector is thickened up to 5 m, then prompt neutron lifetime elongates up to 1 ms (i.e. 2000 times longer!) and becomes comparable with typical values of prompt neutron lifetime in thermal heavy-water reactors CANDU. This drastic elongation of prompt neutron lifetime can be explained by the following considerations. Fast neutrons generated in the reactor core can migrate to ^{208}Pb-reflector, penetrate deeply into reflector, lose their initial energy after many collisions with ^{208}Pb nuclei and, at last, come back to the reactor core after a certain time period (delay). The opportunity for fast neutrons to come back to the reactor core from ^{208}Pb-reflector is provided by weak neutron absorption and effective albedo properties of ^{208}Pb. As these returned neutrons are prompt neutrons on their origination, it may be concluded that the chain fission reaction on prompt neutrons slows down, and, as a consequence, safety of the reactor operation improves.

So, the use of ^{208}Pb as a coolant, neutron reflector and neutron moderator in fast reactors can slow down drastically the chain fission reaction on prompt neutrons and, thus, improve significantly safety of the reactors operation.

3. Radiogenic lead in ADS

Let us note that neutron flux is much higher in ADS compared to thermal reactors that could also be considered for transmutation purposes [14]. Very low cross-sections of neutron absorption by ^{208}Pb together with small step of neutron slowing down can open an opportunity to form zone with high flux of slow neutrons (FSN) in blankets of ADS.

The probabilities for neutrons to avoid absorption in the process of their moderation to thermal energy in ^{208}Pb and in graphite are approximately the same and close to unity. On the contrary, analogous probability in "common" lead is remarkable lower than unity (~0.29). Therefore, the probability for neutrons to be scattered and slowed down by ^{208}Pb nuclei is substantially larger than the probability for neutrons to be absorbed. The inverse situation takes place in the case of "common" lead.

Step of neutron slowing down and cross-sections of neutron absorption by ^{208}Pb nuclei are seventeen times lower than those by graphite nuclei while cross-sections of neutron scattering by ^{208}Pb nuclei are two times larger than those by graphite nuclei. This means that neutron moderation by ^{208}Pb nuclei requires seventeen times larger number of elastic collisions than neutron moderation by graphite nuclei. However, at each neutron collision with ^{208}Pb nuclei probability of neutron absorption is thirty-four times lower than that in neutron collision with graphite nuclei. As a result, the higher FSN can be achieved in ^{208}Pb than that in "common" lead or in graphite.

In order to verify theoretical prerequisites on potential advantages of ^{208}Pb, appropriate numerical studies were carried out to determine radial FSN distributions in ADS blanket composed of ^{208}Pb or graphite. The computations were carried out for infinite plane model of ADS blanket and demonstrated that maximal values of FSN were achieved in ^{208}Pb-blanket (figure 1). As is seen, thermal and slow neutron fluxes in the vicinity of ADS target are remarkably higher in ^{208}Pb-blanket than those in graphite blanket. As far as distance from ADS target increases, this advantage of ^{208}Pb-blanket becomes larger and becomes very quickly. FSN increases for low-energy neutrons because Fermi spectrum with inverse proportionality to neutron energy establishes in low-absorbing medium. So, it may be concluded that, in the case of plane geometry, ^{208}Pb-blanket excels graphite blanket in FSN values at any distance from ADS target.

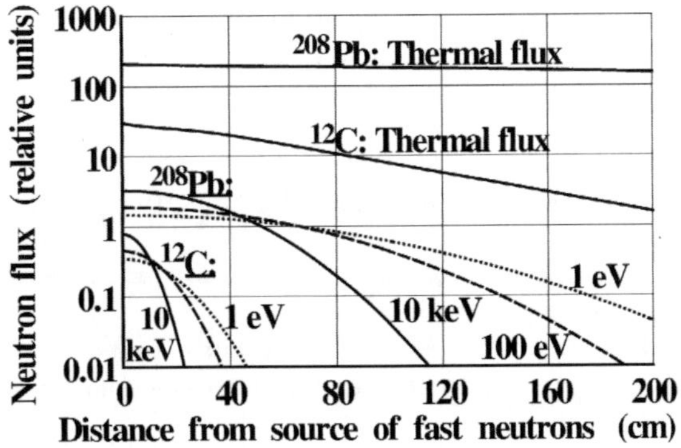

Figure 1. Radial FSN distribution in ADS blanket (infinite plane geometry).

Similar computations were carried out for more real cylindrical and spherical models of infinite ADS blanket (figure 2).

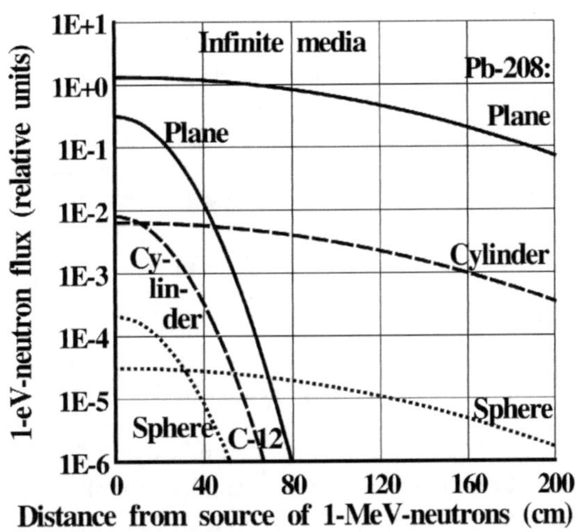

Figure 2. Radial FSN distribution in ADS blanket (plane, cylindrical and spherical geometries).

As is seen, ^{208}Pb-blanket excels graphite blanket in FSN values at any distance from ADS target in plane geometry, at distances longer than 10 cm in cylindrical geometry and at distances longer than 40 cm in spherical geometry. As far as distance from ADS target increases, this advantage of ^{208}Pb-blanket becomes substantially larger. This means that it is possible to form peripheral large-volume zones in ^{208}Pb-blanket with high FSN values.

Thus, isotope ^{208}Pb may be regarded as a candidate on the role of neutron moderator in high-flux ADS blanket with resonance or thermal neutron spectra. ^{208}Pb-blanket can be large enough in volume to place sufficiently large amounts of radioactive wastes for neutron transmutation and to solve problem of neutron leakage.

Let us note that there is one more possible opportunity to use ^{208}Pb for arising controlled fusion chain reaction supported by neutrons in fusion facilities with magnetic plasma confinement [15].

4. Summary

This paper has demonstrated that recovery of the radiogenic lead from thorium and thorium-uranium minerals for needs of nuclear industry could be reasonable because the use of the radiogenic lead as a coolant, neutron moderator and neutron reflector could upgrade safety, neutron-physical and thermal-hydraulic parameters of fast lead-cooled reactors as well as upgrade ADS efficiency on neutron transmutation of radioactive wastes.

References

[1] OECD Nuclear Energy Agency and International Atomic Energy Agency 2014 *Uranium-2014: Resources, Production and Demand*

[2] Polovov I B 2011 *Jaderno-himicheskaja tehnologija torija. Training manual* (Yekaterinburg: Physics and Technology Institute)

[3] Seneda J A, Forbicini O, Queroz S et al 2010 Study on radiogenic lead recovery from residues in thorium facilities using ion exchange and electrochemical process *Progress in Nuclear Energy* **52** pp 304–306

[4] Majorov V G, Nikolaev A I, Zil'berman B Ja and Shherbakov V M 2007 Utilizacija sbrosnyh rastvorov v tehnologii perovskita *Zhurnal prikladnoj himii* **80** №6 pp 899–902

[5] Godoy J M, Godoy M L and Aronne C C 2007 Application of inductively coupled plasma quadrupole mass spectrometry for the determination of monazite ages by lead isotope ratios *Journal of Brazilian Chemical Society* **18** №5 pp 154–162

[6] Nier A O, Tompson R W and Murphey B F 1941 The isotopic constitution of lead and the measurement of geological time *Physical Review* **60** pp 112–117

[7] Holmes A 1918 The pre-Cambrian and associated rocks of the District of Mozambique *Quarterly Journal of Geological Society* **74** pp 31–98

[8] Shherbak N P, Zlobenko V G, Zhukov G V et al 1978 *Katalog izotopnyh dat porod ukrainskogo shhita* (Kiev: Naukova dumka)

[9] Orlov V V 2000 Jevoljucija tehnicheskoj koncepcii bystrogo reaktora. Koncepcija BREST *Trudy Mezhdunarodnogo seminara "Bystryj reaktor i toplivnyj cikl estestvennoj bezopasnosti dlja krupnomasshtabnoj jenergetiki. Toplivnyj balans, jekonomika, bezopasnost', othody, nerasprostranenie"*

[10] Apse V A, Kulikov G G, Shmelev A N and Sirotkin A M 2010 O vozmozhnosti uluchshenija nejtronno-fizicheskih i teplo-gidravlicheskih parametrov jenergeticheskih bystryh reaktorov pri ispol'zovanii radiogennogo svinca v kachestve teplonositelja *Jadernaja fizika i inzhiniring* **1** №5 pp 387–397

[11] Orlov V V, Leonov V N, Sila-Novickij A G et al 2000 Konstrukcija reaktorov BREST. Jeksperimental'nye raboty dlja obosnovanija koncepcii reaktorov BREST. Rezul'taty i plany *Trudy Mezhdunarodnogo seminara "Bystryj reaktor i toplivnyj cikl estestvennoj bezopasnosti dlja krupnomasshtabnoj jenergetiki. Toplivnyj balans, jekonomika, bezopasnost', othody, nerasprostranenie"*

[12] Apse V A and Shmelev A N 2008 *Ispol'zovanie programmy TIME26 v kursovom proektirovanii bystryh reaktorov i jelektrojadernyh ustanovok. Traning material* (Moscow: MEPhI)

[13] Nikolaev M N 1972 *Kompleks programm avtomatizirovannogo rascheta makroskopicheskih konstant (ARAMAKO)* (Obninsk: Institute of Physics and Power Engineering)

[14] Kazansky Yu A and Romanov M I 2015 Transmuting minor actinides with thermal reactor neutrons *Nuclear Energy and Technology* **1** Issue 3 pp 208–212

[15] Shmelyov A N and Kulikov G G 2015 On the physical conditions for arising a controlled fusion chain reaction supported by neutrons in fusion facilities with magnetic plasma confinement *Nuclear Energy and Technology* **1** Issue 3 pp 218–221

Acknowledgments

This research was supported by grant of the President of the Russian Federation for state support of young Russian PhD scientists 2016 (MK-8277.2016.8).

IOP Conf. Series: Journal of Physics: Conf. Series **781**(2017) 012003 doi:10.1088/1742-6596/781/1/01200

Neutronic calculation of fast reactors by the EUCLID/V1 integrated code

D A Koltashev and A A Stakhanova

Nuclear Safety Institute of the Russia Academy of Sciences, 52 Bolshaya Tulskaya, Moscow, 115191, Russia
E-mail: stahanova@ibrae.ac.ru

Abstract. This article considers neutronic calculation of a fast-neutron lead-cooled reactor BREST-OD-300 by the EUCLID/V1 integrated code. The main goal of development and application of integrated codes is a nuclear power plant safety justification. EUCLID/V1 is integrated code designed for coupled neutronics, thermomechanical and thermohydraulic fast reactor calculations under normal and abnormal operating conditions. EUCLID/V1 code is being developed in the Nuclear Safety Institute of the Russian Academy of Sciences. The integrated code has a modular structure and consists of three main modules: thermohydraulic module HYDRA-IBRAE/LM/V1, thermomechanical module BERKUT and neutronic module DN3D. In addition, the integrated code includes databases with fuel, coolant and structural materials properties. Neutronic module DN3D provides full-scale simulation of neutronic processes in fast reactors. Heat sources distribution, control rods movement, reactivity level changes and other processes can be simulated. Neutron transport equation in multigroup diffusion approximation is solved. This paper contains some calculations implemented as a part of EUCLID/V1 code validation. A fast-neutron lead-cooled reactor BREST-OD-300 transient simulation (fuel assembly floating, decompression of passive feedback system channel) and cross-validation with MCU-FR code results are presented in this paper. The calculations demonstrate EUCLID/V1 code application for BREST-OD-300 simulating and safety justification.

1. Introduction

The main goal of development and application of integrated codes is a nuclear power plant safety justification. The codes provide a self-consistent simulation of various physical processes. One of these codes is the EUCLID/V1 [1] integrated code designed for coupled neutronics, thermomechanical and thermohydraulic fast reactor calculations under normal (steady state on the permitted power levels, normal transient conditions) and abnormal operating conditions.

EUCLID/V1 code is being developed in the Nuclear Safety Institute of the Russian Academy of Sciences within the PRORYV (or Breakthrough) project of the Russian Federal Targeted Program "Nuclear Power Technologies of the New Generation for 2010–2015 and until 2020".
The integrated code has a modular structure and consists of three main modules: thermohydraulic module HYDRA-IBRAE/LM/V1, thermomechanical module BERKUT and neutronic module DN3D. In addition, the integrated code includes databases with fuel, coolant and structural materials properties.

Neutronic module DN3D [2] provides full-scale simulation of neutronic processes in fast reactors. Heat sources distribution, control rods movement, reactivity level changes and other processes can be simulated. Neutron transport equation in multigroup diffusion approximation is solved. Neutron field can be calculated for seven radial nodes in each assembly (subunit G7) to take into account absorber heterogeneity in the fuel assembly. This approach provides the control rods worth correct calculation and more detailed simulation of the neutron flux.

Kinetic option is included in DN3D module for calculation the spatial and energy neutron flux distribution and different functionals in the problems with cavities, large gradients of neutron field and strong degree of radiation extinction. Kinetic option is based on the S_N discrete ordinates method. Database of nuclear cross sections is prepared by CONSYST [3] code.

A fast-neutron lead-cooled reactor BREST-OD-300 transient simulation (fuel assembly floating, decompression of passive feedback system channel) and cross-validation with MCU-FR code results are shown in this paper.

2. Coupled calculations of BREST-OD-300 reactor system

Fuel assembly floating modes calculated for one (control system rods operating/failure) and two assemblies floating (control system rods operating/failure). Reactor power, full reactivity and parts of reactivity are considering. The initial state is steady state at nominal power level. The floating velocity equals 5 cm/s. An additional mode is calculated with one assembly floating at speed of 20 cm/s and control system rods are operating.

Figures 1 and 2 present power changes for different floating velocity, figures 3 and 4 presents reactivity and parts of reactivity.

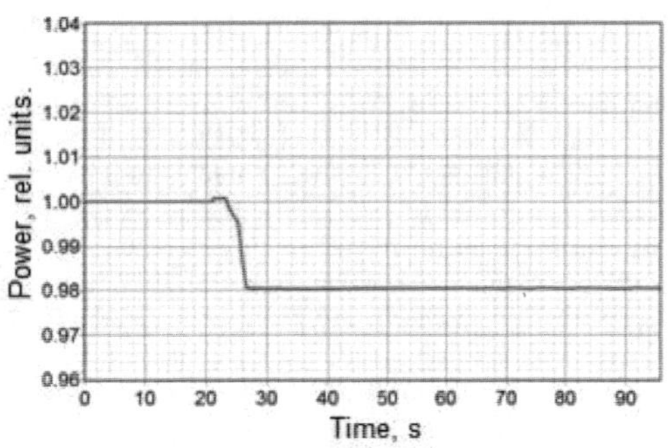

Figure 1. Power changes (floating velocity =5 cm/s).

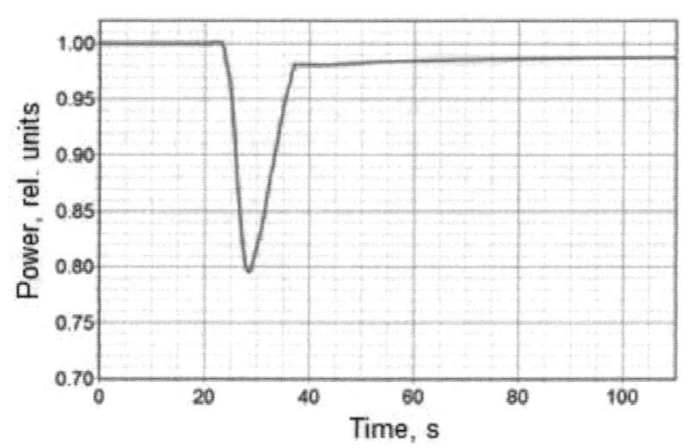

Figure 2. Power changes (floating velocity =20 cm/s).

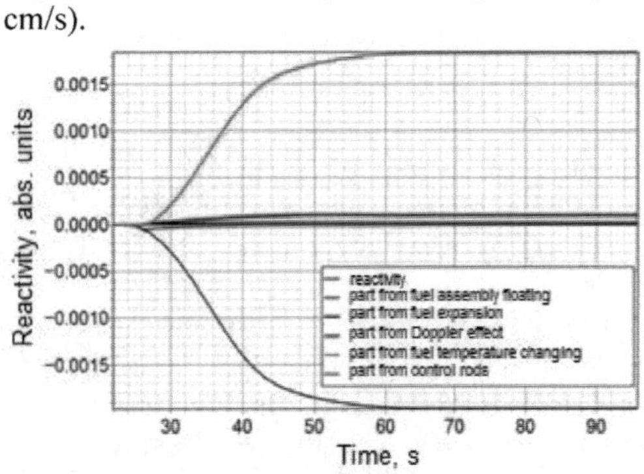

Figure 3. The reactivity and parts of reactivity (floating velocity =5 cm/s).

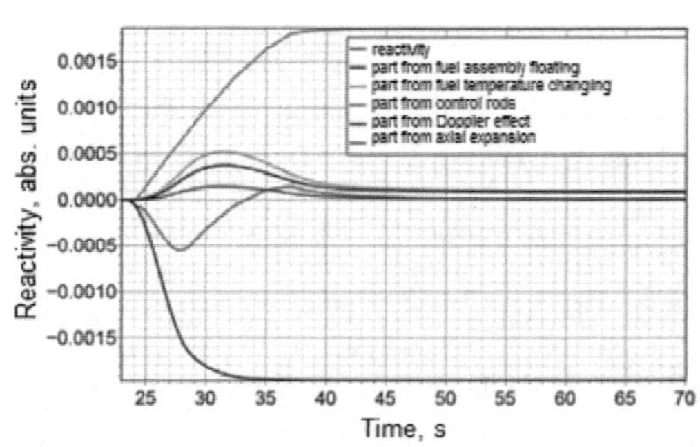

Figure 4. The reactivity and parts of reactivity (floating velocity =20 cm/s).

In case of slow floating (5cm/sec) control system rods can keep power inside the dead zone. In case of faster floating (20 cm/sec), power decreases fast down to 80% level and then control system rods stabilize power at the initial level. Fuel assembly floating insert negative reactivity, which can be compensate by control system rods. Fuel temperature negative feedback is more efficient in case of a fast floating of fuel assembly. The fast power level decrease down to 80% causes fuel temperature change and fuel temperature reactivity effect gives a considerable contribution in reactivity balance.

Decompression of passive feedback system channel was simulated. In this simulation control system rods weren't operating. At the first stage of process the lead level in passive feedback system channel decreases uniformly down to 130 cm, which corresponds to coolant rate decline. At the second stage the lead level increases uniformly from 130 cm up to coolant free level in the reactor. The second stage is actually the decompression of passive feedback system channel.

Changes of power level, reactivity and parts of reactivity are shown in the figures 5 and 6.

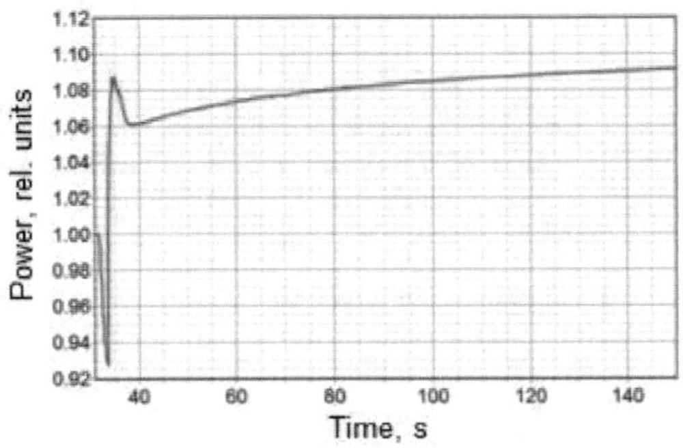

Figure 5. Power changes.

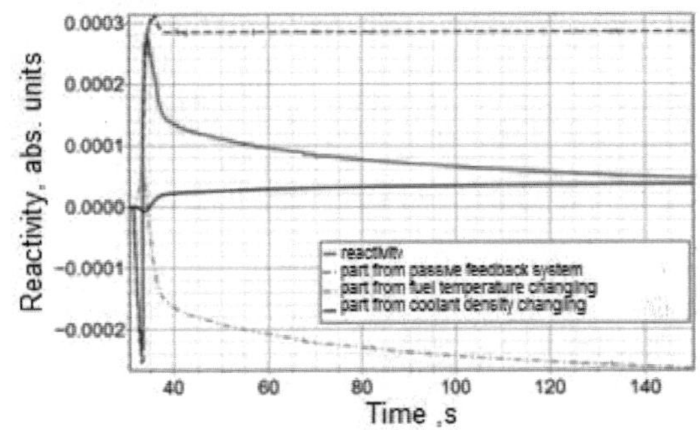

Figure 6. Reactivity and parts of reactivity.

At the first stage, with the lead level in passive feedback system channel decreasing, negative reactivity is inserted, which can be compensated by fuel temperature feedback. At the second stage, with the lead level in passive feedback system channel increasing, negative reactivity from the first stage is compensated by insertion the positive reactivity, which a caused by lead level increasing, and the positive reactivity can be compensated by fuel temperature negative feedback.

3. Cross-validation with MCU-FR code

At the lack of experimental data cross-validation is very important for systems with lead coolant.The BREST-OD-300 reactor core model is calculated with EUCLID/V1 and MCU-FR [4] codes. MCU-FR code is the version of MCU, based on MCU-6 package and designed for calculations of systems with a fast neutron spectrum.

MCU-FR code uses 3D pin-by-pin fuel assembly nodalization and continuous energy cross sections supplied with the code. The cross-sections are prepared from RUSFOND [5] files and temperature of 600 K is included. EUCLID/V1 code uses homogenization of fuel assemblies and cross sections for 26 energy groups, prepared with CONSYST code based on RUSFOND files.

Four different critical states under normal reactor operating conditions are calculated and neutron spectrum, neutron multiplication factor and power distribution are analyzed. Different control rods positions are defined in the states (table 1).

Table 1. Control rods positions.

Variant №	The control rods position (cm)			
	CR1	CR2	RC	PS
1	50	50	60	70
2	50	130	60	70
3	130	130	60	70
4	50	130	60	180

The core of BREST-OD-300 contains protection system (PS) rods, reactivity compensator (RC) rods and control system (CS) rods. The position is set from the bottom of reactor, and control system rods has two groups (CR1 и CR2) with different position for each group.

The neutron spectra for 26 energy groups calculated by EUCLID/V1 and MCU-FR codes are in a good agreement (figure 7) for variants 1 and 4.

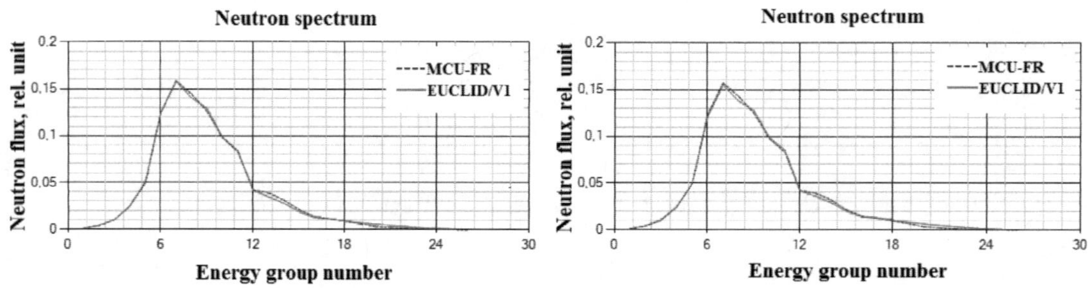

(a) (b)

Figure 7. Neutron spectrum for variants 1 (a) and 4 (b).

Difference of neutron multiplication factors in codes EUCLID/V1 and MCU-FR does not exceed 0.1% (table 2) for all variants. The calculation error in MCU-FR code equals 0.0002.

Table 2. Neutron multiplication factors calculated by EUCLID/V1 and MCU-FR codes.

Variant №	EUCLID/V1	MCU-FR	Difference (%)
1	1.0085	1.0093	0.08
2	1.0065	1.0070	0.05
3	1.0046	1.0053	0.07
4	0.9760	0.9757	0.03

The agreement can be explained by different heterogeneous correction factors in CONSYST code. Power distribution calculations are caused by good agreement of neutron multiplication factors.

Power distributions are considered for variants 1 and 4 with very different control rods positions. Power distributions for variants 2 and 3 are almost the same as for variant 1.

The axial power distributions calculated by EUCLID/V1 и MCU-FR (figure 8) are in good agreement for variants 1 and 4.

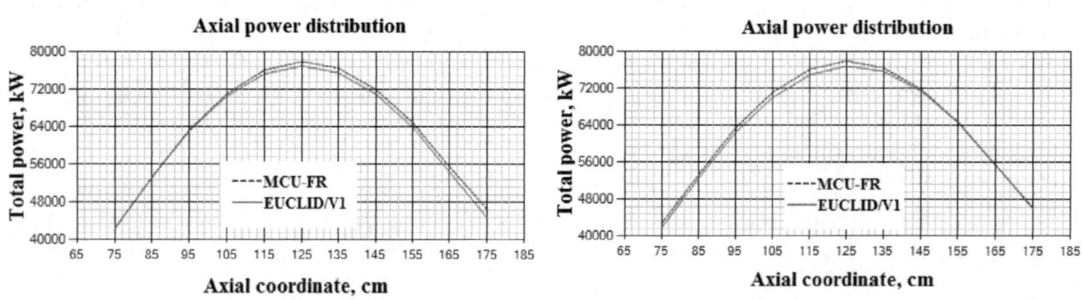

(a) (b)

Figure 8. The axial power distribution for variants 1 (a) and 4 (b).

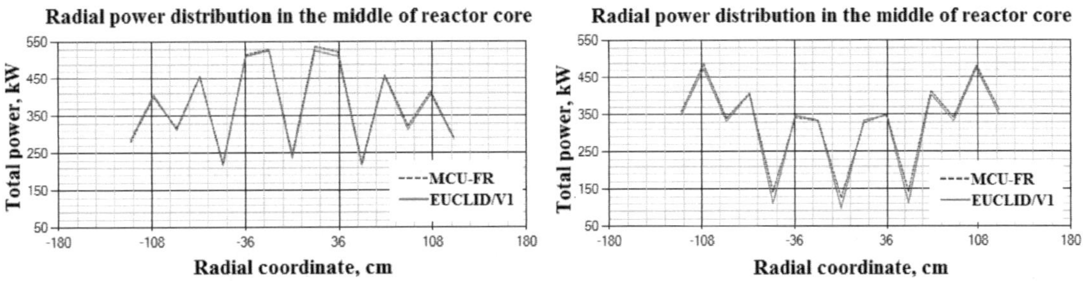

(a) (b)

Figure 9. The radial power distribution in the middle of reactor core for variants 1 (a) and 4 (b).

The radial power distributions calculated by EUCLID/V1 и MCU-FR in the middle of reactor core (figure 9) are in good agreement for variants 1 and 4.

4. Conclusion

The fuel assembly floating and decompression of passive feedback system channel modes for BREST-OD-300 reactor system was simulated with the EUCLID/V1 integrated code. The calculations demonstrate EUCLID/V1 code application for BREST-OD-300 simulating.

The EUCLID/V1 и MCU-FR result analysis demonstrates that EUCLID/V1 code models allow calculating correct neutron multiplication factors and power distributions. The difference of k_{eff} does not exceed 0.1%. The maximum difference in power distributions equals 5%.

References

[1] Belov A A, Vasekin V N, Veprev D P, Veretentsev V A, Veshchunov M S, Mosunova N A, Polous M A, Stakhanova A A, Strizhov V F, Muratov A G and Tukov V V 2013 Calculation of steady-state and transient operation of the lead cooled nuclear reactor by the EUCLID/V1 integrated code *Heavy Liquid Metal Coolants in nuclear technology* (Obninsk, Russia) p 23

[2] Vasekin V N and Stakhanova A A 2013 Neutronic code in a structure of integrated code EUCLID/V1 *XXIV Interdepartmental Seminar "Neutronic problems of nuclear energy"* (Obninsk, Russia)

[3] Manturov G N, Nikolaev M N, Polyakov A Y, Tsibulya A M 1999 Annotation of CONSYST code *Problems of Atomic Science and Technology. Series: Nuclear Constants* **2** p 148

[4] Kalugin M A, Oleynik D S and Shkarovsky D A 2015 Overview of the MCU Monte Carlo Software Package *Annals of Nuclear Energy* **82** pp 54–62

[5] Zabrodskaya S V, Ignatyuk A V, Koshcheev V N, Manochin V N, Nikolaev M N and Pronyaev V G 2007 RUSFOND – Russian national library of evaluated neutron data. *Problems of Atomic Science and Technology. Series: Nuclear Constants* **1–2** pp 3–21

The concerted calculation of the BN-600 reactor for the deterministic and stochastic codes

E V Bogdanova[1] and A N Kuznetsov[1]

[1] Department of theoretical and experimental physics of nuclear reactors, National Research Nuclear University MEPhI (Moscow Engineering Physics Institute), Kashirskoe shosse 31, Moscow, Russia

bogdanovaekaterina.mephi@gmail.com

Abstract. The solution of the problem of increasing the safety of nuclear power plants implies the existence of complete and reliable information about the processes occurring in the core of a working reactor. Nowadays the Monte-Carlo method is the most general-purpose method used to calculate the neutron-physical characteristic of the reactor. But it is characterized by large time of calculation. Therefore, it may be useful to carry out coupled calculations with stochastic and deterministic codes. This article presents the results of research for possibility of combining stochastic and deterministic algorithms in calculation the reactor BN-600. This is only one part of the work, which was carried out in the framework of the graduation project at the NRC "Kurchatov Institute" in cooperation with S. S. Gorodkov and M. A. Kalugin. It is considering the 2-D layer of the BN-600 reactor core from the international benchmark test, published in the report IAEA-TECDOC-1623. Calculations of the reactor were performed with MCU code and then with a standard operative diffusion algorithm with constants taken from the Monte - Carlo computation. Macro cross-section, diffusion coefficients, the effective multiplication factor and the distribution of neutron flux and power were obtained in 15 energy groups. The reasonable agreement between stochastic and deterministic calculations of the BN-600 is observed.

1. Introduction

Verification of modern program codes and methods of fast reactors core calculation is an important step in research activity devoted to possibilities of plutonium usage and minor actinides transmutation. Precision of neutron transfer modeling predictions and related reactivity coefficients should be evaluated.

Such evaluations should be received from comparison of calculations with real reactor conditions. However, the processes in reactor are very complicated. Precise and reliable information about these processes is difficult to obtain. Necessary detectors cannot be positioned everywhere and their indications have significant uncertainties. Moreover, the information about reactor design is not absolutely correct. Small geometry or composition deviations related with technological limits can make significant influence on calculation results.

This situation is simplified by the fact that the multiform process is calculated with complicated program complex, which gives an opportunity to verify parts of this complex separately. Thus, the quality of neutron calculation helps to establish the application of mathematical Monte-Carlo based model. According to the verified program such model has no differences from real reactor and allows obtaining any data necessary for results analyses. Information about cross-sections, reactions rates, fluxes and diffusion coefficients is located in MCU output [1] and can be transferred to some operative algorithm with the results that are possible to compare with corresponding MCU results. Such comparisons for VVER reactors were published earlier [2], [3].

2. Formulation of the problem

In this paper the 2-D layer of BN-600 core from international mathematic test published in IAEA report IAEA-TECDOC-1623 [4] is considered. Cartogram of the core in MCU coding is shown in Figure 1. Core is surrounded by five rows of reflector cells. There are seven fuel assembly types in this configuration.

```
R R R R
R R R R R R R R R R R R R
R R R R R R R R R R R R R R R R
R R R R R R R R R R R R R R R R R R R
R R R R R R R R R R R R R R R R R R R R
R R R R R R R R W W W W W R R R R R R R
R R R R R R W W W U U U U W W W R R R R R R
R R R R R R W W U U U U U U U U U W W R R R R R R
R R R R R R W U U U Z Z Z Z Z Z U U U W R R R R R R
R R R R R W W U U Z Z T Y Y Y T Z Z U U W W R R R R R
R R R R R W U U Z Z Y Y Y Y Y Y Y Y Z Z U U W R R R R R
R R R R W U U Z T Y Y Y Y T Y Y Y Y T Z U U W R R R R
R R R R W U U Z Y Y Y Y Y Y Y Y Y Y Y Y Z U U W R R R R
R R R R W U U Z Y Y Y Y T X X X T Y Y Y Y Z U U W R R R R
R R R R R W U Z Y Y T Y X X X X X X Y T Y Y Z U W R R R R R
R R R R R W U Z T Y Y Y X X X X X X X Y Y Y T Z U W R R R R R
R R R R W U U Z Y Y Y X X X X X X X X Y Y Y Z U U W R R R R
R R R R R W U Z Y Y Y T X X X T X X X T Y Y Y Z U W R R R R R
R R R R W U U Z Y Y Y X X X X X X X X Y Y Y Z U U W R R R R
R R R R R W U Z T Y Y Y X X X X X X X Y Y Y T Z U W R R R R R
R R R R R W U Z Y Y T Y X X X X X X Y T Y Y Z U W R R R R R
R R R R W U U Z Y Y Y Y T X X X T Y Y Y Y Z U U W R R R R
R R R R W U U Z Y Y Y Y Y Y Y Y Y Y Y Y Z U U W R R R R
R R R R W U U Z T Y Y Y Y T Y Y Y Y T Z U U W R R R R
R R R R R W U U Z Z Y Y Y Y Y Y Y Y Z Z U U W R R R R R
R R R R R W W U U Z Z T Y Y Y T Z Z U U W W R R R R R
R R R R R R W U U U Z Z Z Z Z Z U U U W R R R R R R
R R R R R R W W U U U U U U U U U W W R R R R R R
R R R R R R W W W U U U U W W W R R R R R R
R R R R R R R R W W W W W R R R R R R R
R R R R R R R R R R R R R R R R R R R R
R R R R R R R R R R R R R R R R R R R
R R R R R R R R R R R R R R R R
R R R R R R R R R R R R R
R R R R
```

Figure 1. Core cartogram in MCU coding. T - control rod, R - reflector, U - mixed oxide zone, W - high enrichment outer zone, X - low enrichment zone, Y - another low enrichment zone, Z - middle enrichment zone

Nuclear concentrations are described in [4]. Approximate «fission product» is included in these concentrations. Also there is a significant axial leakage in 3-D variant. In 2-D variant the leakage is absent and the fission product was not included in MCU input. Therefore, the model we used turned notably over-critical.

The supposing condition on flux equality to zero on core side surface is difficult too model with Monte-Carlo code in described 2-D tests. In our calculations we changed it by the demand of full neutron absorption on side surface. This surface has complicated geometry. In order to compute it the core was surrounded by the additional row of cells, which were united in special zone. On the surface of this zone full absorption condition can be imposed. This zone is marked with gray color in the Figure 2.

IOP Conf. Series: Journal of Physics: Conf. Series **781** (2017) 012004 doi:10.1088/1742-6596/781/1/012004

After the calculations on MCU code we computed this core with standard diffusion program which used the constants obtained by Monte-Carlo calculations. The mirror reflection condition was stated on upper and lower surfaces and zero flux condition in special zone cells surrounding the reactor was stated on the side surface.

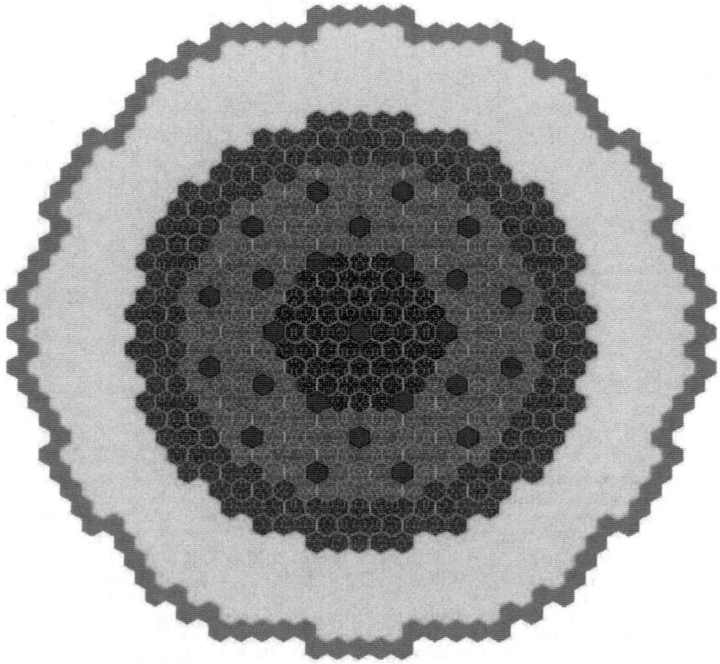

Figure 2. Core in calculation environment.

3. Stochastic and deterministic calculations

In calculation of the described source case 4.8 billion of neutrons were modeled so that the statistic uncertainty of fluxes distribution and reactions rates is about 0.1%. 96 processors were used, therefore the problem was solved in 11 hours.

Macro cross-section and diffusion coefficients in 15 energy groups were obtained for each of 823 assembly cells with MCU code. Boundaries of energy groups are shown in Table 1. Those cell characteristics were transferred to HEM-3 code [5], which made 15-group calculation of K_{eff} and distribution of group neutron fluxes and power that is proportional to MCU registered fission reaction intensity in standard finite-difference homogenous diffusion algorithm.

It is interesting that 40 years ago HEM-3 code was originally made in NRC "Kurchatov Institute" for 2-group calculations of cores with square cells. 25 years ago the ability of multigroup calculations and hexagonal cells were added. After that, the code was rarely used in 1-4 groups mode. About three years ago several cases with up to 12 groups were calculated. In these cases cell characteristics were obtained from whole core calculations with application of Monte-Carlo method and the results of finite-difference calculations were compared to stochastic as benchmarks.

4. Comparison of stochastic and deterministic calculations results

Both power production calculations were normalized by single fission energy. The differences in received distributions are often measured in percent of functional average value in whole calculation field, however in our case the major part of this field is filled with reflector cells. That is the average values are decreased and deviations are noticeably increased. Therefore, deviations are calculated as percent of functional maximum value. The main results are shown in Tables 2, 3.

Table 1. Boundaries of energy groups.

Group	Emin, eV	Emax, eV
1	6.50E+06	1.05E+07
2	4.00E+06	6.50E+06
3	2.50E+06	4.00E+06
4	1.40E+06	2.50E+06
5	8.00E+05	1.40E+06
6	4.00E+05	8.00E+05
7	2.00E+05	4.00E+05
8	1.00E+05	2.00E+05
9	4.65E+04	1.00E+05
10	2.15E+04	4.65E+04
11	1.00E+04	2.15E+04
12	4.65E+03	1.00E+04
13	2.15E+03	4.65E+03
14	1.00E+03	2.15E+03
15	0.00E+00	1.00E+03

Table 2. Effective neutron multiplication factor and maximum of energy discharge in stochastic and deterministic calculations.

	MCU	HEM-3
Effective neutron multiplication factor	1.1912	1.1893
Maximal value of assembly power per 1 neutron	0.001323	0.001330

Table 3. Average and maximum deviation calculations as percent of functional maximum value.

Group	Average deviation, %	Maximum deviation, %
1	0.4	3.2
2	0.5	2.2
3	0.5	2.5
4	0.4	1.6
5	0.3	1.7
6	0.6	1.7
7	1.5	3.5
8	1.7	3.7
9	1.8	3.9
10	2.7	6.1
11	2.7	6.3
12	2.0	4.5
13	4.4	8.8
14	4.6	22.0
15	12.9	28.3

As seen, the differences K_{eff} reach ~ 0.2%. Such and slightly bigger underestimations are observed regularly when MCU is used as benchmark for operative calculations. Hopefully, further experience will bring some clarity to this issue.

The differences in lower groups fluxes as shown in Table 3 are at first glance too big. However, reactor radial flux distribution shows that these fluxes make low contribution on the whole flux. The distribution of fluxes is presented in the Figure 3. According to the reactor radial flux distribution obtained with MCU and HEM-3 (Figure 4, 5) main biases in these fluxes are not in the core so that the importance of these biases is low. Favorable impression of performed calculations is provided by Figure 4, 5.

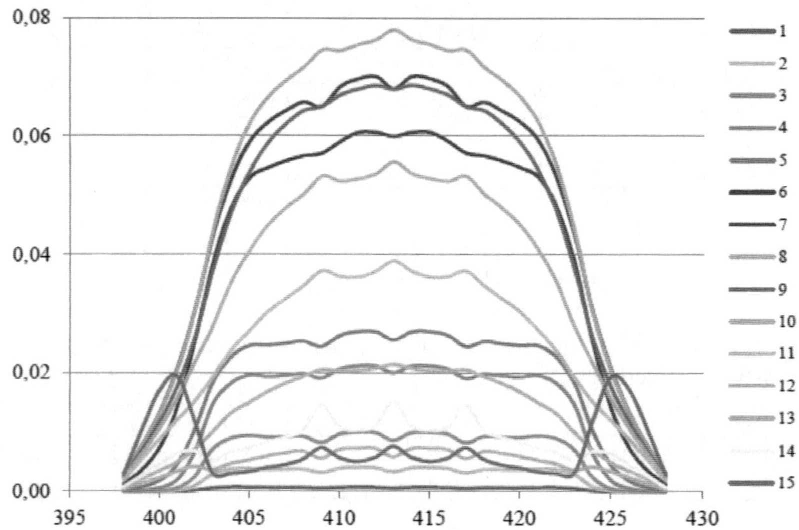

Figure 3. Distribution of group neutron fluxes calculated by MCU in horizontal diameter of reactor.

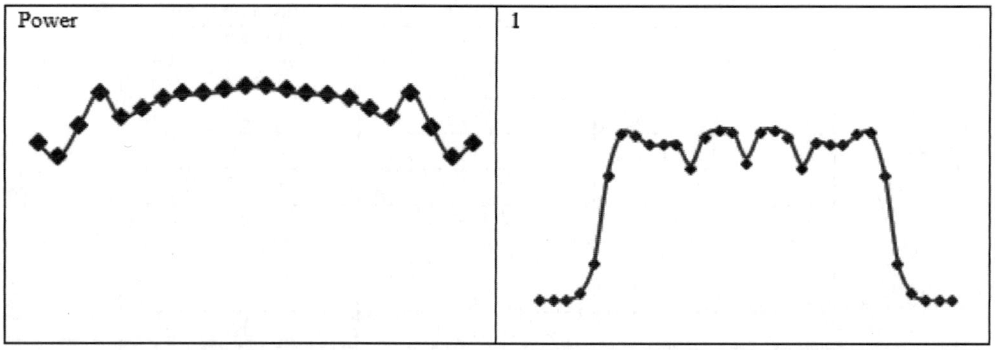

Figure 4. Flux distribution obtained with MCU (♦) and HEM-3(—) in horizontal diameter of reactor on the first energy group and energy production along the horizontal line next to the diameter on Figure 1.

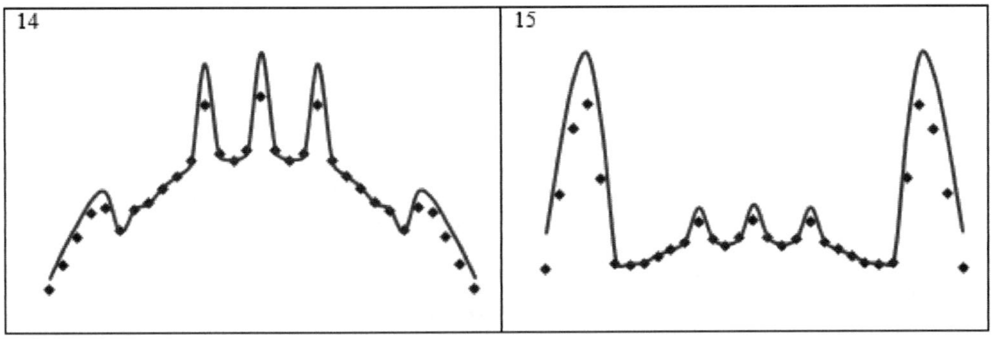

Figure 5. Flux distribution obtained with MCU (♦) and HEM-3 (—) in horizontal diameter of the reactor in the 14 - 15 energy groups.

5. Conclusion

The results of this work show reasonable agreement in stochastic and deterministic calculations of BN-600 reactor. The necessary cell characteristics for deterministic calculation were obtained as a byproduct of stochastic calculation without any subsequent corrections. Further researches are advisable for the issue of agreement improvement possibility with the use of the same group number or with single node of finite-difference mesh for a cell and alteration of these parameters.

References

[1] Alekseev N I, Bolshagin S N, Gomin E A, Gorodkov S S, Gurevich M I, Kalugin M A, Kulakov A S, Marin S V, Novoselsev A P, Oleinik D S, Prianichnikov A V, Sukhino-Fomenko E A, Shkarovskiy D A, Yudkevich M S 2011 Status of MCU-5 *Probl. of At. Sci. and Techn.: Phys. of Nucl. React.* ed 4 pp 4 – 23

[2] Gorodkov S S and Kalugin M A 2014 Applying full multigroup cell characteristics from MCU code to finite difference calculations of neutron field in VVER core *Proc. of the 24th Symp. of AER* Sochi Russian Federation pp 181 - 188

[3] Gorodkov S S, Kalugin M A and Kuznetcov A N 2015 Finite - difference fine mesh 2d calculations of neutron fields in VVER reactors using full sets of multigroup cell characteristic obtained with the MCU package *Proc. of 25th Symp. of AER* Balatongyörök, Hungary pp 571 – 579

[4] BN-600 hybrid core benchmark analyses 2009 IAEA_TECDOC-1623. Vienna

[5] Gorodkov S S 1984 Annotation of the program HEM-3 *Probl. of At. Sci. and Techn.: Phys. and Techn. of the Nucl. React.* ed 85(4) pp 42 – 42

Acknowledgments

The results were received with computational resources of MCC NRC «Kurchatov Institute» (http://computing.kiae.ru/).

Concerning advantages in using ^{208}Pb as such a FR coolant

G Khorasanov*[1], E Zemskov[2] and A Blokhin[3]

*[1]Obninsk Institute for Nuclear Power Engineering, Obninsk, 1 Studgorodok, 249040, Russia
[2]Institute for Physics and Power Engineering, Obninsk, 1 Bondarenko Square, 249033, Russia
[3]Nuclear Safety Institute of Russian Academy of Sciences, Moscow, 52 Bol'shaya Toul'skaya Street, 115191, Russia

*E-mail: khorasanow@yandex.ru

Abstract. In the paper cores of two fast reactors with heavy liquid metal coolant are considered. The first object, RBETS-M, is a project of a medium power, 900 MW, reactor cooled with lead-bismuth. The second object, BRUTS, is a project of an ultra-small power, 0.5 MW, reactor cooled with lead. The results of replacement of their standard coolants with the lead coolant enriched up to 100% with ^{208}Pb are presented. In the RBETS-M core having a large coolant volume share this replacement results in sufficient increasing the share of ^{238}U involved in the fission process and respective decreasing the share of ^{239}Pu and ^{241}Pu burning.

1. Introduction

Presently, several projects dedicated to fast reactors (FRs) with heavy liquid metal coolant (HLMC) are in progress in Russia. Two projects among them are considered in this paper. The first is a medium power reactor, RBETS-M, proposed by the Moscow Kurchatov Institute since 2000 [1] and developed in the frame of international collaboration. Innovation of this reactor lies in using gas lifting and lead-bismuth for core cooling. The second reactor considered is BRUTS, an ultra-small power reactor proposed by the Obninsk Institute for Nuclear Power Engineering in 2015 [2]. Innovation of BRUTS lies in using a coolant from natural lead, ^{nat}Pb, and a core with a stock of reactivity less than beta effective that makes this reactor safe and suitable for education purposes. Earlier, in 2013, a project of ELECTRA, a lead cooled reactor for training and education was published in Ref. 3. The goal of the present paper is to estimate the advantages which can be achieved in these two above mentioned reactors in the case of replacement of their standard coolants, Pb-Bi and ^{nat}Pb, with a coolant from ^{208}Pb which is expensive enough but is possessed of unique consuming properties [4-10].

2. Description of reactor's main parameters

In Table 1 several BRUTS and RBETS-M core parameters are given.

3. Method of calculations

In this work densities of neutron flux in the central parts of BRUTS and RBETS-M cooled with ^{nat}Pb/^{208}Pb and Pb-Bi/^{208}Pb, respectively, were calculated using Monte Carlo code. On this base, several neutron and physical characteristics: mean neutron energy, share of fast neutrons with E_n>0.8 MeV, shares of fissile and low fissile nuclides involved in the fission process were calculated.

Table 1. Main standard core parameters of the reactors BRUTS and RBETS.

Parameter	BRUTS's value	RBETS-M inner core's value
Thermal power [MW]	0.5	900
Fuel	UO_2	(Pu+U)N
Fuel enrichment [%]	19.7 of ^{235}U	13.7 of power grade Pu
Coolant	^{nat}Pb	Pb-Bi
Core diameter [cm]	62	86
Core height [cm]	62	100
Coolant volume share [%]	25.5	62.5

4. Results

In Table 2 the results of calculations are presented.

Table 2. Core neutron and physical characteristics of the reactors BRUTS and RBETS-M cooled with ^{nat}Pb and Pb-Bi (thin lettering) and ^{208}Pb (bold lettering).

Parameter	BRUTS's value	RBETS-M inner core's value
Mean neutron energy [MeV]	**0.5650** / 0.5543	**0.4226** / 0.3992
Variation of mean neutron energy [%]	**+ 1.9304**	**+ 5,8617**
Share of fast neutrons with E_n>0.8 MeV [%]	**0.1869** / 0.1832	**0.1438** / 0.1293
Variation of share of fast neutrons [%]	**+ 2.0197**	**+ 11.2142**
^{235}U share in total number of fissions [%]	**86.0610** / 86.5368	
^{237}Np share in total number of fissions [%]		**0.0236** / 0.0227
^{238}U share in total number of fissions [%]	**13.9390** / 13.4632	**16.6218** / 14.0162
^{238}Pu share in total number of fissions [%]		**0.8599** / 0.8713
^{239}Pu share in total number of fissions [%]		**69.0942** / 71.7217
^{240}Pu share in total number of fissions [%]		**5.5982** / 5.4488
^{241}Pu share in total number of fissions [%]		**5.9602** / 6.1699
^{241}Am share in total number of fissions [%]		**1.0825** / 1.0210
^{242}Pu share in total number of fissions [%]		**0.7594** / 0.7283

From Table 2 follows that using a coolant from ^{208}Pb instead of standard coolants from ^{nat}Pb or Pb-Bi leads to the following results:

- Increasing the mean neutron energy;
- Increasing the share of fast neutrons;
- Decreasing ^{235}U share in total number of fissions in BRUTS;
- Increasing ^{238}U share in total number of fissions in BRUTS;
- Increasing ^{237}Np share in total number of fissions in RBETS-M;
- Increasing ^{238}U share in total number of fissions in RBETS-M;
- Decreasing ^{238}Pu share in total number of fissions in RBETS-M;
- Decreasing ^{239}Pu share in total number of fissions in RBETS-M;
- Increasing ^{240}Pu share in total number of fissions in RBETS-M;
- Decreasing ^{241}Pu share in total number of fissions in RBETS-M;
- Increasing ^{241}Am share in total number of fissions in RBETS-M;
- Increasing ^{242}Pu share in total number of fissions in RBETS-M.

The profit in using ^{208}Pb as such a coolant increases with increasing a volume share of coolant in the core. For medium and high power FRs, the typical coolant volume share in the core is of 60-70%. For

the RBETS-M inner core it is equal to 62.5%, for the small-power reactor BRUTS it is equal to 25.5%.
In Fig.1 shares of ^{238}U and power grade Pu nuclides related to the total number of fissions in RBETS-M inner core cooled with Pb-Bi/^{208}Pb are presented.
Figure 1 demonstrates the advantages in using ^{208}Pb as such a coolant:
- Increase of low fissile nuclides: ^{237}Np, ^{238}U, ^{240,242}Pu, ^{241}Am burning;
- Decrease of valuable fissile nuclides: $^{238, 239, 241}$Pu burning.

Another advantage of ^{208}Pb consists in its low neutron absorbing ability. As it was shown in Ref. 10, one-group cross-section of neutron radiation capture by ^{208}Pb in RBETS-M inner core cooled with ^{208}Pb is equal to 0.93 mbarn, while one-group cross-section of neutron radiation capture by the mix Pb-Bi in the inner core of RBETS-M cooled with Pb-Bi is equal to 3.71 mbarn.

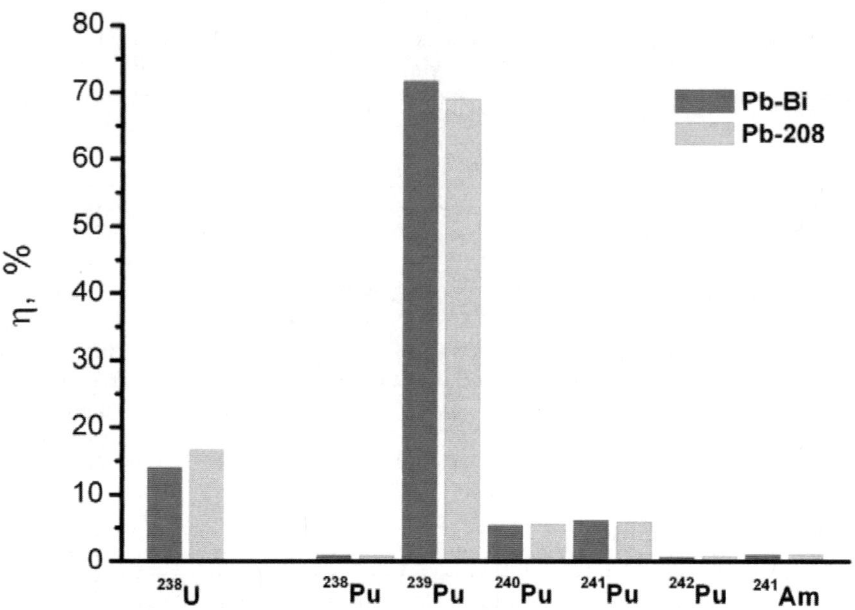

Figure 1. Shares of ^{238}U and power grade Pu nuclides related to the total number of fissions in RBETS-M inner core cooled with Pb-Bi/^{208}Pb.

5. Conclusion

On the base of the paper's results, it can be expected that in small-sized small power (100 MW electric) FR fueled with metallic uranium and power grade plutonium and cooled with innovative coolant from ^{208}Pb (about 150 tons) further increasing core's mean neutron energy and, respectively, increasing shares of ^{238}U, ^{237}Np, ^{240,242}Pu and ^{241}Am fissions in the total fission process can be acquired.

References

[1] Alekseev P N et al 2000 Preprint of the RRC Kurchatov Institute IAE-6179/4 Atomnaya Energiya 2004 97 pp 115-125
[2] Samokhin D S et al 2015 Communications of Higher Schools Nuclear Power Engineering (Izvestiya Visshikh Uchebnikh Zavedenij Yadernaya Energetika) Issue 3 pp 135-141
[3] Wallenius J et al 2013 Proceedings of the International Conference on Heavy Liquid Metal Coolants in Nuclear Technologies HLMC-2013 (Obninsk: IPPE) 1 pp 29-43

[4] Khorasanov G L and Blokhin A I 2013 Application of Stable Lead Isotope Pb-208 in Nuclear Power Engineering and Its Acquisition Techniques ed G L Khorasanov (New-York: Nova Science Publishers) Chapter 1 pp 1-19
[5] Khorasanov G L et al 2009 Nuclear Engineering and Design 239 pp 1703-1707
[6] Yu G and Wang K 2010 Proceedings of PHYSOR 2010 Pittsburgh Pennsylvania USA
[7] Okawa T and Sekimoto H 2011 Progress in Nuclear Energy 53 p 886-890
[8] Shmelyov A N et al 2011 Proceedings of GLOBAL 2011 Makuhari Japan
[9] Khorasanov G L and Blokhin A I 2013 Proceedings of the International Conference on Heavy Liquid Metal Coolants in Nuclear Technologies HLMC-2013 (Obninsk: IPPE) 2 pp 503-508
[10] Khorasanov G 2015 Transactions of the American Nuclear Society 112 Issue 1 pp 803-804

IOP Conf. Series: Journal of Physics: Conf. Series **781** (2017) 012006 doi:10.1088/1742-6596/781/1/012006

On fundamental quality of fission chain reaction to oppose rapid runaways of nuclear reactors

G G Kulikov, A N Shmelev, V A Apse and E G Kulikov

National Research Nuclear University MEPhI (Moscow Engineering Physics Institute), Kashirskoe highway, 31, Moscow, Russia

E-mai: ggkulikov@mephi.ru

Abstract. It has been shown that the in-hour equation characterizes the barriers and resistibility of fission chain reaction (FCR) against rapid runaways in nuclear reactors. Traditionally, nuclear reactors are characterized by the presence of barriers based on delayed and prompt neutrons. A new barrier based on the reflector neutrons that can occur when the fast reactor core is surrounded by a weakly absorbing neutron reflector with heavy atomic weight was proposed. It has been shown that the safety of this fast reactor is substantially improved, and considerable elongation of prompt neutron lifetime "devalues" the role of delayed neutron fraction as the maximum permissible reactivity for the reactor safety.

1. The in-hour equation and its physical sense

Within the frames of one-point kinetic model with six groups of delayed neutrons (DN) without accounting for the reactivity feedbacks, the in-hour equation that links the asymptotic period of the runaway T with the inserted positive reactivity ρ may be written in the following form:

$$\rho = \frac{\Lambda_{prt}}{T} + \sum_{i=1}^{6} \frac{\Lambda_i \cdot \beta_i}{T + \Lambda_i} \qquad (1)$$

where: ρ – the reactor reactivity, i.e. $\rho = (K_{ef} - K_0)\cdot(K_{ef} \cdot K_0)^{-1}$; K_0, K_{ef} – the values of effective neutron multiplication factor in two different reactor states: the unperturbed state and the state perturbed by the reactivity insertion, respectively; Λ_{prt} – means prompt neutron lifetime in the reactors with traditional neutron reflectors; Λ_i – means lifetime of nuclei-emitters of the i-th DN group ($\Lambda_i = 1/\lambda_i$, where λ_i – decay constant of nuclei-emitters of the i-th DN group, Λ_i = 0.3 ÷ 80 s); β_i – effective fraction of the i-th DN group; β - effective DN fraction ($\beta = \sum_{i=1}^{6} \beta_i$). The left term of the in-hour equation is the reactivity that should be inserted to provoke the runaway with a certain value of asymptotic period. The right part of the in-hour equation consists of two summands. The first summand depends only on mean prompt neutron lifetime Λ_{prt}, while the second summand depends only on DN parameters Λ_i, β_i. It seems natural to ascribe the contributions given by prompt and delayed neutrons into the reactor reactivity to the first and second summand, respectively. However, this conclusion contradicts with physical sense because of the following simple reason. Infinitely long asymptotic period of the runaway ($T \rightarrow \infty$) corresponds to the stationary mode of the reactor operation. As a consequence, the contribution from the second summand tends to zero, i.e. there is no any DN contribution into the FCR balance. But it is well-known that delayed neutrons mostly contribute into the reactor reactivity just in the case of the stationary reactor operation mode, and the maximal DN contribution is equal to effective DN fraction β. On the other hand, at rapid runaway, i.e. at very short asymptotic period ($T \to 0$), the so-called DN contribution tends to its maximal value, namely to effective DN fraction β. But this conclusion contradicts with elementary logical consideration. Indeed, at rapid runaway, delayed neutrons have no enough time to contribute into the FCR balance. According to the laws of radioactive decay, DN generation is a time-stretched process. When the most DN fraction appears in the reactor core, the FCR already increases substantially total neutron population. So, DN fraction in total neutron population and DN contribution into FCR balance become negligible. Similar arguments are also correct for prompt fission neurons.

Thus, it may be concluded that the two summands in the right part of the in-hour equation can define the value of **REDUCTION in the contribution fraction** (or **the contribution deficit**) into the reactor reactivity. The reactivity (the left part of the in-hour equation) determines the outside reactivity to be inserted in order to compensate the effects caused by postponed appearance of prompt and DN.

2. Analysis of the in-hour equation for fast and thermal reactors

By using the in-hour equation, dependencies of the reactor reactivity ρ on the asymptotic period of the runaway were calculated for the following two types of nuclear reactors (figure 1): fast lead-cooled reactor BREST (uranium-plutonium nitride fuel, neutron reflector made of natural lead, 50-cm thick) and thermal heavy-water reactor CANDU (uranium oxide fuel). Mean prompt neutron lifetimes were taken as 0.5 μs for BREST reactor and as 1 ms for CANDU reactor.

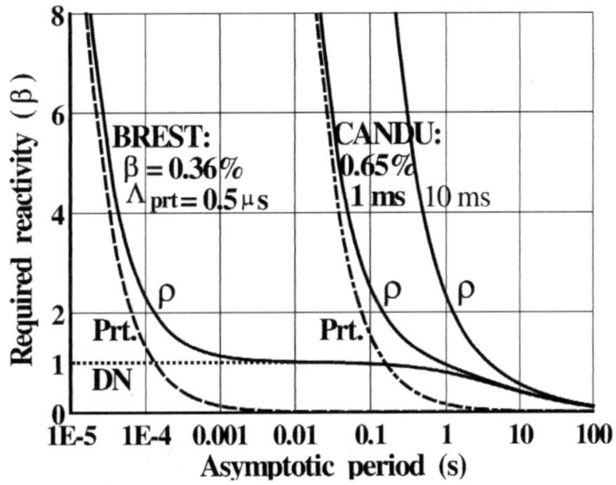

Figure 1. Dependencies of the required reactivity ρ on the asymptotic period of runaway T for BREST-type and CANDU-type reactors.

As is seen, both reactors behave themselves similarly at slow runaways ($T \geq 100\,s$). Very small reactivities (ρ □ β) must be inserted to provide these runaways. It may be said that there are no significant barriers against slow enough runaways.

Rapider runaways (T □ $10\,s$) require inserting the larger reactivities (ρ □ $0.5 \cdot \beta$) because not all delayed neutrons are able to contribute into FCR balance in proper time. So, the missing DN fraction (the DN contribution deficit) must be compensated by insertion of appropriate outside reactivity.

If the runaway rate becomes more large (T □ $1\,s$), then different reactivities must be inserted into different reactor types. In fast BREST-type reactor the runaway requires inserting the reactivity slightly lower than DN fraction ($\rho \leq \beta$). In thermal CANDU-type reactor the runaway requires inserting slightly more reactivity ($\rho \approx \beta$).

This difference may be explained as follows. At relatively short asymptotic period of the runaways (T □ $1\,s$) a significant part of delayed neutrons is not able to contribute into the FCR balance in proper time. The DN contribution deficit appears in both reactor types. In fast reactor, prompt neutron lifetime (0.5 μs) is considerably shorter than the asymptotic runaway period, and all prompt neutrons have enough time to contribute into the FCR balance (the deficit of prompt neutron contribution is absent). It is sufficient to insert the outside reactivity only for compensation of the DN contribution deficit ($\rho \leq \beta$). In thermal CANDU-type reactor with prompt neutron lifetime at the level of 1 ms not all prompt neutrons are able to contribute into the FCR balance in proper time. The deficit of prompt neutron contribution appears at the level of □ $0.1 \cdot \beta$.

The difference between fast and thermal reactors becomes especially large at even more rapid runaways ($T = 0.1 \div 0.001\,s$). Such the runaways in thermal CANDU-type reactor requires inserting very large outside reactivity to compensate significant deficits of contributions from prompt and delayed neutrons. In fast BREST-type reactor all prompt neutrons, as before, are able to contribute into the FCR balance in proper time. So, it is only sufficient to compensate the DN contribution deficit by small additional reactivity ($\rho = \beta + \varepsilon,\ \varepsilon \to 0$). It may be concluded from these considerations that thermal CANDU-type reactor is more protected against very rapid runaways than fast BREST-type reactor because thermal reactor requires the larger outside reactivity.

The protective properties of fast BREST-type reactor become more evident only under extremely rapid runaways ($T < 0.001\,s$), when not all prompt neutrons are able to contribute into the FCR balance in proper time. These extremely rapid runaways can be already regarded as explosive events.

How can we enhance operational safety of fast reactors? As it follows from the in-hour equation, it is necessary to elongate mean prompt neutron lifetime. Neutron moderator can slow down the FCR, but in this case the neutron spectrum in the reactor becomes thermal. Way out of the situation is considered in the next section.

3. Fast reactor with weak neutron absorber as a neutron reflector

3.1. The in-hour equation

Mean prompt neutron lifetime can be elongated by using relatively intense neutron leakage from the reactor core. It is proposed to surround the reactor core by a low-absorbing neutron reflector where neutrons could stay as long as possible. Then, neutrons could come back to the reactor core and contribute into the FCR balance with a certain time delay, thus elongating mean lifetime of prompt neutrons. So, it is necessary to select such a reflecting material that allows neutrons to penetrate as deep as possible and then come back, with a high probability, to the reactor core. As is known, mean neutron displacement in the slowing-down process is equal to $\sqrt{6 \cdot \tau}$, where τ – age of neutrons, and mean displacement of thermal neutrons is equal to $\sqrt{6} \cdot L$, where L – length of diffusion. Material for neutron reflector must satisfy the following two conditions. Firstly, the neutron age must be long enough to provide the deepest penetration of the slowing-down neutrons into the neutron reflector (maximal value of τ). Secondly, in order to provide high probability for the slowed neutrons to come back to the reactor core, the neutron diffusion length must be much larger than square root of the neutron age (L □ $\sqrt{\tau}$). Till now, it was found no one suitable material to be used as a neutron reflector in fast reactors that would satisfy the second requirement, including project of **fast reactor BREST with natural lead as a coolant and neutron reflector**. The required material must be characterized by heavy atomic weight and weak neutron absorption. We propose to use lead enriched with isotope ^{208}Pb as a quite suitable material. Mean neutron displacement during the slowing-down process in ^{208}Pb is equal to ~1.3 m, mean displacement of thermal neutrons is equal to ~8 m. This means that fast leakage neutrons are able, in the process of their slowing-down, penetrate deep enough in the neutron reflector (up to 1.3 m, in average), and then, with a high probability, to come back from deep layers of the neutron reflector into the reactor core. On the one hand, thanks to very low cross-section of neutron absorption by [208]Pb ($\sigma_c^{th} = 0.23\,mb$) [1], mean lifetime of thermal neutrons in infinite [208]Pb is equal to 0.6 s (this value may be considered as a typical delay time of neutrons in the [208]Pb-reflector). On the other hand, heavy atomic weight of [208]Pb allows us to hope that neutrons coming back from the neutron reflector to the reactor core are mainly relatively high-energy neutrons, partially slow neutrons and thermal neutrons only as a minor fraction. So, the reactor could remain to be the fast spectrum reactor.

Those neutrons which are coming back from the neutron reflector are prompt neutrons in their physical essence, but they are delayed neutrons for the FCR balance because these neutrons can contribute into the FCR balance only after some time delay they spent on migration from the reactor core, slowing-down in the neutron reflector and backward diffusion to the reactor core. By analogy with delayed neutrons, the neutrons which are coming back from the neutron reflector may be divided into several groups each of them includes the neutrons coming from the *j*-th layer of the neutron reflector. Then, the following modified in-hour equation can be obtained with accounting for neutrons from the neutron reflector:

$$\rho = \frac{\Lambda_C}{T} + \sum_{i=1}^{6} \frac{\Lambda_i \cdot \beta_i}{T + \Lambda_i} + \sum_{j=1}^{J} \frac{\Lambda_R^j \cdot \rho_R^j}{T + \Lambda_R^j} \tag{2}$$

where: J – the number of spatial layers in the neutron reflector, Λ_R^j – lifetime of neutrons coming back from the *j*-th layer of the neutron reflector, this value is a sum of neutron lifetime in the reactor core before leakage into the neutron reflector, time needed to migrate from the reactor core to the *j*-th layer of the neutron reflector, time of staying in the *j*-th layer of the neutron reflector, time needed to migrate from the *j*-th layer of the neutron reflector to the reactor core and lifetime in the reactor core after coming back; ρ_R^j – the reactivity gain caused by surrounding the reactor core with the *j*-th layer of the neutron reflector: $\rho_R^j = (K_{ef}^j - K_{ef}^{j-1}) \cdot (K_{ef}^j \cdot K_{ef}^{j-1})^{-1}$, K_{ef}^j – effective neutron multiplication factor in the reactor with *j* layers of the neutron reflector. For brevity, this value may be called as the reflector reactivity.

It is noteworthy here that the first summand in the in-hour equation (1) includes prompt neutron lifetime Λ_{prt} in the reactor as a whole. As distinct to equation (1), the first summand in the modified in-hour equation (2) includes neutron lifetime in the unreflected reactor core Λ_C.

If the number of spatial layers tends to infinity, then the discrete form of the in-hour equation (2) transforms into its continuous analog:

$$\rho = \frac{\Lambda_C}{T} + \sum_{i=1}^{6} \frac{\Lambda_i \cdot \beta_i}{T + \Lambda_i} + \int_0^V \frac{(d\rho_R(r)/dv) \cdot \Lambda_R(r) \cdot dv}{T + \Lambda_R(r)} \tag{3}$$

where: V – volume of the neutron reflector; $d\rho_R(r)/dv$ – spatial distribution of the specific reactivity gain caused by the neutron reflector, $\Lambda_R(r)$ – spatial lifetime distribution for neutrons from the neutron reflector.

3.2. Lifetime of neutrons from the neutron reflector

Lifetime of neutrons from the neutron reflector can be calculated by using the balance relationship presented in [2]. For the discrete model of the neutron reflector the balance relationship can be written in the following form: $\Lambda_{prt}^j = (1 - \rho_R^j) \cdot \Lambda_{prt}^{j-1} + \rho_R^j \cdot \Lambda_R^j$, where Λ_{prt}^j – prompt neutron lifetime in the reactor with *j* layers of the neutron reflector, if *j*=0, then $\Lambda_{prt}^0 = \Lambda_{\tilde{N}}$ – neutron lifetime in the unreflected reactor core. For the continuous model of the neutron reflector the balance relationship can be re-written as follows: $\Lambda_R(r) = \Lambda_{prt}(r) + \Lambda_{prt}'(r) / \rho_R'(r)$.

Some dependencies of relevant physical parameters on thickness of [208]Pb-reflector are presented in figure 2 for fast BREST-type reactor. These dependencies were calculated within the frames of spherical continuous model of neutron kinetics. It can be seen that the reactivity gain thanks to the reflector $\rho_R(r)$ grows as the reflector becomes thicker and approaches a saturation level.

Correspondingly, derivative of the reactivity gain decreases and vanishes. Derivative of the reactivity gain $d\rho_R(r)/dr$ defines the growth rate of the contribution given by neutrons from the neutron reflector to the FCR balance (analog of the DN fraction) as the neutron reflector becomes thicker. So, it is natural that derivative of the reactivity gain decreases as the reflector thickness increases because contribution of neutrons coming back from more distant reflector layers into fission chain reaction in the reactor core vanishes.

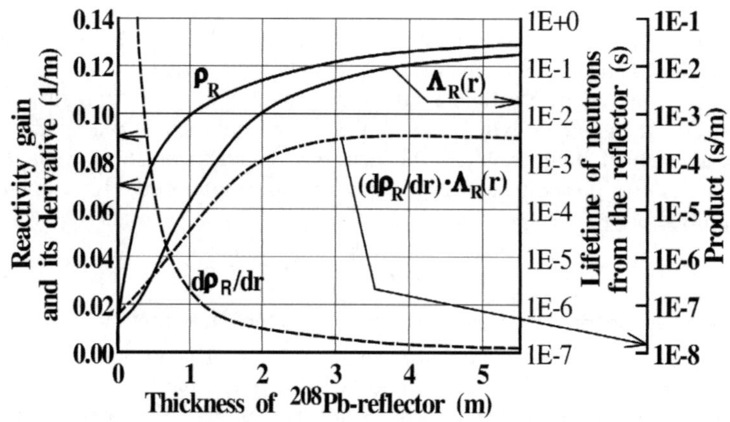

Figure 2. Characteristics of fast BREST-type reactor depending on thickness of ^{208}Pb-reflector.

Those neutrons which are coming back to the reactor core from the deeper layers of the neutron reflector are characterized by the longer lifetime. But the number of these neutrons gradually decreases and, as a result, their contribution into the FCR balance decreases too. The product of two values $\rho_R'(r)\cdot\Lambda_R$, denominator in the third summand of equation (3), is shown in figure 2. The product sharply increases within the range of small thickness, approaches the plateau level and then begins decreasing at large thickness of the neutron reflector. This means that there is a certain effective thickness which should not be exceeded. As is seen from figure 2, total contribution of neutrons from the thick (4 m) neutron reflector into the reactor reactivity reaches a significant value (about $35\cdot\beta$), and their lifetimes cover the range from 0.4 μs to 0.1 s. The reflector with so long neutron lifetime may be regarded as a physically thick medium in contrast to the reflectors with relatively short neutron lifetimes.

3.3. Analysis of the in-hour equation

The spherical continuous model of neutron kinetics was used to determine dependencies of the reactor reactivity and its components, namely the components related with prompt neutrons (Prompt), delayed neutrons (DN), their sum (Prompt + DN) and with neutrons from the reflector, on the asymptotic period of the runaway *T*. These dependencies are presented in figure 3 for fast BREST-type reactor with thick (4 m) ^{208}Pb-reflector.

Since lifetimes of neutrons from the reflector (from 1 μs to 0.1 s) are between lifetimes of prompt and delayed neutrons, under relatively slow runaways (relatively long asymptotic period $T \geq 1\,s$), large fraction of neutrons has enough time to come back from the neutron reflector to the reactor core and contribute into the FCR balance. Therefore, neutrons from the reflector can produce only insignificant effect on the outside reactivity to be inserted into the reactor core (see figure 3). So, the only protective barrier against such runaways in fast reactors is the barrier related with delayed neutrons. However, under rapid runaways ($T = 1 \div 0.001\,s$) the situation changes drastically. Within the range of so rapid and the most dangerous runaways the protective barrier caused by DN effects has exhausted completely its capability because delayed neutrons are not able now to contribute into the

FCR balance in proper time. The outside reactivity compensates the DN contribution deficit. Prompt fission neutrons can introduce their full contribution into the FCR balance. The protective barrier related with prompt neutrons is not working yet. Just within this range of the runaways a very small outside reactivity (much lower than DN fraction) inserted into fast BREST-type reactor without ^{208}Pb-reflector can accelerate the runaway by three orders of magnitude (the asymptotic period reduces from 1 s to 0.001 s, see sum of prompt and delayed neutrons in figure 3). But in this case a favorable effect produced by neutrons from the reflector on the reactor safety becomes definitive. Under so rapid runaways, not all neutrons have enough time to come back from the neutron reflector, and it becomes necessary to compensate their deficit. For example, it is sufficient to insert roughly the same outside reactivity (about DN fraction β) for initiation of the runaway in fast BREST-type reactor without ^{208}Pb-reflector with the asymptotic periods of 0.1 s, 0.01 s and 0.001 s. Initiation of the same runaways in the reactor core surrounded by ^{208}Pb-reflector requires inserting substantially larger values of the outside reactivity ($\sim 2\cdot\beta$, $\sim 4\cdot\beta$ and $\sim 7\cdot\beta$, respectively). The most fraction of these reactivities ($\sim\beta$, $\sim 3\cdot\beta$ and $\sim 6\cdot\beta$, respectively) is needed to compensate the contribution deficit related with neutrons from the reflector, Only about one β is needed to compensate the DN contribution deficit. If the asymptotic period of the runaway decreases, then the protective barrier related with neutrons from the reflector becomes stronger at the degree as the protective barrier related with prompt neutrons. As a result, the reactor resistibility to the runaways becomes substantially stronger too.

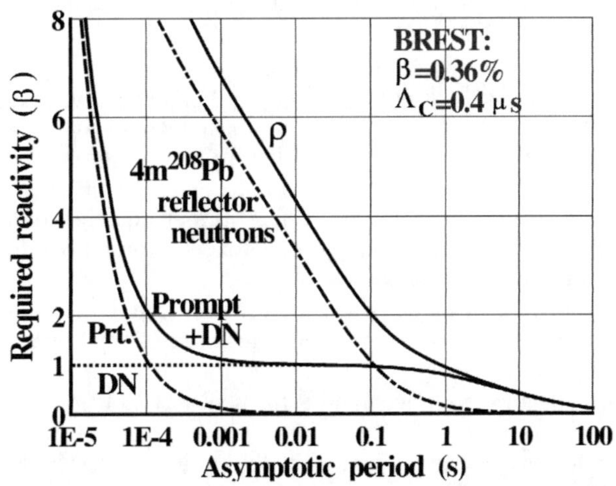

Figure 3. Reactivity and its components as functions of the asymptotic runaway period for fast BREST-type reactor with thick (4 m) ^{208}Pb-reflector.

Fuel rods with time constant about 0.1 s [3, 4] can provide additional protective effect even under so rapid runaways. In this case, heat will be efficiently removed by coolant from fuel rods, and negative reactivity feedback on the coolant temperature can counteract with the runaway.

4. Devaluation of DN fraction β as a maximal permissible reactivity for the reactor safety

Let us consider fast BREST-type reactor with ^{208}Pb-reflector and thermal CANDU-type reactor with prompt neutron lifetime at the level of 1 ms. Both reactors use such fuel rods for which the minimal asymptotic runaway period of 0.1 s is an acceptable value. If total reactivity margin in these reactors is below $2\cdot\beta$, then the runaways with the asymptotic periods shorter than 0.1 s are impossible under any reactivity-induced accidents (see figures 1 and 3). One half of the required reactivity is needed to compensate the DN contribution deficit while the other half is needed to compensate the contribution deficit related with neutrons from the reflector (for fast BREST-type reactor with ^{208}Pb-reflector) or with prompt neutrons (for thermal CANDU-type reactor).

Thus, the generally accepted threshold of the maximal permissible reactivity for safe reactor operation, namely effective fraction of delayed neutrons, can be exceeded, and the reactor operation will remain to be safe.

5. Conclusions

As is well-known, the FCR resistibility to rapid runaways in thermal reactors is substantially stronger than that in fast reactors with traditional neutron reflectors. However, a special neutron reflector can upgrade the FCR resistibility in fast reactors up to the level of the FCR resistibility in thermal reactors. This advantage of thermal reactors does not become apparent during their routine operation. The same situation takes place in fast reactors with a special neutron reflector because both reactor types use fuel rods with relative long time constant (~1-3 s). Under rapid runaways with the asymptotic periods shorter than time constant of these fuel rods the released heat has no enough time to flow down into coolant, and fuel rods can be melted. If fuel rods with relatively short time constants (0.01÷0.1 s) [3-4] are used in thermal and fast reactors with sufficiently long mean prompt neutron lifetime, then an opportunity arises to create conditions for timely actuation of various stabilization feedbacks even under the reactivity jumps above DN fraction β [5]. So, strong FCR resistibility to rapid runaways in fast reactors could be practically used in their routine operation. In particular, fast lead-cooled reactor BREST with a special neutron reflector, i.e. with strong FCR resistibility to rapid runaways, could be regarded as a reactor with inherent safety only if the rapidest acceptable runaway would require inserting the reactivity that would exceed full explicit (and implicit) reactivity margins stored in the reactor. **This property may be considered as an additional protective barrier in the reactor safety system that, as is known, includes fuel meat, cladding of fuel rod, and so on.**

The following main conclusions can be made from the obtained results:

1. New interpretation was proposed for the terms of the in-hour equation. The summands in the right part should be considered as the contribution deficits into the FCR balance from prompt and delayed neutrons, respectively. The left part of the in-hour equation is the reactivity needed to compensate the contribution deficits caused by time delays of prompt and delayed neutrons and to provide the runaway with given asymptotic period.
2. The FCR resistibility to rapid runaways in fast reactors can be significantly strengthened by surrounding the reactor core with the neutron reflector made of weak neutron absorber with heavy atomic weight.
3. The modified in-hour equation with accounting for neutrons from the reflector was derived and analyzed.
4. Neutrons from the reflector can play a significant role in upgrading the FCR resistibility to rapid runaways.

References

[1] Shibata K *et al* 2011 JENDL-4.0: a new library for nuclear science and engineering *Nuclear Science and Technology* **48** 1

[2] Kulikov G G, Shmelev A N and Apse V A 2014 Improving nuclear safety of fast reactors by slowing down fission chain reaction *Int. J. Hindawi Publication Corporation* Nuclear Energy **2014** 373726

[3] Ponomarev-Stepnoj N N *et al* 1999 *Atomic Energy* **86** 6 pp 443–449

[4] Mallone J, Totenmeier A, Shapiro N and Vaidyanathan S 2012 Lightbridge's advanced metallic fuel for light water reactors *Nuclear Technology* **180** pp 437–442

[5] Shmelev A N, Kulikov G G, Apse VA, Ternovykh M Yu, Chubko N V and Kozhahmet B K 2014 *Nuclear Physics and Engineering* (Moscow: NRNU MEPhI Press) **5** 7–8 pp 578–592

Acknowledgments

This research was supported by grant of the President of the Russian Federation for state support of young Russian PhD scientists 2016 (MK-8277.2016.8).

IOP Conf. Series: Journal of Physics: Conf. Series 781(2017) 012007 doi:10.1088/1742-6596/781/1/012007

Low-power lead-cooled fast reactor loaded with MOX-fuel

E R Sitdikov and A M Terekhova

Obninsk Institute for Nuclear Power Engineering, NRNU "MEPhI", 1 Studgorodok, Obninsk 249030, Kaluga Region, Russia

E-mail: emilsitdikov@gmail.com

Abstract. Fast reactor for the purpose of implementation of research, education of undergraduate and doctoral students in handling innovative fast reactors and training specialists for atomic research centers and nuclear power plants (BRUTs) was considered. Hard neutron spectrum achieved in the fast reactor with compact core and lead coolant. Possibility of prompt neutron runaway of the reactor is excluded due to the low reactivity margin which is less than the effective fraction of delayed neutrons. The possibility of using MOX fuel in the BRUTs reactor was examined. The effect of K_{eff} growth connected with replacement of natural lead coolant to 208Pb coolant was evaluated. The calculations and reactor core model were performed using the Serpent Monte Carlo code.

1. Introduction

Existing training simulators are not fully recreating nuclear power plant conditions, so the creation of training reactor will provide an opportunity to gain valuable practical skills. In addition to education, such reactors can be used for scientific and industrial purposes. At present, training reactors already exist in the world, or they are developed. For example, there are SLOWPOKE reactors in Canada [1] and training reactor VR-1 in Czech Republic [2].

Of particular interest are lead-cooled fast reactor projects for training purposes (ELECTRA, BRUTs) [3]. The main features of the BRUTs are: low power – 0,5 MW, the small sized core with a hard neutron spectrum, the low reactivity margin which is less than the effective fraction of delayed neutrons [4].

Purpose of the investigation is to obtain the characteristics of the BRUTs core loaded with MOX fuel and cooled with different lead coolant composition.

2. BRUTs model

2.1. General design layout and reactor core

Pool-type configuration ensuring arrangement within the same reactor vessel of the reactor core with reflectors, as well as upper shielding plug, heat exchangers, coolant treatment equipment for maintaining thermodynamic activity of oxygen dissolved in the coolant, thermal engineering sensors, in-vessel shielding and buffer plenum above the molten lead free surface was suggested for the examined design of BRUTs reactor.

BRUTs reactor vessel represents rugged steel cylindrical containment with elliptical bottom and lid equipped with bores for installation and fastening the retractable in-vessel devices.

Reactor vessel is mounted inside concrete shaft with gap between the vessel and shaft walls designed for arrangement of natural air circulation circuit for removing residual heat through the vessel in passive mode and releasing this heat in atmosphere. Separating barrel dividing cold and hot coolant flows and ensuring installation of the retractable core support barrel with reflectors and support plate and designed as part of the arrangement of the primary coolant circulation loop is available inside the vessel.

It is expected that circulation of coolant in the primary cooling loop will be achieved by natural convection. Selection of natural circulation for the primary cooling loop allows avoiding the problem of development and substantiation of operability of circulation pump for pumping heavy liquid metal coolant. Calculation estimations demonstrate that for achieving natural circulation of coolant in the primary loop it is sufficient to have riser with height of ~3 m above the reactor core [4].

Reactor core consists of seven fuel assemblies, with 252 fuel pins each. Fuel pins are arranged in hexagonal lattice. Technological channels are arranged in the center of each of the fuel assemblies for placement of (rods) of the reactor control and protection system (SUZ) (emergency shutdown system (RO AZ), reactivity compensation system (RO KR) and automatic reactor control system (RO AR)).

The core model used in the investigation has the parameters listed below in Table 1. Cross-section of the BRUTs reactor is shown in Figure 1.

Table 1. Main technical characteristics of the BRUTs reactor core.

Characteristics, measurement unit	Value
Fuel pin diameter along the smooth part, mm	12,7
Fuel pellet diameter, mm	11,5
The gas gap width (helium), mm	0,05
Fuel pin lattice spacing, mm	14
Number of fuel pins in the fuel assembly, items	252
Reactor core diameter, mm	618
Reactor core height, mm	780
Fuel pin cladding material	Steel EP 823
Technological channels outer/inner diameters, mm	48/44
Fuel composition	$PuO_2 + UO_2$
Fuel mass density, g/cm^3	9,2
Fuel volume in the core, %	61
Coolant volume in the core, %	25,5

2.2. Lead coolant

The primary coolant is lead coolant circulated by natural convection. Depending on the type of lead coolant, it may have a different isotopic composition. Natural lead (Pb-nat) includes 4 isotopes: ^{204}Pb (1,5%), ^{206}Pb (23,6%), ^{207}Pb (22,6%), ^{208}Pb (52,3%). Natural lead mass density is 10,686 g/cm^3 (at 600°C), atomic weight – 207,2.

^{208}Pb has the best neutronic properties among listed isotopes. Compared to other isotopes it has the smallest cross-section of the one-group neutron capture and the highest threshold for inelastic neutron scattering [5][6].

IOP Conf. Series: Journal of Physics: Conf. Series **781** (2017) 012007 doi:10.1088/1742-6596/781/1/012007

2.3. MOX-fuel

Current study deals with MOX-fuel with the following composition of civil (reactor-grade) plutonium: ^{234}U – 0,03%, ^{235}U – 0,03%, ^{236}U – 0,04%, ^{237}Np – 0,1%, ^{238}Pu – 1,19%, ^{239}Pu – 63,05%, ^{240}Pu – 21,5%, ^{241}Pu– 4,07%, ^{242}Pu – 4,12%, ^{241}Am – 5,87%.

Previous BRUTs reactor model is used for the study, which was created to calculate the BRUTs core cooled with ^{208}Pb and loaded with UO_2 fuel [7].

Possibility of prompt neutron runaway of the reactor can be excluded by the low reactivity margin which is less than the effective fraction of delayed neutrons. MOX-fuel has β_{eff}=0,0035.

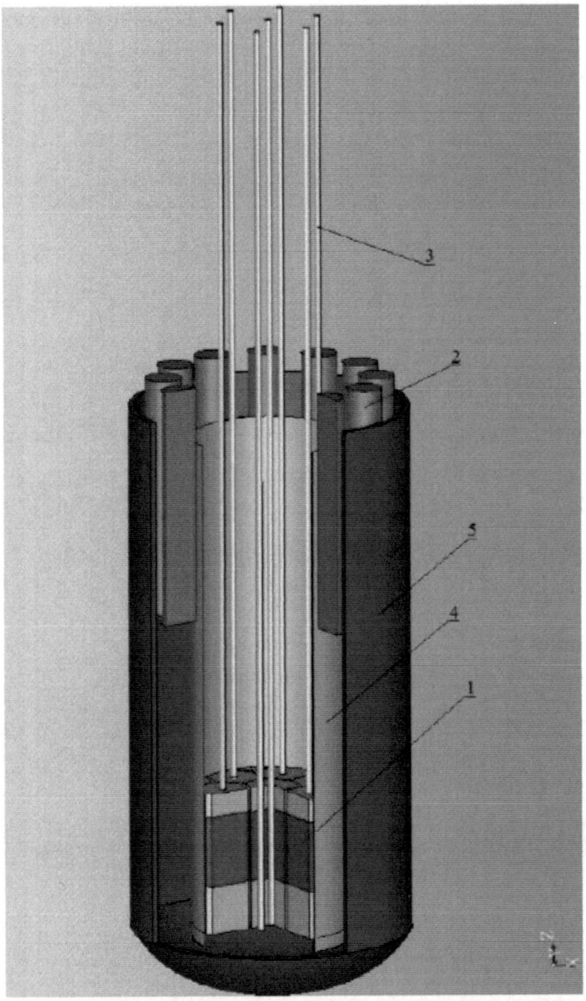

Figure 1. General appearance of the BRUTs reactor:
1 – reactor core;
2 –heat exchanger;
3 – reactor control and protection system drives;
4 – reactor core barrel;
5 – reactor vessel

3. Calculation results

The enrichment of MOX-fuel has been consistently chosen to meet the low reactivity requirement. At constant technical parameters of the core with the use of natural lead coolant (Pb-nat) the desired enrichment was obtained 22.8%, and K_{eff} = 1,00310 ± 0,00042. MOX-fuel with ^{239}Pu = 14,4% and ^{241}Pu = 0,93% is acceptable for use in the BRUTs. All calculations in the study were carried out with Serpent 1.1.17, which is Monte Carlo reactor physics calculation code. It allows calculation of various characteristics of systems containing nuclear fissile material. Serpent calculation code is capable of full-core modeling of research reactors, SMR's, and other closely coupled systems [8].

Replacement of Pb-nat coolant to ^{208}Pb with other constant core parameters gives increased effective multiplication factor equal to $K_{eff} = 1{,}03183 \pm 0{,}00088$, and $\Delta K_{eff} \approx 2{,}86\%$. This effect was examined in previous studies [7]. As compared with uranium oxide fuel MOX-fuel gives stronger effect (the effect for UO_2 was $\Delta K_{eff} \approx 1{,}6\%$).

As the percentage of the coolant in the reactor core is low, increase in K_{eff} is insignificant. In more powerful reactors with greater fraction of coolant in the reactor core will provide greater effect [5]. The same considerations are acceptable for achievement of harder neutron spectrum with ^{208}Pb [9].

Conclusion

The possibility of using MOX fuel in the BRUTs reactor was examined. Fuel enrichment of reactor-grade plutonium was obtained. Low reactivity margin which is less than the effective fraction of delayed neutrons excludes prompt neutron runaway. The effect of K_{eff} growth connected with replacement of natural lead coolant to ^{208}Pb coolant was evaluated – K_{eff} growth on MOX-fuel is higher. Calculations confirm that the use of lead coolant enriched by ^{208}Pb can reduce the initial enrichment of MOX fuel or to reduce the amount of the fuel loading [10].

References

[1] Kennedy G, Pierre J 2002 LEU-fuelled SLOWPOKE-2 research reactors: operational experience and utilisation *Proc. of the Int. Meeting on Reduced Enrichment for Research and Test Reactors* (Bariloche, Argentina)

[2] Rataj J 2010 Extensive utilisation of VR-1 reactor for nuclear education and training *14th Int. topical meeting on Research Reactor Fuel Management (Marrakeh,Morocco)* (Brussels: European Nuclear Society) pp 179-183

[3] Wallenius J, Suvdantsetseg E, Borrot S, Pukari M, Jolkkonen M, Claisse A, Olsson P, Ejenstam J, Szakalos P 2013 ELECTRA: A lead cooled reactor for training and education *Proc. of the 4th Conf. on Heavy Liquid Metal Coolants in Nuclear Tech. (HLMC-2013)* vol 1 (Obninsk: SSC RF-IPPE) pp 29–42

[4] Samokhin D, Khorasanov G, Tormyshev I, Zemskob E, Gostev A, Terehova A, Kuzmichyov S 2015 Low-power lead-cooled fast reactor for education purposes *J. Nuclear Energy and Tech.* JNET01(2015)003

[5] Khorasanov G 2015 Isotopic tailored lead coolant with new consuming properties for fast reactors *Transactions of the American Nuclear Society* Vol. 112 pp 803-804

[6] Blokhin D, Mitenkova E, Khorasanov G, Zemskov E, Blokhin A 2012 Evolution of fast reactor core spectra in changing a heavy liquid metal coolant by molten PB-208 *Proc. Int. Conf. on the Physics of Reactors (Advances in Reactor Physics* vol 2*)* (Knoxville: Curran Associates) pp 1452-1457

[7] Sitdikov E, Terekhova A 2016 *Proc. of the Russian Conf. on Nuclear Power Plants of ultra-small power for the applied and educational purposes* (Obninsk: OINPE NRNU MEPHI) p 18

[8] Leppänen J 2015 *Serpent – a Continuous-energy Monte Carlo Reactor Physics Burn up Calculation Code* (VTT Technical Research Centre of Finland)

[9] Khorasanov G 2013 *Application of Stable Lead Isotope Pb-208 in Nuclear Power Engineering and Its Acquisition Techniques* (New-York: NOVA publishers) p 184

[10] Khorasanov G, Korobeynikov V, Ivanov A, Blokhin A 2009 Minimization of an initial fast reactor uranium–plutonium load by using enriched lead-208 as a coolant *J. Nuclear Engineering and Design* JNED239(2009)009

IOP Conf. Series: Journal of Physics: Conf. Series 781(2017) 012008 doi:10.1088/1742-6596/781/1/01200

Atomistic simulation of defects formation and structure transitions in U-Mo alloys at swift heavy ion irradiation

L N Kolotova[1,2] **and S V Starikov**[1,2]

[1] Joint Institute for High Temperatures of the Russian Academy of Sciences, Izhorskaya 13, Bldg 2, Moscow, Russia
[2] Moscow Institute of Physics and Technology, InstitutskiyPereulok 9, Dolgoprudny, Moscow Region, Russia

E-mail: lada.kolotova@gmail.com

Abstract. At irradiation of swift heavy ions, the track formation frequently takes place in nuclear materials. There is a large interest to understanding of the mechanisms of defects/track formation at this phenomenon. In this work, the atomistic simulation of defects formation and melting in U-Mo alloys at irradiation of swift heavy ions has been carried out. We use the two-temperature atomistic model with explicit account of electron pressure and electron thermal conductivity. This two-temperature model describes ionic subsystem by means of molecular dynamics while the electron subsystem is considered in the continuum approach. The various mechanisms of structure changes at irradiation are examined. In particular, the simulation results indicate that the defects formation may be produced without melting and subsequent crystallization. Threshold stopping power of swift ions for the defects formation at irradiation in the various conditions are calculated.

1. Introduction

At irradiation of swift heavy ions (e.g. Xe ion - the typical fission fragment), the track formation frequently takes place in nuclear materials. The mechanism of the formation and structure of ion track has been a subject of a discussion since the late 50s of the last century [1, 2]. This phenomenon has great significance in radiation material science and nuclear engineering because a typical example of swift heavy ion is a fission fragment in a nuclear fuel (the initial energy of such an ion is about 100 MeV). In addition, the great interest in track formation is caused by the importance of fundamental problems in this field which need to be resolved.

Uranium has received a lot of attention for its unique nuclear properties and its various applications in nuclear industry. A high-temperature γ-phase has the best technical properties for nuclear engineering because it has body-centered cubic (bcc) lattice. However, γ-phase is extremely unstable at low temperature. For this reason, uranium is alloyed with other metals, which have bcc structure. Mo exhibits a high solubility in γ-U. Compared with other high density uranium alloys and compounds, the low-enriched uranium alloys with 6-12 wt.% of Mo have attracted a great deal of attention and are recognized as the most prominent candidates for advanced research and test reactors, because they have a relatively larger γ phase region and present more stable irradiation performance.

Mo is a strong γ-stabilizer which provides stable swelling behavior in U-Pu-Mo fuels, has high thermal conductivity, low thermal expansion, and high melting points [3–5].

Phase and structural transformations during irradiation are important for describing the evolution of the nuclear fuel in the operating conditions and fuel fabrication because the physical properties of U-Mo alloy strongly depend on the material lattice structure. There are a lot of models describing track formation in different materials. In this work the ion track formation in U-Mo alloy is associated with defect formation as it is usually done in pure metals [6, 7]. The mechanism of defect formation during irradiation is usually associated with collision cascades [8–10] or with melting and subsequent crystallization of material [11, 12]. In this work it is shown that several number of defects may be produced without melting and subsequent crystallization.

This work is devoted to theoretical analysis of the phenomena taking place in a nuclear fuel material under operating conditions. The simulation of defects formation and structure transitions in U-Mo alloys at irradiation of swift heavy ions has been carried out using the two-temperature atomistic model [12–14] with explicit account of electron pressure and electron thermal conductivity [12]. The various mechanisms of structure changes at irradiation are examined.

2. Computational technique

The structures of U–Mo alloys and the ion subsystem were described using molecular dynamics simulation with the novel interatomic angular dependent potential [15]. This potential contains an angular-dependent term and was parameterized using the force-matching method [16, 17]. All calculations were performed with the LAMMPS code [18]. The simulation cell had the following size: $160\alpha_0 \times 160\alpha_0 \times 6\alpha_0$, where α_0 is the lattice parameter of a U-Mo alloy structure with the periodic boundary conditions in all directions.

The energy of the swift heavy ions transfers primarily into the energy of the electron subsystem. For a short time, this produces a two-temperature state of matter, where the electron temperature T_e may be several orders higher than the ion temperature T_i. In this case the use of a two-temperature model is necessary. The ionic subsystem was described by the molecular dynamics approach, while the electron subsystem was considered in the continuum approach [12–14]. Energy transport within the electronic subsystem was solved according to the heat diffusion equation with added source terms for heat transfer between the subsystems:

$$C_e \rho_e \frac{\partial T_e}{\partial t} = \nabla(\kappa_e \nabla T_e) - g_p(T_e - T_i) \quad (1)$$

where C_e is the specific heat, $\kappa_e = D_e \cdot \rho_e \cdot C_e$ is the thermal conductivity (ρ_e is an electron density and D_e is an electron diffusivity), g_p is the coupling constant for the electron-ion interaction.

The electronic specific heat is taken in the form:

$$C_e = C_0 + C_1 T_e \quad (2)$$

where C_0 is equal $1.3 \cdot 10^{-10}$eV/(Ke), $C_1 = 1.37 \cdot 10^{-6}$eV/(K^2e).

Relaxation of strong electron excitations generated by swift ions is the main factor that determines the nature of the track regions. Thus the important task for the simulation of track formation is to estimate of the thermal diffusion coefficient D_e for the two-temperature model. It was obtained from the experimental data of the thermal conductivity κ_e [19].

Electronic pressure effects were included in the model to account for the blast force acting on ions because of electronic pressure gradient:

$$\vec{F}_i = -\partial U / \partial \vec{r}_i + \vec{F}_{\text{lang}} - \nabla P_e / n_i \quad (3)$$

where F_{lang} is a force from Langevin thermostat simulating electron-phonon coupling, and $\nabla P_e/n_i$ is the electron blast force. The electronic pressure is taken to be $P_e = B \cdot \rho_e \cdot C_e \cdot T_e$ [12].

3. The track formation

Thermal spike model was used for the description of track formation in this work. In this model the process of swift heavy ions irradiation passes through two main stages. The first stage involves the energy dissipation of electrons and energy transfer by means of electron-electron collisions. So the energy input from swift heavy ions was modeled by setting the initial electron temperature profile. The profile was calculated using the following model:

$$T_e = 300 + T_{max} \cdot exp(-d/r) \qquad (4)$$

The electron temperature profile was set in the plane in the center of the simulation cell, r - track radius, d - distance from the track axis. The track simulations were carried out with the swift heavy ions velocity directed along the [0 0 1] direction of the U-Mo crystalline lattice.

The second stage consists in the movement of atoms due to the electron-phonon interaction, while a high density of nonequilibrium defects is produced.

The various mechanisms of structure changes at irradiation were examined. These mechanisms significantly depend on the initial state of U-Mo alloy and ion energy. The influence of ion energy is usually described in terms of so-called stopping power $S = dE/dx$, which defines the rate of the kinetic energy loss of an ion per unit length along its path. In this work the values of the electron stopping power were ranged from 15 keV·nm^{-1} to 35 keV·nm^{-1}, which are close to the experimental data.

The atomistic simulation of defects formation and melting in U-Mo alloys at irradiation of swift heavy ions has been carried out for different initial temperatures and molybdenum concentrations. The energy input from swift heavy ion results in the increasing of the system temperature near the center of the simulation cell. The simulation results indicate that the defects formation may be produced without melting and subsequent crystallization. Apparently this is due to the low formation energy of self-interstitial atom [15]. Since the structural changes in the alloy occur at the certain temperature, the initial temperature also affects the defects formation in addition to the energy input. Therefore, threshold stopping power of swift ions for the defects formation at irradiation decreases from 33 keV·nm^{-1} to 18 keV·nm^{-1} when the temperature of U-Mo alloy increases from 300 K to 900 K (fig. 1) for U-Mo alloy with 5 atomic percents of molybdenum. Increasing the molybdenum concentration from 5 to 50 atomic percent leads to the increasing of melting temperature of U-Mo alloy (threshold stopping power increases from 22 keV·nm^{-1} to 39 keV·nm^{-1} at $T = 700$ K).

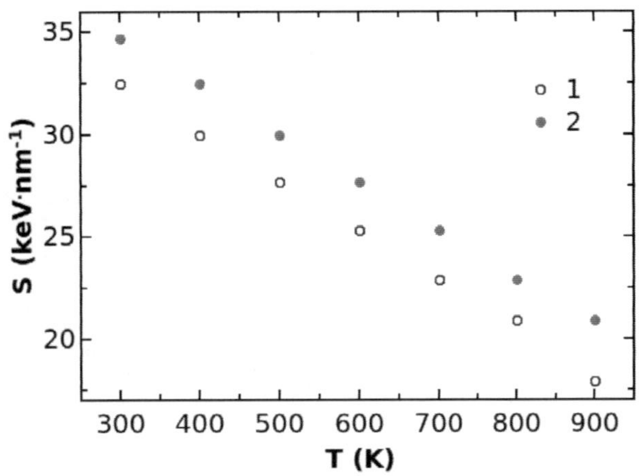

Figure 1. The dependence of threshold stopping power on temperature for U-Mo alloy with 5 atomic percents of molybdenum: 1 - threshold stopping power for defects formation, 2 - threshold stopping power for melting of ion subsystem.

Despite the mechanism of defect formation during irradiation is usually associated with melting and subsequent crystallization of material, the simulation results indicate that the defects formation in U-Mo alloy may be produced without melting and subsequent crystallization. The possible reason for this effect is small energy formation of Frekel pair in U-Mo alloy [15]. Thus threshold stopping power for defects formation at irradiation is 2-3 keV·nm^{-1} less than threshold stopping power for the melting of the ion subsystem (fig. 1).

4. Acknowledgements

The calculations were carried out on the computer clusters MVS-100K of the Joint Supercomputer Center of RAS. The work was supported by President RF Grant MK-7688.2015.8

References

[1] Young D A 1958 *Nature* **182** 375–382
[2] Silk E C H and Barnes R S 1959 *Phil. Mag.* **4** 970–972
[3] Kalashnikov V V, Titova V V, Sergeev G I and SamoilovA G 1959 *At. Energ.* **5** 1315
[4] Landa A, Söderlind P and Turchi P 2011 *J. Nucl. Mater.* **414** 132–137
[5] Kim Y S and Hofman G 2011 *J. Nucl. Mater.* **419** 291–301
[6] Itoh N, Duffy D M, Khakshouri S and Stoneham A M 2009 *J. Phys.: Condens. Matter* **21** 474205
[7] Norman G E, Starikov S V, Stegailov V V et al. 2013 *Contrib. Plasma Phys.* **53** 129-139
[8] Osetsky Y N, Calder A F and Stoller R E 2015 *Current Opinion in Solid State and Materials Science* **19** 277–286
[9] Korchuganov A V, Chernov V M, Zolnikov K P, Kryzhevich D S, Psakhie S G 2016 *Inorg. Mater. Appl. Res.***7** 648–657.
[10] Veshchunov M S, Boldyrev A V, Kuznetsov A V, Ozrin V D et al. 2015 *Nuclear Engineering and Design* **295** 116–126.
[11] Njoroge E G, Theron C C, Malherbe J B, van der Berg N G, Hlatshwayo T T and Skuratov V A 2015 *Nucl. Instr. Meth. Phys. Res.* **354** 249–254
[12] Pisarev V V and Starikov S V 2014 *J. Phys.: Condens. Matter* **26** 475401
[13] Duffy D M and Rutherford A M 2007 *J. Phys.: Condens. Matter* **19** 016207-016218
[14] Rutherford A M and Duffy D M 2007 *J. Phys.: Condens. Matter* **19** 496201-496210
[15] Smirnova D, Kuksin A and Starikov S 2015 *J. Nucl. Mater.* **458** 304–311
[16] Ercolessi F and Adams J B 1994 *Europhys. Lett.* **26** 583
[17] Brommer P and Gahler F 2007 *Modelling Simulation Mater. Sci. Eng.* **15** 295
[18] Plimpton S J 1995 *J. Comp. Phys.* **117** 1
[19] Rest J, Kim Y S, Hofman G L, Meyer M K and Hayes S L 2009 U-Mo Fuels Handbook *Cass Avenue Lemont (IL): Argonne Natoinal Laboratory.*

Improved safety fast reactor with "reservoir" for delayed neutrons generating

G G Kulikov, V A Apse, A N Shmelev and E G Kulikov

National Research Nuclear University MEPhI (Moscow Engineering Physics Institute), Kashirskoe highway, 31, Moscow, Russia

E-mai: ggkulikov@mephi.ru

Abstract. The paper considers the possibility to improve safety of fast reactors by using weak neutron absorber with large atomic weight as a material for external neutron reflector and for internal cavity in the reactor core (the neutron "reservoir") where generation of some additional "delayed" neutron takes place. The effects produced by the external neutron reflector and the internal neutron "reservoir" on kinetic behavior of fast reactors are inter-compared. It is demonstrated that neutron kinetics of fast reactors with such external and internal zones becomes the quieter as compared with neutron kinetics of thermal reactors.

1. The in-hour equation and slowing down of fission chain reaction

Within the frames of one-point kinetic model with six groups of delayed neutrons (DN) without accounting for the reactivity feedbacks, the in-hour equation that links the asymptotic period of the runaway T with the inserted positive reactivity ρ may be written in the following form:

$$\rho = \frac{\Lambda_{prt}}{T} + \sum_{i=1}^{6}\frac{\Lambda_i \cdot \beta_i}{T + \Lambda_i} \qquad (1)$$

where: ρ – the reactor reactivity, i.e. $\rho = (K_{ef} - K_0)\cdot(K_{ef}\cdot K_0)^{-1}$; K_0, K_{ef} – the values of effective neutron multiplication factor in two different reactor states: the unperturbed state and the state perturbed by the reactivity insertion, respectively; Λ_{prt} – means prompt neutron lifetime in the reactors with traditional neutron reflectors; Λ_i – means lifetime of nuclei-emitters of the i-th DN group ($\Lambda_i = 1/\lambda_i$, where λ_i – decay constant of nuclei-emitters of the i-th DN group, Λ_i = 0.3 ÷ 80 s); β_i – effective fraction of the i-th DN group; β - effective DN fraction ($\beta = \sum_{i=1}^{6}\beta_i$).

It is desirable to slow down intensification of the fission chain reaction (FCR) in order to improve the reactor safety. The FCR deceleration means that the larger reactivity ρ should be inserted into the reactor core to provoke the power excursion with the same asymptotic period T. As is seen from the in-hour equation (1), the FCR could be slowed down by elongating mean prompt neutron lifetime. The prompt neutron lifetimes in thermal reactors are longer than those in fast reactors by several orders of magnitude. If the fast reactor core is surrounded by a light neutron moderator, then the prompt neutron lifetime becomes longer, but the fast reactor is transformed into the reactor with thermal neutron spectrum. What should be done to elongate the prompt neutron lifetime and retain high-energy neutron spectrum of fast reactors?

2. The in-hour equation for fast reactor with weak neutron absorber as a neutron reflector

Mean prompt neutron lifetime can be elongated by using relatively intense neutron leakage from the reactor core. It is proposed to surround the reactor core by a low-absorbing neutron reflector where neutrons could stay as long as possible. Then, neutrons could come back to the reactor core and contribute into the FCR balance with a certain time delay, thus elongating mean lifetime of prompt neutrons. So, it is necessary to select such a reflecting material that allows neutrons to penetrate as deep as possible and then come back, with a high probability, to the reactor core. We propose to use lead enriched with isotope ^{208}Pb as a quite suitable material [1-5]. On the one hand, thanks to very low cross-section of neutron absorption by ^{208}Pb ($\sigma_c^{th} = 0.23\, mb$), mean lifetime of thermal neutrons in infinite ^{208}Pb is equal to 0.6 s (this value may be considered as a typical delay time of neutrons in the ^{208}Pb-reflector). On the other hand, heavy atomic weight of ^{208}Pb allows us to hope that neutrons coming back from the neutron reflector to the reactor core are mainly relatively high-energy neutrons, partially slow neutrons and thermal neutrons only as a minor fraction. So, the reactor could remain to be the fast spectrum reactor.

By analogy with delayed neutrons, the neutrons which are coming back from the neutron reflector may be divided into several groups each of them includes the neutrons coming from the *j*-th layer of the neutron reflector. Then, the following modified in-hour equation can be obtained with accounting for neutrons from the neutron reflector:

$$\rho = \frac{\Lambda_C}{T} + \sum_{i=1}^{6} \frac{\Lambda_i \cdot \beta_i}{T + \Lambda_i} + \sum_{j=1}^{J} \frac{\Lambda_R^j \cdot \rho_R^j}{T + \Lambda_R^j} \qquad (2)$$

where: J – the number of spatial layers in the neutron reflector, Λ_R^j – lifetime of neutrons coming back from the *j*-th layer of the neutron reflector, this value is a sum of neutron lifetime in the reactor core before leakage into the neutron reflector, time needed to migrate from the reactor core to the *j*-th layer of the neutron reflector, time of staying in the *j*-th layer of the neutron reflector, time needed to migrate from the *j*-th layer of the neutron reflector to the reactor core and lifetime in the reactor core after coming back; ρ_R^j – the reactivity gain caused by surrounding the reactor core with the *j*-th layer of the neutron reflector: $\rho_R^j = (K_{ef}^j - K_{ef}^{j-1}) \cdot (K_{ef}^j \cdot K_{ef}^{j-1})^{-1}$, K_{ef}^j – effective neutron multiplication factor in the reactor with *j* layers of the neutron reflector. For brevity, this value may be called as the reflector reactivity.

It is noteworthy here that the first summand in the in-hour equation (1) includes prompt neutron lifetime Λ_{prt} in the reactor as a whole. As distinct to equation (1), the first summand in the modified in-hour equation (2) includes neutron lifetime in the unreflected reactor core Λ_C.

If the number of spatial layers tends to infinity, then the discrete form of the in-hour equation (2) transforms into its continuous analog:

$$\rho = \frac{\Lambda_C}{T} + \sum_{i=1}^{6} \frac{\Lambda_i \cdot \beta_i}{T + \Lambda_i} + \int_0^V \frac{(d\rho_R(r)/dv) \cdot \Lambda_R(r) \cdot dv}{T + \Lambda_R(r)} \qquad (3)$$

where: V – volume of the neutron reflector; $d\rho_R(r)/dv$ – spatial distribution of the specific reactivity gain caused by the neutron reflector, $\Lambda_R(r)$ – spatial lifetime distribution for neutrons from the neutron reflector.

3. Fast reactors with different neutron reflectors

The paper considers three variants of the neutron reflectors in fast lead-cooled reactor BREST: the external neutron reflector made of natural lead (50-cm thick), the external neutron reflector made of ^{208}Pb (250-cm thick), the internal neutron reflector (the neutron “reservoir”, 250-cm thick) made of ^{208}Pb. The internal neutron “reservoir” occupies central zone of the reactor core, i.e. the neutron “reservoir” is encircled all round by the reactor core. For simplicity, initial cylindrical models were replaced by equivalent spherical models (see Fig. 1.a, b, c). Initial cylindrical model of the fast reactor core with the internal neutron “reservoir” is presented in Fig. 1.d. Thicknesses of the reactor core and fuel enrichments were chosen by such a way that the reactors would be critical and their breeding ratio would be about 1.05, as in initial model. This means that neutron-physical parameters of the fast reactors with the external and internal neutron reflectors made of ^{208}Pb would be close to the reactors reflected by natural lead.

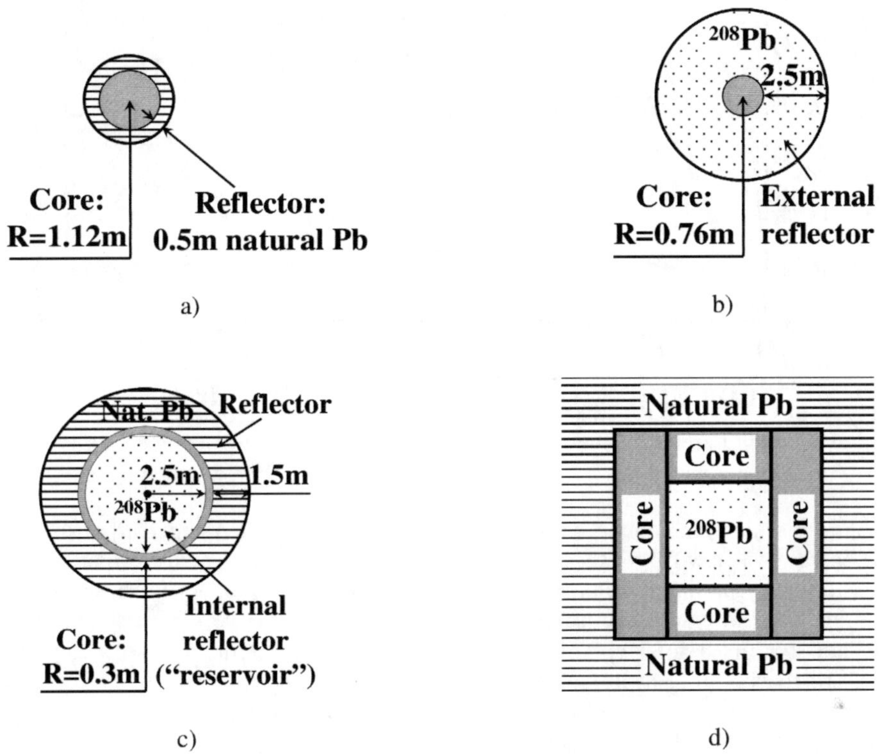

Figure 1. Spherical (a, b, c) and cylindrical (d) models of fast BREST-type reactor.

4. Lifetime and reactivity of neutrons from the neutron reflector

Lifetime of neutrons from the neutron reflector can be calculated by using the balance relationship presented in [3]. For the discrete model of the neutron reflector the balance relationship can be written in the following form: $\Lambda^{j}_{prt} = (1-\rho^{j}_{R})\cdot\Lambda^{j-1}_{prt} + \rho^{j}_{R}\cdot\Lambda^{j}_{R}$, where Λ^{j}_{prt} – prompt neutron lifetime in the reactor with j layers of the neutron reflector, if j=0, then $\Lambda^{0}_{prt} = \Lambda_{N}$ – neutron lifetime in the unreflected reactor core. For the continuous model of the neutron reflector the balance relationship can be re-written as follows: $\Lambda_{R}(r) = \Lambda_{prt}(r) + \Lambda'_{prt}(r)/\rho'_{R}(r)$.

Some dependencies of relevant physical parameters on thickness of ^{208}Pb-reflector are presented in Fig. 2 for fast BREST-type reactor. These dependencies were calculated within the frames of spherical continuous model of neutron kinetics. It can be seen that the reactivity gain thanks to the reflector $\rho_{R}(r)$ grows as the reflector becomes thicker and approaches a saturation level. The term "thickening of the internal reflector" means increment of the reflector thickness from the inner boundary of the annular reactor core to the reactor center.

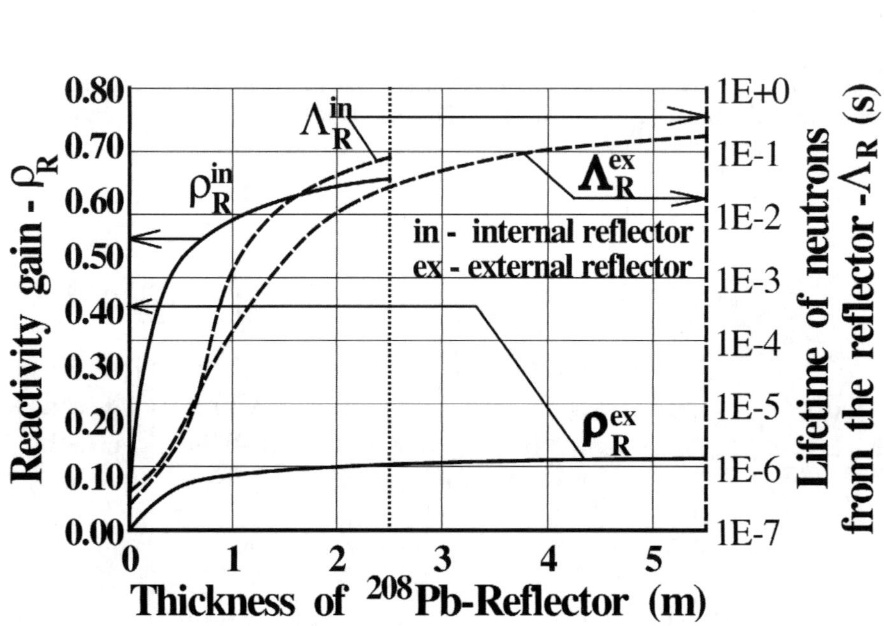

Figure 2. Characteristics of fast BREST-type reactor depending on thickness of external and internal ^{208}Pb-reflectors.

Those neutrons which are coming back to the reactor core from the deeper layers of the neutron reflector are characterized by the longer lifetime. But the number of these neutrons gradually decreases and, as a result, their contribution into the FCR balance decreases too. As is shown in Fig. 2, the neutrons coming from the external reflector and from the neutron "reservoir" (^{208}Pb, 250-cm thick) are able to provide significant total contribution into the reactivity, at the level of 15·β and 100·β, respectively. Lifetimes of these neutrons cover the range from 0.3 μs to 0.02 s and 0.08 s, respectively.

5. Asymptotic process of behavior of fast and thermal reactors

Dependencies of the reactivity required to provoke the power excursion with the asymptotic period T are presented in Fig. 3 for the following cases: fast BREST-type reactor with the external reflectors made of natural lead (50-cm thick) and made of ^{208}Pb (250-cm thick); fast BREST-type reactor with the internal reflector (neutron "reservoir") made of ^{208}Pb (250-cm thick); thermal VVER-type reactor and thermal CANDU-type reactor. As is seen, fast BREST-type reactor reflected by natural lead (50-cm thick) is characterized by rather short prompt neutron lifetime (about 1 μs) and, as a consequence, the most rapid and the most dangerous neutron kinetics – the rapid power excursions can be provoked by relatively small reactivities. Neutron kinetics of fast BREST-type reactor reflected by ^{208}Pb at the power excursions with the asymptotic periods shorter than 0.01 s is comparable with neutron kinetics of thermal VVER-type reactor with prompt neutron lifetimes at the level of 0.1 ms. Fast BREST-type reactor with the internal reflector (neutron "reservoir") made of ^{208}Pb is characterized by even the quieter neutron kinetics than thermal CANDU-type reactor with the longest prompt neutron lifetimes (at the level of 1 ms).

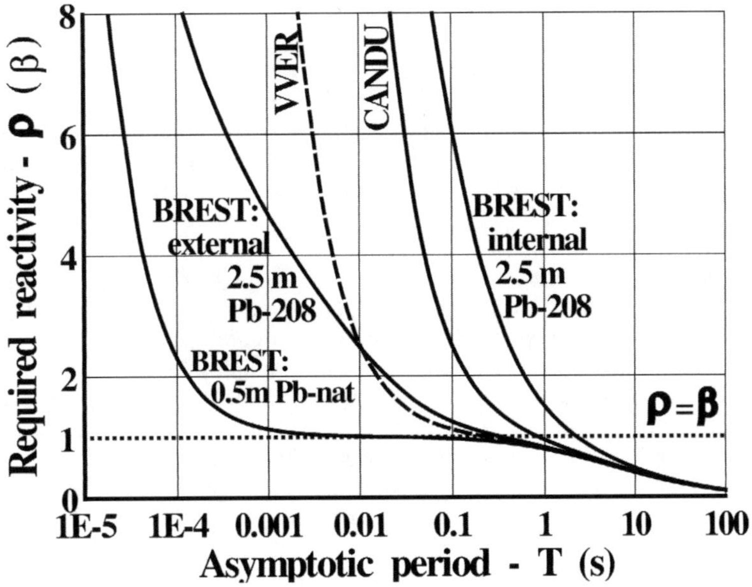

Figure 3. Dependencies of the required reactivity ρ on the asymptotic period of runaway T for BREST-type with different reflectors as well as VVER-type and CANDU-type reactors.

6. Transitional process of behavior of fast and thermal reactors

Time-dependent power evolutions are presented in Fig. 4 for the reactors mentioned above just after the positive reactivity jump (+0.9·β), i.e. before establishing the asymptotic power evolution.

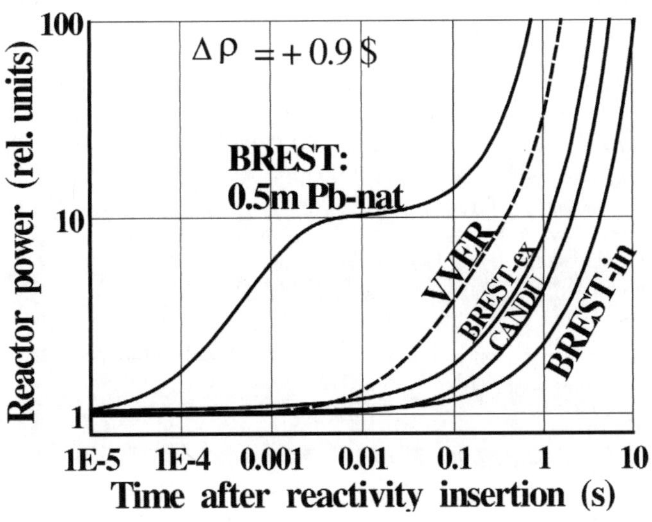

Figure 4. Reactor power as functions of time after reactivity insertion for fast BREST-type reactor with different reflectors as well as VVER-type and CANDU-type reactors.
"BREST" – BREST-type reactor with 0.5 m natural lead external reflector (Fig. 1a),
"BREST-ex" – BREST-type reactor with 2.5 m ^{208}Pb external reflector (Fig. 1b),
"BREST-in" – BREST-type reactor with 2.5 m ^{208}Pb internal reflector ("reservoir") and 1.5 m natural lead external reflector (Fig. 1c, d).

As is seen (Fig. 4), power of fast BREST-type reactor reflected by natural lead (50-cm thick) increases on one order of magnitude instantly, for 10 ms after the reactivity jumped. The same power excursions occurred in thermal VVER-type reactor and CANDU-type reactor for substantially longer time intervals (0.4 s and 2 s, respectively). The same power excursions in fast BREST-type reactor with the external reflector can occur for one second, with the external and internal reflectors – for four seconds. So long power evolutions can produce a favorable effect on the reactor safety because negative reactivity feedbacks on coolant properties can receive enough time to actuate.

7. Conclusions

1. The way is proposed to decelerate neutron kinetics of fast reactors and, thus, improve the reactor safety by using ^{208}Pb as a material of the external neutron reflector and as a material of the internal neutron reflector (the neutron "reservoir" inside of the reactor core).

2. The neutron "reservoir" is more effective than the external reflector because neutrons from the "reservoir" are not lost for the FCR, they come back to the reactor core inevitably. On the contrary, significant neutron fraction can pass through the external reflector and become useless for the FCR.

3. After positive reactivity jump, power of fast BREST-type reactor with the external and internal neutron reflectors made of ^{208}Pb increases substantially slower than power of thermal VVER-type reactor and CANDU-type reactor in transient and asymptotic processes.

References

[1] Shmelev A N, Kulikov G G, Kryuchkov E F, Apse V A, and Kulikov E G **2013** Application of radiogenic lead with dominant content of ^{208}Pb for long prompt neutron lifetime in fast reactor *Nuclear Technoogy* **183** 3 pp 409–426

[2] Kulikov G G, Shmelev A N, Kryuchkov E F, Apse VA, Kulikov E G, and Chekin A A **2013** Advantages of the use of radiogenic lead in fas reactors as a coolant and a reflector; some features of fast kinetics ($\rho \geq \beta$) *Nuclear Physics and Engineering* (Moscow: NRNU MEPhI Press) **4** 5 pp 462–480

[3] Kulikov G G, Shmelev A N, and Apse V A **2014** Improving nuclear safety of fast reactors by slowing down fission chain reaction *Int. J. Hindawi Publication Corporation* Nuclear Energy **2014** 373726

[4] Shmelev A N, Kulikov G G, Apse VA, Ternovykh M Yu, Chubko N V, and Kozhahmet B K **2014** Role of the neutron lifetime in prompt neutron excursions (model analysis) *Nuclear Physics and Engineering* (Moscow: NRNU MEPhI Press) **5** 7–8 pp 578–592

[5] Kulikov G G, Apse V A, Naumov V I, Shmelev A N, Kulikov E G, and Shkodin A O. **2015** Fundamental property of resistibility of fission chain reaction against rapid runaways of nuclear reactors *Nuclear Physics and Engineering* (Moscow: NRNU MEPhI Press) **6** 8 pp 1–15

Acknowledgments

This research was supported by grant of the President of the Russian Federation for state support of young Russian PhD scientists 2016 (MK-8277.2016.8).

Mechanistic approach for nitride fuel evolution and fission product release under irradiation

A P Dolgodvorov and V D Ozrin

Nuclear Safety Institute (IBRAE), Russian Academy of Science, 52 Bolshaya Tulskaya street, Moscow, Russia

E-mail: alexey.dolgodvorov@ibrae.ac.ru

Abstract. A model for describing uranium-plutonium mixed nitride fuel pellet burning was developed. Except fission products generating, the model includes impurities of oxygen and carbon. Nitrogen behaviour in nitride fuel was analysed and the nitrogen chemical potential in solid solution with uranium-plutonium nitride was constructed. The chemical program module was tested with the help of thermodynamic equilibrium phase distribution calculation. Results were compared with analogous data in literature, quite good agreement was achieved, especially for uranium sesquinitride, metallic species and some oxides. Calculation of a process of nitride fuel burning was also conducted. Used mechanistic approaches for fission product evolution give the opportunity to find fission gas release fractions and also volumes of intergranular secondary phases. Calculations present that the most massive secondary phases are the oxide and metallic phases. Oxide phase contain approximately 1 % wt of substance over all time of burning with slightly increasing of content. Metallic phase has considerable rising of mass and by the last stage of burning it contains about 0.6 % wt of substance. Intermetallic phase has less increasing rate than metallic phase and include from 0.1 to 0.2 % wt over all time of burning. The highest element fractions of released gaseous fission products correspond to caesium and iodide.

1. Introduction

Mechanistic modeling of nuclear fuel evolution during irradiation was developed for oxide fuel [1,2]. Similar approaches were realized for uranium-plutonium mixed nitride fuel. Nevertheless mixed nitride fuel has its own unique features that should be taken into account during a description. Nitrogen is less chemical active element than oxygen. This property effects on phase distribution and fission product release. A transport property of nitrogen in nitride fuel differs from that of oxygen in oxide fuel. Mixed nitride fuel includes impurities such as oxygen and carbon as a result of technological producing. There is also a problem with existence of uranium sesquinitride phase and nitrogen potential.

There is a set of works in literature that use thermodynamic modeling for predicting equilibrium molecular and phase distribution [3,4,5]. But this approach can describe fuel burning real situation not very precisely. Because fuel rod and fuel pellets have nonhomogeneous distribution of thermodynamic conditions such as temperature, pressure and fission rate. Thus, fuel processes under irradiation have no equilibrium character. The current work include mechanistic and thermodynamic approaches for a description of fission product transport properties and molecular and phase composition.

2. A brief description of the model

It is supposed that fission products are generated in the (U,Pu)N matrix. Chemically active fission products diffuse to grain boundary according to an equation

$$\frac{\partial e_i(r,t)}{\partial t} = D_i \Delta e_i(r,t) + s_i(t), \tag{1}$$

where $e_i(r,t)$ is the concentration of the ith chemical active element, D_i is the diffusion constant that was taken from [6], s_i is a source. Intragranular transport of fission gas Xe and Kr is described by bi- and multi-modal models [1]. The most important chemically active elements are combined into 20 groups: Cs, I, Mo, Ru, Rh, Pd, Tc, Ba, Sr, Zr, La, Ce, Eu, Nd, Nb, Sb, Te, N, O, C, were O and C are admixtures. Due to lack information about O and C behavior in (U,Pu)N matrix, it is assumed that there is a thermodynamic equilibrium with respect to these elements in the system. The system consists of grain, bubbles on grain faces and bubbles on grain edges. Fission products released from grain originate separate solid phases (precipitates) in bubbles on grain faces and optionally on grain edges. There is a follow list of precipitates that is considered: metallic phase (Mo, Tc, Ru, Rh, Pd), nitride phase (Ba_3N_2, Sr_3N_2), cesium iodide (CsI), telluride phase (BaTe, Cs_2Te), intermetallic phase (Ru_3U, Rh_3U, Pd_3U), oxide phase (UO_2, PuO_2, BaO, SrO and other), grey phase ($BaUO_4$, $SrUO_4$, Cs_2UO_4 and other), carbide phase (UC, BaC_2, SrC_2 and other) and gas phase.

Boundary conditions for the system of equations (1) are defined by thermodynamic equilibrium within grain and gas phase (bubbles on grain face) [1]. The law of mass action is used

$$A_i = K_i \prod_{j=1}^{n_e} \left(A_j\right)^{b_{ij}}. \tag{2}$$

Chemical activity of ith solid component is determined as

$$A_i = \gamma_i x_i \tag{3}$$

and of gas component as

$$A_i = \frac{x_i p}{p^0} = \frac{n_i RT}{V p^0}, \tag{4}$$

where γ_i is activity coefficient, x_i is mole fraction, p is pressure, $p^0 = 1$ atm, T is temperature, V is volume and R is the universal gas constant.

For describing nitrogen behavior in (U,Pu)N matrix, a chemical potential was constructed

$$\mu_{\mathrm{N(c)}}(x,T) = \frac{1}{2} G^0_{\mathrm{N_2(g)}} + \frac{1}{2} RT \ln \overline{p}_{\mathrm{N_2(g)}}(0,T) + RT \begin{cases} ax, & 0 \le x \le a^{-1} \\ 1 + \ln(ax), & x > a^{-1} \end{cases}, \tag{5}$$

where $G^0_{\mathrm{N_2(g)}}$ is Gibbs energy of pure substance, $\overline{p}_{\mathrm{N_2(g)}}$ is partial pressure at zero stoichiometry deviation [7], a is a parameter.

Chemical solver was tested and results were compared with the one from literature (table 1).

3. Calculation of a phase composition and fission product release

Calculation of uranium-plutonium mixed nitride fuel pellet burning was conducted. Impurities of oxygen and carbon were included. Temperature was taken 1200 K and fission rate 10^{19} $m^{-3}s^{-1}$. The most massive secondary phases are shown in the figure 1. The oxide phase that includes rather massive elements (U, Pu, Zr, La, Ce and other) has the highest fraction near 1 % wt. The metal phase is on the second place and has near 0.5 % wt at burn-ups more than 10 %. Calculations show that almost all carbon at high burn-ups is concentrated in fuel matrix.

Table 1. Equilibrium phase distribution, parameter in nitrogen potential $a = 10^{12}$, wt %.

Component	*T* = 1500 K, burnup 11.4 %		*T* = 1400 K, burnup 10 %	
	[3], without impurities	This work, without impurities	[5], with oxygen impurity	This work, with oxygen impurity
(U,Pu,M)N	93.7	92.4	81.6	83.5
U_2N_3	0	0	10.5	9.4
$(Ru,Rh,Pd)_3U$	3.32	1.66	1.57	1.62
Ba_3N_2	0.25	0	0	0
Sr_3N_2	0.16	0	0	0
Cs_2Te	0.006	0.0003	0.29	0.66
BaTe	0.35	0.46	0.19	0
Mo	0.92	0.75	0.86	0.74
Tc	0.26	0.27	0.22	0.26
UO_2	–	–	0	0.27
BaO	–	–	0	0.45
SrO	–	–	0.15	0.17

Element fraction of released chemical active fission products from a pellet is shown in the figure 2. The highest fractions correspond to cesium and iodide.

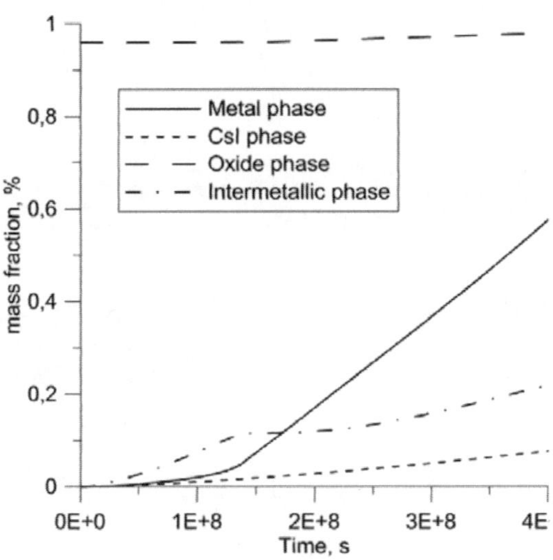

Figure 1. Mass fractions of secondary phases of burning mixed nitride fuel.

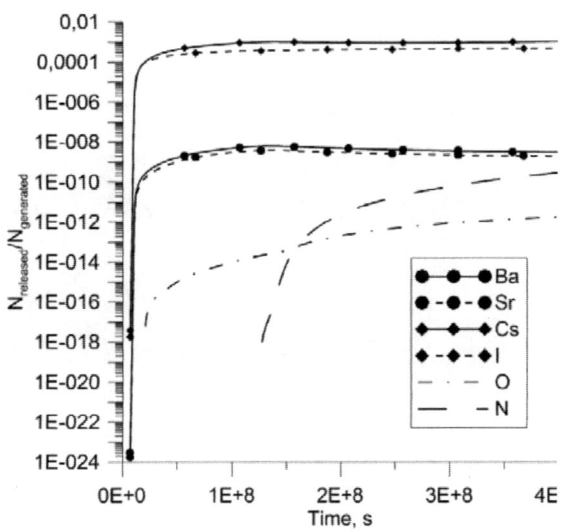

Figure 2. Element fractions of released fission products.

In conclusion, chemical part of the model was constructed and tested and gives reasonably correct results as it was compared with the other works. On the next step of research considered approaches should be verified on experimental data.

References

[1] Veshchunov M S, Ozrin V D, Shestak V E, Tarasov V I, Dubourg R and Nicaise G 2006 *Nucl. Eng. Des.* **236** 179
[2] Veshchunov M S, Dubourg R, Ozrin V D, Shestak V E and Tarasov V I 2007 *J. Nucl. Mater.* **362** 327
[3] Arai Y, Maeda A, Shiozawa K and Ohmichi T 1994 *J. Nucl. Mater.* **210** 161
[4] Lyubimov D Yu, Androsov A V, Bulatov G S and Gedgovd K N 2013 *Atomic Energy* **114** 243
[5] Lyubimov D Yu, Deryabin I A, Bulatov G S and Gedgovd K N 2015 *Atomic Energy* **118** 32
[6] Thetford R and Mignanelli M 2003 *J. Nucl. Mater.* **320** 44
[7] Hayes S L, Thomas J K, Peddicord K L 1990 *J. Nucl. Mater.* **171** 300

Contribution of neutron-capture reactions in energy release in the fuel core of BN-600

R B Bahdanovich, V I Romanenko and G V Tikhomirov

Department of theoretical and experimental physics of nuclear reactors, National Research Nuclear University MEPhI (Moscow Engineering Physics Institute), Kashirskoe shosse 31, Moscow, Russia

E-mail: RynatB@gmail.com

Abstract. The use of modern computing powers and calculation methods allows to get closer to reality results of modelling, as well as to explore areas inaccessible to the experiment. Until now, the calculation of the energy released from the capture of neutrons in the reactor core has been given little attention. The method for calculation of the effective energy release components in a nuclear reactor allows to specify the values used by engineering programs for capture energy release in fast reactors. The paper presents improved method and the results of calculation of three models of the reactor BN-600. It is shown that the contribution of capture energy release in effective energy release for fresh fuel is equal to 4%, which is more than for VVER reactors. During the calculation we created a simple calculation model of the fast reactor, considering its features.

1. Introduction

Calculation of the effective energy release in a nuclear reactor is very important at the different stages of the design and operation of the reactor facility. Currently, the majority of calculation programs is not entirely correct in taking into account the capture component of the energy release, which is associated with the neutron capture reactions ((n, γ), (n, α), etc.). The value of the capture energy is primarily determined by the nuclide composition of the fuel core and the neutron spectrum in the reactor. Using an approximate value of the effective energy release in the reactor (associated with a drawback in the method of calculation of the capture component) may cause a significant error in the calculation of the power distribution, fuel burn-up and spent fuel characteristics. The accuracy of the calculation of effective energy is important for predicting and reconstruction of the reactors' antineutrino spectrum [1].

The method proposed for the calculation of capture energy contribution in the total energy release in nuclear reactors based on use of high-precision calculation programs, for example, Monte Carlo codes (MCNP, MCU, SERPENT, etc.). These programs allow to conduct precise calculations and obtain results close to the experimental ones.

For the purpose of testing the method were developed three models of BN-600: two models of the equivalent cell and homogeneous model of the fuel core. To obtain nuclear reactions rates were used neutron codes MCU and MCNP. Calculations of all models were made for fresh fuel, without burn-up. All the calculations using MCNP were made in the past [2, 3].

2. Method

This method allows to calculate the mean value of the effective energy released in the fuel core of a nuclear reactor per fission. The accuracy of the obtained values depends on the used evaluated nuclear data libraries and any adopted simplifications.

Average energy release in the fission of a heavy nucleus is ($\boldsymbol{E_f}$):

$$\boldsymbol{E_f = E_k + E_n + E_\beta + E_{\gamma\, inst} + E_{\gamma\, del} + E_\nu} \quad (1)$$

where $\boldsymbol{E_k}$ – kinetic energy of fission products, E_n – kinetic energy of instant and delayed fission neutrons, $E_{\gamma\ inst}$ and $E_{\gamma\ del}$ – energy of instant and delayed gamma quanta's, E_β –kinetic energy of delayed beta particles, E_ν – kinetic energy of antineutrinos.

Part of fission energy which is dissipated in a nuclear reactor is called the effective fission energy ($\boldsymbol{E_{eff}}$):

$$\boldsymbol{E_{eff} = E_f - E_\nu - \Delta E_n} \quad (2)$$

where $\boldsymbol{\Delta E_n}$ – adjustment, which takes into account loss of fission neutrons kinetic energy after absorption, which leads to fission.

Part of the energy is also released during the neutron capture reactions, considering that the effective energy release ($\boldsymbol{E_{efe}}$) in a nuclear reactor is:

$$\boldsymbol{E_{efe} = E_{eff} + E_{capt}} \quad (3)$$

$$\boldsymbol{E_{efe} = E_k + E_n + E_\beta + E_{\gamma\, del} + E_{\gamma\, inst} - \Delta E_n + E_{capt\, inst} + E_{capt\ del}} \quad (4)$$

where E_{capt} – energy released in neutron-capture reactions, sum of instant capture energy ($E_{capt\ inst}$) and delayed capture energy ($E_{capt.\ del}$), released during decay of products of neutron-capture reactions (is not considered in this study). The main contributors in capture energy in fuel core are (n,γ) and (n,α) reactions [1,4].

The proposed method of calculation should be carried out in three stages:

1. Development of the reactor model and selection of important fission and capture reactions. Specification of the model geometry and its material composition can significantly affect the bias of the results.
2. The calculation of the rates of nuclear fission and capture reactions using developed reactor model for precise computer codes. It is necessary to take into account the fission energy dependency on neutron energy (especially for fast reactors).
3. Consistent calculation of the effective energy components, based on the rates of nuclear reactions and fission energy table of values, neutron-capture ((n,γ), (n,α), etc.) and decay energy yields table of values obtained from Nubase2012 and ENDF/B-VII.I [5, 6].

MCU and MCNP codes allow to obtain nuclear reaction rates in a given volume, normalized per one fission neutron. Thus, to calculate the number of fission neutrons involved in the nuclear reaction, the reaction rate should be multiplied by the average number of neutrons generated in one fission. To improve the accuracy of calculation of functionals using MCU code one should use command NBAT with a big amount of independent series (more than 10) [7].

3. Calculation of energy release in BN-600

3.1. Equivalent cell models

The first calculated model – a cell of the BN-600 reactor – is shown on figure 1 (table 1). This model is based on the description of the reactor's fuel rod, it was developed in order to understand the difference between the calculation of energy release components in fast and thermal reactors. Mirror reflection of neutrons is on all borders. To compensate the excessive reactivity boron absorber is put into the coolant (in the amount of $11\%_{wt}$), however, the neutron spectrum is distorted and, for this reason, the model cannot be considered correct. Percentage of the integrated fission rate in the thermal (<0.625 eV), intermediate (0.625 eV - 100 keV) and fast energy regions (> 100 keV) is 0/34/66, respectively. The ratio of fissions of ^{235}U and ^{238}U is 87/13. The resulting value of the capture energy component is 5.9 MeV / fission, or 2.95% of the effective energy release (199.7 MeV / fission, without adjustment ΔE_n) [2, 3].

In the BN-600 there are radial and end blanket zones, which strongly affect the neutron balance in the entire fuel core. Most of the neutrons disappear in radioactive capture on ^{238}U (about 45%). For the BN-1200 the percentage of the integral fission rate in the three regions of energy equal to 11/49/40, respectively. The ratio of fissions of ^{235}U and ^{238}U in the BN-1000 is equal to 79/21 [8].

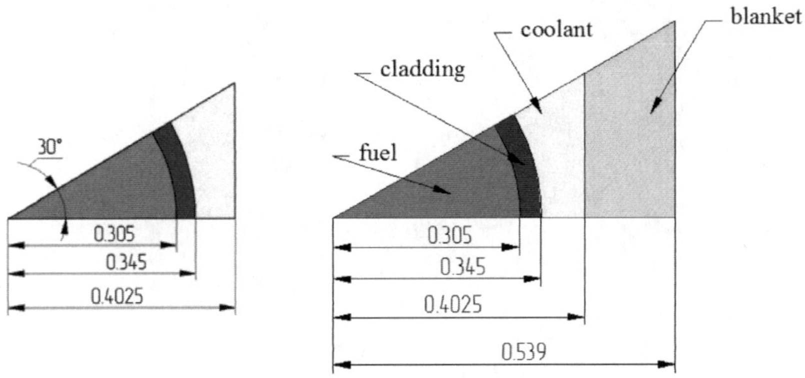

Figure 1. The first and the second (with blanket) model of the BN-600 equivalent cell (cm) [9]

Method to compensate reactivity with boron was replaced by compensation with natural uranium in order to obtain a more realistic neutron spectrum (figure 1, table 1). Percentage of integrated rate of the fission reaction in the three areas of energy equal to 0/50/50, respectively. The ratio of fissions of ^{235}U and ^{238}U in the model is equal to 81/19. The capture energy component for the model with blanket is 7.7 MeV / fission, or 3.8% of the effective energy release (201.7 MeV / fission, without adjustment $\boldsymbol{\Delta E_n}$). When calculating the effective energy release for both models was made an assumption: energy released in fission of ^{235}U is 193.14 MeV, and of ^{238}U is 197.7 MeV.

Table 1. Features of BN-600 model with blanket ** and without it* [10, 11].

Material	Composition	Temperature, K	Density, g/cm^3
Fuel	UO_2, $21\%_{wt}$ ^{235}U	1200	8.40
Cladding (steel ЧС-68)	^{56}Fe+2% Mo+ 15% ^{52}Cr+10% ^{58}Ni	294	7.90
Coolant	Na (Na + ^{10}B)*	500 (294)*	0.84
Blanket**	UO_2, 0,71% ^{235}U	1200	8.40

3.2. The homogeneous model of the reactor core

IOP Conf. Series: Journal of Physics: Conf. Series **781**(2017) 012011 doi:10.1088/1742-6596/781/1/012011

The above results were tested using a closer to real-life model of BN-600 – a homogeneous model of the core with blanket zones. Its description is based on the information available in the open-source literature. The model is a cylindrical reactor core, consisting of 10 homogeneous zones (figure 2) [9, 12-16].

The actual fuel core and blanket were divided into zones, which were then homogenized: LB1 - lower axial blanket in zone with small enrichment, LB2 - lower axial blanket in zone with average enrichment, LB3 - lower axial blanket in zone with large enrichment, F1 - zone with small enrichment (17%$_{wt}$ ^{235}U), F2 - zone with average enrichment (21% $_{wt}$ ^{235}U), F3 - zone with large enrichment (26% $_{wt}$ ^{235}U), UB1 - upper axial blanket in zone with small enrichment, UB2 - upper axial blanket in zone with average enrichment, UB3 - upper axial blanket in zone with large enrichment, RB – radial blanket. The enrichment of depleted uranium in the blanket zones is 0.3% $_{wt}$ ^{235}U.

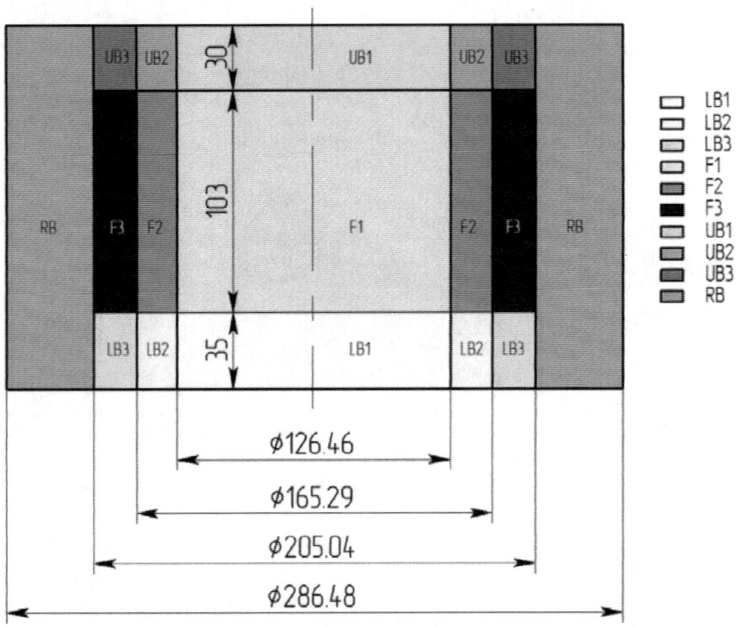

Figure 2. Model of reactor BN-600 homogeneous core with a blanket (cm)

While creating the model we did not take into account the ends of all the rods. Scram control rods and automatic control rods were considered fully withdrawn, fuel assemblies containing them were presented as a wrapping filled with coolant. The porosity of the fuel is 86% (it causes decrease in real density in comparison with the theoretical). Density of fuel was considered to be 9.426 g/cm^3, density of cladding and wrapping – 7.674 g/cm^3, sodium density – 0.841 g/cm^3, boron carbide – 2.510 g/cm^3.

Shim control rods were completely dropped in the upper blanket of small and average enrichment zones (UB1, UB2). Further, in order to make the neutron multiplication factor equal unity, they were immersed in the small and average enrichment zones (F1, F2), which resulted in the increase of atomic concentrations of boron, carbon and decrease of atomic concentration of sodium. Atomic concentrations of steel isotopes in areas with shim rods were calculated with the consideration that the steel cladding and the wrapper of shim rod put into the core on a half of its length. The critical immersion equals to 0.465 of the height of the fuel core (F1, F2).

Temperature of the lower axial blanket taken to be the same as the temperature of coolant at the reactor inlet, temperature of the upper axial blanket taken to be the same as sodium outlet temperature (table 2). Temperature of the radial blanket equal to the average temperature of sodium in the core, temperature of the core materials was taken from BN-600 hybrid benchmark [17].

Table 2. Temperatures of materials in BN-600 reactor model.

Material	Temperature, K
LB1, LB2, LB3	650
F1, F2, F3	1500
UB1, UB2, UB3	823
RB	736

When calculating the effective energy release were defined fission reaction rates for ^{238}U and ^{235}U isotopes in twenty energy groups. Then were calculated fission reaction rate contributions from each energy group in the average effective energy (by weighing). Dependence of the fission energy on neutron energy was taken from ENDF / B-VII.I [6]. Similarly, the adjustment ΔE_n= 0.66 ± 0.16 MeV was calculated (the error is determined by the number of energy groups). It should be noted that the fission reaction on ^{235}U caused primarily by neutrons of [0.01, 7.5] MeV energy, and on ^{238}U – by [0.8, 8.5] MeV neutrons (contribution to the total fission rate more than 0.1%).

Since the MCU code does not allow to calculate the rates of (n, γ), (n, α), etc. reactions separately at the full range of neutron energies, but only in the area of physical sub module FARION (fast energy region, only cross-sections for 300 K are distributed), the simplification was made: we attributed capture reaction rate to the reaction which is significantly prevail in calculation with MCNP code. Thus, we consider the capture reaction as the reaction (n, γ) for all isotopes, except ^{16}O and ^{10}B, which are mainly involved in the reaction (n, α). According to our estimations, this simplification could lead to 2.5% error in the capture energy calculation. The statistical error of the Monte-Carlo method and the error in the value of nuclear reactions energy yield makes a negligible contribution to the error of capture energy component.

Percentage of integrated rates of the fission reaction in the three areas of energy equal to 0/49/51, respectively. The ratio of fissions of ^{235}U and ^{238}U in the model is equal to 88/12. The capture component is 8.2 ± 0.2 MeV / fission (error from simplification), or 4.1% of the effective energy (201.4 ± 0.3 MeV / fission). Comparison with previous model is in the table 3.

Table 3. Results of effective energy release calculation for BN-600 models (MeV / fission).

Model	ΔE_n	E_{capt}	E_{efe}	$\frac{E_{capt}}{E_{efe}}$
First (cell with boron)	--	5.9	199.7	2.95%
Second (cell with blanket)	--	7.7	201.7	3.8%
Third (homogeneous fuel core)	0.66	8.2	201.4	4.1%

As can be seen from the results, model with blanket and the homogeneous fuel core model give similar results. This suggests that the simplified model of the reactor can be used for evaluation of its neutron features and testing of energy release calculation methods. The fraction of ^{235}U fission in full core model is higher than in the cell model with blanket. It can be explained by the fact that the blanket zone is located on the periphery of the fuel core, but not inside the core like in the cell model.

4. Conclusions

This paper presents the results of the use of previously developed method for calculation of the contribution of neutron capture reactions in the effective energy release in the fuel core of the BN-600 reactor. Three reactor models were designed and calculated (cell model, cell model with blanket and homogenous core model).

Calculation shows that the value of the capture energy release in the fresh core of BN-600 reactor (OU_2 fuel) is about 8 MeV or 4% of the effective energy release in the core. One should take into account the presence of blanket zones when calculating the neutron balance in a fast reactor.

Simple model of the reactor cell with a blanket which was developed during this work can be used for estimation of effective energy release components in the BN-600 reactor. The described technique allows to obtain the exact values of the effective energy release and to estimate its dependency on fuel burn and other characteristics of the reactor.

References

[1] Kopeikin V I, Mikaelyan L A and Sinev V V 2004 *Phys. of At. Nuc.* V 67 № **10**

[2] Bahdanovich R B, Tikhomirov G V, Saldikov I S and Kryuchkov E F 2014 *Method for calculation capturing reactions contribution to total energy release in nuclear reactors* Proceedings of PHYSOR 2014: American Nuclear Society's Topical Meeting on Reactor Physics. The role of nuclear reactors towards a sustainable future Kyoto Japan p 7

[3] Bahdanovich R B, Tikhomirov G V, Saldikov I S, Ternovykh M Yu and Gerasimov A S 2015 *Contribution of Neutron Capturing Reactions in Precision Calculations of Total Energy Release in Nuclear Reactors* Proceedings of Global 2015 September 20-24 Paris France

[4] Kramerov A Y and Shevelev Y V 1984 *Engineering calculations of nuclear reactors,* Energoatomizdat Moscow Russia

[5] Pritychenko B, Sonzogni A, Q-value Calculator (QCalc) NNDC, Brookhaven National Laboratory http://www.nndc.bnl.gov/qcalc/

[6] Nuclear Energy Agency, Java-based Nuclear Data Information System (JANIS), available online. URL http://www.oecd-nea.org/janis/

[7] Skorokhodov D N and Tikhomirov G V 2012 The problem of estimating the accuracy of calculation of local tallies by means of the Monte Carlo method *Physics of Atomic Nuclei* V 75 № **14**

[8] Walter A E and Reynolds A E 1981 *Fast breeder reactors* Pergamon Press Inc. New York

[9] Kuznetsov I A and Poplavsky V M 2012 *Safety of NPPs equipped with fast reactors* IzDat Moscow p 70

[10] Judd A M 1984 *Fast Breeder Reactors* Energoatomisdat p 137

[11] Usynin G B and Kusmartsev E V 1985 *Fast Breeder Reactors* Energoatonizdat p 288

[12] Adamov E O 2005 *Mechanical Engineering of nuclear technology* Encyclopedia in forty volumes IV **25** book 1 p 513-520, 584-619

[13] Poltsev V P and Guseva T M 1973 *Operability of fast reactors control rods and the ways of its increase* Scientific Research Institute of Atomic Reactors p 17

[14] Petchenkin V A, Chernov K G, Moiseev A V, Eliseev V A and Konobeev Y V 2013 Features of damaging dose in metals and construction materials under irradiation in the core of the BN and VVER reactors *Nuclear Physics and Engineering* V 4 № **3**

[15] Vasiliev B A, Kuzavkov N and Mishin O V 2011 *Experience and prospects of modernization of the reactor core BN-600* Izvestia Vysshikh Uchebnykh Zawedeniy. Yadernaya Energetika №**1** p 158-168

[16] IAEA 2008 *Thermophysical properties of materials for nuclear engineering: a tutorial and collection of data* Vienna ISBN 978–92–0–106508–7

[17] IAEA 2009 *BN-600 hybrid core benchmark analyses* Vienna ISBN 978–92–0–109409–4 ISSN

1011-4289

Acknowledgments
This work was performed at the Department of Theoretical and Experimental Physics of Nuclear Reactors in the framework of 5-100 project. BASOV and CHERENKOV clusters of MEPhI were used for the preliminary calculations.

Closed fuel cycle with increased fuel burn-up and economy applying of thorium resources

G G Kulikov and V A Apse

National Research Nuclear University MEPhI (Moscow Engineering Physics Institute), 115409, Kashirskoe highway, 31, Moscow, Russia

E-mai: ggkulikov@mephi.ru

Abstract. The possible role of existing thorium reserves in the Russian Federation on engaging thorium in being currently closed (U-Pu)-fuel cycle of nuclear power of the country is considered. The application efficiency of thermonuclear neutron sources with thorium blanket for the economical use of existing thorium reserves is demonstrated. The aim of the work is to find solutions of such major tasks as the reduction of both front-end and back-end of nuclear fuel cycle and an enhancing its protection against the uncontrolled proliferation of fissile materials by means of the smallest changes in the fuel cycle. During implementation of the work we analyzed the results obtained earlier by the authors, brought new information on the number of thorium available in the Russian Federation and made further assessments. On the basis of proposal on the inclusion of hybrid reactors with Th-blanket into the future nuclear power for the production of light uranium fraction $^{232+233+234}U$, and ^{231}Pa, we obtained the following results: 1. The fuel cycle will shift from fissile ^{235}U to ^{233}U which is more attractive for thermal power reactors. 2. The light uranium fraction is the most "protected" in the uranium component of fuel and mixed with regenerated uranium will in addition become a low enriched uranium fuel, that will weaken the problem of uncontrolled proliferation of fissile materials. 3. ^{231}Pa doping into the fuel stabilizes its multiplying properties that will allow us to implement long-term fuel residence time and eventually to increase the export potential of all nuclear power technologies. 4. The thorium reserves being near city Krasnoufimsk (Russia) are large enough for operation of large-scale nuclear power of the Russian Federation of 70 GWe capacity during more than a quarter century under assumption that thorium is loaded into blankets of hybrid TNS only. The general conclusion: the inclusion of a small number of hybrid reactors with Th-blanket into the future nuclear power will allow us substantially to solve its problems, as well as to increase its export potential.

1. Introduction

Closure of nuclear fuel cycle (NFC), the process under practical implementation now in the Russian Federation, can substantially upgrade efficiency of uranium utilization. Therefore, additional involvement of natural thorium resources into nuclear fuel cycle for further enlargement of practically unlimited natural uranium resources looks now as an option with the slightest validity. This option will require large additional expenses for development and industrial mastering of complicated technologies for thorium mining, primary processing, fabrication and handling with other nuclear fuel material. It would be appropriately to note here that till now no one country had decided yet to undertake wide involvement of natural thorium resources into national nuclear power system (NPS), even India with large stockpiles of natural thorium.

So, involvement of thorium resources into NPS in the process of the NFC closure has to be validated by other and very weighty arguments. As is known, neutron-physical parameters of thermal power reactors can be remarkably improved by transition from uranium (^{235}U-^{238}U) fuel to uranium-thorium (^{233}U-^{232}Th) fuel. Just this consideration has been taken into account when "The Strategy on Development of Russian Nuclear Power System in the First Half of the XXI century" [1] was formed. The Strategy defines that closure of uranium-plutonium NFC will be followed by thorium involvement, and thermal reactors will be gradually transferred from uranium-plutonium fuel to uranium-thorium (^{233}U-^{232}Th) fuel.

2. Thorium application for generation of "light" uranium fraction

When the problem of thorium involvement into the NFC to be closed is under discussion, it seems reasonable to give attention to the following important (in a practical sense) circumstance. As is known, from the very beginning of nuclear power development and till nowadays the NPS is based on utilization of "heavy" uranium fraction $^{235+236+238}U$, where ^{236}U is a component of the regenerated uranium.

As far as thorium will be involved into the NFC, then a new (on its isotopic composition) uranium material appears in spent fuel, and the following neutron reactions (Fig. 1) are the channels for its generation:

1) Radiative neutron capture by thorium with generation of fissile isotope ^{233}U.
2) Threshold reactions of high-energy neutrons:
   $^{232}Th(n,2n)^{231}Th(\beta^-)^{231}Pa$; $^{232}Th(n,3n)^{230}Th(n,\gamma)...^{231}Pa$.
3) Radiative neutron capture by protactinium: $^{231}Pa(n,\gamma)...^{232}U$.

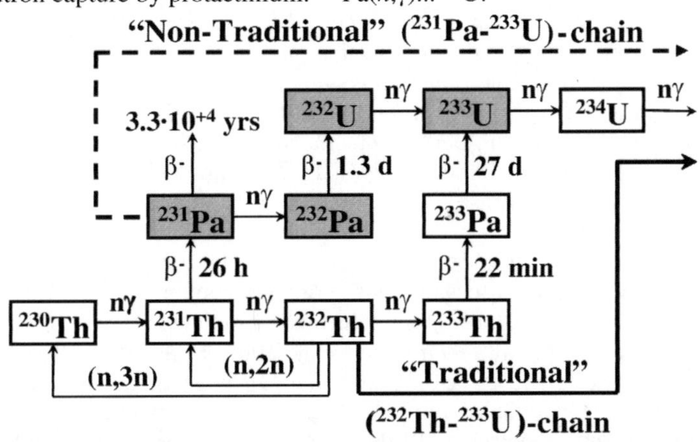

Figure 1. Chains of nuclide transformations in the thorium-uranium fuel.

These reactions can generate, in addition to fissile uranium isotope ^{233}U, protactinium isotope ^{231}Pa and uranium isotope ^{232}U in the neutron-irradiated thorium. The chains of isotopic transitions initiated by neutron-thorium reactions can be grouped by such a way:

1) "Traditional" chain that started from radiative neutron capture by ^{232}Th (the capture channel).
   $^{232}Th(n,\gamma)^{233}Th(\beta^-,T_{1/2}=22')^{233}Pa(\beta^-,T_{1/2}=27\ d)\ ^{233}U(n,\gamma)^{234}U(n,\gamma)...$
2) "Non-traditional" chain that started from threshold $^{232}Th(n,2n)$ and $^{232}Th(n,3n)$ reactions (the threshold channel):
   $^{232}Th(n,2n)^{231}Th(\beta^-,T_{1/2}=26\ h)^{231}Pa(n,\gamma)^{232}U...$
   $^{232}Th(n,3n)^{230}Th(n,\gamma)^{231}Th(\beta^-,T_{1/2}=26\ h)^{231}Pa...$
   $^{231}Pa(n,\gamma)^{232}Pa(\beta^-,T_{1/2}=1.3\ d)^{232}U(n,\gamma)^{233}U(n,\gamma)^{234}U(n,\gamma)^{235}U...$

So, introduction of natural thorium in fresh fuel composition will result in production of the "light" uranium fraction, namely mixture of light uranium isotopes ^{232}U, ^{233}U, ^{234}U with dominant role of well-fissile isotope ^{233}U.

If the "light" uranium fraction is used as a component of fresh nuclear fuel, then the NFC of nuclear power reactors (both thermal and fast reactors) may be based on utilization of either uranium-based fuel or mixed uranium-plutonium fuel compositions for a long time, even for an ultra-long time period taking into account practically unlimited resources of natural uranium and natural thorium. Besides, plutonium and minor actinides can be utilized within the frames of the NFC. Isotope composition of uranium fraction in fresh fuel will contain mixture of light uranium isotopes ^{232}U, ^{233}U, ^{234}U with uranium regenerated from spent LWR fuel.

If the "light" uranium fraction is generated by thorium-based fertile materials in relatively small number of highly efficient producers of fissile materials, then the main park of nuclear power reactors will be fuelled by uranium as before. The hybrid thermonuclear "FUSION-FISSION" reactors with thorium blankets and with plasma parameters already achieved in contemporary experimental fusion installations can be regarded, in the first turn, as such producers. In the second turn, accelerator-driven facilities and fast reactors with thorium blankets can be also used to produce the "light" uranium fraction but they are characterized by relatively low production potential.

Thus, the closed (^{232}Th-^{233}U-^{235}U-^{238}U-Pu) NFC with multiple fuel recycle and with feeding by the "light" uranium fraction can form the closed (^{232}Th-^{232}U-^{233}U-^{234}U-^{235}U-^{236}U-^{238}U-Pu) NFC with multi-isotope (^{232}U-^{233}U-^{234}U-^{235}U-^{236}U-^{238}U) uranium fraction containing practically all uranium isotopes. As a consequence, the NFC "center of gravity" can be shifted from the currently used ^{235}U towards the use of ^{233}U in the future.

The "light" uranium fraction, in addition to content of very attractive fissile uranium isotope ^{233}U, can offer the following significant feature from nuclear non-proliferation point of view. If the "light" uranium fraction would be mixed with uranium regenerate extracted from spent LWR fuel, then some kind of low-enriched uranium fuel could be produced for nuclear power reactors. This is a very auspicious factor for weakening the problem of unauthorized proliferation of fissile materials. By the way, radioactivities of three light uranium isotopes (^{232}U, ^{233}U and ^{234}U) are more intense than those of other uranium isotopes. The more intense radioactivity of such uranium fuel can additionally enhance its proliferation resistance.

When natural ^{230}Th thorium is irradiated by high-energy neutrons, the lighter nuclides ^{231}Pa and ^{230}Th [2, 3] are generated in addition to the "light" uranium fraction (^{232}U, ^{233}U, ^{234}U). Protactinium isotope ^{231}Pa can be directly extracted from neutron-irradiated thorium with application of radiochemical technologies. Under further neutron irradiation light thorium isotope ^{230}Th captures neutron, converts to ^{231}Pa that can be also extracted from neutron-irradiated thorium.

It would be wrong to estimate build-up of ^{231}Pa under neutron irradiation of thorium as a generation of useless by-product and neutron poison. ^{231}Pa, being admixed to fresh uranium fuel of nuclear power reactors, can initiate the chain of isotopic transformations $^{231}Pa \rightarrow {}^{232}U \rightarrow {}^{233}U$. As a result, ^{231}Pa-bearing fuel acquires a new valuable quality, namely stable neutron-multiplying properties under high and ultra-high values of fuel burn-up [3]. The opportunity arises to provide long and ultra-long fuel lifetimes, strengthen autonomy of nuclear energy sources and heighten export potential of nuclear power technologies.

Involvement of natural thorium into the closed NFC can promote resolving some principal problems with which nuclear power will face in the foreseeable future. Some of these problems have become apparent already, some other bottlenecks of large-scale NPS with the closed NFC are already foreseen.

There are three, at least, difficulties in further development of global nuclear power. The first problem is a necessity to reduce scope of operations in back-end NFC part via substantially higher values of fuel burn-up in nuclear power reactors. The second problem is a necessity to enhance NFC protection against unauthorized proliferation of fissile materials and, thus, heighten export potential of nuclear technologies. The third problem is a necessity to make radioactive wastes harmless via neutron transmutation of long-lived fission products and minor actinides, i.e. their conversion into short-lived and stable isotopes. At present, these problems produce only negative effects on public opinion concerning the future destiny of large-scale nuclear power technologies.

3. Specific features of thorium resource problem in Russia

The following two specific features are usually mentioned in the connection with involvement of thorium, one else natural fertile material in addition to uranium, into the NPS. The first feature consists in the necessity of developing other fuel technologies as compared with the existing uranium technologies due to quite different material properties of thorium and uranium. The second feature is related with thorium mining and primary processing. The question arises: up to what scales the industrial plants on mining and primary processing of natural thorium ores should by deployed? Generally speaking, till now no one country in the world has developed thorium technologies up to an industrial and, therefore, very expensive scale.

Current applications of natural thorium are limited by its usage in chemical and metallurgical industries, in electronics and medicine [4]. According to various publications and evaluations, total annual consumption of natural thorium in the world covers the range from several dozens of tons up to 200–300 tons.

Thorium is being mined now mainly as a by-product of rare-earth elements. Thorium is extracted from mineral ores during their primary processing and stockpiled for potential applications in the future. Besides, thorium is a by-product of mining and primary processing of uranium and mixed uranium-thorium ores [5–7].

However, when considering the problem, some peculiarities in mining, primary processing and energy utilization .of natural uranium and thorium resources should be taken into account. Large scale of uranium mining industry is mainly defined by low content of the only natural fissile isotope ^{235}U in uranium (~0.71%). About 200 tons of natural uranium must be mined in order to produce one ton of ^{235}U with application of isotope separation technologies.

Since effective fissile isotope, analogue of ^{235}U in natural uranium, is absent in natural thorium, thorium involvement into the closed NFC does not require the same scale of thorium mining industry. Scale of thorium mining industry may be defined by thorium consumption needed to produce large enough amount of well-fissile isotope ^{233}U. Anyway, scale of thorium mining industry is mainly defined by the NFC structure and by strategies of thorium utilization.

The following simple assumptions may be used to evaluate minimal thorium consumption in the closed NFC. As is known, one typical LWR with electric power of 1 GWe incinerates annually about 960 kg of fissionable materials. These 960 kilograms include ~580 kg ^{235}U (i.e. about 60%). If useless transformation of ^{235}U into ^{236}U via radiative neutron capture is taken into account, then total amount of the eliminated ^{235}U may be evaluated as ~800 kg $^{235}U/(GWe\cdot year)$.

Total power of Russian nuclear power plants is planned to be at the level of 50-60 GWe in the middle of the XXI century. If ^{233}U is substituted for ^{235}U, then, under the same other conditions, thorium consumption for production of ^{233}U will be laughably low, at the level of 40-50 tons annually. So small value of thorium consumption will require no additional efforts on exploration of new thorium deposits and deployment of thorium mining plants in Russia.

Nowadays, available stockpiles of thorium-bearing monazite concentrate (82 000 tons with mean thorium content of 4.7%, i.e. ~3800 tons of natural thorium, [8]) are stored in the vicinity of Krasnoufimsk (Sverdlovsk Region, Russia). Even so modest resources of natural thorium are large enough to satisfy demands of the large-scale Russian NPS (50-60 GWe in the middle of century) during the next 70 years.

In addition to thorium consumption for ^{233}U generation in nuclear power reactors, it is necessary to evaluate thorium amounts needed to fill up blankets of hybrid fusion facilities, and thorium amounts stored at spent fuel reprocessing plants. As is known, radiochemical reprocessing of irradiated fertile materials can be performed without cooling-time at all, as it was shown for hybrid fusion facility with molten-salt blanket [9]. This means that thorium amounts stored at reprocessing plants can be evaluated as negligibly small.

As to thorium amounts needed to fill up thermonuclear neutron source (TNS) blankets, they can be sufficiently large. Some relevant information was taken from the project of pilot hybrid fusion facility with uranium blanket developed by the National Research Center – Kurchatov Institute in the 1980s as a plutonium-production plant [10]. Metal uranium blanket (20-cm thick) was placed just behind the first wall of plasma chamber, and contained molybdenum alloy with depleted uranium. Main parameters of the fusion facility: thermonuclear power – 578 MWt; neutron load on the first wall - 0.8 MW/m^2; plutonium production rate - 4.0 tons per year. Total uranium amount loaded into the TNS blanket was evaluated as 1040 tons.

If the same thorium amount is taken to load into the TNS blanket, then currently available resources of natural thorium (~3800 tons) with accounting for thorium consumption for ^{233}U production (~40-50 tons annually) are large enough to provide operation of two TNS for about 30 years. Afterwards, either additional thorium mining or thorium purchasing will be required.

However, these evaluations are correct only for the case when no thorium amounts are introduced into fresh fuel compositions of nuclear power reactors. Thorium must be loaded into blankets of highly efficient producers of ^{233}U. Just this variant of Russian NPS structure may be realized, if conceptual design of hybrid TNS with thorium blanket will be worked out. Then, involvement of so effective producers of ^{233}U (even at relatively low number of these producers) into the closed NFC can provide Russian NPS with artificial ^{233}U-based fuel for a very long time period. In this case, multi-isotope uranium compositions with dominant content of ^{233}U and ^{238}U can be used as a fuel of thermal nuclear reactors.

Nuclear nonproliferation problem is considered in [11]. If several percents of ^{231}Pa and ^{237}Np are introduced into fresh (^{233}U-^{238}U) fuel compositions, then, under neutron irradiation in nuclear power reactors, uranium isotope ^{232}U can be produced in amounts comparable with amount of ^{233}U via the following chain of isotopic transitions $^{231}Pa(n,\gamma)^{232}Pa(\beta)^{232}U$ [12]. Similarly, plutonium isotope ^{238}Pu can be produced in amounts comparable with amount of ^{239}Pu via the chain of isotopic transitions $^{237}Np\ (n,\gamma)^{238}Np(\beta)^{238}Pu$ [13]. Both these isotopes (^{232}U and ^{238}Pu) are able to enhance proliferation resistance of fissile, weapon-usable materials.

4. Conclusion

The paper analyzes potential involvement of hybrid fusion facilities with thorium blanket into the closed (U-Pu) NFC of future Russian NPS for production of the "light" uranium fraction (^{232}U-^{233}U-^{234}U) and ^{231}Pa that can lead to the formation of advanced (^{231}Pa-^{232}Th-^{232}U-^{233}U-^{234}U-^{235}U-^{236}U-^{238}U-Pu) fuel cycle. The following main results are listed below:

1. Nuclear fuel cycle will be shifted from fissile isotope ^{235}U towards ^{233}U, more attractive isotope for thermal nuclear reactors.
2. The "light" uranium fraction is the most proliferation-resistant fuel component. The "light" uranium fraction, being mixed with uranium regenerate, becomes some kind of low-enriched uranium, i.e. proliferation resistance can be additionally strengthened.
3. Introduction of ^{231}Pa into fresh fuel compositions can stabilize their neutron-multiplying properties and, thus, can create conditions for ultra-long fuel lifetimes, upgrade autonomy of nuclear energy sources and heighten export potential of nuclear power technologies.
4. Natural thorium resources currently available in Russia are large enough to provide operation of large-scale (50-60 GWe) Russian NPS for about 30 years under assumption that thorium is loaded into blankets of hybrid TNS only.

References

[1] 2000 *Proc. Int. Conf. Strategy of Development of Nuclear Power Generation of Russia in the First Half of the XXI-st Century. Main Provisions (Moscow)* (Moscow: Minatom Press)
[2] Marin S V and Shatalov G E 1984 *Atomic Energy* **56** pp 289–291

[3] Shmelev A N, Kulikov G G, Kulikov E G and Apse V A 2014 *New potentsiale role of fusion hybrids in nuclear non-proliferation ensuring within the frames of the international closed nuclear fuel cycle with application of thorium technologies* (Moscow: NRNU MEPhI Press) p 120
[4] Gavrilin V I, Zubkov L D and Petrova N V 1998 *Mineral raw materials, thorium. Reference book* (Moscow: ZAO "Geoinformmark" Press) p 27
[5] Kac Dzh, Siborg G and Mors L 1991 *The chemistry of the actinide elements* vol **1** (Moscow: "Mir" Press) p 525
[6] Godoy J M, Godoy M L D P and Aronne C C 2007 Application of inductively coupled plasma quadrupole mass spectrometry for the determination of monazite ages by lead isotope ratios *J. Brazil Chemical Society* **18**
[7] Holmes A 1918 The pre-cambrian and associated rocks of the district of Mozambique *Quarterly J. Geological Society* **74** pp 31–98
[8] Avaiable at: http://profbeckman.narod.ru/RH0.files/25_1.pdf
[9] Velikhov E P *et al* 2014 *VANT* Thermonuclear fusion **4**, 37 pp 3–8
[10] Gornostaev B D *et al* 1978 *Proc. 2nd Soviet-US Seminar (*Moscow: IAE, Atomizdat Press) pp 94–122
[11] Avrorin E N and Chebeskov A N 2015 Fast reactors and nuclear nonproliferation problem *Nuclear Energy and Technology* **1** pp 1–7
[12] Kulikov G G, Shmelev A N, Geraskin N I, Kulikov E G and Apse V A 2016 Advanced nuclear fuel cycle for the RF using actinides breeding in thorium blankets of fusion neutron source *Nuclear Energy and Technology* **2** pp 147–150
[13] Fedorov M I, Dyachenko A I, Balagurov N A and Artisyuk V V 2015 Formation of proliferation-resistant nuclear fuel supplies based on reprocessed uranium for Russian nuclear technologies recipient countries *Nuclear Energy and Technology* **1** pp 111–116

Radioactive waste from decommissioning of fast reactors (through the example of BN-800)

A A Rybin and O A Momot

Obninsk Institute for Nuclear Power Engineering of National Research Nuclear University «MEPHI», Studgorodok, 1, Obninsk, Kaluga region, 249040, Russian Federation

E-mail: alexeyrybin94@gmail.com

Abstract. Estimation of volume of radioactive waste from operating and decommissioning of fast reactors is introduced. Preliminary estimation has shown that the volume of RW from decommissioning of BN-800 is amounted to 63,000 cu. m. Comparison of the amount of liquid radioactive waste derived from operation of different reactor types is performed. Approximate costs of all wastes disposal for complete decommissioning of BN-800 reactor are estimated amounting up to approx. $145 million.

1. Introduction

The experience of decommissioning of the fast neutron reactor and complex reprocessing of big volume of radioactive waste (RW), including liquid-metal sodium, is quite limited in the world's practice. There are only two fast neutron reactors in operation: BN-600 and BN-800. Reprocessing of the used liquid sodium coolant was performed at experimental facilities in a limited way. Consequently, analyses of the volume, content, activity level of RW and costs of RW disposal will be a prototype for handling RW of liquid sodium cooled fast neutron power reactors, e.g. BN-600, BN-800, BN-1200, and may be applied for the operating water cooled nuclear power plants as well as in projects of new power units.

Purpose of the work: to analyse the volume, content and activity level of operational and decommissioning RW of fast neutron reactors; to compare RW characteristics of the thermal neutron and fast neutron reactors, to evaluate RW peculiarities of fast neutron reactors; costs of wastes disposal for reactor decommissioning.

2. Analysis of radioactive waste

At present the process of decommission of different types of reactors becomes complicated because in case of design of most of them the stage of decommission wasn't considered. It is well known that one of sources of radioactive waste and a source of potential danger to the population and the environment are the neutron induced activity (neutron activation) of the equipment, protective concrete constructions and other materials.

So, under the influence of neutrons the equipment, reactor construction and protective construction materials become radioactive. Activated materials are the reactor vessel and internal components,

IOP Conf. Series: Journal of Physics: Conf. Series **781** (2017) 012013 doi:10.1088/1742-6596/781/1/012013

pipelines, the equipment and elements which are in reactor space. Important feature of the activated materials is that they aren't decontaminated.

It is established that during decommissioning nuclear power plants up to 50% of the volume of radiation protection, lining and sealing of coatings and structures are activated, i.e. they will belong to the category of radioactive waste not subjected to decontamination. It is worth noting that the containment is a concrete monolith, and performs two functions: protective and bearing, so when removing it is very difficult to separate the active and inactive parts, which is why the volume of radioactive waste due to the induced activity is increased many times [1].

Evaluation of RW peculiarities of fast neutron reactors was performed through comparison of activity, structure, chemical composition and volume of liquid and solid radioactive waste (LRW and SRW) of BN and VVER. The volume of RW accumulated depends not only of the number of power plan units, their power and duration of operation, but also of the reactor type. Thus LRW volume comparison was conducted based on the amounts specified for 1 reactor·year and for 1 MW_{el}·year [2]. For estimation of LWR volume for BN-600 and BN-800 reactors, these reactors characteristics were compared to those of BN-350. The data on yearly average LRW volume are presented in Table 1. The data for BN-350, BN-800 and VVER-1000 is referenced as well for illustrative purposes.

Chemical composition, specific activity, and radionuclide composition of LRW of fast neutron reactors is little different to those of water cooled thermal neutron reactors. Thus, the same methods can be used for LRW reprocessing as for thermal neutron reactors.

Table 1. LWR volume for Nuclear Power Plants with reactors of different types.

Specified Volume	Reactor Type			
	BN-350	BN-600	BN-800	VVER-1000
m^3/(reactor·year)	50-75	100-150	120-180	220-350
m^3/(MW_{el}·year)	0.2-0.3	0.22-0.33	0.17-0.25	0.4-0.6

As table 1 shows, 4000-6000 m^3 of LRW can be accumulated during operation of BN-600 (30+10 years); and 5400-8100 m^3 of LRW can be accumulated during operation of BN-800 (45 years).

SRW content is similar for the fast neutron and thermal neutron reactors are comparable. Volumes of SRW of different activity level correlate almost equally excluding the fact that HLW ratio for BN reactors is a little higher in comparison with VVER SRW.

It is worth considering that the volume of SRW significantly decreases during operation due to organizational-technical measures and decontamination of some amount of the waste. However, higher volume of SRW accumulated during operation is typical for fast neutron reactors, namely 1.5-3.5 higher than for VVER reactor type and are comparable to SRW for RBMK reactor type.

A big amount of SRW forms upon decommissioning of the Nuclear Power Plants. For instance, from 20000 to 30000 tons of SRW forms upon decommissioning of the average 1000 MW reactor [3]. According to preliminary estimates, about 11180 tons of SRW (from which 36% are low-level, 41% are intermediate-level and 23% are high-level) will be produced upon decommissioning of BN-350 reactor. We can estimate from the above that the volume of SRW produced upon decommissioning of BN-800 reactor, the data on SRW volume is presented in Table 2.

Distinctive SRW is the main peculiarity of RW of BN reactors: sodium metal, highly-active adsorbers and cold traps containing sodium. It is also important to consider that the volume of decommissioning SRW and correlation of SRW of different groups on activity and especially HLW depends on the accepted concept of decommission – immediate or deferred dismantling. The volume of SRW will be higher in case of immediate dismantling compared to deferred one.

Table 2. Amount of SRW upon decommissioning of BN-800.

Amount of SRW, m^3			
Low-level waste	Intermediate-level waste	High-level waste	Total SRW:
12240-14688	13940-16728	7820-9384	34000-40800

3. Costs of RW handling upon Decommission

Upon development of Decommission preparation programs it is necessary to consider the costs of waste handling and implementation of measures for ensuring radioactive and ecological safety. Costs of RW and radioactive materials handling (approx. 30-50%) and disassembly of radioactive equipment are significant part of the total Decommissioning costs [4].

International experience of reactor decommission shows that costs connected with decommission may amount up to 37% [5] on estimation of the leading Russian specialists [6] costs of decommission of one reactor may amount to half a cost of its construction (i.e. $ 1-1.5 billion).

At the time of power unit decommission for determination of activity levels and volumes of radioactive waste it will be necessary to know composition of all constructional and protective materials near-reactor zone which during all period of operation were subject to radiation. It is also necessary to note that there is a practical possibility of lowering of activation levels of materials by a purposeful choice at design and construction stages of the least activated components and substances that will lead to lowering of the neutron induced activity activity, and finally – to lowering of volumes of radioactive waste. Such approach is also recommended by the International Atomic Energy Agency (IAEA) [7-8].

Utilization and disposal of the RW produced during operation as well as RW produced during decommission is required upon decommission of the reactor unit.

In summary, approximate volume of RW upon decommission for low-level waste will sum up to 29000 m^3, intermediate-level waste – 25000 m^3, high-level waste – 8500 m^3. Costs of RW disposal upon total decommission of BN-800 reactor unit, based on 2016 RW disposal rates [9], will sum up as follows: low-level waste – 1100 million rubles, intermediate-level waste – 3318 million rubles, high-level waste – 4819 million rubles. Therefore approximate cost of disposal of all waste types will amount to 9,25 billion rubles ($ 145 million).

According to the Ministry of Energy of the Russian Federation the construction cost of the power unit No. 4 of the Beloyarsk NPP with the fast neutrons reactor BN-800 is estimated at 145,649 billion rubles [10 Based on the calculations, the cost of disposal of all kinds of waste upon the BN-800 decommissioning will be about 6.4% of the costs for the construction of the power unit. However, it necessary note that the treatment of radioactive waste during decommissioning is the collection, sorting, processing, conditioning, transportation of waste and disposal. These items of expenditure in the calculations is not taken into account except for disposal, but they can have a serious impact on the final cost of RW management during decommissioning.

4. Conclusion

Based on the results the conclusion can be made that content of LRW and SRW as well as their activity levels in fast neutron and thermal neutron reactors are almost similar. It was also determined that the volume of LRW in BN reactors is approx. twice less than those of VVER reactor; and volume of SRW is higher and is dependent from the accepted concept of decommission. SRW peculiarities require special decisions on their handling as well as development of special technologies of its reprocessing. Also do not forget that the modern approach to the design of nuclear power plants requires consideration of the stage of decommissioning at the stage of designing, with the aim of reducing the radiation exposure on the population and the environment, mainly by reducing the

IOP Conf. Series: Journal of Physics: Conf. Series **781** (2017) 012013 doi:10.1088/1742-6596/781/1/012013

volume of radioactive waste and ensure a more secure environment, the dismantling of buildings and structures.

Approximate cost of RW disposal will amount to 9.25 billion rubles ($145 million).

References

[1] Engovatov I A 2011 *Vestnik MSUCE* vol 8 p 325-330

[2] Skvortsov A I 2007 Handling of Radioactive Waste from Decommissioning of Fast Reactors: Author's abstract … PhD in Technical Sciences (Sankt-Petersburg) p 24

[3] Skvortsov A I 2007 Handling of Radioactive Waste from Decommissioning of Fast Reactors:Thesis. … PhD in Technical Sciences (Sankt-Petersburg) p 186

[4] Zhurbenko E A and Tsofin V I 2009 *Issues of Nuclear Science and Technics. Series: NPP Safeguard* **25** 140-144

[5] Tupchienko V A and Shmakov O G 2015 *Strategies of Development of Economics of Russia: Problems and Challenges* (Moscow) p 105-122

[6] Krasyukova N L 2011 *Economic Analyses: theory and practice* **47** 39-45.

[7] IAEA 1999 *Decommissioning of Nuclear Power Plants and Research Reactors* (Vienna, IAEA)

[8] IAEA 2007 *Decommissioning Strategies for Facilities Using Radioactive Material. Safety Reports Series* (Vienna, IAEA) **50**

[9] Order of the Ministry of Natural Resources of Russia as of 13.03.2013 N 89 "On Initial Set-up of Rates for Radioactive Waste Disposal".

[10] The cost of building fast neutron reactor BN-800 / http://www.atomic-energy.ru/news /2016/01/20/62632.

The remote methods for radwaste and SNF control

O Ivanov, V Stepanov, A Danilovich and V Potapov

NRC Kurchatov Institute 1, Kurchatov Sq, 123182, Moscow, Russia

olegivanov@mail.ru

Abstract. With the examples of developments carried out in the Kurchatov Institute and by the world leaders in the field the presentation considers the devices and methods to obtain remotely information on the distribution of radioactivity in radwaste and SNF. It describes the different types of light portable gamma cameras. The application of scanning spectrometric systems is considers also. The methods of recording UV radiation for detection of alpha contamination with the luminescence of air are presented. We discuss the scope and tasks that can be solved using remote and non-destructive methods.

1. Introduction

Contactless obtaining the distributions of gamma sources in the form of images as a method of research in medicine and astronomy applied to the 50's. Medical devices called gamma camera [1], astronomical – gamma-telescopes. The first portable gamma camera appeared in the nuclear industry as a finder of lost active sources in the hot chambers [2, 4, 5]. For these applications it is necessary to combine the optical image of detected gamma-emitting objects with the location of gamma sources. Accidents at nuclear power plants and the preparation for decommissioning of nuclear facilities stimulated the work on the further development of such tools [3, 6, 7, 10, 11, 22]. From the time of the accident at the Chernobyl nuclear power plant in Russia began to be used the name gammavizor [10-12]. On the world market there are first commercially produced cameras [8,9,24]. The last two decades were time of the research and development to create a lighter and more comfortable in use cameras [13, 14, 19-21, 27-29]. Finally, in recent years, very compact and lightweight cameras [24 - 26] weighing about 2-3 kilograms arisen. Progress in these studies is associated with the use of new semiconductor position-sensitive detectors and imaging on the detector by means of coded apertures.

Developers offer a very wide range of applications portable gamma camera: from the search of lost sources, obtaining the distribution of radioactivity in the objects to protect and control leakage of X-ray inspection. Not all the proposals have an implementation in a regular, everyday work with the devices. It may be noted the successful use of gamma cameras in Russia during the inspection of the objects of nuclear legacy [15] and during the work on the decommissioning of research nuclear reactor [16].

Gamma cameras for nuclear industry are based on different imaging principles – pinhole, coded mask, Compton camera. They use different detectors as a spectrometric [6, 9, 23], and without energy resolution [2, 8, 11, 13-14]. Different cameras may be required for different applications. For example, an effective method of scanning space spectrometry detector collimated with subsequent synthesis of images, sometimes also called gamma-ray imaging, and the gamma camera device. It should be noted first implementation of a portable gamma camera based on the principle of the Compton camera [26].

This camera is characterized by the fact that a measurement in a single position determines distribution of the flow of gamma radiation throughout the solid angle of 4π.

For widespread use of gamma cameras to work at nuclear facilities must be the recognition of the importance of using them solved problems and of course, price reduction, which is now a relatively large due to the high prices of portable position-sensitive detectors.

The following describes the principle of obtaining the gamma images in a variety of portable gamma-cameras, considered the device camera operating parameters, are examples of the gamma images. We consider the scanning system and the system discussed the scope and tasks that can be solved with the new devices.

2. Imaging principles

For the formation of gamma-image on the detector of portable gamma-camera the principle of the camera obscure, or a method of coded apertures is used. Coded aperture (mask) is a set of transparent and opaque to gamma radiation elements arranged in a particular pattern [3]. It comprises a base structure and an extra portion of the mask pattern which defines the basic design of the structure. Family -based mask patterns URA type often used in practical applications due to the advantages set inherent to said coding apertures family.

Scheme of measurements with using a coded mask is shown in Fig 1. Emitting object X gives the detector image M, which displays modulated (coded) aperture A flux of radiation from the source M = A * X. To determine the source image must hold its restoration using the inverse matrix C of the matrix apertures C = A- 1. Then X ~ C * M.

If the measurement is required field of view (in angular coordinates) θ x θ sterradian , with an angular resolution - - $\delta\theta$, then the number of elements of the mask (rank masks) will be $N = (\theta/\delta\theta)^2$. If the number of open cells (holes) in the mask n, the transparency mask is $\rho = n / N$.

What are the possible detectors for use in portable systems? There are two types of position-sensitive detectors which are small enough for use in portable systems: Scintillating with different photo registration (PMT or image intensifier + CCD) and semiconductor based detectors of modern semiconductors operating at room temperature - pixel and stripped detectors based on CdTe. First detectors have a large size and complex management and the second in until recently were not very accessible . Recently, the international research consortium develops detector Medipix2[4] made for him power and control through the USB interface, and this detector has become attractive for portable gamma camera. For CdTe detector with a thickness of 2 mm the sensitivity is evaluated using a simple model of image acquisition [6].

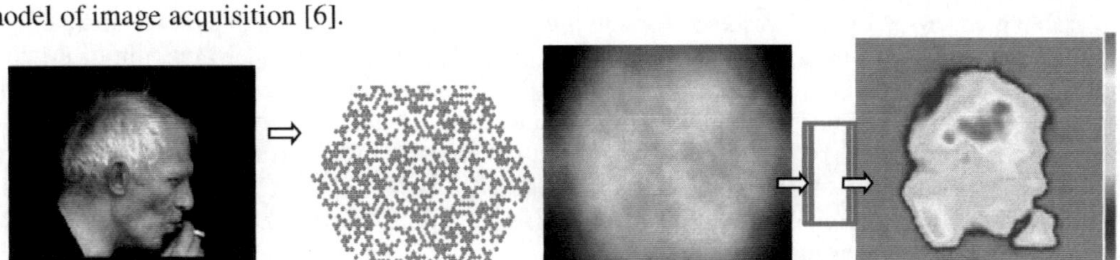

[object] [aperture (mask)] [decoding software]
[shadow-gram on detector] [image restored]

Figure 1. The scheme of gamma-imaging using coded aperture

Gamma radiation of the source is modulated by the mask, and a position-sensitive detector detects that shadow picture - shadowgram, encoded image. The signal from the detector is read out in digital form into the computer. The computer produce a restoration of the image - the original distribution of gamma radiation from the source. This image is displayed on the screen, and to identify the sources of gamma-rays the obtained image is necessarily superimposed on an optical image of the scene. As a result, the synthetic image presents distribution of "activity."

Modern compact position-sensitive detectors have good stability, so after spending the necessary calibration, for the obtained images angular distribution of the dose rate of the incident gamma radiation can be derived. This dose gives an estimate of power from individual partial objects at the location of the camera. Well, then, while continuing to analyze the obvious, knowing (from the spectrum, according to preliminary information) emitting nuclides and spending measure the distance to objects can be calculated activity. The partial dose rate is very important information for the dosimetric characterization of premises, staff work areas.

In many practical problems enough information about the location of the source or nature of the distribution of radioactivity in large objects without quantitative assessments.

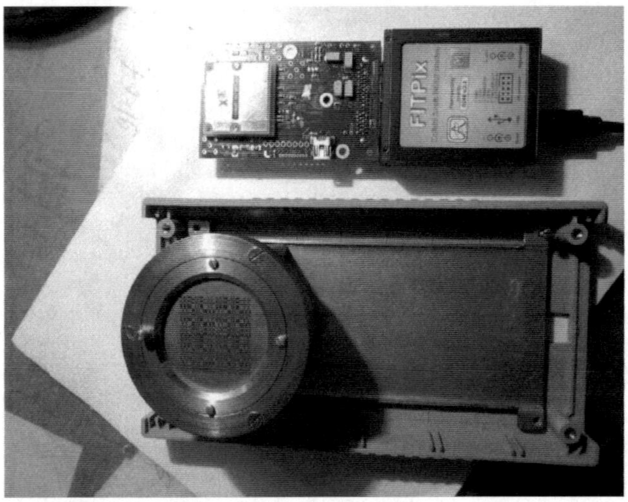

Figure 2. Detector Medipix2 with communication module FITPix before installing into the detector housing with block of coded mask.

Figure 3. The gamma camera which is autonomously set for remote measurements.

2.1. The results of applications

The portable gamma-ray imager was used for solving of different tasks during activity on dismantling of research reactors, remediation of sites, radwaste storage. These tasks include: Location of primary source terms; shielding design and optimization; finding discrete radioactive particles; locating of gamma-ray sources in shipping containers; locating and tracking crud in pipes and valves; tracking sources through time; identifying fuel failures; clean-up verification; determining spatial extent of contamination; filling in blank areas of traditional survey maps; emergency response; decommissioning and decontamination control; differentiating scatter and contamination; after operation contamination control.

In the fig. the results of gamma-ray imaging of packages with HLRW are presented. Main gamma emitter in these images is Co-60.

Figure 4. Gamma-ray imaging of packages with HLRW. Main gamma emitter is Co-60.

3. Scanning spectrometric systems

The scanning collimated spectrometric system is presented in fig.5.

Figure 5. Scanning collimated spectrometric system. Main parts: 1-collimated detector, 2-video camera, 3-control unit, 4-rotator

Maps of activity distribution measured with scanning collimated spectrometric system are presented in fig 6 and 7.

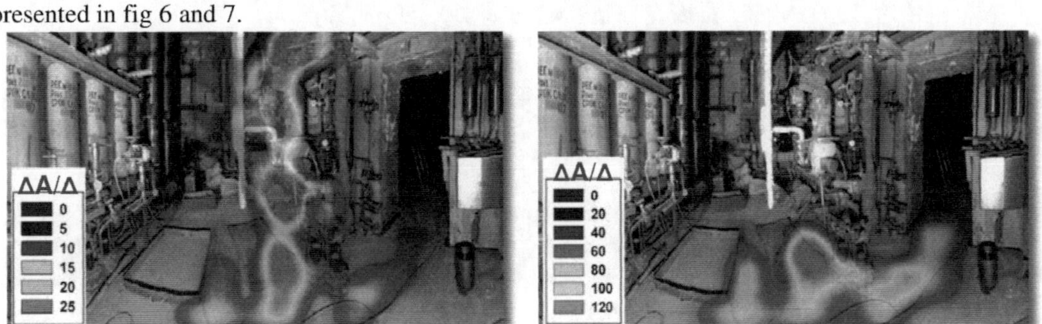

Total activity of ^{60}Co: $2.12 \cdot 10^{9}$ Bq Total activity of ^{137}Cs: $5.9 \cdot 10^{9}$Bq

Figure 6. Results of gamma-ray scanning in a room with contaminated facility. Main gamma emitters are Co-60 and Cs-137.

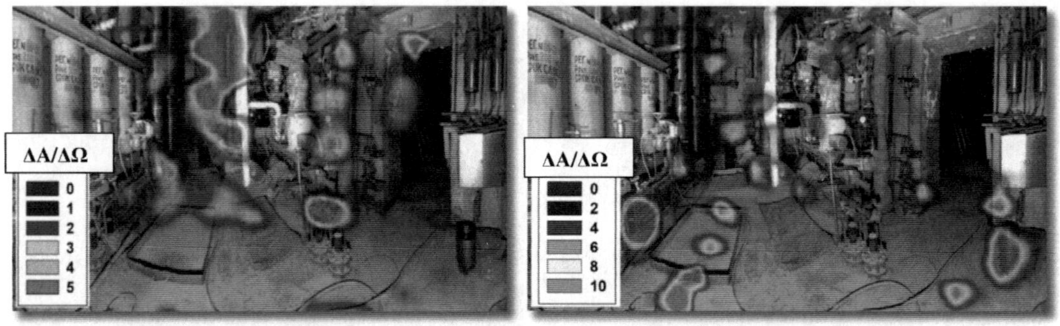

Total activity of ^{60}Co: $3.67 \cdot 10^{8}$ Bq Total activity of ^{137}Cs: $2.59 \cdot 10^{8}$ Bq

Figure 7. New distribution of gamma-activity in a room with contaminated facility if main sources are removed in dismantling works. Results of modelling.

4. Remote monitoring of alpha-ray contamination

Remote monitoring of alpha-ray contamination is very important problem for activities connected with decommission of installation of nuclear fuel processing. There are [34-37] some approaches for optical detection of alpha contamination based on measurements of fluorescence of atmospheric air in near ultraviolet (wavelength – 240 – 390 nm) region. Relative intensities of luminescence lines of air for excitation by Cm-244 can be seen in fig.8. CEA group at Marcoule is developing imaging systems based on this method [35, 36]. Large problem in this approach is solar (or other sources) UV background. New UV cameras are suitable for imaging of α-contamination by fluorescence of atmospheric air the as they are so called solar blind (SB)– they are sensitive to UV this wavelength less 290 nm only and are non- sensitive to solar UV.

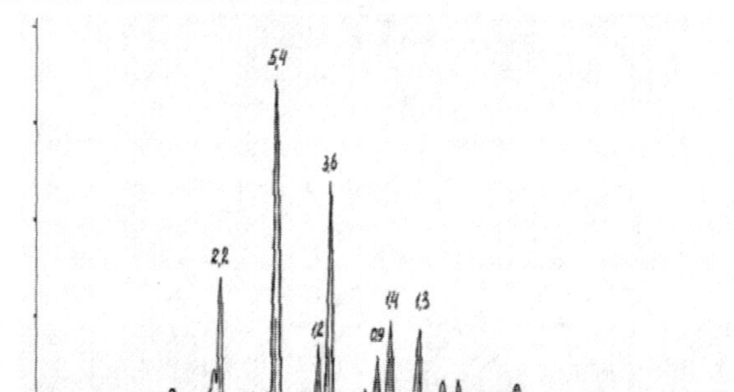

Figure 8. Luminescence spectra of air for excitation by Cm-244 source with activity 100 mCi. Numbers under spectral lines are their corresponding luminescence x10^{5}.

4.1. *Application of DayCor SBUV camera*

DayCor SuperB UV camera (fig. 9) has sensitivity $3x10^{-18}$ W/cm^{2} enables detection and displaying corona emission as weak as 1.5 pC at distance 8 m, and capture moving targets without smearing the output image. As discharge energies are relatively high this camera work in TV frame rate mode and have integration up to ~1 sec only. Our estimations [37] show that, that such sensitivity is enough for visualization of strong enough alpha sources without additional updating of industrial SBUV cameras. Thus for a point source of alpha radiation (typical energy of Ea = 5MeV) located on distance 3 m activity for which it can be found out is equal 10^5Bq. Homogeneous surface contamination, which can be detected is 300 Bq/cm^2

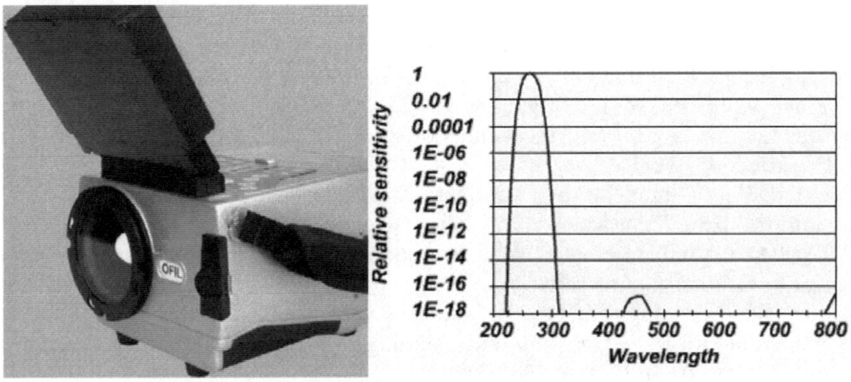

Figure 9. Photo of camera DayCor SuperB and dependence of relative sensitivity of its UV detector on wavelength [38]

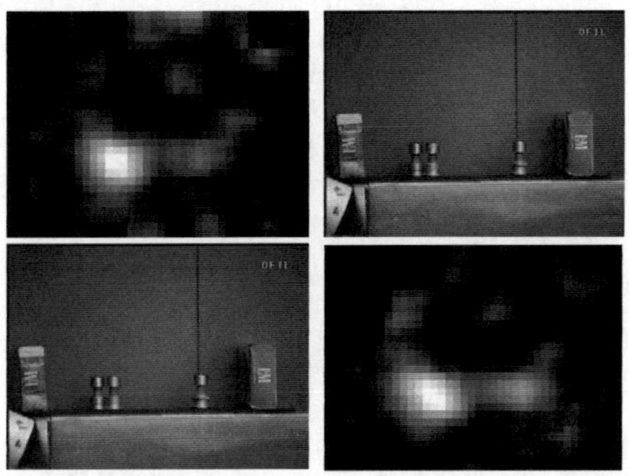

Figure 10. Images of 3 alpha sources of ~ 50 kBq each. Exposure - 10000 sec.

5. Conclusions

In result of the study we conclude that technology of industrial Solar blind UV cameras may be used for remote mapping of alpha contamination with reasonable surface activity. First UV images of alpha sources at day light conditions were obtained

Industrial SBUV cameras CoroCam and DayCor family are suitable for alpha contamination detection and mapping. For a point source of alpha radiation (typical energy of rays Ea = 5 MeV) located on distance 3 m activity for which it can detected is $\approx 10^5$Bq. Homogeneous surface contamination, which can be revealed for 300 Bq/cm^2.

At measurements with accumulation of a signal, for exposition 600 sec it is possible to expect, that for a point source of alpha radiation (Ea = 5MeV) with activity 3.10^4 Bq located on distance 3 m the activity will be detected. Homogeneously contaminated surface can be revealed for surface activity 100 Bq/cm^2 .

References

[1] H. Anger. A new instrument for mapping gamma-ray emitters. Biology and Medicine Quarterly Report UCRL, 1957, 3653: 38. (University of California Radiation Laboratory, Berkeley)

[2] G. Imbard, H. Carcreff "Development of a prototype gamma camera for use in decommissioning facilities", Waste Management 1995, Tucson, Arizona, March 1995.

[3] Z.He, S.V.Guru, D.K.Wehe, G.F.Knoll, A.Truman, D.Ramsden, Portable wide-angle γ-ray vision system", IEEE Transactions on Nuclear Science, vol. 42, No. 4, pp. 668-674, 1995.

[4] Migliaccio, R., Petito D., Nutter V., and Smith D. Gamma ray imaging in nuclear power plants. -Nuclear Plant Journal. 1996, March-April, v. 14, n 2, pp.205-212.

[5] D. Ramsden, A. J. Bird, M. J. Palmer, and P. T. Durrant. Gamma-Ray Imaging Systems for the Nuclear Environment. – In: Proceedings of Remote techniques for hazardous Environment, BNES, London, 1996.

[6] S.V.Guru, Z.He, D.K.Wehe and G.F.Knoll, A portable gamma camera for radiation monitoring," IEEE Transactions on Nuclear Science, vol. 42, No. 4, pp. 940-945, 1995.

[7] C. Le Goaller, G. Imbard et al., "The development and improvement of the Aladin gamma camera to localise gamma activity in nuclear installations," European Commission, Nuclear Science and Technology EUR18230 EN, 1998.

[8] AIL, GammaCam - [on line] https://www.dndkm.org/DOEKMDocuments/GetMedia/ Technology/141-1840-GammaCam.pdf

[9] RMD, RadCam -[on line] http://rmdinc.com/wp-content/uploads/2014/10/RMD-RadcamII.pdf

[10] A.G.Volkovich, O.P.Ivanov, V.E Stepanov, et al. Application of gammavizor for reactors examination. Atomic Energy. 1995. v. 79. n. 5. pp. 367-370. (In Russian).

[11] A.G.Volkovich, A.S.Danilovich, O.P.Ivanov, et al. Gammavizor – automated system for imaging of radioactive objects. Experimental Technique. 1996. No.3. pp.131-135. (In Russian).

[12] Ivanov O.P., Chesnokov A.V., Sudarkin A.N., Stepanov V.E., Urutskoev L.I. History of development and application of gamma-ray imagers in Russia since 1986. NIM A. 1999. v. 422. n 1-3. pp. 677-682.

[13] Gmar M., Gal O., Le Goaller C. Ivanov O. P., Stepanov V.E., et. al. Development of coded-aperture imaging with a compact gamma camera. - IEEE Trans. on Nucl. Science 2004, v. 51; n 4; p. 1682-1687.

[14] Ivanov O.P., Potapov V.N., Schmidt C., et. al. Development of new version of RayMoS gamma-camera with coded mask. - Abstract Book of IEEE NSS/MIC, Dresden, Germany 19 - 25 October 2008 p.37.

[15] K.E.Ivanov, N.N.Ponomarev-Stepnoi, Y.A.Teterin, A.Y.Teterin, V.V. Kharitonov. Remote measurement of radiation dose rate at the site of temporary storage of solid radioactive waste. Atomic Energy. 2008, v. 105, n. 2. pp. 103-107. (In Russian)

[16] V.G.Volkov, O.P.Ivanov, V.I. Kolyadin, et al. Treatment methods of high-level waste during decommissioning of research reactors RFT and MR. Atomic Energy. 2013. v. 115. n. 5. pp. 271-275. (In Russian)

[17] G.A.Fedorov, S.A.Tereshchenko. Computer emission tomography. - M .: Energoatomizdat, 1990.-184 pp. (In Russian)

[18] G.A.Fedorov. Integrated-coded registration system of ionizing radiation. ANRY 2008, no.1 (52), pp. 32-47. (In Russian)

[19] Ivanov O. P. , Stepanov V. E., Sudarkin A. N., and Urutskoev L. I. Different methods of image reconstruction for portable X-Ray and gamma-ray imager with coded aperture. – In Proc. of IEEE Nucl. Science Symp. 1997, November 9-15 1997, Albuquerque, New Mexico. v. 2, pp.1586-1590.

[20] A.N.Sudarkin, O.P.Ivanov, V.E.Stepanov, et al, “High-Energy Radiation Visualizer (HERV). A New System for Imaging in X-Ray and Gamma-Ray Emission Regions,” *IEEE Trans. on Nuc.Sci.* v.43, N4, p.2427, 1996.

[21] A.N. Sudarkin, O.P. Ivanov, V.E. Stepanov and L.I. Urutskoev, "Portable gamma ray imager and its application for the inspection of the near-reactor premises contaminated by radioactive substances," *Nucl. Instrum. Meth.,* vol. A 414, pp. 418-426, 1998.

[22] P.T. Durrant, M. Dallimore, I.D. Jupp and D. Ramsden, "The application of pinhole and coded aperture imaging in the nuclear environment," *Nucl. Instrum. Meth.*, vol. A 422, pp. 667-671, 1999.
[23] O. Gal, C. Izac, F. Jean, F. Laine, C. Leveque, A. Nguyen, "CARTOGAM - a portable gamma camera for remote localization of radioactive sources in nuclear facilities," *Nucl. Instrum. Meth.*, vol. A 460, pp. 138-145, 1999.
[24] Система iPIX [on line] http://www.canberra.com/products/insitu_systems/gamma-imaging-systems.asp
[25] Ivanov O, Semin I.,. Potapov V, et al. Ultra-Light Gamma-Camera for Security and Emergency Situation. Proceedings of WM2014 conference [online] www.wmsym.org/archives/2014/papers/14066.pdf
[26] Polaris-H: gamma-ray camera [on line] http://www.h3dgamma.com/applications.html
[27] O. P. Ivanov, V. E. Stepanov, V. G. Volkov, et al, "New Portable Gamma-Camera for Nuclear Environment and its Application at Rehabilitation Works," Book of abstracts IEEE NSS-MIC Conference, Rome, Italy, October 16-23, 2004, p.89.
[28] O.P. Ivanov, A.N. Sudarkin, V.E. Stepanov and L.I. Urutskoev, "Portable Digital X-Ray and Gamma-Ray Imaging with Coded Mask -Performance Characteristics and Methods of Image Reconstruction," *Nucl. Instrum. Meth.*, vol. A 422, pp. 729-734, 1999.
[29] O. Gal, B. Dessus, F. Jean, F. Laine, C. Leveque, "Operation of the CARTOGAM portable gamma camera in a photon-counting mode," *IEEE Trans. Nucl. Sci.*, vol. 48, no. 4, pp. 1198-1204, Aug. 2001.
[30] Medipix2 [on line]: http://medipix.web.cern.ch/medipix/pages/medipix2/medipix2.php
[31] X-ray Imaging Europe GmbH [on line]: http://www.xi-europe.de/
[32] ADVACAM products [on line]: http://www.advacam.com/en/pro
[33] S. R. Gottesman and E. E. Fenimore. New family of binary arrays for coded aperture imaging. Applied Optics, Vol. 28, Issue 20, pp. 4344-4352 (1989)
[34] Baschenko SM. Remote optical detection of alpha particle sources. Journal of Radiological Protection 2004; 24: 75-82.
[35] Lamadie F, Delmas F, Mahe C, Girones P, Le Goaller C, Costes JR. Remote alpha imaging in nuclear installations: New results and prospects. IEEE Transactions on Nuclear Science 2005; v.52: pp.3035-3039.
[36] C. Mahe, F. Lamadie, C. Le Goaller. Recent Advances in Low-Level Nuclear Measurements at the CEA– WM2009 Conference, March 1-5, 2009, Phoenix, AZ 9212
[37] O P Ivanov, A S Danilovich, V E Stepanov, SV Smirnov, A Volkovich. Visualization of Radioactive Sources without Gamma-Radiation with UV Imaging Systems, Paper no. ICEM2009-16145 pp. 321-325, ASME 2009 12-th International Conference on Environmental Remediation and Radioactive Waste Management (ICEM2009) October 11–15, 2009, Liverpool, UK
[38] DayCor cameras family of OLIF [online] http://www.ofilsystems.com/ and http://www.daycor.com/

X-ray fluorescence-based method for the quantitative determination of uranium in the aqueous solutions

S Dubrovka, S Chursin and V Verkhoturova

National Research Tomsk Polytechnic University, Lenin Avenue, 30, 634034, Tomsk, Russia

E-mail: spd1@tpu.ru

Abstract. Currently, one of the important issues in the field of nuclear technology is providing special handling with respect to nuclear materials, due to their energy and commercial significancy, as well as their potential radiation contamination threat. There is a necessity to have information about the full qualitative and quantitative composition of the sample as a part of special handling with nuclear materials. Spectrometric methods solve this problem effectively. One of these methods is the X-ray fluorescence analysis, which is fast, nondestructive and environmentally friendly with a high accuracy and reproducibility of the results. Development of uranium quantitative determination method in aqueous solutions to solve the problems of accounting and control of nuclear materials is the subject of research in this article. The development of the uranium concentration determination method in the aqueous solutions of uranyl nitrate $UO_2(NO_3)_2$ was carried out using Spectroscan MAKC-G – wavelength dispersive crystal diffraction XRF spectrometer.

1. Introduction

The significant efforts are made today in the world to provide special handling with respect to nuclear materials, as many of them have commercial and energy values. Also, there is a real threat of using nuclear materials for military or terrorist purposes. Moreover, the possibility of radiation contamination is of high importance. As it is known, special handling is aimed at providing the safety of nuclear materials and full information about them. It consists of three main components – the physical protection, accounting and control of nuclear materials [1].

From the system of accounting and control of nuclear materials point of view, it is always necessary to have full information about the qualitative and quantitative composition of a substance, as nuclear materials in any compound should be registered, starting from the minimum amount, and should be divided into categories. The issues concerning qualitative and quantitative composition of sample are resolved by spectrometric methods effectively. One of these methods is the X-ray fluorescence analysis, which is fast, nondestructive and environmentally friendly method with a high accuracy and results repeatability.

The subject of the research in this paper is to develop a method for determination of uranium concentration in the aqueous solutions to solve the problems of accounting and control of nuclear materials using MAKC-G analytical complex Spectroscan.

Spectroscan MAKC-G is a portable X-ray fluorescence wavelength dispersive crystal diffraction scanning spectrometer designed to determine the content of chemical elements from Ca to U of the substances in solids, liquids or powders, in solutions and thin films, deposits on filters.

2. The experimental evaluation of the method accuracy

With spectrometer Spectroscan MAKC-G standard multi-element sample was measured, in order to verify the settings and calibrations of device, as well as for accuracy evaluations.

With the qualitative analysis it was able to determine exactly the composition of the sample. Then, a quantitative analysis was made. The results are shown in table 1.

Table 1. The measurement results of the standard multi-element sample.

Components	Abundance ratio, %					
	Experimental values					Average values
	1	2	3	4	5	
Ca	20.01	20.37	20.35	20.13	20.20	20.210±0.068
Ti	28.38	28.21	28.29	28.60	28.12	28.320±0.082
Cr	4.22	4.11	4.09	3.98	4.24	4.129±0.047
Fe	2.88	2.81	2.79	2.76	2.86	2.819±0.022
Ni	2.05	2.06	2.04	2.04	2.08	2.055±0.007
Zn	4.04	4.06	4.04	4.07	4.07	4.058±0.007
Sr	2.31	2.30	2.32	2.32	2.32	2.315±0.003
Pb	2.36	2.34	2.34	2.35	2.35	2.345±0.052
H*	0.07	0.07	0.07	0.07	0.07	–
C*	0.70	0.70	0.70	0.70	0.70	–
O*	32.98	32.98	32.98	32.98	32.98	–

* – filling component, cannot be measured using this XRF spectrometer

The results of the qualitative and quantitative analysis of the values correspond to the passport. The measurement error is introduced in the second decimal place.

3. Preparation of uranyl nitrate aqueous solutions

The studies were conducted to measure aqueous solution of uranyl nitrate $UO_2(NO_3)_2$. This uranium connection was chosen for several reasons. Firstly, it is readily soluble in water and, therefore, it is safer to handle (as opposed to acid solutions). Secondly, uranyl nitrate is widely used at different stages of the nuclear fuel cycle, from the ore opening-up to the extraction separation of uranium and plutonium in spent nuclear fuel [2]. The step of prospecting and ore extraction is of particular interest, since the use of XRF analysis at this stage would greatly speed the process and replace longer, expensive and destructive chemical methods for qualitative and quantitative determination of the sample composition.

The process of sample preparation consists of several stages: mass calculation of uranyl nitrate dry powder to obtain the desired concentration of uranium in the solution; weighing of uranium powder using electronic scales with accuracy to three decimal places; dry uranyl nitrate dilution with water of 100 ml volume and mixing thoroughly; moving an aqueous solution into a special cell, which is supplied equipment to the spectrometer, with volume of 15 ml; the cell must be closed with the Mylar film and put through the adapter into the holder.

IOP Conf. Series: Journal of Physics: Conf. Series **781** (2017) 012015 doi:10.1088/1742-6596/781/1/012015

4. The experimental calibration curves

Several series of measurements of uranyl nitrate with concentration varying from 1 to 100 g/l. were made during the research. Measurement time is 8 minutes. The studies were conducted on two analyzing crystals – LiF(200) and C002. Characteristic lines of uranium were detected in the first and the second order of reflection on the analyzing crystal LiF(200) and in the second order on C002. In total, the five characteristic lines of the uranium, which are clearly visible in the measurement of all prepared solutions, have been selected for measurements. The calibration characteristics were obtained in the research result. These dependences are complex linear functions, which consist of two lines with transition between them, as it is shown in Figures 1 – 5.

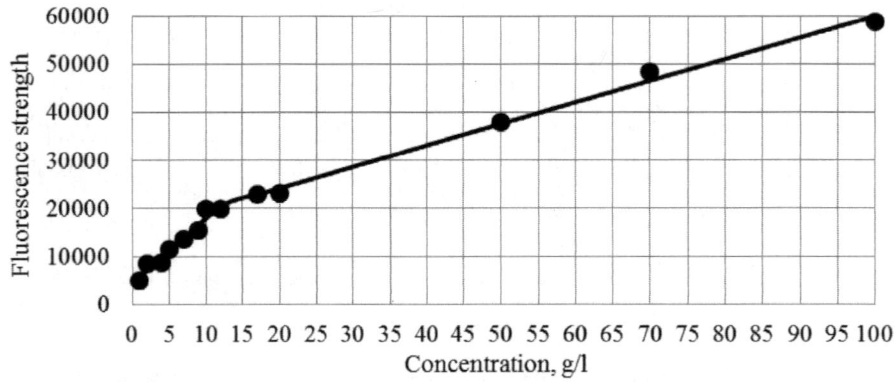

Figure 1. The calibration characteristic on a crystal LiF of 1 order reflection with a wavelength of 910 mA.

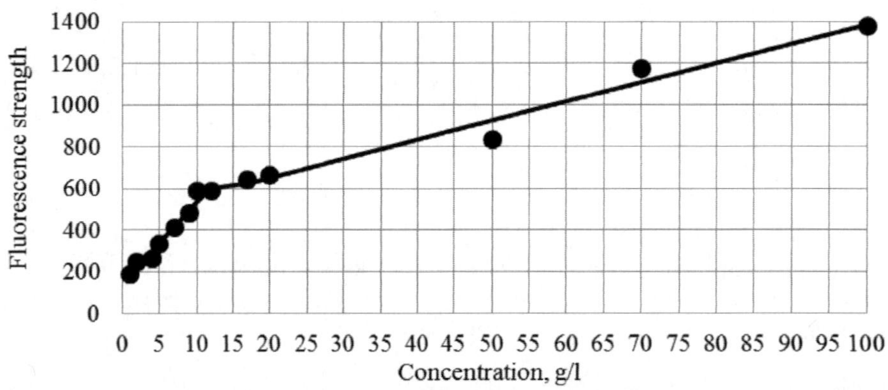

Figure 2. The calibration characteristic on a crystal LiF of 2 order reflection with a wavelength of 720 mA.

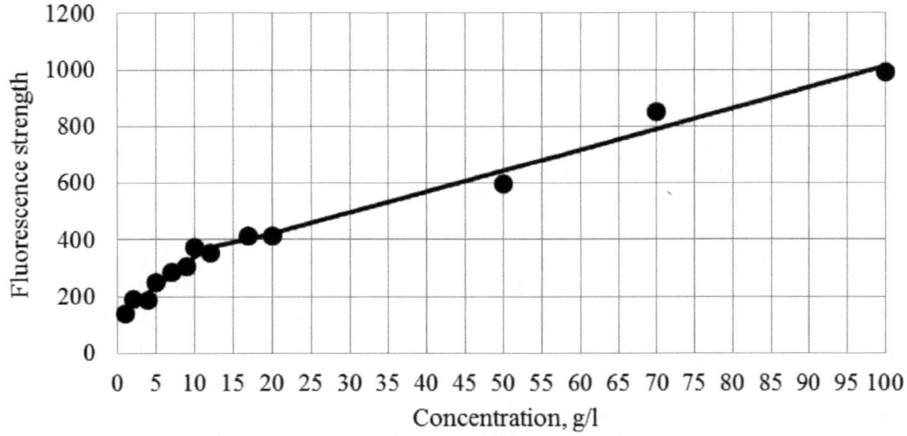

Figure 3. The calibration characteristic on a crystal LiF of 2 order reflection with a wavelength of 754 mA.

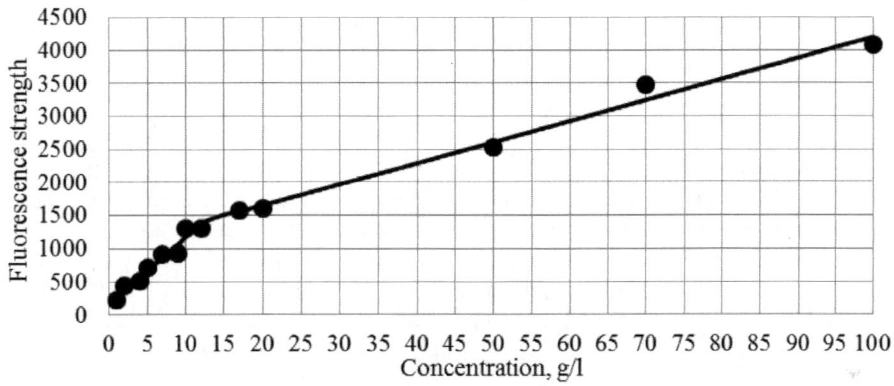

Figure 4. The calibration characteristic on a crystal LiF of 2 order reflection with a wavelength of 910 mA.

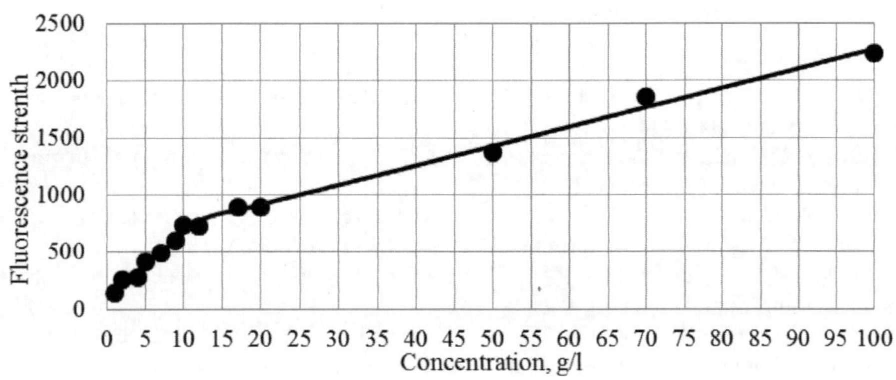

Figure 5. The calibration characteristic on a crystal C002 of 2 order reflection with a wavelength of 910 mA.

5. Analysis and testing of the method

Taking into account the dependencies between the concentrations of uranium in solution and fluorescence strength, it can be seen that a kink between two straight falls on a point with a value of 15 g/l, both on the analyzing crystal LiF and on the pyrolytic graphite in all investigated orders of reflection. Decomposition of calibration curves in two lines is due to effects of fluorescence quenching and self-shielding in the samples.

The studies were conducted with aqueous solutions of uranyl nitrate, and, as it is known, the water includes oxygen, which is the most prominent dynamic fluorescence quenching agent and is able to quench the fluorescence of almost all known fluorophores. With the decrease of uranium concentration in the solution the fluorescence strength decreases, but the concentration of oxygen on the contrary increases, consequently, the influence of the dynamic fluorescence quenching process increases. It is also important to note that even conducting rapid measurement (which was used in the work), the sample is heating. The dynamic quenching depends on diffusion, which, in turn, depends on the temperature. The increasing temperature increases diffusion coefficient. In addition, the quenching rate constant increases as a result of increasing temperature.

Moreover, the self-shielding effect influences on the results of measurements, because with increasing concentrations of uranium in solution, its solubility decreases and the probability of radiation's absorption in sample increases [3, 4].

Possibility of using the developed method was tested on the two standard samples, which have uranium concentrations 60 and 40 g/l. The maximum error in determining the uranium concentration was 2%.

6. Conclusion

During the research, by measuring the standard multi-elements sample it has been proved that the measurement results obtained by this XRF spectrometer have a high precision. Also, the possible use of this spectrometer to determine the elemental composition of the samples and quantitative determination of uranium in accounting and control of nuclear materials was proved.

Within the framework of the objectives a series of the experiments to determine the interconnection between the various uranium concentrations in aqueous solutions and their secondary radiation. As a result, the method for the quantitative determination of uranium in an aqueous solution has been developed and, later, successfully tested using analytical complex Spectroscan MAKC-G. The measurement error doesn't exceed 2 %.

The using of this method at different stages of the nuclear fuel cycle will simplify main manufacturing processes and will make them faster and cheaper. Also, it should be noted that this method does not depend on the chemical compounds of uranium. It is only necessary for the compound to be able to dissolve in water.

References

[1] V I Boyko and M E Silaev 2011 Methods and instruments for measuring the of nuclear and other radioactive materials (ISTC) p 356
[2] I I Zherin, G N Amelina 2010 Chemistry of thorium, uranium and plutonium (Tomsk: Tomsk Polytechnic University Press) p 147
[3] D Reilly, N Ensslin, H Smith, Jr. 1991 Passive Nondestractive Assay of Nuclear Materials, ed S Kreiner (Washington: U.S. Government Printing Office) p 700
[4] A V Bahtiarov. S K Savel'ev 2014 X-ray fluorecence analysis of minerals (Saint Peterburg: Saint Peterburg State University Press) p 132
[5] N G Chernorukov, O V Nipruk 2012 Theory and practice of X-ray fluorecence analisys (Nizhny Novgorod: Nizhny Novgorod State University Press) p 57
[6] L A Shikin 2009 X-ray fluorecence analysis of environmental objects (Vladimir: Vladimir State University) p 65
[7] I P Zhizhin 2002 X-ray fluorescence spectrometers "Spectroscan MAX" series. Analytical characteristics (Ekaterinburg: Ural Federal University Press) p 463 – 469
[8] R V Abrazheev, O.V. Nipruk, N.S. Godovanova 2012 X-ray fluorescence determination of Germanium, Arsenic and uranium (OOO ID Voda: chemistry and ecology Press) p 76 – 80
[9] A A Maslov, G V Kalyatskaya 2007 Technology of uranium (Tomsk: Tomsk Polytechnic University Press) p 97
[10] N S Turaeva, I I Zherin 2005 Chemistry and technology of uranium (TSNIIATOMINFORM) p 407

Nondestructive Determination of ^{235}U Enrichment of Uranium Samples in the Presence of Actinides (^{232}Th, ^{236}U, ^{237}Np)

Yu V Stogov

Department of Theoretical and Experimental Physics of Nuclear Reactors, National Research Nuclear University MEPhI (Moscow Engineering Physics Institute), Kashirskoe highway, 31, Moscow, Russia

YVStogov@mephi.ru

Abstract. The measurements were performed using a planar Ge detector. The multi-group analysis (MGA) for the determination isotopic abundances in low enriched uranium samples is applied in this study. In order to perform the analysis of complex gamma-spectrums, were taken a number of gamma-spectrums of single reference samples, containing ^{235}U, ^{238}U, ^{232}Th, ^{236}U, ^{237}Np with declared isotopic abundance. The gamma-ray spectrums in the range of low energy X-ray and gamma-ray peaks of these samples were obtained and analyzed.

1. Introduction

The determination of ^{235}U enrichment is of great importance of nuclear technology. The MGA for uranium (MGAU) method for non-destructive assay of ^{235}U enrichment does not require the use of any nuclear material standards [1].

The MGAU-code is based on the analysis of the 89-120 keV region of Ge detector spectrum. There are some limitations of using of MGAU-code: the resolution at $E\gamma$=122 keV of Germanium detector less than 700 eV; it requires from the daughter isotopes be in activity equilibrium with the ^{235}U and ^{238}U parent isotopes (it require up to six months after chemical separation); the accuracy of the method diminishes when the thickness of sample container becomes excessive; the abundances of actinides (^{232}Th, ^{236}U, ^{237}Np, ^{239}Pu) and other gamma-emitter nuclides (fission products, et all) must be very little. The MGAU-code was designed to operate with little or no user interaction, that is as «complete secrecy»-code (we can not change the parameters of the code). Hence the conclusion: it is necessary to perform a preliminary analysis of the gamma-ray spectrums of the samples for the presence of «interfering» impurities - gamma emitters (actinides) [1-2].

The first task of this investigation is to perform the analysis of gamma-spectrums of single reference samples containing ^{235}U, ^{238}U, ^{232}Th, ^{236}U, ^{237}Np with declared isotopic abundance.

The second task – using the MGAU-code to obtain the results of the ^{235}U enrichment of samples: a) «pure» U-samples and b) «dirty» U-samples, containing different concentrations of actinides (^{232}Th, ^{236}U, ^{237}Np). Before each using of the MGAU-code it is necessary to analyze gamma-spectrum of sample.

The third task – to obtain the results of the ^{235}U enrichment sample using the gamma-spectrum of investigated sample without use of MGAU-code. In order to decide this task the «peak-ratio» technique was applied [3] in the narrow 89-99-keV energy range. The peak 89.95-keV in spectrum of the sample is a measure of the ^{235}U concentration (X-ray Th X $_{K\alpha2}$). The intensities of 92.4 and 92.8-

keV lines of ^{234}Th are used as a measure of ^{238}U concentration of the sample. The 93.35-keV thorium $K\alpha_1$ line was used as a measure of ^{235}U concentration. The contribution of 93.35-keV line was taken into account by using the special procedure.

2. Experimental description

2.1. Equipment

Spectrums were taken in planar Ge detector with resolution 460 eV at E_γ=122 keV (full width at half maximum, FWHM). The multichannel digital analyzer DSA-1000 Canberra and Genie-2000 (including MGAU-code) soft were used. The spectrums were stored in 4096 channels with a gain of 0.75 keV/channel. During the measurements there was performed the control of the position of the X-ray and gamma-ray peaks at energies of 63.1 and 185.72 keV, respectively belonging ^{234}Th and ^{235}U. The maxima of the peaks were situated respectively in the channels (844±1) and (2476±4).

2.2. The Gamma and X-rays Nuclear Data

The Gamma and X-rays nuclear data are presented in Table 1.

Table 1. Gamma and X-rays nuclear data [4].

	Energy (keV)	Photon emission probability (%)
γ-ray(^{235}U/^{231}Th)	89.95±0.2	1.00±0.6
Th $_{K\alpha2}$	89.957	28.2±0.6
Pa $_{K\alpha2}$	92.282	28.3±0.6
γ-ray (^{238}U/^{234}Th)	92.38	2.8±0.3
γ-ray (^{238}U/^{234}Th)	92.80	2.8±0.3
γ-ray (^{235}U/^{231}Th)	93.02	0.047±0.006
Th $_{K\alpha1}$	93.35	46.0±0.9
U $_{K\alpha3}$	93.844	0.098±0.003

2.3. The Procedure of the Experimental Investigation

2.3.1. The Investigation of Influence of Different Concentrations of ^{232}Th and ^{236}U Presence in UO_2-Samples on the Results of ^{235}U Enrichment Determination by MGAU-code

The ^{235}U enrichment of the reference UO_2-sample (^{235}U enrichment x=6.50 wt%) was determined by MGAU-code (Table 2).

Table 2. The results of ^{235}U enrichment using MGAU-code analysis for UO_2-sample (x=6.50 wt%).

Nuclide	Weight fraction of nuclide, (%)	Derivation
^{234}U	0.045	0.010 (22.2 %)
^{235}U	6.352	0.084 (1.3%)
^{238}U	93.603	0.084 (0.1%)

The gamma-spectrum of the reference UO_2-sample (^{235}U enrichment x=6.50 wt%) was taken. The peak 89.95-keV in spectrum of this sample is a measure of the ^{235}U concentration (X-ray Th X $_{K\alpha 2}$) (Figure 1).

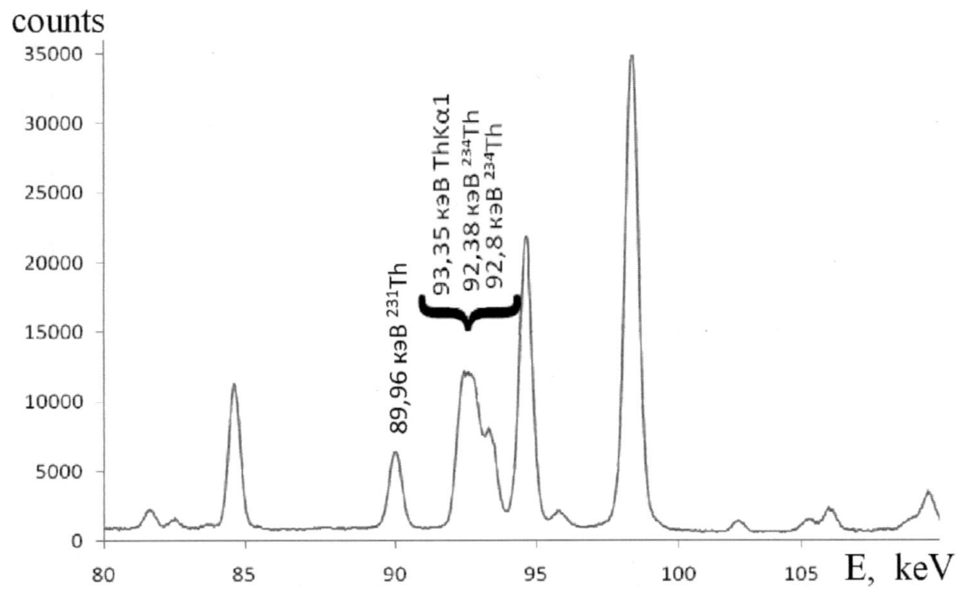

Figure 1. Gamma and X-ray spectrum of enriched uranium sample (x=6.50 wt%).

The gamma-spectrum of ^{232}Th sample was taken. In this spectrum the peak of sample 89.95-keV is a measure of ^{232}Th concentration.

The count rate in peaks 89.95 keV in both samples was determined (measuring peak intensity in counts/s). The time of measuring ^{232}Th sample spectrum, corresponding the increasing the value of count rate in peak 89.95 keV in the spectrum of UO_2-sample (^{235}U enrichment x=6.50 wt%) is obtained.

Using the gamma-spectrum of UO_2-sample contained in the memory of multi-channel analyzer, the UO_2-sample was removed from Ge - detector and then the ^{232}Th-sample was put on the detector. After the spectrum measurement of ^{232}Th-sample during the necessary time the count rate in the peak 89.95-keV increased by 1%. That corresponds to an increase of weight fraction ^{232}Th in the mixture of (UO_2+^{232}Th). A new treatment of obtained spectrum was performed by MGAU-code, ect. Step by step we obtain the response function of MGAU-code of increasing 89.95 keV count rate peak by gamma-emitter (^{232}Th-sample). This count rate depends from parameter P:

$$P = \frac{S^{90\,keV}(UO_2(6.5\%) + {}^{232}Th)}{S^{90\,keV}(UO_2(6.5\%))} \quad (1)$$

The dependence of ^{235}U enrichment (obtained using MGAU-code analysis) from parameter P is presented on Figure 2. The number of the «big points» on the Figure 2 corresponds that there are no message from MGAU-code «gamma-emitting nuclides are present in investigated sample» on measure of increase of parameter P before certain moment. The analogical procedure was performed with ^{236}U sample. The results are presented in Figure 3 in dependence of the parameter $P_1 = \frac{S^{90\,keV}(UO_2(6.5\%) + {}^{236}U)}{S^{90\,keV}(UO_2(6.5\%))}$.

IOP Conf. Series: Journal of Physics: Conf. Series **781** (2017) 012016 doi:10.1088/1742-6596/781/1/012016

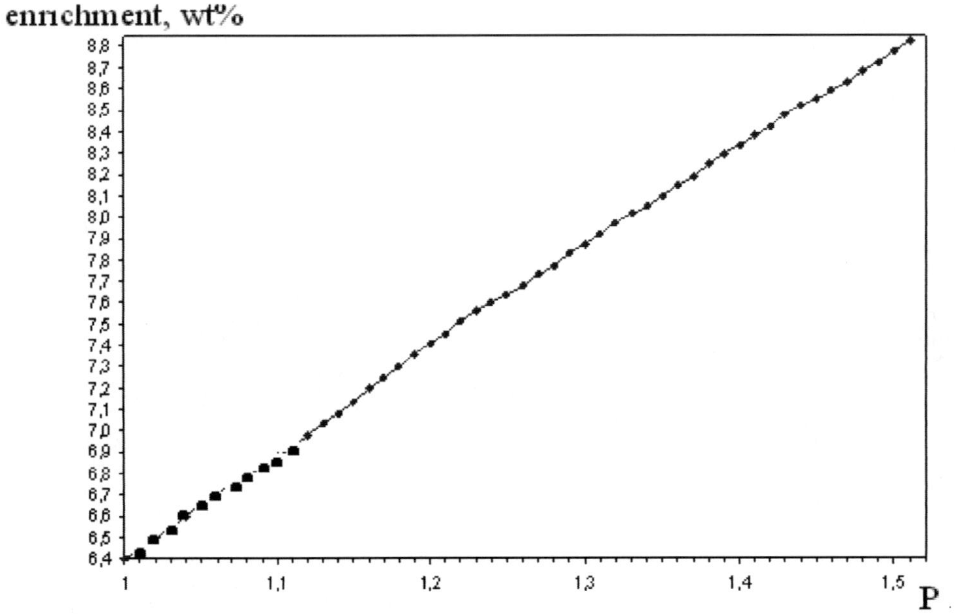

Figure 2. Dependence of ^{235}U enrichment (obtained using MGAU-code analysis) from parameter P: ^{232}Th gamma-emitting nuclide is present in investigated sample.

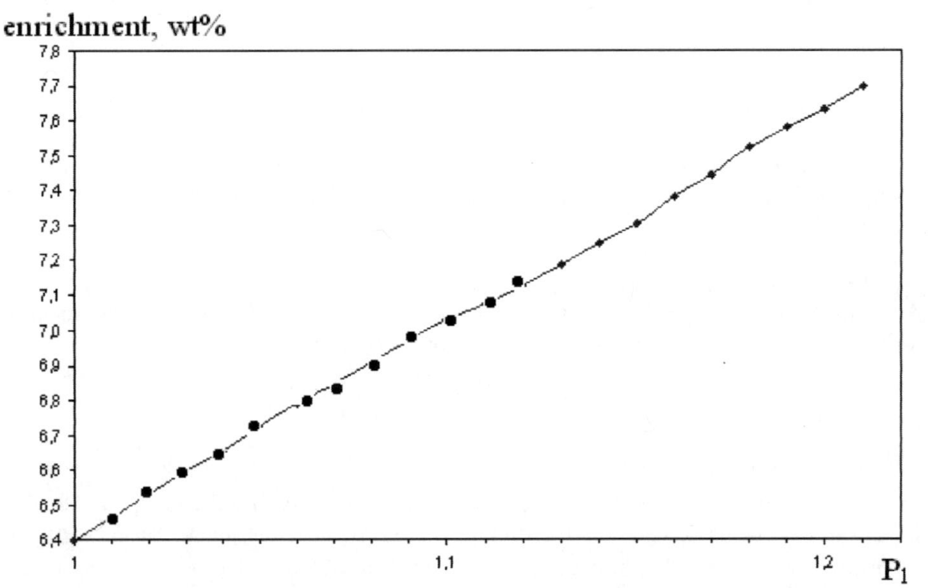

Figure 3. Dependence of ^{235}U enrichment (obtained using MGAU-code analysis) from parameter P_1: ^{236}U gamma-emitting nuclide is present in investigated sample.

2.2.2. The «Peak-Ratio» Technique for ^{235}U Enrichment Determination in Samples (Without Using of MGAU-Code)

On the Figure 4 is presented the gamma and X-ray spectrum of depleted uranium sample (x=0.0025 wt%). On the Figure 5 is presented the gamma and X-ray spectrum of enriched uranium sample (x=98 wt%)

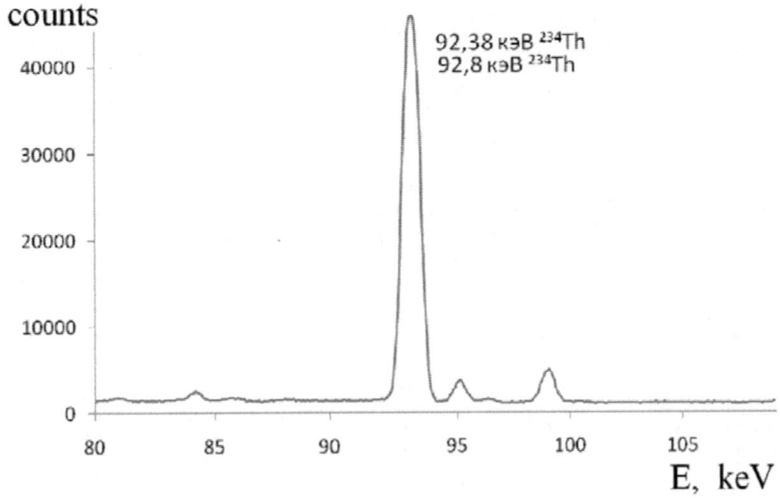

Figure 4. Gamma and X-ray spectrum of depleted uranium sample (x=0.0025 wt%).

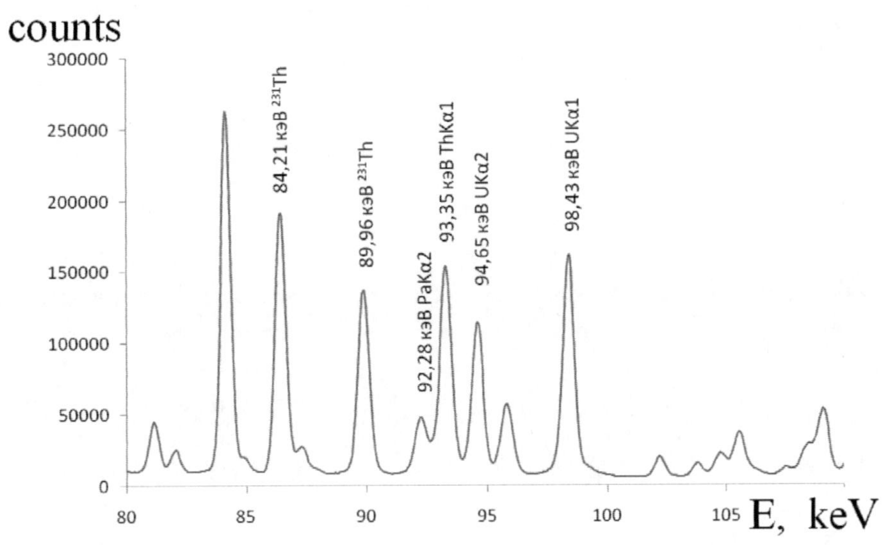

Figure 5. Gamma and X-ray spectrum of enriched uranium sample (x=98 wt%).

The «strip» function of the multichannel analyzer was used as follows: 1) store an «enriched» spectrum of uranium enriched to 99.5 % in ^{235}U; 2) obtain the sample spectrum, then normalize the «enriched» spectrum to the 89.96 keV X-ray of the sample spectrum; and 3) subtract the normalized «enriched» spectrum from the sample spectrum. The resultant stripped spectrum contains the cleanly resolved ^{238}U gamma-ray doublet at 92-keV peak from the stripped spectrum and the 89.96-keV X-ray peak area from the original sample spectrum are used to obtain the ^{238}U/^{235}U ratio.
A relationship between enrichment, X, and the ^{238}U/^{235}U ratio, R, can be obtained as follows:

$$X = \frac{100 \cdot w.f.^{235}U}{w.f.^{234}U + w.f.^{235}U + w.f.^{236}U + w.f.^{238}U} \qquad (2),$$

where $w.f.^{234}U$, $w.f.^{235}U$, $w.f.^{236}U$, $w.f.^{238}U$ - weight fraction of ^{234}U, ^{235}U, ^{236}U, ^{238}U, respectively
Substitutive the following approximation into the above equation:

$$w.f.^{234}U + w.f.^{236}U = 0.01 \cdot w.f.^{235}U \qquad (3).$$

Than rearrange the terms to obtain:

$$X = \frac{100}{1.01 + \frac{w.f.^{238}U}{w.f.^{235}U}} = \frac{100}{1.01 + K \cdot R},$$

where $R = \frac{S^{92\,keV}}{S^{90\,keV}}$; $K = 5{,}46$ is an experimental constant. The constant K is determined in by iterative way to give the best fit to the calibration data where $R = \frac{S^{92\,keV}}{S^{90\,keV}}$. The relationship between peak area ratios S^{92keV}/S^{90keV} and ^{235}U enrichment are presented in figure 6.

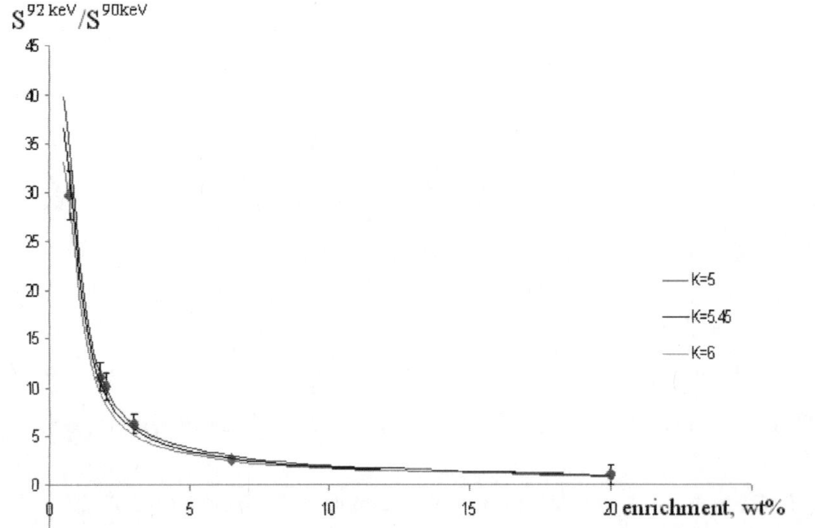

Figure 6. (color online) Relationship between peak area ratios S^{92keV}/S^{90keV} and ^{235}U enrichment.

3. Conclusions

Using the high resolution planar Ge detector it is possible to take the gamma-spectrums of «clean» uranium samples and «unknown» «dirty» uranium samples, containing some concentrations of actinides. Using a single reference samples, containing ^{232}Th, ^{236}U, ^{237}Np etc. with declared isotopic abundance, it is possible to increase the count rates in corresponding gamma-spectrum peaks of «unknown» «dirty» uranium samples and obtain the response function of MGAU-code from the ^{235}U enrichment (using MGAU-code) in dependence of parameter **P**. At last, now it is possible to estimate the number of steps, that corresponds the increasing of **P** on 1%, when the MGAU-code message «gamma-emitters are present in sample» will be obtain.

The «peak-ratio» technique description, presented in this paper, allows explaining the procedure of ^{235}U enrichment determination of uranium samples without using of MGAU-code and its allows obtaining the ^{235}U enrichment of investigated samples.

The study of nondestructive determination of ^{235}U enrichment of uranium samples in the presence of actinides will be continued.

4. References

[1] Gunnik R *et al* 1994 A New Analysis Code for Measuring U-235 Enrichments in Arbitrary Samples *IAEA Symposium on International Safeguards, Vienna, Austria. Lawrence Livermore National Laboratory. Report UCRL-JC-114713 Livermore, California*

[2] Yucel H 2007 The Applicability of MGA Method for Depleted and Natural Uranium Isotopic Analysis in the Presence of Actinides (^{232}Th, ^{237}Np, ^{233}Pa and ^{241}Am) *Applied Radiation and Isotopes* **65** p 1269-1280

[3] Passive Nondestructive Assay of Nuclear Materials 1991 NUREG/CR-5550 LA-UR-90-732 Edited by: Reilly D, Ensslin N and Smith H

[4] Firestone R B, Shirley V S 1996 Table of Isotopes. *Wiley, New-York*

IOP Conf. Series: Journal of Physics: Conf. Series **781** (2017) 012017 doi:10.1088/1742-6596/781/1/012017

The basic features of a closed fuel cycle without fast reactors

E A Bobrov, P N Alekseev and P S Teplov

National Research Center "Kurchatov Institute",
1, Akademika Kurchatova pl., Moscow, 123182, Russia

E-mail: Bobrov_EA@nrcki.ru

Abstract. In this paper the basic features of a closed fuel cycle with thermal reactors are considered. The three variants of multiple Pu and U recycling in VVER reactors was investigated. The comparison of MOX and REMIX fuel approaches for closed fuel cycle with thermal reactors is presented. All variants make possible to recycle several times the total amount of Pu and U obtained from spent fuel. The reported study was funded by RFBR according to the research project № 16-38-00021

1. Introduction

Basic strategy of Russian Nuclear Energetic is propagation of closed fuel cycle on the base of fast breeder and thermal reactors. The strategy can help to solve such systematic problems as the huge quantity of accumulated spent nuclear fuel in the storages and the limited inventory of cheap natural uranium for fuel production, and to increase the economic attractiveness of the nuclear industry. There is the program which based on the development of fast nuclear reactors in Russia, but this technology is not ready for global implementation. The main element of the nuclear power fleet in Russia today is VVER reactors. The first stage for closed fuel cycle can be done with applying thermal reactors. It will help to decrease the amount of spent nuclear fuel in storages, reduce natural uranium consumption and develop modern reprocessing technologies.

This paper shows the three concepts of U and Pu multiple recycling in VVER type reactors. The first two variants differ with feeding fissile material. The first is the standard REMIX fuel [1-3] approach and the second is close to MOX fuel where regenerated uranium used instead of depleted uranium and the feeding fissile material is reactor grade plutonium. It makes possible to recycle several times the total amount of Pu obtained from spent fuel. The main difference in Pu recycling is the concept of 100% or partial fuel assemblies (FAs) loading of the core. The third variant is heterogeneous composition of enriched uranium and MOX fuel pins in the FA. These three variants of the full core loadings are balanced on zero Pu accumulation in the fuel cycle. The all Pu from the spent fuel of the core loading is used to produce new fuel for the next loading. This approach makes it possible to compare physical and economical aspects of the three variants of Pu multiple recycling in VVER core.

The neutron-physics calculations were performed by the Consul code package [4]. All calculations were performed for the standard VVER-1000 FA [5] configuration. Duration of the fuel campaign is 4 years (4X300 EFPD) with the average burnup 49.3 MW•day/kg_{HM}.

2. The fuel cycle concepts

The first variant observed in the paper based on enriched uranium addition to the recycled mixture of U and Pu. Basically the enrichment of feeding U supposed to be lesser then 20% of ^{235}U, but it will be impossible to rich 100% usage of spent fuel for the next loading in that case. The selected in the investigation first variant presumes enrichment of feeding U in the range of 50-55% of ^{235}U to achieve parameters given in the task. That is standard REMIX fuel concept. The resulting mixture consists of 3.8% of ^{235}U and 1.2% of Pu. The Pu and ^{235}U content growth with recycling number. This variant assumes 100% loading of the core with REMIX FAs.

The second variant based on Pu addition to the mixture. The necessary amount of Pu received from the pre-recycled FA with standard UOX fuel. It is possible to achieve 100% usage of spent fuel in this case. The second variant needs additional reprocessing of UOX fuel with full separation of plutonium fraction. The main difference from the standard MOX fuel is the presence of regenerated U instead of depleted one. The resulting mixture consists of 0.8% of ^{235}U and 9% of Pu. The investigation of FA depletion was done under assumption that MOX FA surrounded with UOX FAs to take into account spectral effects. The standard construction of VVER-1000 FA with 312 fuel pins was chosen for the investigation of burnup properties of new fuel compositions for the first and second variants. No burnable absorbers or Pu content profiling were taken into account.

The third variant based on the facts that during reprocessing the full FA is cut and melted down and that REMIX technology allows getting the mixture with any Pu content. The main idea was to separate UOX and MOX fuel pins in the REMIX FA to achieve better fission properties for ^{235}U. The variant of heterogeneous fuel pin positioning for VVER FA is presented in Figure 1. This concept close to CORAIL FA design for Pu multirecycling in PWR [6,7]. The main difference is MOX fuel pin amount in FA and regenerated U presence in fuel composition which helps to reduce Pu content in MOX fuel pins.

The presented FA consists of 78-90 MOX fuel pins (25%) with 0.8% of ^{235}U and 4.5% Pu and 234-222 UOX fuel pins with 4.6% of ^{235}U. Pu content in MOX fuel pin, total number and positioning of these pins in FA were chosen to meet the following tasks:

- the average burn-up of MOX and UOX fuel pins are equal;
- peaking factor does not exceed 1.2 (assembly calculation).

Total amount of Pu and MOX pins will increase with recycling number. The investigation doesn't assume the usage of regenerated uranium for UOX pin manufacturing, and they have standard design.

3. Main conclusions

The proposed above approach to the reprocessing technology has potential to improve economic parameters by reducing the number of process steps during SNF reprocessing and fuel manufacturing (reducing the several stages of fuel powder and pellets fabrication as compared to MOX fuel fabrication). New technology process can provide better quality for the mixed uranium plutonium pellet fabrication. Low Pu content in the core has negligible influence on safety parameters of the NPP.

The three options of REMIX FA fabrication proposed in the paper. All considered variants assume Pu multiple recycling. The Pu content and isotopic quality in the fuel matrix stabilized with growth of recycle number for the most variants. The better results in natural uranium savings can be achieved for the standard REMIX(UOX) approach. The regenerated materials usage in thermal power reactors gives not more than 30% saving of natural uranium consumption.

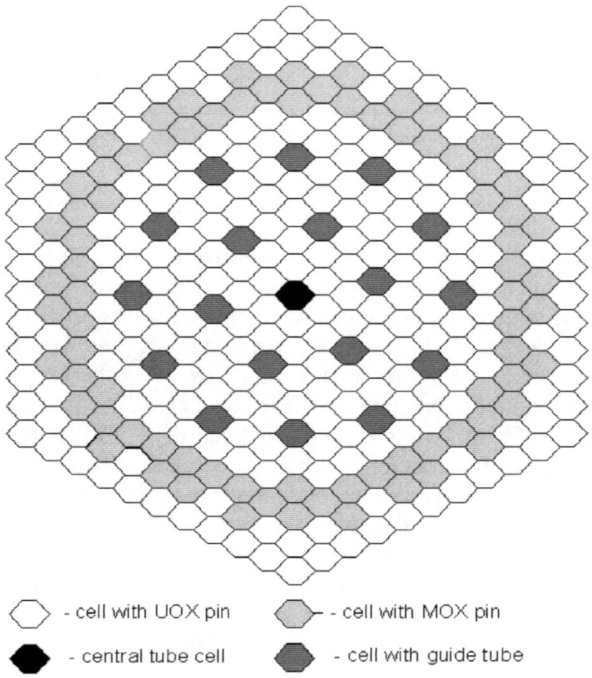

Figure 1. The REMIX (het) FA configuration for heterogeneous fuel pin positioning

Estimation of technical and economic assessment presented in the paper has demonstrated that the use of REMIX technology in the closed fuel cycle is more expensive than the open fuel cycle with direct SNF disposal. The Pu positioning in the expensive MOX FA gives better results from the economical point of view but it is worse for multiple recycling. The main idea is to place expensive fuel in the expensive fuel pins or FAs so the integral FA cost for the loading will decrease.

It is important to note that the uranium multiple recycling in the REMIX fuel form or using the reenrichment process leads to the uranium isotope composition degradation. The 236U and 232U concentrations in the fuel are increasing and the regenerated uranium treatment becomes more complicated.

References

[1] Pavlovichev A. M., Pavlov V. I., Semchenkov Yu. M., Kudryavtsev E. G., Fedorov Yu. S., Bibichev B. A. 2006 *Neutron-physical characteristics of a VVÉR core with 100% load of reprocessed uranium and plutonium fuel* Atomic Energy, 101 863-868

[2] Pavlovichev A. M., Pavlov V. I., Semchenkov Yu. M., Kudryavtsev E. G., Fedorov Yu. S., Bibichev B. A., Zil'berman B Ya. 2008 *Neutron-physical characteristics of a WWER-1000 core with 100% fuel load consisting of a mixture of recovered uranium and plutonium and enriched uranium* Atomic Energy 104 257-261

[3] Fedorov Yu.S., Baryshnikov M.V., Bibichev B.A., Zilberman B.Ya., Kryukov O.V., Khaperskaya A.V. 2013 *Multiple recycle of REMIX fuel based on reprocessed uranium and plutonium mixture in thermal reactors* Proc of Global 2013, Salt Lake City, Utah

[4] Chibinyaev A.V., Teplov P.S. 2007 *CONSUL – code package for comprehensive LWR core calculations* Proc of ICAPP 2007 Nice, France, May 13-18

[5] Vygovskyy S.B, Ryabov N.O., Semenov A.A., Chernov E.V., Bogachek L.N. 2011 *The physical and structural characteristics of nuclear power plants with VVER* Textbook

Moscow MEPhI pp. 376 - ISBN 978-5-7262-1458-0
[6] Grouiller J.P., Doriath J.Y., Vasile A., Zaetta A. 2001 *Different possible scenarios for plutonium recycling in PWRs* Global 2001, Paris (France), INIS-FR-1314
[7] Youinou G. and Vasile A. 2005 *Plutonium multirecycling in standard PWRs loaded with evolutionary fuels* Nuclear SCIENCE and engineering 151 pp. 25–45

Acknowledgments

The authors are grateful for the help to Dudnikov A.A. and Chibinyaev A.V.
The reported study was funded by RFBR according to the research project № 16-38-00021.

Influence of uncertainties of isotopic composition of the reprocessed uranium on effectiveness of its enrichment in gas centrifuge cascades

A Yu Smirnov[1], A R Mustafin[1], V A Nevinitsa[2], G A Sulaberidze[1], A A Dudnikov[2] and V E Gusev[1]

[1] National Research Nuclear University MEPhI (Moscow Engineering Physics Institute), Kashirskoe shosse 31, Moscow, 115409, Russia
[2] National Research Center "Kurchatov Institute", Moscow, Russia

E-mail: AYSmirnov@mephi.ru

Abstract. The effect of the uncertainties of the isotopic composition of the reprocessed uranium on its enrichment process in gas centrifuge cascades while diluting it by adding low-enriched uranium (LEU) and waste uranium. It is shown that changing the content of ^{232}U and ^{236}U isotopes in the initial reprocessed uranium within 15% (rel.) can significantly change natural uranium consumption and separative work (up to 2-3%). However, even in case of increase of these parameters is possible to find the ratio of diluents, where the cascade with three feed flows (depleted uranium, LEU and reprocessed uranium) will be more effective than ordinary separation cascade with one feed point for producing LEU from natural uranium.

1. Introduction

In the context of increasing needs of the nuclear power in the fuel for thermal reactors and the gap between production and consumption of natural uranium [1], the reprocessed uranium and plutonium are considered as an alternative to natural raw materials, and partly involved in the light water reactors' fuel cycle. In addition to that there are schemes of using reprocessed uranium and plutonium together [2-5] and separately [6-9].

In the case of separate use of these materials for the production of nuclear fuel, reprocessed uranium must first be enriched in ^{235}U by means of gas centrifuge cascades [10-16]. Moreover, besides saving natural resources and reducing the volume of radioactive waste, reprocessed uranium fully meets the requirements for nuclear non-proliferation treaty (NPT) [17-18]. But enrichment of this material in separation cascades is impeded due to the presence of ^{232}U и ^{236}U [6], concentrations of which are strictly limited to fulfill the requirements of radiation safety in the manufacture of fuel elements and to preserve the neutron-physical characteristics of the fuel.

For today, a number of methods to enrich the reprocessed uranium in gas centrifuge cascades have been proposed [10-16]. They could be based on an ordinary three-flow centrifuge cascades (one feed point, one product out (enriched output), one waste out (depleted tailings outlet)) or on use of cascades with additional feed point of reprocessed uranium or two additional points (reprocessed and waste uranium) and in some cases, additional product out. At the same time the main advantage of centrifuge cascades with additional feed points is to minimize separative work, due to mixing of flows with different contents of ^{235}U. However, the use of the ordinary centrifuge cascade is technologically easier.

Computational experiments show that cascade with three feed points (waste uranium, LEU, reprocessed uranium) is one of the most effective schemes due to the simultaneous input of LEU and reprocessed uranium which leads to significant decrease in concentrations of harmful isotopes (^{232}U, ^{236}U) while increasing the concentration of ^{235}U isotope [15]. In particular, it is shown that this scheme may be more efficient than cascade with two feed points (natural and reprocessed uranium) [11] or cascade with three feed points (waste uranium, natural uranium, reprocessed uranium) [13]. In addition, preliminary economic calculations have shown that this cascade can be more effective than ordinary separation cascade [19]. But it should be noted that in all calculations was considered spent fuel from Russian type of light water reactors, VVER-1000, and suggested that the concentration of isotopes in it are constant. At the same time, alterations in the concentrations of the unwanted isotopes (^{232}U, ^{236}U) in reprocessed fuel, which needs to be enriched, can have a significant impact on the key parameters of the separation cascade (the consumption of diluent per unit of product, separative work). That must be considered in the further elaboration of the possibility of involving reprocessed uranium in the production of fuel for light-water reactors.

This paper deals with the analysis of the impact of uncertainties of isotopic composition of the reprocessed uranium on efficiency of its further enrichment in gas centrifuge cascades with three feed points. Estimates of changes in consumption of natural uranium and in separative work are made by varying the ^{232}U and ^{236}U concentrations in the initial reprocessed uranium. As the object of study a mathematical model of the quasi-ideal cascade [20] with three feed points is considered [13].

2. Mathematical model

Let us consider cascade with three feed points (Fig. 1.). The cascade has product out (P) and waste out (W) with a concentration C_i^P , C_i^W (from now on, the concentration is expressed as a percent by mass; $i=\overline{1,m}$, i - index number of component of separated mixture; m - the total number of components of the separated mixture). F1, F2 and F3 are flows with concentrations $C_i^{F_1}$, $C_i^{F_2}$, $C_i^{F_3}$ into feed points of waste uranium, reprocessed uranium and low-enriched uranium respectively. Detailed description of this mathematical model is given in [13].

The calculations were performed for the problem statement as follows: the concentrations C_n^P, C_n^W and relations F_2/F_3, F_1/F_3 are given (n –the index number of the target component). Through calculations is necessary to compute remaining parameters of the cascade, including the concentrations of other components in outlets and the number of stages in the cascade, etc. The problem is interesting from a practical point of view, because in this case clearly defined product quality that allows you to compare the effectiveness of different ways to produce commercial reactor-grade LEU.

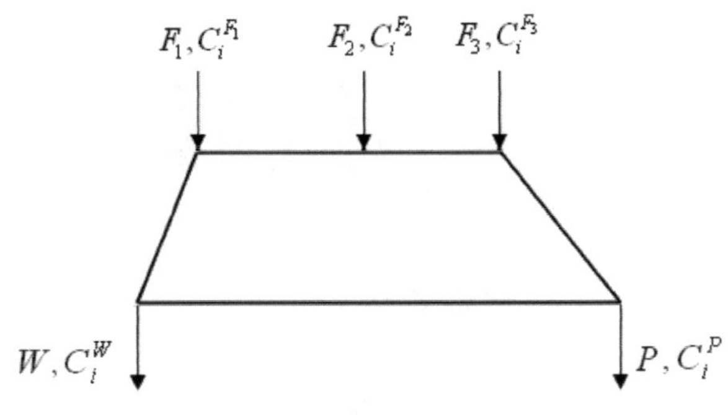

Figure 1. The schematic drawing of the cascade with three feed flows to enrich recycled uranium.

3. Results and discussion

As an example consider the enrichment of spent fuel from VVER-1000 with a burnup level of 40 GWd/t [15]. The isotopic composition of the material is shown in Table 1.

Table 1. The isotopic composition of spent fuel from Water-Water Energetic Reactor, VVER-1000.

Component	^{232}U	^{233}U	^{234}U	^{235}U	^{236}U	^{238}U
Concentration, %	$2.98{\cdot}10^{-7}$	$5.81{\cdot}10^{-7}$	$1.91{\cdot}10^{-2}$	0.901	0.573	The rest

As a calculation model is considered a particular case of "quasi-ideal" cascade - *R*-cascade [16] or cascade with non-mixing condition for chosen pair of components, in this case ^{235}U and ^{238}U isotopes. In the calculations the following parameters were given: concentration of ^{235}U in product was 4% (excluding additional enrichment in order to compensate the effect of ^{236}U), concentration of ^{235}U in waste was 0.1%, the reactivity compensation coefficient was 0.29. The concentration of ^{232}U in the product was limited by value $2{\cdot}10^{-7}$% [18]. As diluents were considered the waste uranium with 0.25% of ^{235}U and LEU with 1.5% of ^{235}U. As a comparison has been considered ordinary R-cascade to enrich natural uranium to the same concentration of ^{235}U.

For these conditions were calculated cascade parameters with different concentrations of ^{232}U and ^{236}U isotopes in reprocessed uranium, which varied within the range of 15% (relat.). Moreover, concentrations of ^{238}U were calculated by subtracting the sum of all concentrations from unity for each set of concentrations in order to satisfy the condition of equality of total concentration to unity.

Three-dimensional surfaces (Fig. 2-4) show the consumption of natural uranium, separative work and relative costs of Low-enriched uranium production versus concentrations of ^{232}U, ^{236}U isotopes in reprocessed uranium. The relative costs of Low-enriched uranium production was calculated as follows:

$$S = A_C \cdot Z + A_{F_1} \cdot F_1 + A_{F_2} \cdot F_2 + A_{F_0} \cdot F_0 , \qquad (1)$$

where *Z* is the total number of separation units, in this case centrifuges, necessary to enrich uranium to the given concentrations of ^{235}U in product and in waste. For cascades with three feed points, *Z* also takes into account the number of separation units needed for producing LEU diluent. A_C -factor which takes into account the costs associated with the work of gas centrifuges; F_0- consumption of natural uranium for producing commercial reactor-grade LEU. A_{F1}, A_{F2}, A_{F0} – prices of waste, reprocessed and natural uranium, respectively.

Properly speaking, the value *S* is not "the relative costs of Low-enriched uranium production" itself, but is a value determining relative costs to a large degree [21]. The first term of (1) is proportional to the cost of separative work, and the others represent prices of input material. The calculations assumed that the average price of natural uranium approximately 9 times higher than the average cost of waste uranium [22], and 3 times less than the cost of the uranium hexafluoride produced from reprocessed uranium. [8]. In the case of calculating the value of *S* for the ordinary cascade, values *F1* and *F2* is set to 0.

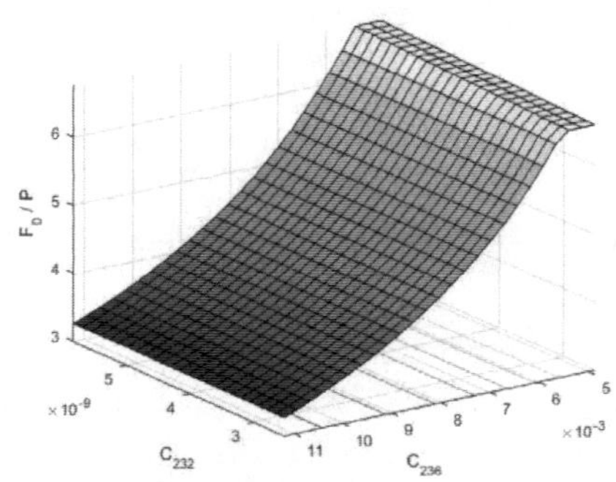

Figure 2. Consumption of natural uranium versus concentrations of ^{232}U, ^{236}U in reprocessed uranium.

IOP Conf. Series: Journal of Physics: Conf. Series **781** (2017) 012018 doi:10.1088/1742-6596/781/1/012018

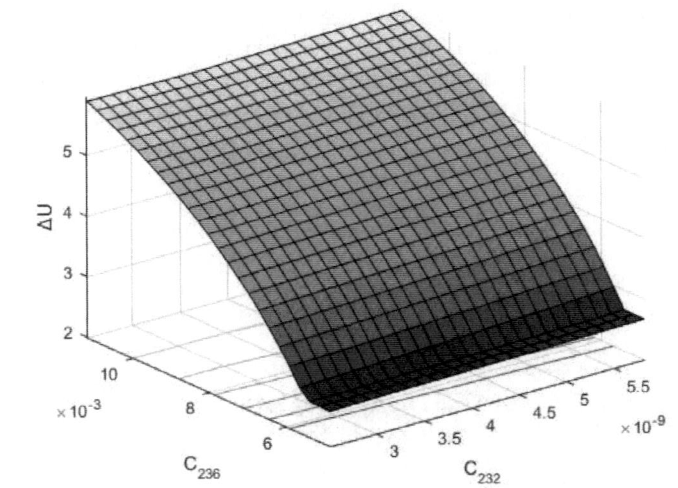

Figure 3. Separative work versus concentrations of ^{232}U, ^{236}U in reprocessed uranium.

Analysis of surfaces (Fig. 2-4) shows that uncertainties of isotopic composition of the reprocessed uranium can have a significant impact on consumption of natural uranium, on separative work and, as a result, on relative costs of Low-enriched uranium production. Therefore, this factor must be considered in a feasibility study of various options of reprocessed uranium enrichment in the gas centrifuge cascades. However, the relative costs of Low-enriched uranium production on the same or even lower level comparing to enrichment of natural uranium can be achieved by varying the cascade parameters.

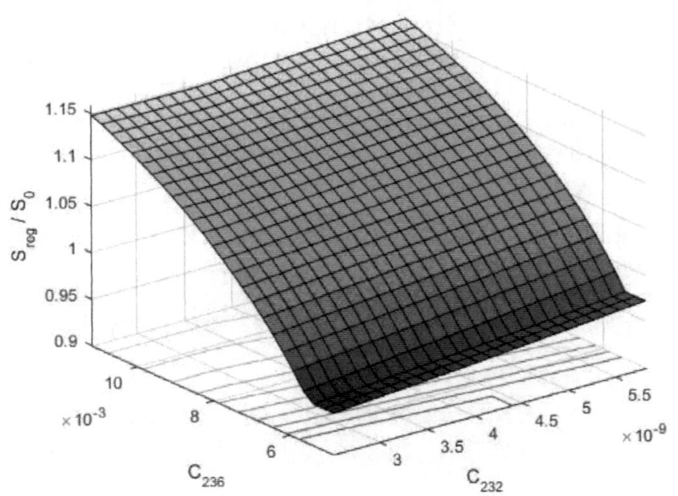

Figure 4. Relative costs of Low-enriched uranium production versus concentrations of ^{232}U, ^{236}U in reprocessed uranium.

4. Conclusion

The impact of uncertainties of isotopic composition of the reprocessed uranium on effectiveness of its enrichment in gas centrifuge cascades with simultaneous dilution with waste and low-enriched uranium is considered. It is shown that changing the content of ^{232}U and ^{236}U isotopes in the initial reprocessed uranium within 15% (rel.) can significantly change natural uranium consumption and separative work (up to 2-3%). However, even in case of increase of these parameters is possible to find the ratio of diluents, where the cascade with three feeds (depleted uranium, LEU, reprocessed uranium) will be more effective than ordinary separation cascade with one feed flowing for producing LEU from natural raw material.

References

[1] Andrianova E A, Davidenko V D, Tsibul'skii V F 2015 *At. Energy* **118** 301–6
[2] Pavlovichev A M et al 2006 *At. Energy* **101** 407–413
[3] Zilberman B Ya et al 2012 *At. Energy* **113** 383–391
[4] Fedorov Yu S et al 2005 *At. Energy* **99** 136–141
[5] Pavlovichev A M et al 2008 *At. Energy* **104** 195–8
[6] Smirnov A Yu et al 2012 *Phys. At. Nuclei* **75** 1616-25
[7] Coleman J.R., Knight T.W. 2010 *Nucl. Eng. Design* **240** 1028–1032
[8] Butler G G, Wilcox P 1988 *Nucl. Engineer* **28** 186-90
[9] *Hida K, Kusuno S, Seino T* 1986 *Nuclear Technology* **75** 148-159
[10] Palkin V A 2013 *At. Energy* **115** 32–7
[11] Sulaberidze G A, Borisevich V D, Xie Q X 2006 *Theor. Found. Chem. Eng.* **40** 5-14
[12] Palkin V.A. 2010 *Prosp. Mater.* **8** 11–4
[13] Smirnov A Yu, Sulaberidze G A 2014 *At. Energy* **117** 44–51
[14] Sulaberidze G A, Borisevich V D, Xie Q X 2004 *In: Proc. of IX All-Russia (International) Scientific Conference "Physical and Chemical Processes on Selection of Atoms and Molecules"* (Russia, October 4-8, 2004) p 70
[15] Smirnov A Yu, Sulaberidze G A 2015 *Phys. Proc.* **72** 132-5
[16] De la Garza A, Garrett G A and Murphy J E 1961 *Chem. Eng. Sci.* **15** 188–209
[17] Abbas K et al 2013 *Esarda Bulletin* **49** 75-81
[18] Alekseev P N et al 2012 *Phys. At. Nuclei* **73** 2264-2270
[19] Smirnov A Yu et al 2016 *IOP Conf. Ser.* **751** 012005
[20] Sazykin A A 2000 In the book: ISOTOPES. Properties. Production. Application. Ed. V.Yu. Baranov Moscow. IzdAt. 72-108 (in Russian)
[21] Smirnov A Yu, Borisevich V D, Sulaberidze G A 2012 *Theor. Found. Chem. Eng.* **46** 373-8
[22] Varela C, Akapo D, Schneider E A et al. 2009 *In: Proc. of GLOBAL-2009* 9463

Acknowledgements

This research is supported by the grant No.14.Y30.16.6284-MK of the President of the Russian Federation for young scientists.

The scheme for evaluation of isotopic composition of fast reactor core in closed nuclear fuel cycle

I S Saldikov[1], M Yu Ternovykh[1], P A Fomichenko[2] and A S Gerasimov[3]

[1] National Research Nuclear University "Moscow Engineering Physics Institute" (NRNU MEPhI), 31 Kashirskoe highway, Moscow 115409, Russian Federation
[2] National Research Center "Kurchatov Institute", 123183, Kurchatov square, 1, Moscow, Russia.
[3] Institute for Theoretical and Experimental Physics, 117218, Bolshaya Cheremushkinskaya street, 25, Moscow, Russia.

Corresponding author's e-mail address: ISSaldikov@mephi.ru

Abstract. The PRORYV (i.e. «Breakthrough» in Russian) project is currently under development. Within the framework of this project, fast reactors BN-1200 and BREST-OD-300 should be built to, inter alia, demonstrate possibility of the closed nuclear fuel cycle technologies with plutonium as a main source of power. Russia has a large inventory of plutonium which was accumulated in the result of reprocessing of spent fuel of thermal power reactors and conversion of nuclear weapons. This kind of plutonium will be used for development of initial fuel assemblies for fast reactors. To solve the closed nuclear fuel modeling tasks REPRORYV code was developed. It simulates the mass flow for nuclides in the closed fuel cycle. This paper presents the results of modeling of a closed nuclear fuel cycle, nuclide flows considering the influence of the uncertainty on the outcome of neutron-physical characteristics of the reactor.

1. Introduction

The code REPRORYV (REcycle for PRORYV), which simulates the process of loading, processing, and handling of fuel assemblies in the cores of BN-600, BN-800, BN-1200 [1, 2, 3] and other advanced fast reactors was developed to support the PRORYV project. The code can be used to analyze closed nuclear fuel cycles. It allows to study the possibility of the self-sufficient mode of operation in fast reactor. Similar work on the modeling of neutron-physical characteristics of the stationary states of the reactor core with detailed scheme of refueling of fuel assemblies was performed for the BN-600 reactor and was presented in [4]. A third-party software tool is used to calculate fuel radiation and depletion on each step of fuel burnup. We can use these data to analyze external fuel cycle stages. Any product can be used as the third-party code, which meets the requirements to get the needed functionality.

Russian code JARFR [5] is applied at present in the REPRORYV. The code is designed for evaluation of neutron-physical characteristics of fast reactors in the multi-group diffusion approximation [6]. It released with application constant support based on the library of nuclear data, BNAB 93.01 [7]. JARFR can model the entire active core with separate fuel assemblies. This program enables to set various properties of different fuel assemblies in the core of the considered reactor.

IOP Conf. Series: Journal of Physics: Conf. Series **781** (2017) 012019 doi:10.1088/1742-6596/781/1/012019

Thus, the program JARFR allows to take into account the history of irradiation of fuel assemblies, their unique isotopic composition at any given point of time.

We study the influence of isotopic composition uncertainties to the final result (multiplication factor, the amount of plutonium in the reactor) in 60 years of work on the power. Fuel is (Pu-U-N).

Special input file of JARFR code is required for correct operation of REPRORYV code. Results are calculated by using JARFR code on each step based on this input file. To analyze the closed nuclear fuel cycle it is needed to specify some input parameters in REPRORYV input file: cooling time, reprocessing time, fuel reload layout, distribution of isotopes into different groups, and conditions of processing for each of the groups.

Fuel reload layout indicates an order in which fuel assemblies need to be unloaded (Figure 1). The user can specify the part of the core which is unloaded each year. Detailed example of fuel reload layout is considered below.

```
                        0,
                        1,
                      2,  3,
                    4,  0,  5,
                  6,  1,  2,  3,
                4,  5,  6,  1,  0,
              2,  3,   0, 4,  5,  6,
            0,  1,  2,  3,  4,  5,  6,
          1,  2,  3,  4,  0,  5,  6,  1,
        2,  3,  4,  5,  6,  1,  2,  3,  4,
      5,  6,  1, 2,  3,  4,  5,   6,  1,  2,
    3,  4,  5,  6,  1,  2,  3,  4,  5,  6,  1,
  0,  0,  2,  3,  4,  5,  6,  1,  2,  3,  4,  0,
0,  0,  0,  0,  0,  0,  5,  6,  0,  0,  0,  0,  0,
0,  0,  0,  0,  0,  0,  0,  0,  0,  0,  0,  0,  0,  0,
0,  0,  0,  0,  0,  0,  0,  0,  0,  0,  0,  0,  0,
            0,  0,  0,  0,  0,  0,  0,
```

Figure 1. Fuel reload layout (60 degrees symmetry sector).

It is important to separate isotopes correctly to the groups and then operate with groups of isotopes on the stages of reprocessing. User may choose one of three options for each group as follows:

- Do not change a group concentrations, i.e. retain the mass of group as it was unloaded;
- Remove the group after recycling (assign concentrations to zero);
- Assign the initial value of the mass of the group, i.e. add or remove mass of the group so that the mass would be the same as at the first loading. When this option is selected, concentrations of each isotope can be changed while the whole mass should be the same as at the initial step.

Thus, the user can allocate separate groups of plutonium isotopes (to change mass to its original value, "removing the excess of plutonium"), uranium (to add mass to the initial level by adding ^{238}U), structure materials (to leave them as they were) and actinides (remove them from reprocessed fuel, or leave them in case of the post-combustion scenario).

Various reload layouts may be entered to agree with stated designers specifications. The effect of plutonium losses on fuel reprocessing stages is studied. Little loss of plutonium is allowable under equipment pollution in processes of separation of fuel fractions and fission products. To specify the input data of REPRORYV, the data from publications on the isotopic composition of fuel, energy release in the core composition and structural materials [8] were used in this work.

2. Uncertainties and errors of calculations

Advantage of the code REPRORYV is the possibility of direct calculation of the errors in the concentrations on the basis of errors in the input data. For this purpose, third-party code of nuclide kinetics BPSD (Burning and Poison calculation System with Deviation) [9] is used in REPRORYV code. The BPSD code was developed in the Institute of the Problems of the Development of Nuclear Power of the Russian Academy of Sciences (IBRAE RAS). This code allows to calculate any chains of nuclide transitions both for actinides and fission products. Library RUSFOND (IPPE) simulating up to about 1500 fission products and up to 40 actinides is used as database for BPSD code. Feature of the program is the direct calculation of the above-described errors: errors of the used numerical methods and errors of the input data, which are set inside the program. The result containing numerical values of the errors allows the user to take into account in their work the possible inaccuracies associated with the simulation, to get the confidence interval, the maximum deviation value, etc.

The user of the BPSD code has to assign the input data for the beginning of every step of burnup that is shown in figure 2. These input data are:

• Concentrations of actinides.
• Multi-group (26 groups) cross sections of fission, radioactive capture, (n,2n) and (n,3n) reactions.
• Neutrons spectrum (26 groups).

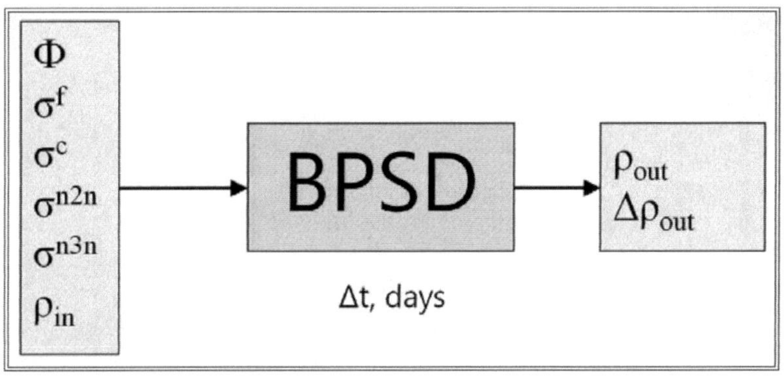

Figure 2. Simplified scheme of the BPSD program with the output of errors of the concentrations.

This module allows turn on function of estimation of uncertainties of the mass for any time of reactor operation into the REPRORYV code. These uncertainties allow estimate the error of the resulting functionals, such as reactivity and density effects of reactivity. In this paper the effect of uncertainty on the final result of the simulation is considered.

3. Processing

Simulation of the recycling process in the REPRORYV code is based on the assumption that the output spent fuel should consist of two fractions: the product and the waste Scheme of processing in REPRORYV code taking into account uncertainties in isotopic composition is shown in figure 3.

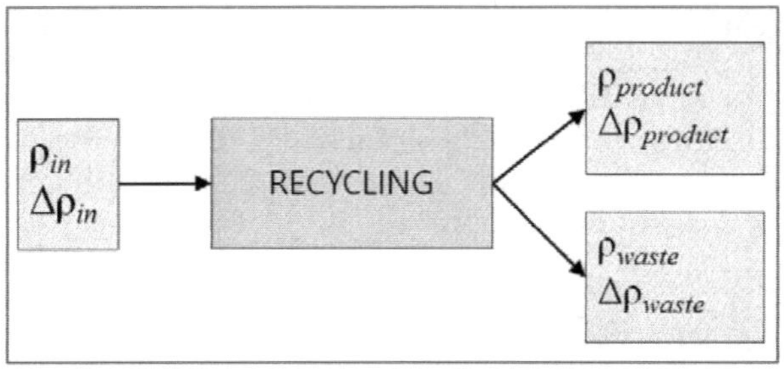

Figure 3. Scheme of processing in REPRORYV code taking into account uncertainties in isotopic composition.

In general case, the schema of processing is nonlinear and depends on many factors. Some of the software codes specifically designed to simulate the process of fuel reprocessing, separation of individual sections of processing lines, calculation of the processing coefficients relying on the chemical reactions, refining, etc. In the REPRORYV code, all processing steps are replaced by one common stage, after which it is already clear what part of the isotopes is the product and what part of the isotopes is waste. The general scheme of closed nuclear fuel cycle simulated in the task by the code REPRORYV is shown in figure 4.

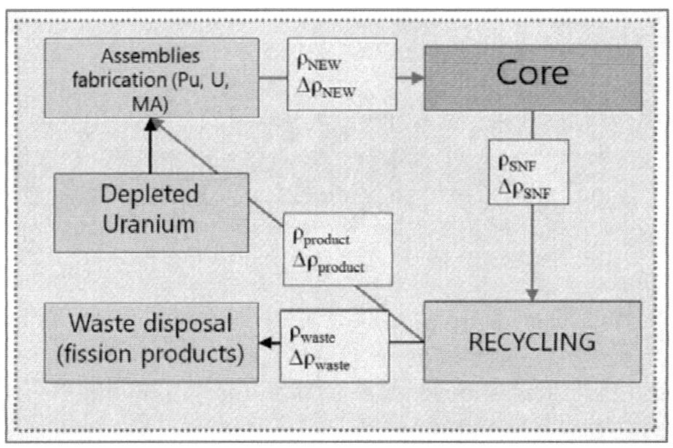

Figure 4. The general scheme of closed nuclear fuel cycle simulated in the task by the code REPRORYV.

This simplified scheme allows us to specify vector of coefficients of processing for each isotope, removing the complexity of the scheme of spent fuel reprocessing. After processing the waste is directed to the burial, and the product is used to create new fuel loading for the reactor. Vector of coefficients of the separation of the product from the waste for the major isotopes (~20-30 isotopes) is assigned as the parameters of the separation process. Values of these coefficients of the separation were obtained with account of losses and errors. The following equations are used at each step of processing to obtain the new concentration values:

- $\vec{\rho}_{product} = \vec{k}_{recl} \times \vec{\rho}_{in}$;
- $\vec{\rho}_{waste} = \vec{k}_{waste} \times \vec{\rho}_{in}$;
- $\vec{k}_{recl} + \vec{k}_{waste} + \vec{d} = \vec{1}$;
- $\vec{k}_{recl} > 0$; $\vec{k}_{waste} > 0$; $\vec{d} > 0$.

Where $\vec{\rho}_{in}$ is vector of concentrations of isotopes at the beginning of the procedure of processing;

$\vec{\rho}_{product}$ is vector of concentrations of isotopes in the product after processing;

$\vec{\rho}_{waste}$ is vector of concentrations of isotopes in the waste after processing;

$\vec{k}_{recl}$ is vector of coefficients of processing. Each component of this vector contains coefficients of processing for each of the major isotopes. Coefficient of processing means the share of the transition of each isotope in the product;

$\vec{k}_{waste}$ is vector of the coefficients of the transition to waste. Each component of this vector contains coefficients of transition to the waste for each of the major isotopes. Coefficient of transition to waste means the share of the transition of each isotope into the waste;

$\vec{d}$ is vector of the share of losses in the processing specified by the user.

It is obvious that each element of these vectors is greater than zero, and the sum of these vectors is equal to a unit vector (the law of conservation of mass).

There are certain difficulties in obtaining accurate values of losses at each operation of the processing. However, assignment of approximate values dives a possibility for the application of the algorithm with accurate data, which can be obtained experimentally directly in the processing plant.

After starting the program with all assigned required characteristics of the fresh fuel, irradiation parameters, cooling time, etc., the output file will contain the isotopic composition of the fuel for loading into a new reactor, its mass, neutron-physical characteristics of all active core and its parts, the errors of the obtained values.

The general scheme of operation of the program REPRORYV is presented in figure 5.

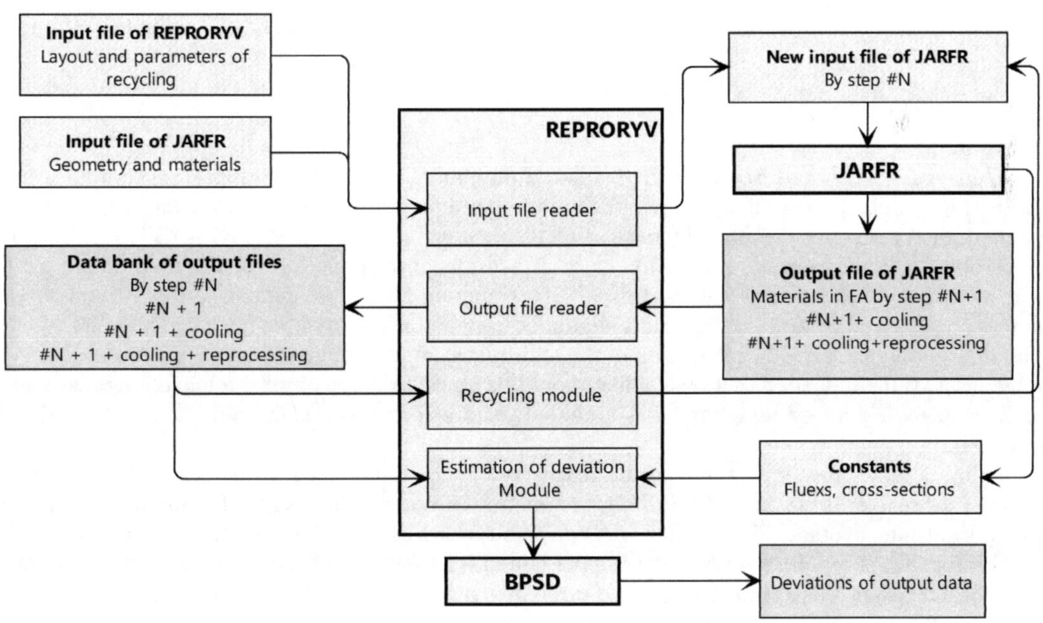

Figure 5. The general scheme of operation of the program REPRORYV.

IOP Conf. Series: Journal of Physics: Conf. Series **781** (2017) 012019 doi:10.1088/1742-6596/781/1/012019

4. Core reloading layout

Reloading layout of the fuel assemblies in the BN reactor core was accepted as follows. It was assumed that plutonium content was 12.8%. Plutonium isotopic composition corresponded to that of discharged from the spent nuclear fuel of VVER-440 reactors [10]. Initially new fuel assemblies were loaded in the core. In the first step of reloading after 330 days of operation, 12 fuel assemblies of 72 (1/6 part of the core) marked by index of "1" (first lot of fuel) in Figure 1 were unloaded and placed into intrareactor storage for cooling until the activity of these fuel assemblies would decrease to the level when they could be reprocessed. This time of cooling was evaluated as two years in our case. New fuel assemblies with the original composition were placed to the positions of unloaded fuel assemblies during following 35 days. The time of operation during 330 days and time for reloading of 35 days were assumed for all steps of reloading. The total time was equal to one year. In the second step of reloading after next 330 days of operation, fuel assemblies marked by index of "2"(second lot of fuel) were unloaded for cooling and replaced by new fuel assemblies during 35 days. Then reactor operated again during 330 days and so on. After two years of cooling, fuel assemblies of type "1" that have been loaded first, were sent for recycling. Several variants of reprocessing were considered. Actinides are removed from the fuel. In the process of separation of fission products, a little amount of plutonium could be extracted along with fission products. The amounts of plutonium as impurity to separated fission products of 0.1% were considered. The time of reprocessing was assumed as one year. In the third step of reloading the fresh fuel assemblies were place to position "3" and the reactor operated for following 330 days. Finally in the fourth step of reloading, recycled fuel which was reprocessed over the last year was placed to position "4" (on the contrary to the fresh fuel as in the previous steps), composed of the first lot of fuel. Then, in following steps of reloading, fuel assemblies that have been unloaded and reprocessed after the second step (second lot) would be placed in position "5", after the third step - in position "6", etc. Thus, the reactor comes to steady state mode of operation, if the vector of plutonium (the fraction of each plutonium isotope in the mass of plutonium) at each following step of the refueling of the active core remains constant (self-sufficient plutonium mode)

5. Results

Geometry parameters and isotopic composition of the model of high-power reactor core was assigned in JARFR to obtain the results on multiplication factor and mass of plutonium at the end of the time interval. Fuel reloading layout and modes of recycling were specified in REPRORYV code. Each fuel assembly was homogenized in the active core. All results correspond to fuel density 11.5 g/cm^3. The reactor was set to the critical state (Keff =~ 1) by changing plutonium content in the active core only in the initial period of time so the results should be viewed only from a methodological point of view. The methodological error of calculation of the multiplication factor in each case was equal to 10^{-5}. The rule "mass of the unloaded fuel assemblies should be equal to the mass of the loaded fuel assemblies in each of the cycles of refueling" was accepted for the formation of the new fuel composition by processing.

Figure 6 shows the dependence of the reactivity of the reactor core (in percent) on time (the life cycle of the reactor is 60 years) for different estimation of error in the results. The middle graph in this figure shows the average value, which was evaluated by REPRORYV code using the built-in module JARFR on each step. The top graph is an upper bound for the effect of errors in the mass of plutonium to Keff. To obtain this result the absolute error of plutonium was added to its entire mass on each iteration step. Obviously, the more reactor works, the larger the absolute deviation from the mean of Keff. The same procedure was done for the lower chart: here the error value of plutonium mass was subtracted from the entire mass of plutonium in each iteration step. The relative difference between the maximum and minimum value of Keff in relation to the average was 0.7%. So such an effect can be expected, if you taking into account uncertainties in the mass of plutonium.

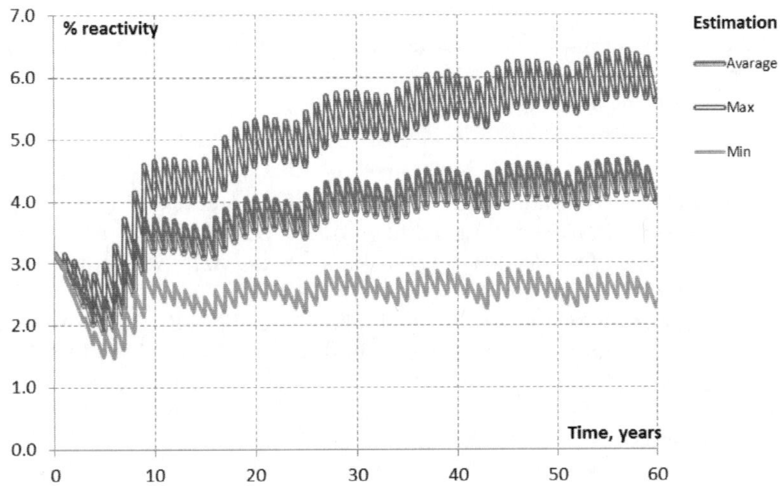

Figure 6. The dependence of reactivity of the core on time (the life cycle of the reactor is 60 years) for different estimation of error in masses of plutonium in fuel.

In each fuel batch after 9 years it is necessary to add a smaller amount of plutonium to support the operation of the reactor (for the same mass of each fuel assembly, see table 1). From 4-th cycle the "excess" of plutonium accumulated in an amount of not more than 1%. So we can say that the reactor came to "steady-state" when the mass change of Plutonium between two steps is almost zero.

Table 1. Change (%) of the mass of plutonium during its production in each fuel batch during the reactor life cycle

	Cycle number						
Batch #	1	2	3	4	5	6	7
1	-	-3.0	-1.9	-1.2	-0.7	-0.4	-0.3
2	-	-2.5	-1.6	-1.1	-0.6	-0.4	-0.2
3	-	-2.6	-1.7	-1.1	-0.6	-0.4	-0.2
4	-	-2.8	-1.9	-1.1	-0.6	-0.4	-0.2
5	-	-1.9	-1.5	-0.8	-0.6	-0.3	-0.3
6	-	-1.7	-1.3	-0.7	-0.5	-0.3	-
7	-	-1.7	-1.2	-0.7	-0.4	-0.2	-
8	-	-1.5	-0.9	-0.7	-0.3	-0.3	-
9	-	-1.7	-1.0	-0.7	-0.3	-0.3	-

6. Conclusion

The paper deals with evaluating of fuel handling in fast reactors in CNFC. To solve these problems, the REPRORYV program code is developed. It simulates the nuclide streams in out-of-reactor stages of the closed fuel cycle. Existed verified code JARFR is used for the calculation of neutron-physical characteristics of the core. Various options for nuclide streams are considered with a representative full-scale model of a fast reactor with sodium-cooled high-power. The changes of multiplication factor

IOP Conf. Series: Journal of Physics: Conf. Series **781** (2017) 012019 doi:10.1088/1742-6596/781/1/012019

and mass of plutonium in the closed nuclear fuel cycle for Russian fast high-power reactor with use of the REPRORYV code are evaluated. The maximum and minimum estimation of the effect of errors in the mass of plutonium to the multiplication factor was considered. The effect is equal to 0.7%.

This work was carried out within the framework of 5-100.

References

[1] Poplavsky V, Tsiboulia A, Khomyakov Yu, Matveev V, Eliseev V, Tsikunov A, Vasiliev B, Belov S, Farakshin M Core design and fuel cycle of advanced fast reactor with sodium coolant. *International Conference on Fast Reactors and Related Fuel Cycles - Challenges and Opportunities* - December 7-11 2009 Kyoto Japan

[2] Vasilyev B, Shepelev S, Ashirmetov M, Poplavsky V BN-1200 Reactor Power Unit Design Development. International Conference on Fast Reactors and Related Fuel Cycles: Safe Technologies and Sustainable Scenarios Paris France 4-7 March 2013

[3] Vasiliev A, Fomichenko P, Alekseev P, Nevinnitsa V, Belov S, Vasiliev B, Evsees V, Farakshin M Preprint Kurchatov Institute IAE-6275/4 (2003)

[4] Nevinitsa A I, Matveev V I, Stogov V Yu, Cherny V A Problems of Atomic Science and Engineering Series: Physics of Nuclear Reactors 2 pp 53-57 (2011)

[5] Yaroslavtseva L N The JAR Code System for Nuclear Reactor Neutronics Calculations. VANT issue 8 (37) 1983 pp 41-43

[6] Yaroslavtseva L N Problems of Atomic Science and Engineering, Series: Physics and Technology of Nuclear Reactors 8(37) pp 41-43 (1983)

[7] Manturov G N, Nikolayev M N, Tsybulya A M Problems of Atomic Science and Engineering, Series: Nuclear Constants 1 pp 4-12 (1996)

[8] R. Prodanova. Calculation of isotope composition of WWER- 440 spent fuel assembly by the NESSEL-NUKO code system on the basis of the ISTC burn up credit project data. Institute for Nuclear Research and Nuclear Energy Bulgarian Academy of Sciences

[9] A. Belov, E. Seleznev. Atomic Energy, 108, 321-324 (2010).

[10] S. B. Belov, A. V. Kiselyov, E. V. Marova, M. R. Farakshin, M. N. Zizin, P. A. Fomichenko, K. F. Raskach, M. Yu. Semyonov, Problems of Atomic Science and Engineering, Series: Physics of Nuclear Reactors, 4, 77-82 (2014).

On some perturbation theory singularities for reactivity effects calculations

B D Abramov

Leypunsky Institute for Physics and Power Engineering (IPPE), Obninsk, Russia.

E-mail: abramov@ippe.ru

Abstract. It is shown in this paper that traditional formulas for reactivity effects calculation based on diffusion approximation small-perturbation theory sometimes need updating. The corresponding specifications are provided. These are caused by the unbalance of neutron "leakages" between subdomains (zones, cells, fuel subassemblies, etc.) of the reactor as a result of perturbation. The unbalance occurs because of the jumps of relative increments of diffusion coefficients on the subdomain interfaces.

1. Introduction

As it is known from [1 – 8], reactor neutronics analysis in diffusion approximation is based on conditional critical equation

$$-\nabla D\nabla\Phi + C\Phi = F\Phi / k_{eff}, \tag{1}$$

where C and F – operators defined by the following formulas:

$$D = 1/3\Sigma_{tr}, \quad C = \Sigma - K_s, \quad F = K_f,$$
$$K_b\Phi = \int dE' \omega_b(x,E,E')\Phi(x,E'), \quad b = s,f$$

on $\Phi(x,E)$ functions obeying (1) in reactor domain G and, for example, zero boundary condition on its (extrapolated) boundary S.

Effects $\delta\rho = \rho' - \rho$ of reactivity $\rho = 1 - 1/k_{eff}$ caused by the following type small perturbations

$$D' = D + \delta D, \quad C' = C + \delta C, \quad F' = F + \delta F, \quad \Phi' = \Phi + \delta\Phi, \quad k'_{eff} = k_{eff} + \delta k_{eff}$$

are usually calculated using well known small perturbation theory formula [1,2,4]

$$\delta\rho \approx -\sum_i \sum_n 3\frac{(J_i^*, \delta\Sigma_{tr} J_i)_n}{(\Phi^*, F\Phi)} + \sum_n \frac{(\Phi^*, (\delta F / k_{eff} - \delta C)\Phi)_n}{(\Phi^*, F\Phi)}, \tag{2}$$

where $(\,,)_n$ – symbol of integration on $x \in G_n$ and E ; $(\,,) = \sum_n (\,,)_n$; $\sum_n$ – summation over subdomains (zones, cells) of the reactor $G_n \subseteq G$; Φ' – solution of perturbed equation

$$-\nabla D'\nabla\Phi' + C'\Phi' = F'\Phi' / k'_{eff} , \quad (3)$$

Φ^* – solution of unperturbed adjoint equation

$$-\nabla D\nabla\Phi^* + C^*\Phi^* = F^*\Phi^* / k_{eff} , \quad (4)$$

J_i, J_i^* – projections of $J = -D\nabla\Phi$ and $J^* = D\nabla\Phi^*$ vectors on axes ($i = 1,2,3$ in Cartesian coordinate system, etc.); $\sum_i$ – summation over projections.

This formula follows (3) and (4) taking into account the following relationship:

$$-\sum_i (\Phi^*, \nabla_i D'\nabla_i \Phi')_n + \sum_i (\Phi', \nabla_i D \nabla_i \Phi^*)_n =$$
$$= \sum_i \{3(J_i^*, \delta\Sigma_{tr} J_i')_n + \int_{S_n} (\vec{i} \cdot d\vec{S}) \int dE\, (\Phi^* J_i' + \Phi' J_i^*)\} \quad (5)$$

and approximation of perturbed fluxes and currents by those unperturbed:

$$\Phi' \approx \Phi, \quad J' = -D'\nabla\Phi' \approx -D\nabla\Phi = J , \quad (6)$$

which results in nulling the last member in (5), where $(\vec{i} \cdot d\vec{S}) = (\vec{i} \cdot \vec{n}_s) dS$, $\vec{n}_s$ – external normal to S_n boundary of G_n cell, $(\vec{i} \cdot \vec{n}_s)$ – cosine of angle formed by $\vec{i}$ and $\vec{n}_s$.

2. New approach

If however approximation of fluxes and their gradients is used instead of (6):

$$\Phi' \approx \Phi, \quad \nabla_i \Phi' \approx \nabla_i \Phi , \quad (7)$$

then less known formula can be derived [7] instead of (2):

$$\delta\rho \approx -\frac{\sum_i \sum_n [(\delta D \nabla_i \Phi^*, \nabla_i \Phi)_n - \int_{S_n} (\vec{i} \cdot d\vec{S}) \int dE\, \delta D \Phi^* \nabla_i \Phi]}{(\Phi^*, F\Phi)} +$$
$$+ \sum_n \frac{(\Phi^*, (\delta F / k_{eff} - \delta C)\Phi)_n}{(\Phi^*, F\Phi)} \quad (8)$$

differing from (4) by that possible jumps of $\delta D / D$ function on S_n, S_m interfaces of adjacent subdomains (cells) G_n, G_m are taken into account.

If there are no jumps, then, in the common assumption of continuity of conjugate fluxes Φ^* and normal components of J currents on the cells interfaces, (2) agrees with (8), since additional members in (8) describing neutron "leakages" through the cells interfaces are canceled on when being summed over cells in (8). However, generally, these formulas are different, thus giving rise to the question of which one of these formulas is more exact.

In this regard the alternative presentation of relationship (8)

$$\delta\rho \approx \sum_i \sum_n \frac{(\Phi^*, \nabla_i \delta D \nabla_i \Phi)_n}{(\Phi^*, F\Phi)} + \sum_n \frac{(\Phi^*, (\delta F / k_{eff} - \delta C)\Phi)_n}{(\Phi^*, F\Phi)}, \tag{9}$$

should be noted, which confirms its validity in that (9) goes to (in case of multigroup approximation and withdrawal of breakdown into subareas) expression (6.81) from [4], which in this case is equivalent to (2).

3. Verification

Since (2) and (9) are generally different from each other, then it is necessary to know which one of them is more true or precise.

Let us take as examples the following boundary value problems for diffusion equations:

$$(-D\frac{d^2}{dz^2} + \Sigma_a)\Phi = \nu\Sigma_f \Phi / k_{eff}, \tag{10}$$

$$(-D'\frac{d^2}{dz^2} + \Sigma_a)\Phi' = \nu\Sigma_f \Phi' / k'_{eff} \tag{11}$$

in the layer $z \in (-h, h)$ with constant coefficients and zero conditions with $z = \pm h$.

The solutions of the problems are as follows:

$$\Phi(z) = \Phi'(z) = \cos Bz, \quad B = \pi / 2h,$$
$$k_{eff} = \frac{\nu\Sigma_f}{\Sigma_a + DB^2}, \quad k'_{eff} = \frac{\nu\Sigma_f}{\Sigma_a + D'B^2}, \tag{12}$$

therefore, the correct value of reactivity effect in this example is calculated by the following formula:

$$\delta\rho = -\,\delta D B^2 / \nu\Sigma_f. \tag{13}$$

Integrating (10) and (11) on $z \in (-h', h')$, $h' < h$ with weight functions Φ' and $\Phi^* = \Phi$, respectively, the desired analogue of (9) in this problem can be derived as follows:

$$\delta\rho = -\frac{\delta D B^2}{\nu\Sigma_f}\frac{(\cos Bz, \cos Bz)'}{(\cos Bz, \cos Bz)'} = -\frac{\delta D B^2}{\nu\Sigma_f}, \tag{14}$$

which also gives the correct effect value irrespective of chosen size $0 \le h' \le h$ of subdomain G', where $(,)'$ – symbol of integral on $z \in G'$.

This result confirms correctness of (8) and (9). In contrast to them, corresponding analogue of equation (2) gives the following result:

$$\delta\rho = -\frac{\delta DB^2}{\nu\Sigma_f}\frac{(\sin Bz, \sin Bz)'}{(\cos Bz, \cos Bz)'} = -\frac{\delta DB^2}{\nu\Sigma_f}\frac{(1-t^{-1}\sin t)}{(1+t^{-1}\cos t)}, \qquad (15)$$

coinciding with correct one if only $h' = 3h/4$, where $t = 2Bh' = \pi h'/h$. If $0 \le h' < 3h/4$, then the desired effect value is underestimated by (15) right down to zero (if $h' \to 0$), and if $h' > 3h/4$, then the value is overestimated, although by a factor of not more than $\pi/(\pi-1)$ (if $h' = h$).

The above results show that (2) is generally less precise as compared to (8) and its equivalent (9).

3. Other results

The similar conclusions are presented in [7] as applied to more comprehensive analysis of reactivity effects caused by thermal deformation of reactor cells. Results given in [7] are synthesis and specification of those presented in [4-6] for the case of different dimensions, configuration and mutual arrangement of cells. These results are different from data of [8] obtained using different method of taking into account the above neutron leakages between subdomains.

4. Conclusion

The paper is devoted to updating of traditional diffusion approximation relationships for evaluation of reactivity effects. The updating involves taking into account additional terms (missing in traditional formulas) related to the unbalance of neutron leakages between reactor subdomains (zones or cells) in the course of perturbation. Another approaches see in [9,10].

5. References

[1] Marchuk G 1961 Methods of analysis of nuclear reactors, Moscow, Gosatomizdat.
[2] Shikhov S and Troyansky V 1983 Theory of nuclear reactors, Moscow, Energoatomizdat.
[3] Bell G and Glasston S 1973 Nuclear reactor theory, Moscow, Atomizdat
[4] Zizin M 1978 Fast reactor neutronics analysis, Moscow, Atomizdat.
[5] Shikhov S and Shmelev S 1068 Taking account of the influence of an arbitrary change in the dimensions on the critical mass of a fast reactor using perturbation theory. *Physics of Nuclear Reactors,* Atomizdat, Moscow, pp. 67-85.
[6] Abramov B 1988 Calculation of Reactivity Effects in Deformed Zones of Nuclear Reactors. *Atomic Energy*, **85**(6): pp. 853-859.
[7] Abramov B and Raskatch K 2015 On some methods of evaluation of reactivity effects in deformed reactor cores. *Issues of Atomic Science and Technology, Series: Nuclear Data for Reactors*, pp. 18-32.
[8] Reed M, Smith K and Forget B 2014 The "virtual" density theory on neutronics: a generic method for geometry distortion reactivity coefficients, "*Physor 2014*", on CD-ROM.
[9] Larsen E and Pomraning G 1981 Reactivity perturbation theory. *Nuclear Science and Engineering,* **77,** pp. 415-425.
[10] Pomraning G 1983 Perturbation theory in the diffusion approximation. *Nuclear Science and Engineering*, **83,** pp. 72-74.

IOP Conf. Series: Journal of Physics: Conf. Series **781** (2017) 012021 doi:10.1088/1742-6596/781/1/012021

On monitoring and forecasting of graphite stack temperature in transient modes

A M Zagrebaev[1], N V Ovsyannikova[1] and R N Ramazanov[12]

[1]National Research Nuclear University MEPhI (Moscow Engineering Physics Institute), Department of Cybernetics, 115409, Russia, Moscow, Kashirskoe highway, 31

[2]E-mail: tem@gmx.com

Abstract. The paper presents a method of monitoring and forecasting of the graphite stack temperature of the RBMK reactor in transient modes. The method is based on processing the in-core information about macro-distribution and mathematical model of distribution of temperature changes of the graphite stack in the reactor core. It is shown that the use of archival neutron field monitoring data allows determining the graphite stack temperature in the on-line mode.

1. Introduction

One of the requirements for the control process of the RBMK reactors is the effective and safe operation of the reactor under changes in power level [1-3]. In this regard, the task of monitoring and forecasting of changes of the operational reserve in space and time is relevant, as the operational reactivity margin determines the reactor maneuverability (the operational reactivity margin below refers to the value $K(\vec{r},\tau)-1$, i.e. the excess of birth rate over one).

The change of the propagation coefficient in the RBMK reactors due to the influence of feedback can be represented as follows [3, 4]:

$$\Delta K(\vec{r},\tau)=\Delta K_{\varphi}(\vec{r},\tau)+\Delta K_{F}(\vec{r},\tau)+\Delta K_{C}(\vec{r},\tau)+\Delta K_{X}(\vec{r},\tau), \quad (1)$$

where $\Delta K_{\varphi}(\vec{r},\tau),\Delta K_{F}(\vec{r},\tau),\Delta K_{C}(\vec{r},\tau),\Delta K_{X}(\vec{r},\tau)$ is the change of the propagation coefficient due to the following reactivity effects: steam, fuel and moderator temperature, and xenon concentration, respectively.

In this article the problem of the forecasting of the propagation coefficient changes under the moderator temperature changes is considered. It is assumed that the following relationship is true:

$$\Delta K_{C}(\vec{r},\tau)=\alpha_{C}(\vec{r},\tau)\Delta T_{C}(\vec{r},\tau), \quad (2)$$

where $\alpha_{C}(\vec{r},\tau)$ is the graphite reactivity coefficient (a known function).

Thus, under the assumption made above, the task of monitoring and forecasting of the reactivity changes is reduced to the problem of monitoring and forecasting of the graphite temperature changes in transient modes of the reactor.

2. Mathematical model

To describe the behavior of macro-distribution of the graphite temperature changes in the RBMK reactor core we use the following mathematical model, which is applied in the study of RBMK reactors dynamic characteristics [5-7]:

$$\begin{cases} C_V \dfrac{\partial \Delta T_C}{\partial \tau} == -\beta \Delta T_C + A\Sigma_f \Delta \Phi(\vec{r},\tau); \\ \Delta T_C(\vec{r},0) = 0. \end{cases} \tag{3}$$

where $\Delta T_C(\vec{r},\tau) = T_C(\vec{r},\tau) - T_C(\vec{r},0)$ is the deviation of the graphite stack temperature at the time τ in the coordinate $\vec{r}$ from the initial value of the graphite stack temperature (further this value for the sake of brevity is called macro-distribution of graphite temperature); C_V is heat capacity of graphite; $\Sigma_f(\vec{r})$ is fission cross section; β is the proportionality coefficient, depending on the conditions of heat transfer from the graphite to the coolant; A is the proportionality coefficient between the changes of energy release density and the changes of the graphite temperature; $\Delta\Phi(\vec{r},\tau) = \Phi(\vec{r},\tau) - \Phi(\vec{r},0)$ is deviation of neutron flux distribution at time τ from the initial value.

The model (3) does not account for the leakages of heat over the height and radius of the reactor. Let us present the equation (3) in a dimensionless form. To do this, we introduce a new variable

$$\Theta(\vec{r},\tau) = \frac{\Delta T_C(\vec{r},\tau)}{\overline{T_C - T_S}}, \tag{4}$$

where $\overline{T_C - T_S}$ is the average temperature difference in the reactor in the steady state at nominal power level.

$$\beta(\overline{T_C - T_S}) = A\overline{\Sigma_f \Phi},$$

where T_S is the water saturation temperature under the working pressure in the channel.

Then the original problem (3) can be written as:

$$\begin{cases} \dfrac{\partial \Theta}{\partial \tau} == -\dfrac{1}{\tau_C}\Theta + \dfrac{1}{\tau_C} K\varphi(\vec{r},\tau); \\ \Theta(\vec{r},0) = 0. \end{cases} \tag{5}$$

3. Solution

The solution of the problem (5) presents no difficulties on condition that the values τ_C and $K\varphi = \Sigma_f \Delta\Phi / \overline{\Sigma_f \Phi}$ are given.

$$\Theta(\vec{r},\tau) = \frac{K}{\tau_C} \exp\left(-\frac{\tau}{\tau_C}\right) \int_0^{\tau} \varphi(\vec{r},t) \exp\left(\frac{t}{\tau_C}\right) dt \tag{6}$$

However, with a real RBMK reactor the problem of definition of macro-distribution of the graphite temperature using the model (5) is complicated by the following circumstances:

1) the parameter τ_C may be a function of coordinates and significantly depend on the operating conditions, in particular, the composition of the gas mixture of the graphite stack [3, 5, 9];
2) the experimental information on the energy release distribution in the reactor core is available only in a finite number of coordinates (at the locations of sensors) and is presented in the form of currents of sensors at discrete moments of time;
3) the impact of a large number of perturbing factors leads to the fact that the distribution of energy release under the reactor power change may differ from the forecast. Strictly speaking, this difference is random and from a mathematical point of view is described with a random function.

IOP Conf. Series: Journal of Physics: Conf. Series **781** (2017) 012021 doi:10.1088/1742-6596/781/1/012021

Regarding the first comment let us assume that the time constant of the graphite τ_C does not depend on the coordinate and can be determined experimentally. The rest of the comments can be considered as follows.

Deviations of currents of discretely located sensors of energy release from the initial values are approximated with a set of test functions according to the methodology [8, 10]. This approach to determination of the graphite temperature macro-distribution can be justified by the fact, proven by the archival experimental information, that the distribution of the graphite temperature changes is a fairly smooth function.

The readings of the thermocouples in transient modes along with the readings of energy release sensors allow to determine the conversion factor *K*. Indeed, supposing the macro-distribution $\varphi^*(\vec{r},\tau_i)$ is restored at discrete moments τ_i according to the readings of in-core sensors, let us interpolate the macro-distribution $\varphi^*(\vec{r},\tau_i)$ in time using the stepwise interpolation method.

$$\varphi^*(\vec{r},\tau)=\varphi^*(\vec{r},\tau_i), \quad \tau_i \le \tau \le \tau_{i+1}.$$

The interval between the time counts is considered to be constant: $\tau_0=\tau_{i+1}-\tau_i$. Then the expression for the macro-distribution of the graphite temperature changes is as follows:

$$\delta\Theta = K\frac{\tau_0}{\tau_C}\sum_{i=0}^{N-1}\varphi^*(\vec{r},\tau_i)\exp\left(\frac{\tau_{0i}-\tau}{\tau_C}\right), \tag{7}$$

where $N=\frac{\tau}{\tau_0}$ is the number of time counts per τ. The estimated value of graphite temperature changes in the locations of the sensors is

$$\delta\Theta(\vec{r}_{th_j}) = K\frac{\tau_0}{\tau_C}\sum_{i=0}^{N-1}\varphi^*(\vec{r}_{th_j},\tau_i)\exp\left(\frac{\tau_{0i}-\tau}{\tau_C}\right), \tag{8}$$

where $\vec{r}_{th_j}$, *j*=1,...,24 are locations of the thermocouples in the reactor core.

On the other hand, using the thermocouples readings we can detect changes in the temperature $\Theta(\vec{r}_{th_j})$.

The conversion factor *K* can be defined, for example, from the following requirement:

$$\sum_{j=1}^{24}\left[\left(\delta\Theta(\vec{r}_j)-\Theta(\vec{r}_j)\right)^2\right]=\min. \tag{9}$$

With this in mind, the forecast for the change of graphite temperature in the reactor core can be carried out according to the formula:

$$\Theta(\vec{r},\tau)=\frac{K}{\tau_C}\exp\left(-\frac{\tau}{\tau_C}\right)\int_0^{\tau}\varphi(\vec{r},t)\exp\left(\frac{t}{\tau_C}\right)dt, \tag{10}$$

where $\varphi(\vec{r},t)$ is the proposed change of the energy release macro-distribution in the transient mode.

It should be mentioned that the expression (10) allows to restore a detailed distribution of the graphite temperature $T(\vec{r},t)=T_0(\vec{r},t_0)+\delta\Theta(\vec{r},t)$, on condition that the graphite temperature distribution at the initial moment of time $T_0(\vec{r},t_0)$ and the behavior of the neutron flux density $\varphi(\vec{r},\tau)$ during the interval $(2\div 3)\tau_C$ preceding time t are known.

4. On accuracy of monitoring macro-distribution of the graphite stack temperature

To evaluate errors in graphite temperature macro-distribution let us assume the following [8]:

1) the random error of restoration of the neutron flux space distribution does not depend on the neutron flux density;
2) the systematic error of the restoration of neutron flux density is absent;
3) the space interpolation errors are not time-correlated, i.e.

$$K_{\Delta}(t,t') = \begin{cases} 0, & when\ t \neq t'; \\ D, & when\ t = t'. \end{cases}$$

In this case it is possible to show that the error of monitoring the behavior of macro-distribution of graphite stack temperature under the steady mode of measurements is determined by the period of getting the sensors readings τ_0, by the parameters of the neutron flux correlation function and variance of the error of restoration of neutron flux space distribution, and does not exceed 4% in stable mode.

5. Conclusions

The methodology of monitoring and forecasting of the graphite temperature changes in the RBMK reactor core is proposed.

The methodology is based on using both experimental information (current of sensors and thermocouples in the graphite stack) and the mathematical model of the distribution of the graphite temperature changes in the reactor core. The mathematical model is obtained considering the number of assumptions, namely: the reactor height and radius heat leaks are not considered, the graphite time constant is considered to be constant within the reactor core. The algorithms for the monitoring and forecasting of the graphite stack temperature changes in the reactor core are easy for computer implementation and require small amount of RAM. Estimation of the errors of the graphite stack temperature forecasting and monitoring allows us to conclude that the proposed algorithms can be used to forecast and monitor with a reasonable degree of accuracy.

References

[1] Aleksandrov S, Postnikov V, Sviridenkov A, Yurkin G and Yakunin I 2011 Effectiveness analysis of the reactor operator's adviser computer code articles *Atomic Energy* **111** Nov.

[2] D'Auria F, Gabaraev B, Soloviev S, Novoselsky O, Moskalev A, Uspuras E, Galassi G, Parisi C, Petrov A, Radkevich V, Parafilo L, Kryuchkov D 2008 Deterministic accident analysis for RBMK *J. Nucl. Eng. Des.* **238 (4)** pp 975–1001

[3] IAEA 2005 Accident analysis for nuclear power plants with graphite moderated boiling water RBMK reactors *Safety Report Series* **No. 43** (Vienna) ISBN 92-0-112804-5

[4] D'Auria F, Soloviev S, Malofeev V, Ivanov K, Parisi C 2008 The three-dimensional neutron kinetics coupled with thermal-hydraulics in RBMK accident analysis *J. Nucl. Eng. Des.* **238 (4)** pp 1002–25

[5] D'Auria F, Gabaraev B, Radkevich V, Moskalev A, Uspuras E, Kaliatka A, Parisi C, Cherubini M, Pierro F 2008 Thermal-hydraulic performance of primary system of RBMK in case of accidents *J. Nucl. Eng. Des. 238 (4)* pp 904–24

[6] D'Auria F, Novoselsky O, Safonov V, Uspuras E, Galassi G, Cherubini M, Giannotti W 2008 Thermal-hydraulic performance of confinement system of RBMK in case of accidents *J. Nucl. Eng. Des.* **238 (4)** pp 925–39

[7] Knoglinger E, Wölfl H and Kaliatka A 2015 Heat transfer in the core graphite structures of RBMK Nuclear Power Plants *J. Nucl. Eng. Des.* **293** pp 413–35

[8] Hitchcock A 1960 *Nuclear Reactor Stability* (London: Temple Press)

[9] Tkachev V, Mikhailov D, Zheltukhin K, Afanasiev A, Tkacheva N 2011 Modelling the deformation of RBMK graphite columns surrounding a damaged cell with a ruptured channel pipe taking aging into account *Atomic Energy* **109 No. 4**

[10] Zagrebayev A, Ovsyannikova N, Ramazanov R, Mil'to N 2015 Reconstruction of missing height-sensor readings on the basis of archival neutron field monitoring data *Atomic Energy* **117**

Use of molybdenum as a structural material of fuel elements for improving the safety of nuclear reactors

A N Shmelev and B K Kozhahmet

Department of theoretical and experimental physics of nuclear reactors, National Research Nuclear University MEPhI (Moscow Engineering Physics Institute), Kashirskoe highway, 31, Moscow, Russia

E-mail: ANShmelev@mephi.ru

Abstract. Main purpose of the study is justifying the use of molybdenum as a structural material of fuel elements for improving the safety of nuclear reactors. Particularity of used molybdenum is that its isotopic composition corresponds to molybdenum, which is obtained as the tailing during operation of the separation cascade for producing a material for medical diagnostics of cancer.
When performing the study the neutron-physical properties of isotopes of natural molybdenum (nuclear data library JENDL-4.0) and thermal properties of metallic molybdenum were used.
The following results were obtained:
1. A method for reducing the thermal constant of fuel elements for light water and fast reactors by using dispersion fuel in cylindrical fuel rods containing, for example, granules of metallic U-Mo-alloy into Mo-matrix was proposed.
2. The necessity of molybdenum enrichment by weakly absorbing isotopes was shown.
3. Total use of isotopic molybdenum will be more than 50%.
A method for reducing the thermal constant of the fuel elements, allowing us to increase the safety of light water and fast nuclear reactors by using dispersion fuel in cylindrical fuel rods containing, for example, granules of metallic U-Mo-alloy into Mo-matrix with enrichment by weakly absorbing isotopes of molybdenum is proposed.

1. Introduction

It is well known fact that the safety of nuclear reactor in the case of introduction of reactivity comparable with delayed neutrons fraction depends largely on the properties of fuel and materials constituting fuel element [1]. Sharp reactivity increase initiates neutron flash, which is suppressed by means of feedback due to fuel heating and increase of neutron absorption by fertile nuclide (^{238}U or ^{232}Th) thanks to the Doppler effect. If there is enough time for heat transfer from fuel to coolant then there would be feedback caused by heating of the coolant. This second feedback depends largely on thermal-physical characteristics of the fuel element and materials constituting it.

As is known, a refractory material based on molybdenum is characterized by good thermal-physical properties [2]. Therefore, such a material appears to be attractive when using dispersion fuel elements with good heat-conducting molybdenum matrix. However, there are some difficulties to use such a material in the reactor core. Firstly, molybdenum of a natural isotopic composition is quite a strong absorber of neutrons. Secondly, it is necessary to take into account its compatibility with fuel material on the one hand and with coolant on the other hand. Resolving these issues is the subject of the present paper.

2. Natural isotopic composition of molybdenum and some peculiarities of its nuclear-physical properties

Zirconium alloys are used as a basic structural material in the core of thermal reactors. The advantage of zirconium alloys is small neutron capture cross-section in the thermal energy range. At the same time molybdenum of a natural isotopic composition is characterized by a significantly larger neutron capture cross-section in this energy range (see table 1). It can be seen that neutron capture cross-section at thermal point (E_n = 0.025 eV) of molybdenum of a natural isotopic composition is 13 times larger than that of zirconium [3].

Table 1. Isotopic composition and radiative neutron capture cross-section of natural zirconium and molybdenum, as well as of different isotopes of molybdenum at thermal point (E_n = 0.025 eV).

Nuclide, atomic number of molybdenum isotope	Natural composition [%]	σ_c [barn] (E_n = 0.025 eV)
Zr_{nat}	–	0.196
Mo_{nat}	–	2.57
92	14.8	0.061
94	9.25	0.339
95	15.9	13.6
96	16.7	0.447
97	9.55	2.49
98	24.1	0.132
100	9.63	0.194

It means that technology of isotopic enrichment should be applied in order to use molybdenum as a construction material in thermal reactors. It can be seen from the table that isotopes ^{92}Mo and ^{94}Mo account for almost a quarter of the natural material and are located on the "light" end of isotopes natural mixture. Their mixture provides about the same neutron capture cross-section as natural zirconium does. Isotope ^{95}Mo provides a dominant contribution to the total capture cross-section, the atomic weight of which is intermediate in molybdenum isotopic composition. Use of ^{98}Mo and ^{100}Mo is not excluded, of course. Dependence of capture cross-section of these light and heavy isotopes of molybdenum on neutron energy is shown in figures 1 and 2 (here reactor energy range is considered). One can see that capture cross-section is generally inversely proportional to neutron velocity, and, generally speaking, the resonance integral is more than an order of magnitude smaller than that of ^{238}U [3].

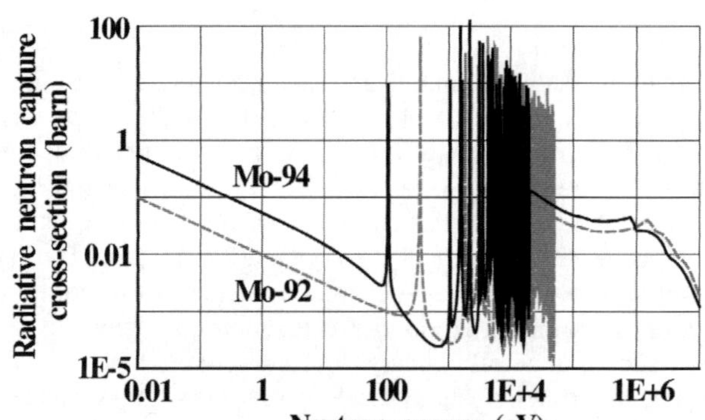

Figure 1. Radiative neutron capture cross-section of molybdenum isotopes ^{92}Mo and ^{94}Mo on neutron energy (nuclear data library JENDL-4.0).

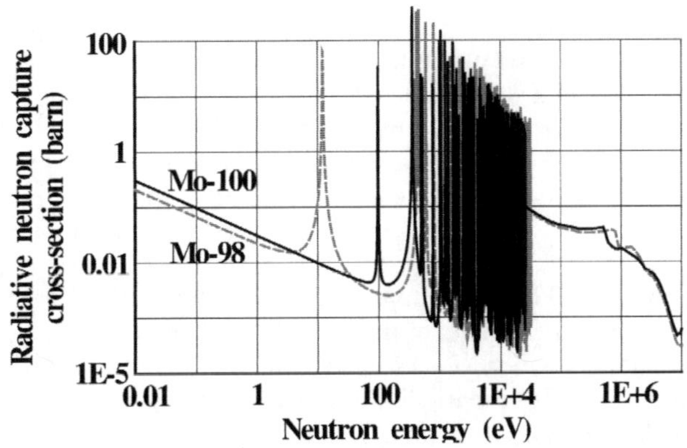

Figure 2. Radiative neutron capture cross-section of molybdenum isotopes ^{98}Mo and ^{100}Mo on neutron energy (nuclear data library JENDL-4.0).

3. Isotopic enrichment of molybdenum for medical purposes and for reducing its radiative neutron capture

As is known, the isotopic enrichment is a high-level technology requiring the use of multi-step separation cascades, and therefore is very expensive. However, for the case of production of enriched molybdenum there is one important factor that can significantly facilitate the solution of this issue.

There is already a commercial enrichment of molybdenum [4] with the production of the desired isotopes ^{98}Mo and ^{100}Mo for medical diagnosis of early formation of cancerous tumors. Currently, the diagnosis is the most advanced in the world in the field of medicine and, therefore, the need for these heavy isotopes of molybdenum is high and continues to rise.

Keeping this fact in mind, it can be assumed that the process of obtaining molybdenum enriched by the light isotopes $^{92+94}Mo$ can be successfully combined with the process of producing heavy isotopes ^{98}Mo and ^{100}Mo. Heavy isotopes are the product at one end of enrichment cascade, while light isotopes are the product at the other end. Of course, the structure of the cascade requires to be changed to get rid of the strong neutron absorber - isotope ^{95}Mo, atomic weight of which is intermediate in a series of atomic weights. Since this isotope is in the middle of the spectrum of the atomic weights of molybdenum natural isotopes (see table 1), a cascade of isotope separation with additional selection in the middle of the cascade should be used. This problem is considered in [4].

4. Thermo-physical parameters of metallic molybdenum, important for use as a structural material of the fuel element

Table 2 shows some thermo-physical parameters of molybdenum and zirconium that are important for heat transfer in the fuel element [2, 3, 5].

Table 2. Nuclear-physical and thermo-physical properties of fuel and structural materials.

Material	Density [g/cm^3] (20°C)	$T_{melting}$ [°C]	Heat capacity [10^7 J/(m^3·K)]	Thermal conductivity [W/(m·K)] (1000°K)	Time constant of cylindrical fuel element (rod) τ_{th} [s] (d = 9.1 mm)
Zr	6.5	1855	0.23	21.5	0.36
Mo	10.22	2623	0.30	112	0.20
U-9%Mo	17.6	1300	0.35	40	0.36
Dispersion fuel[a]	–	–	0.32	71.5	0.26

[a] Granules – U-9%Mo; matrix – Mo; the proportion of granules V_f = 0.5.

Apart from the fact that molybdenum has a melting point substantially greater, it also has a more than 5 times greater coefficient of thermal conductivity. Model fuel element of metallic molybdenum is characterized by a time constant of transferring heat to the environment [1] which is almost two times smaller than that for zirconium. These attractive properties of molybdenum can be used to create a new concept of a fuel element for thermal and fast reactors. This structural material is compatible with the well-known metallic uranium-molybdenum fuel good thermal properties of which are shown in table 2.

5. Rod fuel element with a small time constant τ_{th} for fast reactor and thermal reactor

As is known, fuel material of fuel elements of the world's first nuclear power plant built in the USSR (Obninsk) in 1954, was an alloy U-9%Mo. Molybdenum was chosen not only because it is able to stabilize γ-phase of uranium, but also because its alloy with uranium is characterized by a high thermal conductivity and high nuclear density of uranium (see table 2).

It can be seen that among materials presented in table 2, for creating a fuel element it is preferable to use dispersion fuel containing granules of a metal alloy U-9%Mo, dispersed into the molybdenum matrix. To improve the thermal contact of fuel with cladding, the latter also should be produced, for example, from Mo-based alloy [6]. It is known that molybdenum and its alloys are compatible with both aqueous coolant (in thermal reactors), and liquid metals Na, Pb and Pb-Bi [7, 8] in fast reactors.

Experimental studies [9] performed with a model nuclear fuel have confirmed the possibility of creating such a dispersion fuel (U-Mo - fuel granules, Mo - matrix) both in terms of compatibility over a wide temperature range of fuel granules, matrix and cladding, and at high fuel burn-up.

Since the thermal conductivity of fuel granules (U-9%Mo) is 2.8 times smaller than that of a molybdenum matrix, the average thermal conductivity of dispersion fuel material depends to a significant extent on the proportion of fuel granules, and on their forms. Assuming that fuel granules have a spherical shape, when their share $V_f = 0.5$ in dispersion fuel material, the average coefficient of thermal conductivity of the fuel rod is 71.5 W/(m·K) [5], and the time constant of the fuel element with a diameter d = 9.1 mm will be $\tau_{th} = 0.26$ sec (see table 2).

6. Prospects for the use of dispersion fuel elements with molybdenum as a structural material in fast reactors

As for the apparent attractiveness of the possibility of using molybdenum and its alloys as a structural material in fast reactors (fuel - uranium-molybdenum alloy, the matrix material - molybdenum in dispersion fuel elements, cladding material - possibly also refractory material), the need for use of molybdenum enriched by light isotopes still requires to be considered. However, it is important to note that in a core the combination of a high melting point liquid metal coolant (e.g., Pb, Pb-Bi et al.) and the fuel element with a cladding based on a refractory (molybdenum-based) material [6] and with a molybdenum matrix can significantly increase the stability of the core with respect to the possibility of a crisis in the heat transfer at a jump of reactivity. Using uranium-molybdenum fuel fits well into the concept of protected fuel cycle based on a mixture of (^{233}U+^{238}U) [10].

If two heavy isotopes $^{98+100}Mo$ would be used for medicine purposes (these isotopes account for about 1/3 of the total of molybdenum), and two light isotopes $^{92+94}Mo$ would be used for reactor purposes (these isotopes account for about 1/4 of the total of molybdenum), then the total use of the isotopic molybdenum would be over 50%. The rest of the molybdenum without damage may be used in the national economy.

7. Summary

A method for reducing the time constant of the fuel elements allowing us to increase the safety of light water reactors and fast reactors by using dispersion fuel in cylindrical fuel elements containing, for example, granules of metallic U-Mo-alloy into Mo-matrix with enrichment by weakly absorbing molybdenum isotopes was proposed. At the same time the use of the isotopic molybdenum would be more than 50%.

IOP Conf. Series: Journal of Physics: Conf. Series **781** (2017) 012022 doi:10.1088/1742-6596/781/1/012022

References

[1] Hummel H H and Okrent D 1970 *Reactivity Coefficients in Large Fast Power Reactors* (La Grange Park, Illinois: American Nuclear Society)

[2] Kalin B A, Platonov P A, Tuzov Yu V, Chernov I I and Shtrombakh Ya I 2012 *Fizicheskoe materialovedenie: Uchebnik dlya vuzov, Vol. 6, Konstruktcionnye materialy yadernoy tekhniki* (Moscow: NRNU MEPhI)

[3] Grigor'ev I S and Meylikhov E Z 1991 *Fizicheskie velichiny: Spravochnik* (Moscow: Energoatomizdat)

[4] Smirnov A Yu and Sulaberidze G A 2014 Sravnenie sposobov obogascheniya promezshutochnykh komponentov v kaskadakh iz odinakovogo chisla razdelitel'nykh komponentov *Atomnaya energiya* **117** №5 pp 274–279

[5] Baranov V G, Godin Yu G, Tenishev A V, Khlunov A V and Novikov V 2012 *Fizicheskoe materialovedenie: Uchebnik dlya vuzov, Vol. 7, Yadernye toplivnye materialy* (Moscow: NRNU MEPhI)

[6] Dolgov A B, Novikov V V and Ivanov A S 2013 Novoye toplivo dlya VVER *Atomniy ekspert* **6-7** pp 30–34

[7] Sorokin A P, Kalyakin S G and Kozlov F A 2014 Visokotemperaturnaya yadernaya energotekhnologiya na osnove bistrikh reaktorov s natrievim teplonositelem dlya proizvodstva vodoroda *Atomnaya energiya* **116** №4 pp 194–283

[8] Gromov B F, Demishonkov A P and Toshinskiy G I 1993 Sposob zaschity konstruktcionnykh materialov ot korrozii pri povyshennykh temperaturakh v zhidkom svintce, vismute i ikh splavakh *Patent of RF* №2066710 Class C23F 11/00 Date of filing 10.03.1993

[9] Shornikov D P 2008 Distinctive Features of Interaction of the U-Mo Fuel Composition Fission Products and the Al Matrix with Barrier Coating *Proc. of the 12th International Conference on Research Reactor Fuel Management* pp 26–29

[10] Shmelev A N, Kulikov G G, Kulikov E G and Apse V A 2014 *O potentciale gibridnykh (sintez-delenie) narabotchikov topliva dlya yadernykh reactorov: stabilizirovannye razmnozhayuschie svoystva, glubokoe vygoranie, zaschischennoe toplivo* (Moscow: NRNU MEPhI)

Acknowledgment

The authors thank the staff of the Department №9 "Physical Problems of Materials Science" of the National Research Nuclear University MEPhI Dr. A.V. Tenishev and Dr. D.P. Shornikov for useful consultations.

IOP Conf. Series: Journal of Physics: Conf. Series 781 (2017) 012023 doi:10.1088/1742-6596/781/1/0120

Optimization of the energy complex "NPP-accumulator" in case of force majeure

G Zaluzhnaya and A Zagrebaev

National Research Nuclear University MEPhI (Moscow Engineering Physics Institute), Kashirskoe shosse 31, 115409 Moscow, Russia

Abstract. We consider a problem of optimization of NPP with accumulator operation mode in case of force majeure. A mathematical formulation and solving of problem of energy output's time behavior is provided. A mathematical formulation and solving of problem of energy's optimum allocation to consumers with different priorities. Mathematically, the problem reduces to linear programming problem. We received that optimal time behavior is uniform energy output, and one should start with consumer with highest priority.

Introduction

There are special Nuclear Power Plants for remote areas, where NPP is the only source of energy for quite a big area (for example, Bilibinskaya NPP in Russia [1]). In case of force majeure, there is a need in an instant stop of NPP. This will lead to economic and social losses.

We suppose to establish an accumulator in NPP to solve these problems. A need to store energy for a long time without big losses requires using of accumulators with low energy loss, such as described in [2-5]

Here we consider the problem of optimal distribution of energy in case of force majeure, when the NPP is shut down but still there are consumers that need electricity.

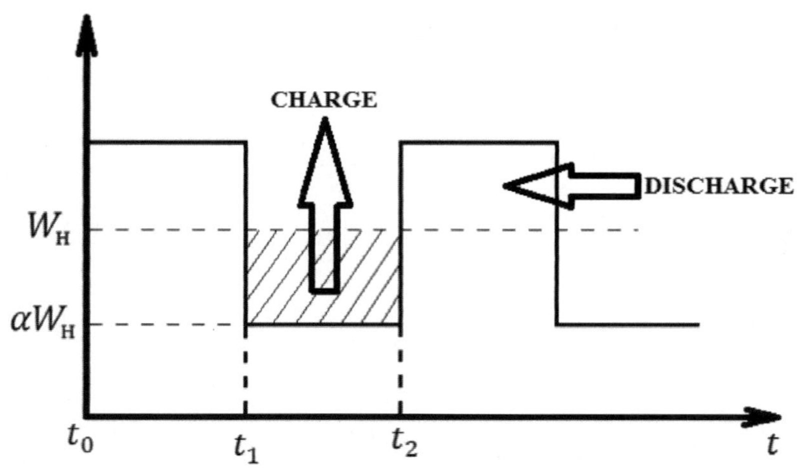

Figure 1. Energy consumption in relation to time.

Fig.1 illustrates energy consumption in relation to time [6]. Energy is demanded more in day time, than at night, therefore, power units have to work in a varying duty. This leads to waste of fuel and power units' reduction of serviceability [7]. Accumulator introduction into the system can solve this problem: power units will work at constant duty, accumulator will be charged with energy excess at night, and it will discharge during day time to cover peak ratings.

The problem is posed this way: there are a NPP with an accumulator and several consumers of energy provided by NPP (for example, factories etc.) A threat is announced at the time zero (t=0) for indefinite period τ. This period is a random value, which has density function $f(\tau)$. Therefore, the NPP is stopped, but there is some charge in accumulator. It is required to find the best way to provide energy to consumers, considering that lack of energy leads to economic loss determined by loss function.

This problem's solution was divided on two parts: optimization of energy output's time behavior and searching of energy's optimum allocation to consumers depending on their loss functions.

1. Optimization of energy output's time behavior

Minimum of average loss is searched to solve the first problem.

Total loss of one consumer is

$$c(\tau)\int_0^{\tau}\big(W_{\text{н}} - W(t)\big)dt \tag{1}$$

($c(\tau)$ is some loss function here, its general form is shown on fig.2).

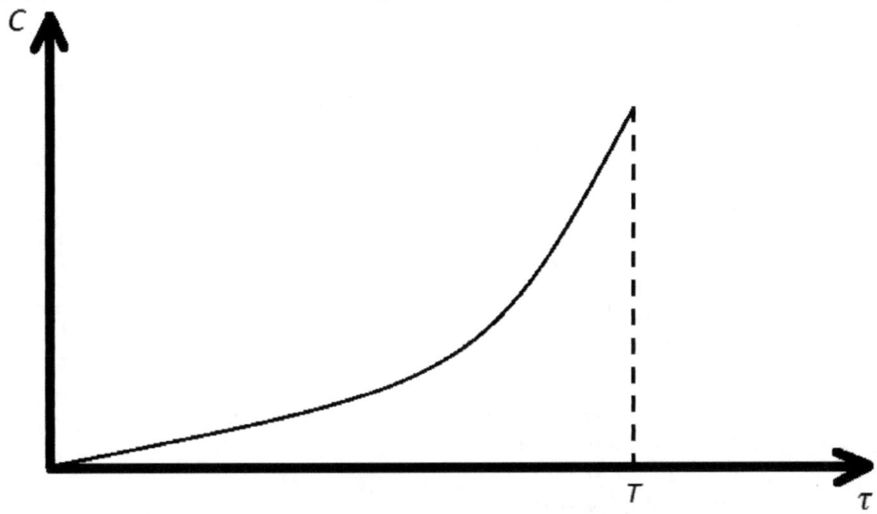

Figure 2. General form of loss function.

Therefore, considering that $f(\tau)$ is density function of duration of force majeure, we obtain average loss (it is average value of (1)):

$$\int_0^{T} f(\tau)c(\tau)\int_0^{\tau}\big(W_{\text{н}} - W(t)\big)dtd\tau \tag{2}$$

In addition to this, we have to set some limitations on $W(t)$:

$$\begin{cases} 0 \le W(t) \le W_0, \\ \int_0^{\tau} W(t)dt \le \int_0^{T} W(t)dt \le Q_A. \end{cases} \tag{3}$$

W_0 here is power that was provided to consumer before force majeure, Q_A is accumulator's charge.

$$Q_A = \eta(t_2 - t_1)(1-\alpha)W_{\text{н}} \tag{4}$$

η – accumulator's coefficient of efficiency, $t_2 - t_1$ – system's working time at night, α – portion of NPP's power at night time, $W_{\text{н}}$ – NPP's power.

It is shown that if we expanse $W(t)$ into power series, this problem is reduced to linear programming problem, which has an optimal solution $W(t) = const$ (uniform energy output). This result will be taken in account while solving the second problem.

2. Searching of energy's optimum allocation to consumers

2.1. Problem definition

To solve the second problem, we have to search for

$$\min_{W_j} \sum_{j=1}^{M} (W_{sj} - W_j) \int_0^T C_j(\tau) f(\tau) \tau d\tau \quad (5)$$

considering that limitations on $W(t)$ are:

$$\begin{cases} 0 \le W_j \le W_{sj}, \\ \sum_{i=1}^{M} W_j \tau \le \sum_{i=1}^{M} W_j T \le Q_A. \end{cases} \quad (6)$$

This problem reduces to problem of searching

$$\max_{W_j} \sum_{j=1}^{M} W_j R_j \quad (7)$$

Where

$$R_j = \int_0^T C_j(\tau) f(\tau) \tau d\tau \quad (8)$$

and ratio between this values is determined by loss function.

2.2. Problem solving

This problem has analytical solution: first, we have to provide energy to consumer, which loss function is maximum [8].

We consider this problem:

$$\begin{cases} f = \max_{W_j} \sum_{j=1}^{M} W_j R_j , \\ 0 \le W_j \le W_{sj}, \sum_{j=1}^{M} W_j = W_A; \end{cases} \quad (9)$$

where $W_A = {}^{Q_A}/_{T}$. We consider that $R_1 > R_2 > \cdots > R_M$ (10) without loss of generality.

Let us prove that if $W_A \le W_{s1}$ then optimum solution is

$$W_1 = W_A, W_2 = \cdots = W_M = 0 \quad (10)$$

and

$$f = W_1 R_1 \quad (11)$$

Now we consider any other allocation of W_j.

Let $W_1' = W_A - \Delta W, W_2' \ne 0, \ldots, W_M' \ne 0,$ and $W_2' + \cdots + W_M' = \Delta W$ (which follows from (9)).

Then

$$f' = (W_A - \Delta W) + R_2 W_2' + \cdots R_M W_M' \quad (12)$$

We have to compare f and f'; if $f > f'$ it means that allocation (11) is optimum.

Let us consider $f - f'$:

$$f - f' = R_1 \Delta W - R_2 W_2' - \cdots R_M W_M' \quad (13)$$

It follows from (10) that:

$$R_M W_2' + \cdots R_M W_M' < R_2 W_2' + \cdots R_M W_M' < R_2 W_2' + \cdots R_2 W_M' \quad (14)$$

And, considering (15), we have:

$$f - f' > R_1\Delta W - R_2(W_2' + \dots + W_M') = R_1\Delta W - R_2\Delta W = (R_1 - R_2)\Delta W > 0 \quad (15)$$

It follows from (16) that any other allocation gives us $f > f'$, and original allocation (11) is optimum. Then, if there is some charge remained, it has to be allocated between the rest of consumers according to aboon: we have to start from consumer with maximum loss function.

3. Optimization effect

To count optimization effect, we have to compare to energy allocations: optimal and non-optimal. Let us sort consumers from lowers R_j to greatest. Then losses in optimal allocation will be:

$$\nu_1 = \left(W_A - \sum_{i=1}^{M-1} W_i\right) R_M + \sum_{j=M+1}^{N} W_j R_j \quad (16)$$

(accumulator's power is enough for M consumers)

And in case of non-optimal allocation, if accumulator's power is enough for P consumers, total loss is:

$$\nu_2 = \left(W_A - \sum_{i=N}^{N-P+2} W_i\right) R_{N-P+1} + \sum_{j=1}^{N-P} W_j R_j \quad (17)$$

Then optimization effect can be counted as

$$\nu = \frac{\nu_2 - \nu_1}{\nu_2} = 1 - \frac{\nu_1}{\nu_2} \quad (18)$$

This formula is difficult to analyse, because we have too many parameters: we can simplify it considering that accumulator's power is only enough for consumers with numbers $1 \div M$ or with numbers $N - P \div N$, and $W_j = W$. Then

$$\nu = 1 - \frac{\sum_{j=M+1}^{N} W_j R_j}{\sum_{j=1}^{N-P} W_j R_j} = 1 - \frac{\sum_{j=M+1}^{N} R_j}{\sum_{j=1}^{N-P} R_j} \quad (19)$$

For example, if we consider that there are only two consumers, and $W_1 = W_2 = W_A$, then

$$\nu = 1 - \frac{(W_A - W_1)R_1 + W_2 R_2}{(W_A - W_2)R_2 + W_1 R_1} = 1 - \frac{R_2}{R_1} \quad (20)$$

4. Future works and conclusions

To summarize, we received that optimal time behavior is uniform energy output, and one should start with consumer with highest priority. We will continue solving the problem optimization of the energy complex "NPP-accumulator" in case of force majeure, and we are going to solve this problem considering that accumulator can be used to satisfy NPP's own needs (for example, cooling water outlet, if energy generator is broken, as it was at Fukushima ([9-10])). Also, we have to take in account that the value of accumulator's charge can also be a random value, because we do not know exactly, in what moment the NPP was stopped. It would be interesting to understand, how big should be accumulator's capacity, that it would be enough to cover possible needs in case of force majeure, but not very expensive in regular regime.

References

[1] Dolgov V V October 1997 Bilibinsakya Nuclear Power Plant: 23 years operation in the specific conditions of the Russian far north-east *Nuclear Engineering and Design* 173 pp 87-97

[2] Baklushin R P 2011 Exploitation of NPP NRNU MEPHI

[3] Mohsen Saadat, Farzad A Shirazi. Perry Y Li 1 January 2015 Modeling and control of an open accumulator Compressed Air Energy Storage (CAES) system for wind turbines Applied Energy 137 pp 603–616
[4] Donald Bender 2016 Flywheels Storing Energy pp 183–201
[5] Stolyarevsky A Ya 2005 Chemic-thermal cycles and energy storage machines International Scientific Journal for Alternative Energy and Ecology 3 pp 45-58
[6] Nuzhdin, Prosvirnov Most effective ways of energy storage and prospects of using energy storage technology in atomic industry VNIIAES
[7] Jacob Edwards, Hitesh Bindra , Piyush Sabharwall October 2016 Exergy analysis of thermal energy storage options with nuclear power plants Annals of Nuclear Energy 96 pp 104–111
[8] Judin, Goldstein 1961 Problems and methods of linear programming Moscow
[9] Ashraf Labib, M.J. Harris January 2015 Learning how to learn from failures: The Fukushima nuclear disaster Engineering Failure Analysis 47 pp 117-128
[10] Atsufumi Yoshizawa, Kyoko Oba, Masaharu Kitamura 2016 Lessons Learned from Good Practices During the Accident at Fukushima Daiichi Nuclear Power Station in Light of Resilience Engineering IFAC-PapersOnLine 49 pp 245–250

IOP Conf. Series: Journal of Physics: Conf. Series **781** (2017) 012024 doi:10.1088/1742-6596/781/1/012024

Equalization of energy density in boiling water reactors (as exemplified by WB-50). Development and testing of WB -50 computational model on the basis of MCU-RR code

Yu B Chertkov, V V Disyuk, E Yu Pimenov and N V Aksenova

Tomsk Polytechnic University, 30, Lenina avenue, Tomsk, 634050, Russia

E-mail: chertkov@tpu.ru, valek_diss@mail.ru, pimenov-eduard@mail.ru, polozova15@tpu.ru

Abstract. Within the framework of research in possibility and prospects of power density equalization in boiling water reactors (as exemplified by WB-50) a work was undertaken to improve prior computational model of the WB-50 reactor implemented in MCU-RR software. Analysis of prior works showed that critical state calculations have deviation of calculated reactivity exceeding ±0.3 % (ΔKef/Kef) for minimum concentrations of boric acid in the reactor water and reaching 2 % for maximum concentration values. Axial coefficient of nonuniform burnup distribution reaches high values in the WB-50 reactor. Thus, the computational model needed refinement to take into account burnup inhomogeneity along the fuel assembly height. At this stage, computational results with mean square deviation of less than 0.7 % (ΔKef/Kef) and dispersion of design values of ±1 % (ΔK/K) shall be deemed acceptable. Further lowering of these parameters apparently requires root cause analysis of such large values and paying more attention to experimental measurement techniques.

1. Influence of raw data uncertainty on the dispersion of results

Uncertainties in the raw data are determined by deviations of parameters describing material composition and geometry of the reactor core from their nominal values and by positioning allowance values for the reactor's elements, as well as by measurement errors.

The following uncertainties were analyzed for their influence over dispersion of the calculation results:

- non-uniform distribution of uranium (burnup) over the volume of fuel assembly;
- uncertainty of concentration and distribution of boron in wrapper tubes;
- uncertainty of thickness and composition of deposits on the fuel elements;
- uncertainty of chemical composition of boric acid water solution;
- uncertainty in manual regulator position and gap position between absorbing and fuel parts of the regulator;
- uncertainty in distribution of ^{10}B along the height of absorbing parts of the manual regulators (during the most of the run the manual regulators are partially submerged into the reactor core).

Non-uniform distribution along the hight of fuel assemblies is defined by two causes: non-uniformity of initial distribution of uranium along the height of fresh fuel elements and fuel assemblies (non-uniform size and density of uranium dioxide pellets) and non-uniform ^{235}U burnup along the height and radius of the assemblies.

According to [3] production of fuel elements for water-cooled reactors, the same as used in the VK-50 reactor, provides the following allowance values:

- deviation of ±0.055mm is allowed for fuel elements with average outside diameter of 9.135mm;
- deviation of ±0.045mm is allowed for pellets of uranium dioxide with average outside diameter of 7.755mm;
- while average density of uranium dioxide in pellets is 10.6 g/cm^3, allowable minimum is 10.3 g/cm^3 and allowable maximum is 10.8 g/cm^3.

The last two parameters help estimate non-uniformity coefficient of uranium mass distribution (not accounting for enrichment tolerance) along the height and section of a fresh fuel assembly with the value of ~1.03. Dispersion of calculated reactivity due to dispersion in fuel with such non-uniformity coefficient may reach (depending on reactor loading) ±0.1 % (ΔKef/Kef).

While the average outside diameter of fuel elements is 9.135 mm, maximum diameter may reach 9.19 mm (as per specifications). Deposits increase the outside diameter of the element and thus decrease the flow passage for reactor coolant. According to [3], registered outside diameter does not exceed 9.173 mm; while with deposits the outside diameter of the fuel element may reach 9.19 mm. Dispersion in calculated reactivity due to change in wrapper tube outer diameter in the range of 9.08 - 9.19 mm does not exceed ±0.1 % (ΔKef/Kef).

At boron reactivity coefficient equal to 4 %/(g/l) the error of ±0.01 g/l in its concentration leads to reactivity error of ±0.04 % (ΔKef/Kef).

Change of distance between the absorbing and fuel parts of the manual regulator in the range from 20 to 34 cm leads to changes in reactivity not exceeding ±0.02 % (ΔKef/Kef).

Change in manual regulator suspension height with respect to reactor core by ±5 cm leads to changes in reactivity not exceeding ±0.01 % (ΔKef/Kef).

Lowering concentration of ^{10}B in absorbing parts of manual regulator by 10% leads to increase in reactivity by 0.04 % (ΔKef/Kef).

Non-uniform distribution of ^{235}U in irradiated fuel assembly is largely determined by non-uniform burnup distribution along the height and radius of the assemblies, which is determined in calculations with the help of software packages BIPR-K and MCU. Estimation of error in fuel distribution along height and radius of a fuel assembly due to inaccuracy of burnup description is a complex task. In general, if the material distribution calculation accuracy along the height and radius is 5 %, then dispersion of calculated reactivity may reach (depending on reactor loading) ±0.2 % (ΔKef/Kef) due to non-uniform fuel distribution.

Analysis of these results shows:

- the most important are effects linked to geometric characteristics, fuel concentrations and fuel distribution calculation accuracy; joint inaccuracy in calculated reactivity due to these factors comprises up to 0.5 % (ΔKef/Kef);
- in the reactor core loaded with fresh fuel assemblies the uncertainties of fuel content may reach ±0.3 % (ΔKef/Kef); but these errors lead to systematic error in calculated values for critical assembly experiments and on 12th run of the reactor; they cannot explain random variation of calculated values; however, during modeling of such experiments the standard mean-square deviation between calculated and experimental values is on average 0.65 % (ΔKef/Kef);
- in the reactor core loaded with irradiated and fresh fuel assemblies the calculated accuracy values due to fuel composition reach up to ±0.5 % (ΔKef/Kef).

Projected boron content in wrapper tubes of the fresh fuel assemblies is (0.07-0.10) per cent by weight. If we select the average boron content (0.085 % wt), as a basic value, then its changes within the stated range lead to dispersion of calculated reactivity of ±0.6 % (ΔKef/Kef).

Changes in boron content within the design range (0.07 - 0.10) % wt lead to changes in reactivity from +0.8 % (ΔKef/Kef) to -0.35 % (ΔKef/Kef). For the experiments described above the average calculated reactivity values were within the range from -0.21 % (ΔKef/Kef) to +0.74 % (ΔKef/Kef). That is, in the same range.

New fuel loading scheme for a run with partial overloads contains another disturbing factor: fresh fuel assemblies are typically loaded in the third and fourth rows of reactor core; at that, power (and, consequently, thermal neutron flux density) of surrounding fuel assemblies relative to its power may vary from being approximately 1.5 times higher to being 2-3 times lower. Such wide range of power variation in surroundings shall lead to high non-uniformity of boron burnup in the wrapper tubes and shall be accounted for (especially orientation!) during rearrangements of fuel assemblies in the reactor core in overload.

2. Refinement of computational model of WB-50 reactor by increasing the number of material zones for fuel along the reactor's height

The first variant of the WB-50 reactor computational model [1] allowed describing each of the reactor's fuel assemblies with a single material zone, while energy release was calculated for each fuel element separately.

2.1. The first variant of the WB-50 reactor computational model

The first variant of the WB-50 reactor computational model [1] allowed describing each of the reactor's fuel assemblies with a single material zone, while energy release was calculated for each fuel element separately. Boron burnup in all the wrapper tubes was described as three zones:

- boron in a fresh fuel assembly;
- boron in an irradiated fuel assembly after the first run;
- boron in a fuel assembly that underwent more than 2 runs (^{10}B burnt up).

2.2. The second variant of the computational model

The second variant of the computational model allowed determining changes in burnup within a single fuel assembly (20 material zones along the assembly height). For other assemblies the fuel was described with one individual zone. Boron burnup in all the wrapper tubes was described as three zones as well. Such computational model allows calculating changes in material composition of an assembly and element-wise calculation of power density as they are changing through the run.

The second variant of the computational model was used to obtain dependency of energy density distribution of ^{235}U in one of the central operating fuel assemblies on its average burnup at constant material content of other assemblies of the reactor. At that, dependencies were obtained between the energy density distribution along the assembly's height and density of principal fissile isotopes (^{235}U + ^{239}Pu) and average burnup.

The previous computational model was refined to take account of detailed calculation of ^{10}B burnup and its distribution along the fuel assembly height depending on the duration of reactor's operation: SA wrapper tube was divided into 20 material zones along its height and ^{10}B burnup was calculated separately for different zones.

Accounting for obtained dependency of ^{10}B distribution along the height of assembly on time, as well as (factory) non-uniformity, significantly influences the results of computational modeling of experimental critical conditions. Additionally a perturbation shall be mentioned that is introduced into the computational modeling of reactor overload due to lack of accounting for orientation of rearranged fuel assemblies with different ^{10}B content in different flanks of the wrapper tubes in their new position in the reactor core.

2.3. The third variant of the computational model

The third variant of the computational model was created to describe each of the reactor's fuel assemblies by 2 material zones (top and bottom halves of the assembly). Burning boron in all wrapper

IOP Conf. Series: Journal of Physics: Conf. Series **781** (2017) 012024 doi:10.1088/1742-6596/781/1/012024

tubes was still described with three zones, however there are plans for future modifications with several material zones describing boron content in the wrapper tubes (from 5 to 10 along the height). This variant will allow more accurate consideration of boron burnup along the fuel assembly height and will hopefully increase accuracy of reactivity calculations for high concentrations of boric acid in the reactor water. From this perspective, it is hard to expect that mean-square deviation of reactivity calculations for critical states will be less than 0.7 % (ΔKef/Kef), and dispersion of calculated values will be less than ±1 % (ΔKef/Kef).

3. Conclusion

In conclusion, the following should be noted:

- uncertainty of raw data determined by characteristics of fuel elements and fuel assemblies with fresh fuel may explain only relatively small deviations of calculated values from experimental ones, such deviations (±0.4 % (ΔKef/Kef)) are typical in calculations for other reactor units;
- a single factory non-uniform distribution of boron along the wrapper tubes can explain deviations of calculated values from experimental results on critical assembly and fresh reactor core in the 12th run;
- during calculations for WB-50 reactor it is necessary to describe ^{10}B along the height and azimuth of the assembly in more detail; when describing wrapper tubes with a single material zone their burnup time is unjustifiably lowered two- or three-fold;
- to create a more accurate computational model of the WB-50 reactor it seems sufficient to introduce at least 2 material zones for description of fuel burnup along the height and at least 5 material zones to cover ^{10}B distribution in the wrapper tubes.

All these issues were largely considered while preparing the third variant of the WB-50 computational model.

References

[1] Yescherkin V, Starkov V, Chertkov Yu et al 2005 Estimation of possibility to use 3.6 % enriched fuel in WB-50 reactor *Research report* Registry no. O-5682

[2] Vaneev Yu, Antonov S, Bulycheva L et al 2004 Development and verification of MCU-based models of the VK-50 reactor active zone at its loading with fresh fuel assemblies *Research report Control* no. O-5542

[3] Scheglov A, Proselkov V, Enin A 1991 Results of statistical treatment of design and technological parameters of WWER-1000 fuel element *Preprint* of Kurchatov Atomic Energy Institute reg. no. IAE-5334/4

IOP Conf. Series: Journal of Physics: Conf. Series **781** (2017) 012025 doi:10.1088/1742-6596/781/1/012025

Validation of deterministic and Monte Carlo codes for neutronics calculation of the IRT-type research reactor

M V Shchurovskaya, V P Alferov, N I Geraskin and A I Radaev

National Research Nuclear University MEPhI (Moscow Engineering Physics Institute), 31 Kashirskoe shosse, 115409 Moscow, Russia

E-mail: mvshchurovskaya@mephi.ru

Abstract. The results of the validation of a research reactor calculation using Monte Carlo and deterministic codes against experimental data and based on code-to-code comparison are presented. The continuous energy Monte Carlo code MCU-PTR and the nodal diffusion-based deterministic code TIGRIS were used for full 3-D calculation of the IRT MEPhI research reactor. The validation included the investigations for the reactor with existing high enriched uranium (HEU, 90 w/o) fuel and low enriched uranium (LEU, 19.7 w/o, U-9%Mo) fuel.

1. Introduction

Monte Carlo neutron transport codes are currently widely used for the operation and safety analysis of research reactors. Diffusion codes are also used, especially for routine core follow calculations. To demonstrate the quality of these codes, the validation against experimental data is necessary. If the special qualification process is developed including detailed description of the experimental data and the test problems, this process can be used simultaneously for the validation of Monte Carlo and diffusion codes. Such approach allows carrying out code-to-code comparison and gives additional information for the analysis. We have developed the validation process for the IRT-type research reactor. The continuous energy Monte Carlo code MCU-PTR [1] and the nodal diffusion-based deterministic code TIGRIS [2] were validated against experimental data of the IRT MEPhI research reactor and on the basis of the comparison with the calculated results obtained using different codes for test problems and real core configurations. The validation included the investigations for the reactor with existing high enriched uranium (HEU, 90 w/o) fuel and low enriched uranium (LEU, 19.7 w/o, U-9%Mo) fuel [3].

2. Validation methodology

MCU-PTR code and TIGRIS code validation included: comparison with the results of the test problems calculation obtained using the other codes; comparison with the results of IRT MEPhI reactor reference cores calculation obtained using the other codes; comparison with the operational measured data of the IRT MEPhI reactor and the results of the calculation of these experiments obtained using the other codes.

The test problems consider simplified configuration of the core and the reflector, only fuel assembly (FA) is described in detail. The test problems have not the IRT MEPhI reactor specific features.

Three reference cores with real geometry of the reflector and experimental channels were considered. The first reference core is IRT MEPhI current core, the other reference cores are the cores with IRT MEPhI current reflector but HEU fuel is replaced with LEU fuel (fresh or with typical burnup). The reference cores and the test problems were used for the comparison with the results calculated using different codes. The calculated results comparison was carried out for the identical fuel isotopic composition.

The comparison with the operational measured data of the IRT MEPhI reactor was based on the modeling of the reactor operation history (burnup process, reloadings, control rod calibrations).

The calculations for the code validation were performed for real and model cores with existing IRT-3M fuel assemblies with HEU fuel as well as for model cores with IRT-3M fuel assemblies with LEU U-9%Mo-Al fuel. Calculated results for both types of fuel show that HEU and LEU FA have similar neutronics parameters and the conclusions made for one type of fuel are correct for another. In the test problems with depletion in the most cases LEU fuel was considered since due to higher plutonium buildup the isotope inventory vs. burnup is more complicated for LEU fuel. In the test problems concerned with control rod (CR) absorber burnup HEU fuel was considered since these test problems simulate the real reactor operation. In the test problems with fresh cores performed for the comparison of the calculations conducted by Monte Carlo codes with different cross-section libraries and diffusion code both HEU and LEU fuel were considered.

3. Models and codes

Full 3-D model for the IRT MEPhI reactor was developed using MCU-PTR code, including fuel assemblies, reflector blocks, control rods, main structural components, horizontal beam tubes and vertical irradiation channels. Investigations to prove the choice of spatial nodalization for the power density and the burnup distribution calculation were carried out [4].

The TIGRIS code is intended for steady-state neutronics calculation and for the burnup calculation of pool-type research reactors using three-dimensional (x-y-z geometry) few group diffusion approximation. Spatial neutron distribution is resolved by means of polynomial nodal method.

4. Calculated results

4.1. Test problems

The calculation of the set of the test problems for the IRT-type research reactor with a tube-type LEU (U-Mo) and HEU fuel, a light water moderator and a beryllium reflector was performed using MCU-PTR and two other continuous-energy Monte Carlo codes. The results of the test problems calculation are presented in [5]. For the fresh cores the discrepancy in neutron multiplication factor between the results of MCNP with ENDF/B-VII cross section library and the results of MCU-PTR with the ACE/MCU library is from 0.03%Δk/k to -0.34%Δk/k for HEU and LEU fuel. A good agreement among the MCU-PTR and the other Monte Carlo codes results was observed for the test problem with burnup. The increase in the discrepancy among the codes in neutron multiplication factor after the first burnup cycle with LEU fuel was found to be less than 0.2%Δk/k.

The calculation of the set of static test problems using TIGRIS code showed that the deviation in neutron multiplication factor from MCU-PTR results is ~-1%Δk/k. The maximum deviation from the precision codes in FA power is less than 16%. The calculation of the depletion test problem showed that reactivity calculated using TIGRIS code has approximately constant bias from MCU-PTR results. So, reactivity vs. time modeling during burnup using TIGRIS code is satisfactory.

4.2. Reference cores

Calculation was performed for three reference cores. Table 1 presents three cases with different CR (regulating rod AR and 3 shim rods KC) positions for the reference HEU and LEU cores. The calculation of these cases enables to estimate criticality, excess reactivity and shutdown margin.

Table 1. CR positions for the reference cores calculation (scram rods withdrawn).

Core	Case	CR position (mm)			
		AR	KC-1	KC-2	KC-3
HEU operational	#1	250	0	0	393
	#2	0	0	0	0
	#3	580	580	580	580
LEU fresh	#1	250	0	180	580
	#2	0	0	0	0
	#3	580	580	580	580
LEU operational	#1	250	0	0	390
	#2	0	0	0	0
	#3	580	580	580	580

Table 2 presents the results of the calculation of the reference cores using TIGRIS code and MCU-PTR code with different CR positions (case #1, #2, #3) described in Table 1. The last column of Table 2 shows the total worth of shim rods and regulating rod.

Table 2. Calculated criticality (#1), excess reactivity (#2) and shutdown margin (#3) for the reference cores.

Core	Code	ρ (%Δk/k)			ρ(#2) - ρ(#3) (%Δk/k)
		#1	#2	#3	
HEU operational	MCU-PTR[a]	0.32	5.52	-12.18	17.7
	TIGRIS	-0.2	4.4	-11.8	16.2
LEU fresh	MCU-PTR	0.18	10.11	-5.68	15.8
	TIGRIS	-0.8	9.0	-6.2	15.1
LEU operational	MCU-PTR	0.62	5.13	-10.48	15.6
	TIGRIS	-0.1	4.1	-10.4	14.6

[a] Standard deviation <0.0002.

For the cases #1 and #2 the difference between the reactivities calculated using TIGRIS code and MCU-PTR code is from -1 to -0.5%Δk/k. The total CR worth calculated using TIGRIS code is 4-9% less than that calculated using MCU-PTR code.

4.3. IRT MEPhI operational data

The calculations of the states with measured critical CR positions for Xe-free cores at the Beginning of Cycle (BOC) and at the End of Cycle (EOC) were performed. Table 3 presents critical CR position and calculated reactivity. The calculations using TIGRIS code and MCU-PTR code are presented.

Table 3. Results of criticality calculation for Xe-free cores at BOC and EOC.

Cycle		CR position (mm)				ρ (%Δk/k)		Δ (%Δk/k)
		AR	KC-1	KC-2	KC-3	TIGRIS	MCU-PTR[a]	
#113	EOC	250	0	0	373	-1.44	-0.91	-0.53
#114	BOC	200	0	219	592	-0.35	0.04	-0.39
	EOC	250	0	81	592	-0.65		
#115	BOC	270	0	0	590	-0.61		
	EOC	250	0	0	515	-0.88		
#116	EOC	250	0	0	385	-1.01	-0.45	-0.55
#117	EOC	250	0	0	360	-1.35		
#118	BOC	250	0	155	592	-0.34	0.02	-0.37
	EOC	231	0	93	592	-0.50		
#119	BOC	248	0	0	489	-0.13	0.35	-0.49
#120	BOC	250	0	25	592	-0.14	0.54	-0.67
	EOC	250	0	0	418	0.05	0.75	-0.69
#121	BOC	250	0	0	393	-0.18	0.42	-0.60

[a] Standard deviation is ±0.00014.

The maximum negative discrepancy with the experiment for both codes is observed for the EOC #113 and #116. It could be explained by the fact that the cores at the end of the cycles #113 and #116 are the cores with maximum burnup: EOC #113 – core-averaged burnup is 33%, 4 FA have deep burnup (50.3%, 51.4%, 54.8%, 56.4%), EOC #116 – core-averaged burnup is 31.8%, 4 FA have deep burnup (48%, 52%, 54.9%, 55.3%). The other cycles have core-averaged burnup of 26-29%. For the cores with large burnup the error of the power measurement has the maximum impact on the calculated results. The maximum positive discrepancy is observed for the cycles #119-121. It could be explained by the fact that starting from the cycle #118 the duration of shutdown periods increased and the error in beryllium poisoning calculation has the maximum impact on the calculated results.

It can be seen from Table 3 that the difference between TIGRIS and MCU-PTR results is from -0.4 to -0.7%Δk/k. This difference is approximately the same as for the reference cores and is less than for the test problems due to the error cancellation.

The detailed simulation of CR calibrations was performed [6,7]. For the IRT MEPhI, the discrepancy between the measured and calculated integral reactivity worth was found to be less than 10% for the shim rods and less than 15% for the regulating rod [7].

The brief summary of the results of MCU-PTR code and TIGRIS code validation based on the comparison with IRT MEPhI operational data is presented in Table 4.

Table 4. Comparison of the calculated and measured results for the IRT MEPhI reactor.

Parameter	Deviation from the experiment	
	MCU-PTR	TIGRIS
Criticality (%Δk/k)	-0.9…0.8	-1.4…0.1
Excess reactivity (%Δk/k)	0.1…0.9	-0.3…0.6
Reactivity worth of shim and scram rods (%)	<10	<10
Reactivity worth of regulating rod (%)	<15	<15

5. Conclusion

The continuous energy Monte Carlo code MCU-PTR and the nodal diffusion-based deterministic code TIGRIS were validated for the IRT MEPhI reactor calculation. The validation was based on the calculation of the test problems for model core configuration with HEU and LEU fuel, reference cores with HEU and LEU fuel, and IRT MEPhI cores with HEU fuel. Detailed comparison with the measured data and code-to-code comparisons were performed. The range of the discrepancy between the calculated and measured results was determined. A satisfactory agreement among the considered codes and among the calculated and measured results is observed. The calculated reactivities for the measured critical CR positions are between the ±1%Δk/k range and ±1.5%Δk/k range for MCU-PTR code and TIGRIS code, respectively.

References

[1] Kalugin M A, Oleynik D S and Shkarovsky D A 2015 Overview of the MCU Monte Carlo software package *Annals of Nuclear Energy* **82** 54–62

[2] Shchurovskaya M V and Alferov V P 2006 Calculation and experiment to determine the operational characteristics of a research reactor *Atomic Energy* **101** 706–13

[3] Alferov V P, Kryuchkov E F, Portnov A A and Shchurovskaya M V 2012 Analysis of the technical feasibility of converting the IRT MIFI reactor to low-enrichment fuel *Atomic Energy* **112** 368–74

[4] Radaev A I and Schurovskaya M V 2015 Substantiation of parameters of the geometric model of the research reactor core for the calculation using the Monte Carlo method *Physics of Atomic Nuclei* **78** 1227–37

[5] Alferov V P *et al.* 2015 Comparative validation of Monte Carlo codes for the conversion of a research reactor *Annals of Nuclear Energy* **77** 273–80

[6] Naymushin A, Chertkov Yu, Shchurovskaya M, Anikin M and Lebedev I 2016 Modeling of operating history of the research nuclear reactor *IOP Conf. Series: Materials Science and Engineering* **135** 012032

[7] Shchurovskaya M V *et al.* 2016 Control rod calibration simulation using Monte Carlo code for the IRT-type research reactor *Annals of Nuclear Energy* **96** 332–43

Acknowledgments
The work was supported by the Argonne National Laboratory under contract No. 0J-30402. The work was also partly supported by the Ministry of Education and Science of the Russian Federation under Project 3092 in the framework of the basic part of the state task for scientific activities for educational organizations.

Development of coupled neutronics/thermal–hydraulics test case for HPLWR

P Pham, I D Gamtsemlidze, R B Bahdanovich, S P Nikonov and A D Smirnov

National Nuclear Research University MEPhI (Moscow Engineering Physics Institute), Department of Theoretical and Experimental Physics of Nuclear Reactors, 115409 Kashirskoe shosse 31, Moscow, Russia

E-mail: phuphamphys@gmail.com

Abstract. The High-Performance Light Water Reactor (HPLWR) is the European concept of a supercritical water reactor (SCWR) which is one of the most promising and innovative designs of the Generation IV nuclear reactor concepts. The thermal–hydraulics behavior of supercritical water is significantly different from water at sub-critical pressure because of the difference in the specific heat value. Coupled analysis of HPLWR assembly neutronics and thermal–hydraulics has become important because of the strong influence of the water density on the neutron spectrum and power distribution. Programs MCU (Monte-Carlo Universal) and ATHLET (Analysis of Thermal-hydraulics of Leaks and Transients) were used for better estimation of power and temperature distribution in HPLWR assembly.

1. Introduction

Supercritical water-cooled reactor (SCWR) is the nuclear reactor using supercritical pressure water as coolant. The SCWR is considered as one of the most promising Generation IV reactors because of its simplicity, high thermal efficiency, and nearly fifty years of industrial experience from thermal-power stations with a supercritical water cycle [1]. Based on existing designs, they can be divided into two main categories of the conceptual designs of SCWRs: (1) pressure vessel concepts being studied mostly in Japan and European Union are analogous to conventional LWRs, and (2) pressure tube or channels containing fuel bundles concepts being focusing in Canada and Russia are analogous to conventional CANDU and RBMK nuclear reactors.

High Performance Light Water Reactor (HPLWR) is the European conceptual design of the SCWR. Distinction between the HPLWR and the LWRs is that the HPLWRs use supercritical pressure water (25 MPa) with very high water temperature increase in outlet (500°C) from inlet (280°C), so that some components of LWRs are not required anymore, for example, steam separators and dryers in BWR; steam generators and pressurizer in the PWR.

In the papers [2], [3], [4] several designs for the SCWR assemblies based on square and hexagonal geometry were proposed. Several partners of the HPLWR project [12] studied the hexagonal assembly design [2] and concluded that the hexagonal assembly design has the major weaknesses: the under-moderation, which causes reactivity swing; the high enrichment; and three different enrichment fuel pins within the subassembly making a complicated mechanical structure.

Design of a square assembly containing a central moderator box and two rows of fuel rods being proposed in [4] was proved that this design compared with an hexagonal assembly with one and two rows of fuel rods is better for minimizing the structural material, for optimizing the moderator to fuel and for achieving a more uniform radial power distribution in a cross section of the assembly [12]. The design concept will be described in more detail further.

Neutronics/thermal–hydraulics coupling schemes having been developed sparingly for HPLWR were introduced in [13] and [14]. For the same input parameters calculations give similar results to these schemes. It is necessary to extend development of neutronics/thermal–hydraulics coupling schemes to predict basic reactor physics design parameters for HPLWRs. In this work a coupling scheme MCU/ATHLET for prediction of the power and temperature distribution of HPLWR fuel assembly was developed. The square assembly as described in [4] is considered. The principal aim of this study is to consider the temperature distribution, the pin power distribution, the maximum power peaking factor of HPLWR fuel assembly obtained by MCU/ATHLET.

2. Fuel assembly design

HPLWR assembly selected in the current analysis is the square fuel assembly with two rows of fuel rods and a central moderator box as described in [4]. The assembly consists of 40 fuel rods of 8mm outer diameter and a single central water box replacing 9 fuel rods (table 3, Figure 2). Because HPLWR reactor core consist of three regions evaporator, first superheater, second superheater with 52 clusters (the cluster is comprising nine fuel assemblies in a 3×3 arrangement) for each region as shown in fig 1. So that we will consider assembly model in three regions with 3 different temperature ranges (Figure 3; tables 1, 2). Due to the strong variation of coolant density along active core height, assembly is divided in 10 axial layers identified as locations 1, 2, 3, 4, 5, 6, 7, 8, 9, 10 respectively (table 3; Figure 4).

Table 1. Temperature-density distribution of coolant in three regions: evaporator, first superheater, second superheater.

Axial layer	Lower Position, cm	Evaporator		First superheater		Second superheater	
		T, °K	ρ, (g/cm^3)	T, °K	ρ, (g/cm^3)	T, °K	ρ, g/cm^3)
1	0	583.0	0.724092	706.0	0.12383	706.0	0.12383
2	42	591.9	0.705878	701.2	0.12479	713.4	0.11779
3	84	600.8	0.686021	696.4	0.13121	720.9	0.11173
4	126	609.7	0.66406	691.7	0.13652	728.3	0.11083
5	168	618.6	0.639275	686.9	0.14076	735.8	0.10233
6	210	627.4	0.610422	682.1	0.14666	743.2	0.09787
7	252	636.3	0.574954	677.3	0.15652	750.7	0.09340
8	294	645.2	0.526317	672.6	0.17089	758.1	0.08893
9	336	654.1	0.428571	667.8	0.19126	765.6	0.08447
10	378	663.0	0.235714	663.0	0.23571	773.0	0.08000

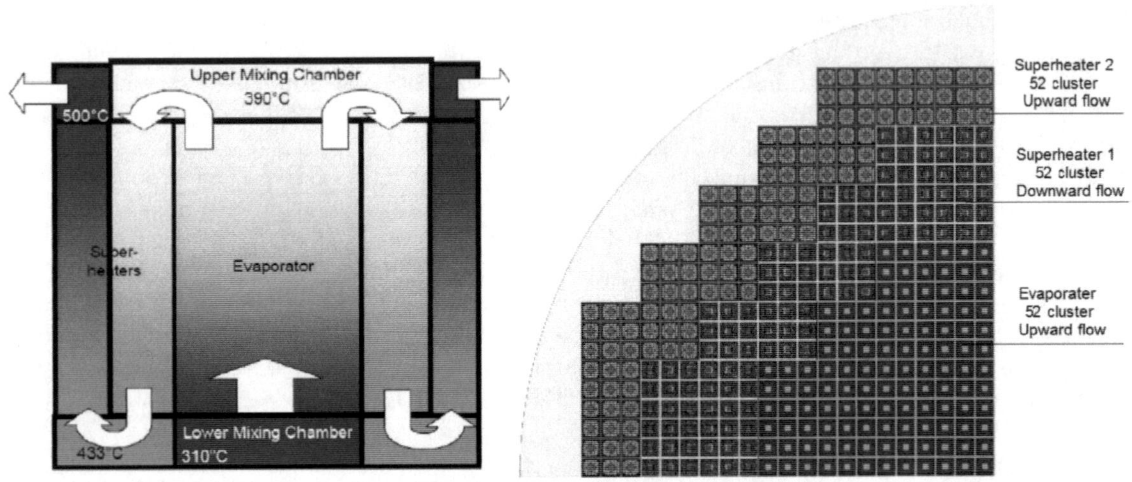

Figure 1. Arrangement of evaporator, first superheater, second superheater and assembly clusters in the HPLWR three-pass core [5].

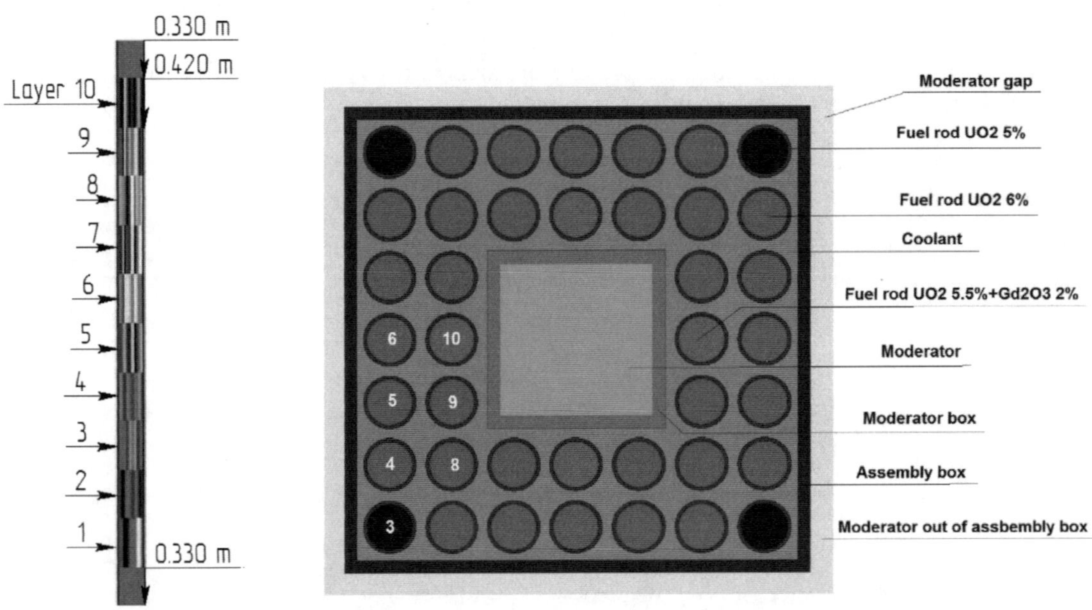

Figure 2. Cross sections of HPLWR assembly.

Thermal hydraulics characteristics are described in table 3, in first iteration of MCU calculation the fuel temperature was constant given 1227 ^{0}K and moderator temperature out of assembly – the outlet temperature of the moderator box about 633 ^{0}K. Because of symmetry of assembly, so that we only calculate one eighth of assembly (rods 3, 4, 5, 6, 8, 9, 10).

Journal of Physics: Conference Series **781** (2016) 012026 doi:10.1088/1742-6596/781/1/012026

Table 2. Temperature-density distribution of moderator and fuel claddings in three regions: evaporator, first superheater, second superheater.

Axial layer	Moderator		Evaporator	First superheater	Second superheater
			Fuel cladding temperature		
	T, °K	ρ, (g/cm³)	T, °K	T, °K	T, °K
1	633.0	0.58929	593.0	716.0	716.0
2	626.9	0.61238	601.9	711.2	723.4
3	620.8	0.63251	610.8	706.4	730.9
4	614.7	0.65053	619.7	701.7	738.3
5	608.6	0.66694	628.6	696.9	745.8
6	602.4	0.68208	637.4	692.1	753.2
7	596.3	0.69618	646.3	687.3	760.7
8	590.2	0.70941	655.2	682.6	768.1
9	584.1	0.72189	664.1	677.8	775.6
10	578.0	0.73374	673.0	673.0	783.0

Table 3. HPLWR fuel Assembly design data [6].

Assembly design data	Value	Units
Fuel active length	420	cm
Inlet temperature	280	°C
Outlet temperature	500	°C
Fuel pin diameter	0.69	cm
Outer cladding diameter	0.8	cm
Cladding thickness	0.05	cm
Pitch of fuel rods	0.92	cm
Water box side length	2.69	cm
Water box side length thickness	0.2	cm
Outer side length box of assembly	7.05	cm
Length box of assembly	0.3	cm
½ gap around the assembly	0.5	cm
Fuel	UO_2	(-)
Enrichment	5, 6, 5.5	%
Density	10.15	g/cm³
First fuel temperature	1227	°K
Cladding and structural material	SS347	(-)
Density	8.027	cm
Burnable absorber material	Gd2O3	(-)
Enrichment	2	%
Fluid pressure	25	MPa
Flow direction in water boxes	downwards	(-)
Coolant direction in Evaporator	upwards	(-)
Coolant direction in first superheater	downwards	(-)
Coolant direction in second superheater	upwards	(-)
Average core power density	57.3	MW/m3
Average coolant mass flux	1600	kg/m2s

Journal of Physics: Conference Series **781** (2016) 012026 doi:10.1088/1742-6596/781/1/012026

3. Brief description of MCU and ATHLET codes

The codes used for calculating power and temperature distribution of HPLWR in this work were MCU and ATHLET.

MCU (Monte Carlo Universal) is neutronics code based on numerical Monte Carlo method which simulates the transport of neutrons, gamma rays and electrons with different energies. The code takes into account the effects of the continuous changes in the particle energy in collisions, as well as a continuous or step-energy dependence of the cross sections. The MCU accuracy is determined primarily used evaluated nuclear data libraries. MCU system allows to calculate the real geometry and material composition of the complex at various temperatures. It contains module sections for recalculating cross sections in the thermalization and resolved resonance regions, depending on the temperature of the material (using the Breit-Wigner formalism or Adler-Adler). It can be used for multi-processor calculations. [7]

Thermal-hydraulic system code ATHLET (Analysis of Thermal-Hydraulics of Leaks and Transients) developed Gesellschaft für Anlagen- und Reaktorsicherheit (GRS mbH) and was originally intended for the analysis of the entire spectrum of leaks and transients in PWRs and BWRs. However, experience has shown that ATHLET is applicable for western reactor designs as well as for Russian VVER and RBMK reactors. ATHLET consists of several modules that allow to describe the various phenomena in the operation of a light water reactor: thermo-fluid dynamics (TFD); heat transfer and heat conduction (HECU); neutron-kinetics (NEUKIN) for describing the point and the one-dimensional kinetics; module for operation of the equipment (GCSM) and a module for the numerical integration of FEBE for implementation the fully implicit scheme. Other independent modules (three-dimensional neutron kinetics, describe the behavior of containment, etc.) can be connected through the main interface.

The TFD – the fundament module that is based on the use of substantially five equations (equations of conservation of mass and energy for the liquid and vapor phase separately and the general equation for the angular momentum of mixture with drift flow) or six equations (equations of conservation of mass, energy, and pulse for the liquid and vapor in the description of the two-fluid model) with a wide range of closing relations. In addition, the module allows to simulate the behavior of non-condensable gases, nitrogen dissolution and describe the boron transport system. Reactor coolant system is modeled by a compound of basic thermodynamic objects (TFO), where the detail of modeling of the studied installation is limited, in principle, only the possibility of computing - the amount of memory and speed. In addition, TFD module is capable for specific descriptions of transverse coolant flow in system of related parallel channels. [8]

A flowchart describing the coupling process between neutron and thermal hydraulic codes MCU/ATHLET is shown on Figure 3. Nomenclature: TFi - fuel temperatures [^{0}K], TCi - cladding temperatures [^{0}K], TMi - moderator temperatures [^{0}K], TMbi - moderator's box temperatures [^{0}K], TCoi - coolant temperatures [^{0}K], TABi - assembly's box temperatures [^{0}K], TOMi - moderator temperatures out of assembly [^{0}K], ρMi - moderator density in moderator's box [g/cm^3], ρOMi - moderator density out of assembly box [g/cm^3], ρCoi - coolant densities [g/cm^3], Qi - local power [KWt], i - layer subscripts.

Coolant flow scheme in ATHLET for HPLWR thermal hydraulics analysis is shown on Figure 4. The coupling procedure carries on until a reasonably converged power profile is achieved. Difference in local powers less than 0.5% in two successive iterations is applied to dampen out the fluctuations and get a converged profile more quickly.

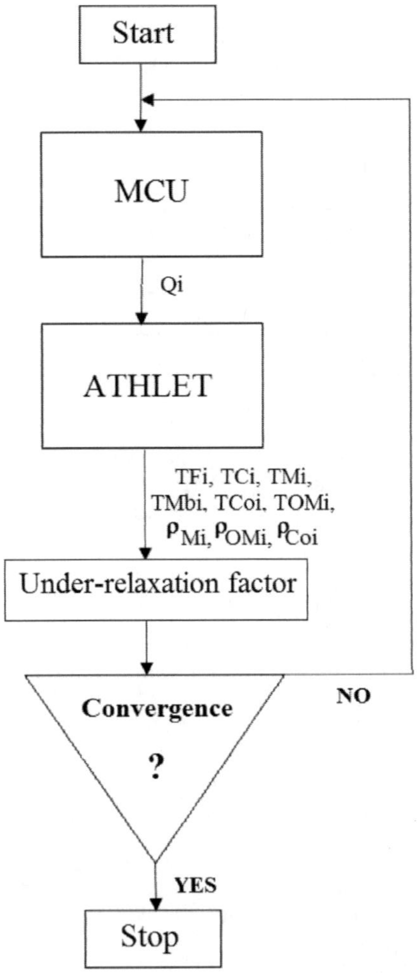

Figure 3. MCU/ATHLET flowchart.

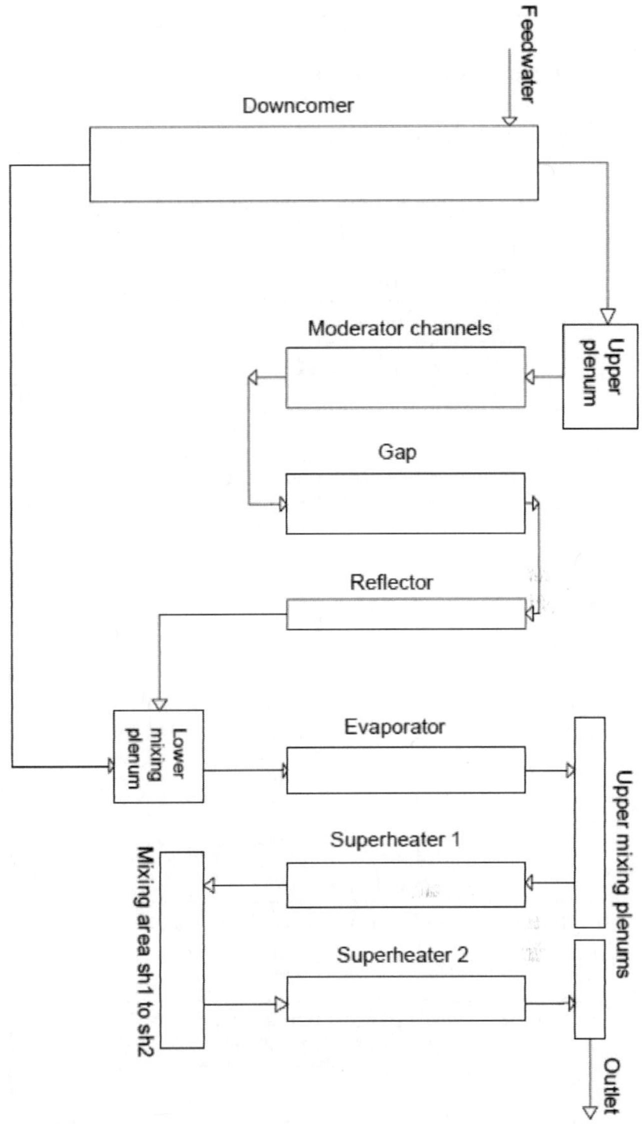

Figure 4. Coolant flow scheme.

4. Preliminary results

The results of neutron calculations in the first iteration are shown in Table. 4, Figures 5, 6, 7, and 8. The value of infinite multiplication factor of assembly in evaporator is greater than that in first superheater and second superheater as shown in table 4. This is explained by smaller coolant temperature and density in evaporator as compared to that in first superheater and second superheater. This also indicates that the coolant temperature coefficient of reactivity is negative.

Table 4. The infinite neutron multiplication factor of assembly in three region.

Assembly in:	K infinite	Std.Dev.
Evaporator	1.115	0.000498
First superheater	1.082	0.000504
Second superheater	1.078	0.000463

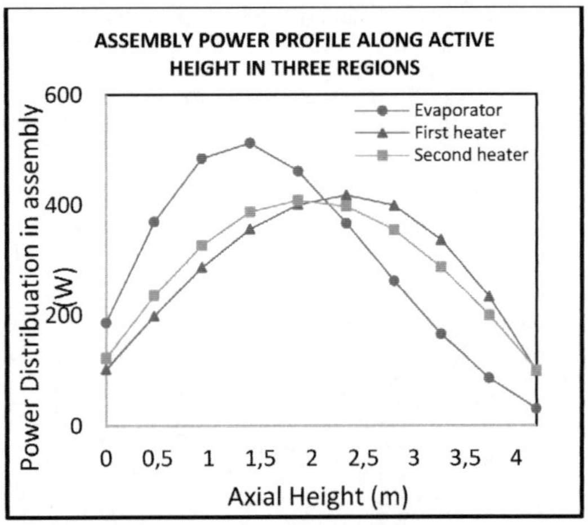

Figure 5. Assembly liner power along active height in three regions.

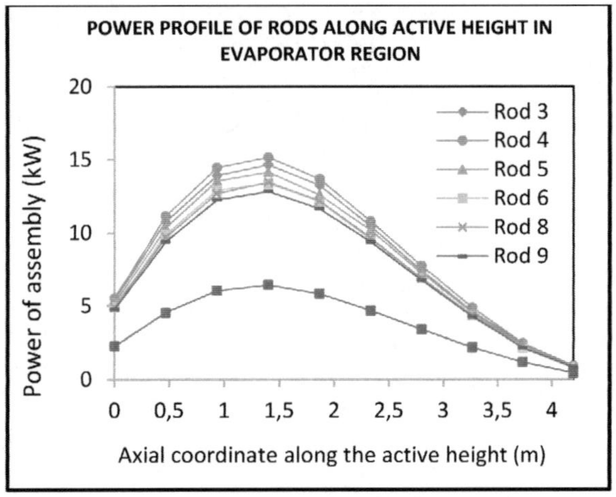

Figure 6. Liner power of rods along active height in evaporator region.

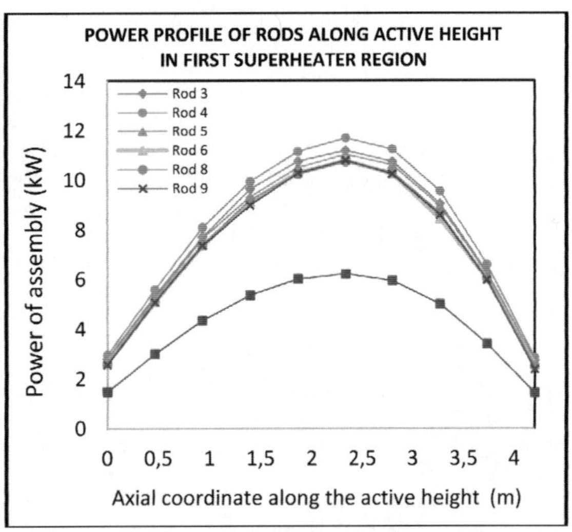

Figure 7. Liner power of rods along active height in first superheater region.

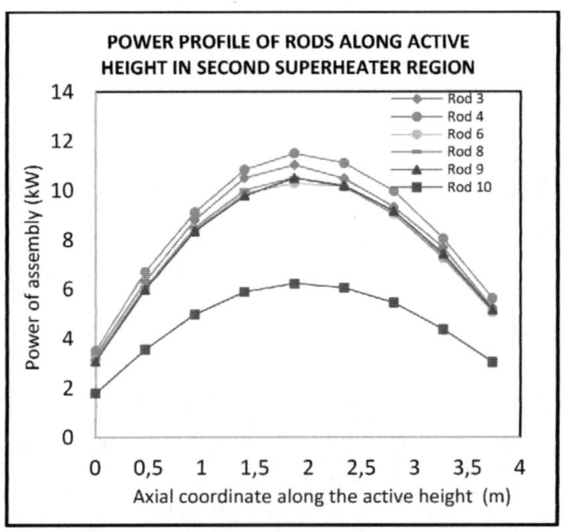

Figure 8. Liner power of rods along active height in second superheater region.

5. Conclusions

The coupled scheme MCU/ATHLET has been developed for analysis of HPLWR reactor behavior. Code MCU were used to calculate the power distribution in HPLWR assembly which is put into ATHLET input file to obtain the corresponding thermal–hydraulic parameters. The obtained thermal–hydraulic parameters are used to generate a new MCU input file for the next calculation. This procedure is repeated until the convergence is achieved.

Three types of HPLWR assembly were calculated with MCU code. The results indicate that in fuel rods with gadolinium (Figure 7-9, rod 10) axial power distribution is closer to uniform and has smaller average power value than in other types of fuel rods.

References

[1] Buongioron, J., 2003. Generation IV R&D Activities for the Development of the SCWR in the U.S, Progress Report for the FY-03. Idaho National Engineering and Environmental Laboratory

[2] Dobashi, K., Oka, Y., Koshizuka, S., 1998. Conceptual design of a high temperature power reactor cooled and moderated by supercritical light water. Annals of Nuclear Energy 25, 487–505.

[3] Yamaji, A., Oka, Y., Koshizuka, A., 2005. Three-dimensional core design of high temperature supercritical-pressure. Nuclear Science and Technology 42, 8–19

[4] Hofmeister, J., Waata, C., Starflinger, J., Schulenberg, T., Laurien, E., 2007. Fuel assembly design study for a reactor with supercritical water. Nuclear Engineering and Design 237, 1513–1521.

[5] S. Tashakora,*, A.A. Salehia,b, G. Jahanfarniaa, A. Abbaspour Tehrani Fard, Thermal–hydraulic analysis of HPLWR fuel assembly cluster, J. of Supercritical Fluids 77 (2013) 91–99

[6] Thomas Schulenberg // Jörg Starflinger (eds.), HIGH PERFORMANCE LIGHT WATER REACTOR Design and Analyses, KIT Scientific Publishing 2012, ISBN 978-3-86644-817-9

[7] Gurevich MI, Shkarovsky DA Monte Carlo neutron transport calculation Meto-house on the MCU program: Textbook. M .: MEPhI. - 154.

[8] G. Lerchl, H. Austregesilo, ATHLET Mod2.2 Cycle B, User's Manual, GRS 2011.

[9] Alejandra-Maribel Barragán-Martínez, Cecilia Martin-del-Campo, Juan-Luis François, Gilberto Espinosa-Paredes, MCNPX and HELIOS-2 comparison for the neutronics calculations of a Supercritical Water Reactor HPLWR, Annals of Nuclear Energy 51 (2013) 181–188.

[10] Thomas Schulenberg, Laurence K.H. Leung, Yoshiaki Oka, Review of R&D for supercritical water cooled reactors, Progress in Nuclear Energy xxx (2014) 1-18.

[11] Igor Pioro, Sarah Mokry, Thermophysical Properties at Critical and Supercritical Conditions, University of Ontario Institute of Technology.

[12] Squarer, D., Schlenberg, T., Struwe, D., Oka, Y., Bitterman, D., Aksan, N., Maraczy, C., Kyrki-Rajamaki, R., Souyri, A., Dumaz, P., 2003. High performance light water reactor. Nuclear Engineering and Design 221, 167–180.

[13] Waata, C.L., 2005. Coupled Neutronics/Thermal–Hydraulics Analysis of a High Performance Light-Water Reactor Fuel Assembly. Report, Forschungszentrums Karlsruhe, Germany.

[14] Reiss, S., Feher, S., Czifrus, S., 2008. Coupled neutronics and thermohydraulics calculations with burn-up for HPLWRS. Progress in Nuclear Energy 50, 1–10.

[15] Fischer C, et al. (2009) Kinetic mechanism for single-stranded DNA binding and translocation by Saccharomyces cerevisiae Isw2. Biochemistry 48(13):2960-8

Acknowledgments

This work was supported by Competitiveness Program of National Research Nuclear University MEPhI.

The relationship between hydraulic resistance and heat transfer in the crossflow-line of staggered tube bundles

V N Fedoseev[1], A S Korsun[1], M I Pisarevsky[2] and Y N Balberkina[3]

[1]Docent, National Research Nuclear University "Moscow Engeneering Physics Institute, Moscow, Russia
[2] Engeneer, National Research Nuclear University "Moscow Engeneering Physics Institute, Moscow, Russia
[3]Student, National Research Nuclear University "Moscow Engeneering Physics Institute, Moscow, Russia

E-mail: MIPisarevskij@mephi.ru

Abstract. The paper develops the idea that the semi-empirical formula for calculating heat transfer in turbulent flow in pipes, expressed in terms of ratio of the dynamic speed to the average flow velocity in the pipe, are universal and can be used for the calculation of heat transfer in channels with complex geometry, where separated flows take place. A relationship between flow resistance and heat transfer at cross turbulent flow in line and staggered bundles of rods was found. The resulting formulas generalize experimental data available in the literature better than the empirical formulas.

1. Introduction

One of the urgent tasks of heat transfer theory and hydrodynamics is to establish correlations between hydraulic resistance and heat transfer for specific cooling surface. Within the framework of the semiempirical Prandtl theory of turbulence this problem is almost solved for smooth direct channels, and in the case of complex shapes with loose-channel eddy currents (crossflow tube bundles, ball filling, the porous material, etc.), where the friction loss of pressure constitute a small part of the total pressure loss for pumping coolant, this question requires serious research. Different approaches to solve this problem are available in literature publications [1-5]. The universal relationship between hydraulic resistance and heat and mass transfer in porous media proposed and substantiated in [1-4]. This universal relationship summarizes a large body of experimental data obtained by different authors for many structures with a deviation of no more than 30%. So, the heat transfer and hydrodynamics in the pebble bed and granular layers crossflow tube bundles, grid inserts, perforated plates packets are bind by the same formula. This suggests that advanced in [1-4] hypothesis of the universal nature of the heat transfer laws, derived from semi-empirical theory, is correct. However, the universal formula is not sufficiently accurate for reliable thermal-hydraulic calculations. That's why, further studies are required to clarify the heat transfer model for each type of porous structures. In particular, proposed in [1-4] universal formula is suitable for cross flow staggered bundles, but it does not describe satisfactorily correlation between hydraulic resistance and heat transfer at crossflow line bundles. This work is aimed to further study of the interrelations of hydraulic resistance and heat transfer in tube bundles in crossflow within the framework of the proposed in [1-4] approach.

2. The interrelation between flow resistance and heat transfer in turbulent flow in smooth straight pipes

In [6-9] showed that in turbulent flow in pipes in the area of stabilized heat transfer for constant heat load heat transfer is well described obtained by theoretical calculation formula

$$Nu = \frac{(\zeta/8)\,\mathrm{Re}\,\mathrm{Pr}}{k_1 + k_2\sqrt{\zeta/8}\,(\mathrm{Pr}^{2/3} - 1)} \tag{1}$$

where $Nu = \alpha d/\lambda$, d - tube diameter; α – heat transfer coefficient; λ – thermal conductivity coefficient of the coolant; $\zeta = (1.8\lg\mathrm{Re} - 1.64)^{-2}$ - hydraulic resistance coefficient of the pipe, $k_1 = 1.07$, $k_2 = 12.7$. Equation (1) is valid for Re numbers from 10^4 to $5 * 10^5$ and Pr numbers from 0.5 to 2000. An important characteristic of the coolant velocity distribution near the pipe wall is a dynamic rate v_*, which is related to the average velocity and drag coefficient $v_*/v = \sqrt{\zeta/8}$. Let's mark the ratio of the dynamic speed to the average speed through $\varphi = v_*/v$, and assume that k_1=1 and k=k_2. Then the formula (1) can be reduced to the form:

$$St = \frac{\varphi^2}{1 + k\varphi(\mathrm{Pr}^{2/3} - 1)} \tag{2}$$

where $St = Nu/(\mathrm{Re}\,\mathrm{Pr})$ - the Stanton number. In [1-4],is assumed that the equation (2) is universal and applicable to the calculation of the average heat transfer in complex configuration channels with separated flows. Thus the main objective is to look at the relationship between the dynamic speed and the average speed of the flow in the channel, i.e. expression relating φ to the characteristics of averaged flow. In [1-4] proposed to determine the speed of a dynamic power through the coolant pumping costs. Dynamic speed associated with power costs from dimensional considerations for pumping expression

$$v_* = c\left(\frac{v\upsilon\nabla P}{\rho}\right)^{0.25} \tag{3}$$

where υ - kinematic viscosity coefficient, c – proportionality factor, which depends in general on the geometry of the channel and the number of Re. For any channel we have the relation:

$$\varphi = \frac{v_*}{v} = c\left(\frac{\zeta}{2\,\mathrm{Re}\,\Pi}\right)^{0.25} \tag{4}$$

In [1-4] on the basis of experimental data analysis it was found that for many complex heat exchange surfaces (granular layers, crossflow staggered bundles, of mesh inserts, packets perforated plates) factor "c" can be assumed to be constant and equal to about 3. The expression (2) becomes a universal for all considered in [1-4] surfaces, and the whole array of studied experimental data is summarized with 30% accuracy.

Set out above model describes a limiting case of separated flows in the winding channels, when the coolant is constantly changing its direction of motion and intensively mixed in the communicating channel. For example, in crossflow line bundles, where portions of rectilinear motion of the coolant can be identified in the annulus, the assumption that c = 3 is false. Calculation of heat transfer by the formula (2) with reasonable accuracy requires knowledge of «c» channel coefficient dependence on the geometry and coolant flow rate for each type of heat-exchange surfaces.

3. Line and staggered tube bundles.

In [10-12] presented experimental data on heat transfer and hydraulic resistance in the tube bundles in crossflow and empirical formulas are offered for engineering calculations. Note that all investigated range of Reynolds numbers for line bundles and staggered bundles is divided into three sections. And each site has its own formula for calculations. Wherein calculations on recommended formulas are in good agreement with the experimental data only for the bundles with a large value of porosity ($\Pi \geq 0.3$). And for the dense packing of tube bundles $\Pi \prec 0.3$ the results of calculations and the experimental data disagree strongly (up to 70 %).

For heat transfer calculations by means of equation (2) it is necessary to know the parameter φ, which in turn depends on the constant "c". Processing of the experimental data from [11] gives the following dependences:

for line bundles: $k = 11.13$

$$c = \begin{cases} 0.76 + 1.93\Pi, \Pi \geq 0.30 \\ (0.97\Pi - 0.03)\mathrm{Re}^{0.157}, \Pi \prec 0.30 \end{cases} \quad (5)$$

for staggered bundles: $k = 8.3$

$$c = \begin{cases} 1.6 + 0.32\Pi, \Pi \geq 0.30 \\ 3.5 + 2.17\,\mathrm{Re}\cdot 10^{-6}, \Pi \prec 0.30 \end{cases} \quad (6)$$

From (5) and (6) it follows that for crossflow in line and staggered bundles with greater porosity $\Pi \geq 0.3$ coefficient "c" is a function of the geometric parameters of the bundle and does not depend on the Reynolds number. In dense bundles of tubes with $\Pi = 0.14 - 0.3$ dependence "c" of the Reynolds number appears. Calculation formula using a set of dependencies for the "c" in good agreement with the experimental data [11] (Figure 1-2).

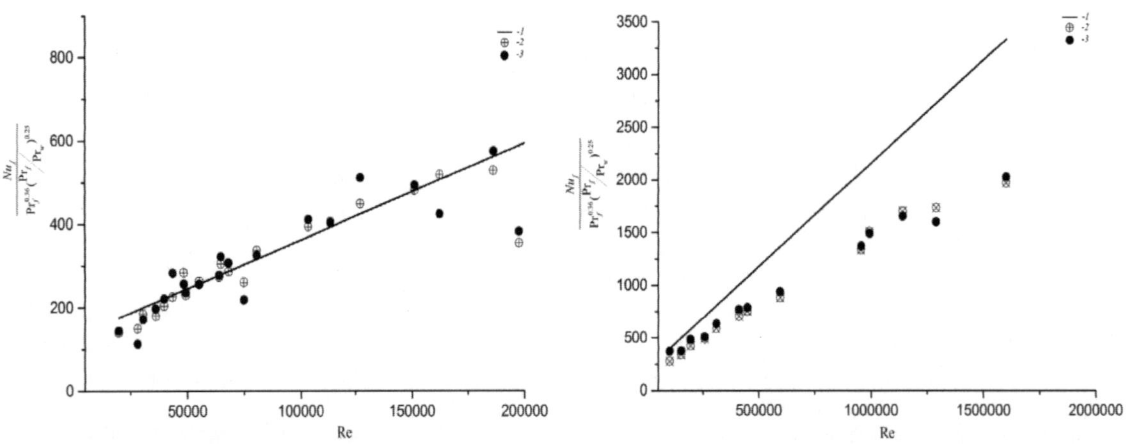

a) b)

Figure 1. The dependence of heat transfer from the complex number Re for a corridor beam with high porosity (a) and low porosity (b): 1- calculation guidelines [4], 2- formula calculation (1), 3- experimental values.

IOP Conf. Series: Journal of Physics: Conf. Series **781** (2017) 012027 doi:10.1088/1742-6596/781/1/012027

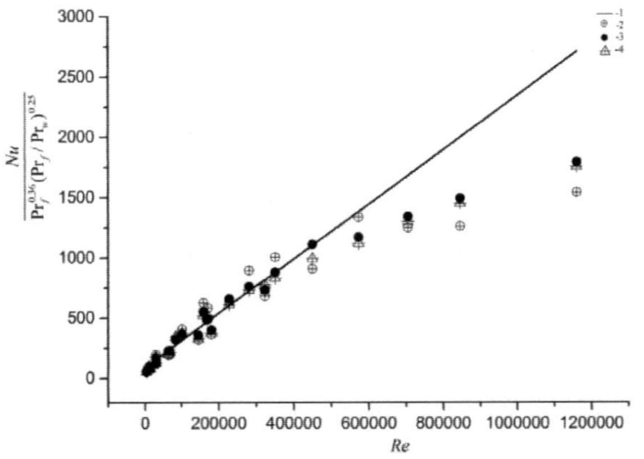

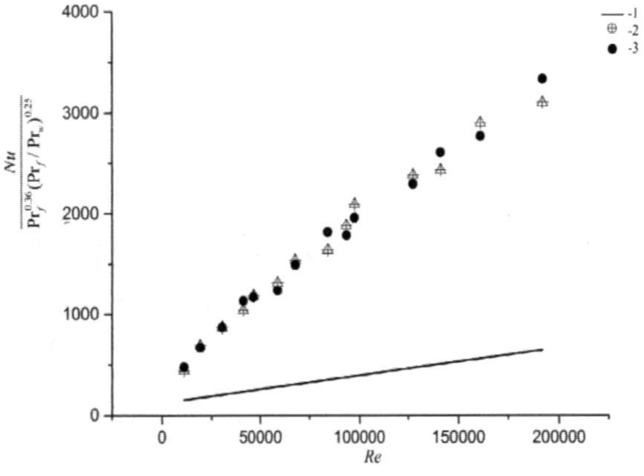

Figure 2. The dependence of heat transfer from the complex number Re for staggered with high porosity (a) and low porosity (b):1- calculation guidelines [4], 2- formula calculation (1), 3- experimental values.

4. Conclusion

The results obtained demonstrate the possibility of applying the given approach to the calculations for average heat transfer in channels of different shapes according to the flow resistance of these channels. It also demonstrates the universality of the laws of heat transfer in a turbulent flow.

References

[1] Kokorev L S *et al* 1986 *MEPHI* New approach to the calculation of heat transfer in porous mediums **20** 1 (In russian)

[2] Kokorev L S *et al* 1987 *High Temperature* On the relationship between flow resistance and heat transfer in porous media **6** 92 (In russian)

[3] Kharitonov V V and Fedoseyev V N 1990 *High Temperature* The generalization of the experimental data on heat transfer in a streamlined cross-beams twisted tubes **3** 180 (In russian)

[4] Fedoseyev V N *et al* 1987 *Thermal Engineering* The universal dependence of heat transfer from the pressure gradient in porous mediums p 61-64 (In russian)

[5] Isachenko VP *et al* 1981 Teploperedacha - M Energoizdat (In russian)

[6] Petukhov B S and Kirillov V V 1958 *Thermal Engineering* To the question of heat transfer in a turbulent flow of fluid in the tubes №4 (In russian)

[7] Petukhov B S and Popov V N 1963 *High Temperature* Theoretical calculation of heat transfer and friction in turbulent flow in pipes of an incompressible fluid with varying physical properties t1 №1 (In russian)

[8] Maxine P L *et al* 1976 *Heat and mass transfer-V Minsk* The calculation of turbulent heat transfer in a stabilized flow in pipes (In russian)

[9] Petukhov B S *et al* 1986. Heat transfer in nuclear power plants. Energoatomizdat p 233 (In russian)

[10] Zukauskas A 1982 *Science* Convective transfer in heat exchangers p 470 (In russian)

[11] Zukauskas A and Makarevicius V 1968 *«MINTIS» Vilnius* Heat transfer in banks of tubes in crossflow of fluid p 192 (In russian)

[12] Martynenko O G 1987 *etc.-M: Energoatomisdat* Handbook of heat exchangers In 2 V 2 (In russian)

Minimization of the energy loss of nuclear power plants in case of partial in-core monitoring system failure

A M Zagrebaev[1], R N Ramazanov[12] and E A Lunegova[1]

[1]National Research Nuclear University MEPhI (Moscow Engineering Physics Institute), Department of Cybernetics, 115409, Russia, Moscow, Kashirskoe highway, 31

[2]E-mail: tem@gmx.com

Abstract. In this paper we consider the optimization problem minimize of the energy loss of nuclear power plants in case of partial in-core monitoring system failure. It is possible to continuation of reactor operation at reduced power or total replacement of the channel neutron measurements, requiring shutdown of the reactor and the stock of detectors. This article examines the reconstruction of the energy release in the core of a nuclear reactor on the basis of the indications of height sensors. The missing measurement information can be reconstructed by mathematical methods, and replacement of the failed sensors can be avoided. It is suggested that a set of 'natural' functions determined by means of statistical estimates obtained from archival data be constructed. The procedure proposed makes it possible to reconstruct the field even with a significant loss of measurement information. Improving the accuracy of the restoration of the neutron flux density in partial loss of measurement information to minimize the stock of necessary components and the associated losses.

1. Introduction

The safe and efficient operation of a nuclear reactor requires the continuous monitoring of the reactor core state. One important task is the control of neutrons flux density. The control is necessary because a critical power ratio can vary greatly with small fluctuations of process parameters. Special sensors of in-core monitoring system (ICMS) [1] located along the core height are used to obtain information about the flux density. Pressurized water reactors VVER have assemblies of seven self-powered neutron detectors (SPNDs) [2], distributed uniformly over the core height. In the practice of operating nuclear reactors there may occur situations when some sensors fail. If a sensor fails, then the use of its indications is 'prohibited', e.g. assembly of SPNDs in VVER reactors is banned if three or more sensors fail [3]. Failure of the ICMS increases error of the neutron flux determination [4]. In such cases it is necessary either to reduce the power of the reactor, or to replace the failed part of the ICMS. The replacement is possible only when the reactor is stopped, moreover, for this purpose at the plant should be a certain reserve of necessary components. There is an opportunity to increase the assembly operation time. The missing measurements information can be restored by mathematical methods and premature replacement of the failed sensors can be avoided, which is also expedient economically. For this, it is proposed that missing sensor indications be calculated by approximation the known points. Harmonic functions are used for the recovery of neutron flux density distribution, but these functions are not optimal. Approximation is suggested to be made as an expansion to the series of functions,

IOP Conf. Series: Journal of Physics: Conf. Series **781** (2017) 012028 doi:10.1088/1742-6596/781/1/012028

adjusted in a certain way using the archive data. The proposed technique enables to use accumulated data from an archive to restore the neutron field even after a serious loss of measurement information.

2. Minimization of the energy loss

Consider the simple situation for VVER, when the moment of failure is random. Until the maximum burnup fuel assembly located in the reactor core for several campaigns. Thus, the campaign is determined by the reactivity margin. It is therefore possible extension of the campaign through the use of reactivity temperature effect [5]. In this case either the reactor power is reduced by a predetermined amount by the operator, or is transferred to block spontaneous power reduction mode. For simplicity we consider the first option. Then it is possible to estimate losses due to shut down the reactor for replacement of the failed part of ICMS and by reducing the power. This makes it possible to calculate the boundary point in time t_b, after which advantageously reduces power:

$$t_b = T - \frac{\tau W + \Delta W (T - T_{low})}{\delta W} \tag{1}$$

T – the estimated time of the reactor campaign; T_{low} – the time of the planned power reduction at the end of the campaign; τ – downtime required for replacement; W – nominal reactor power; ΔW – the value of power reducing to work at the end of the campaign; δW – value of the required power reduction in case of partial failure of the ICMS.

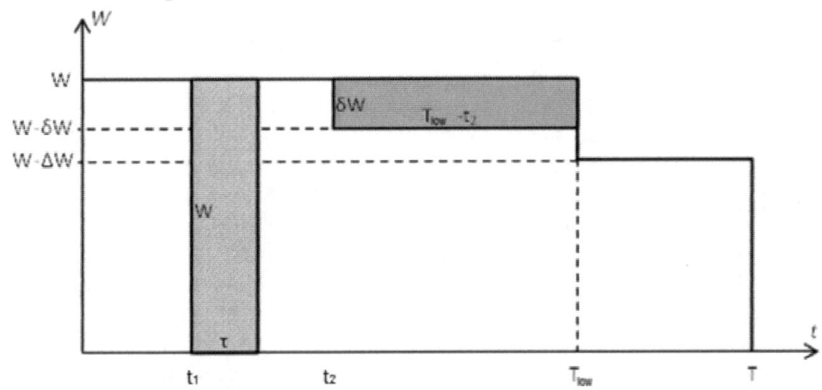

Figure 1. Loss in case of shutdown and in the case of reducing power.

This value can be regarded as a time for the unit should be stored ICMS stock components. Most often a nuclear power plant is equipped with the same type of units. Therefore, at the moment t_b for the first reactor stock can be transferred to another reactor. Then it is possible to plan the start-up of reactors with time t_b offset. This approach is justified, because a separate reserve for each block leads to economic losses. Estimate what stock to have on station of the N reactors of type VVER in order to minimize losses. The minimum number of spare sets will be equal to $[Nt_b/T] + 1$, where $[x]$ is the integer part of x. It is seen that the smaller t_b, the smaller the reserve needed for operation of nuclear power plants. The reduction t_b allows to lower losses. And to reduce t_b it should be possible to reduce δW which depends on the accuracy of the neutron flux density restoration in partial loss of measurement information.

Since the time of the failure is a random variable, we can estimate the average losses for both cases.

$$\overline{Q} = \int_0^T Q(t) f(t) dt \tag{2}$$

$$Q_1 = \begin{cases} \tau W, t < T_{low} \\ \tau (W - \Delta W), t > T_{low} \end{cases} \tag{3}$$

$$Q_2 = \begin{cases} \delta W(T_{low} - t) + (\delta W - \Delta W)(T - T_{low}), t < T_{low} \\ 0 + (\delta W - \Delta W)(T - T_{low}), t > T_{low} \end{cases} \quad (4)$$

$\overline{Q}$ – average losses, Q_1 – losses during sensor replacement, Q_2 – losses as a result of the power reduction, $f(t)$ – probability density function.

Take as $f(t)$ is a uniform density. Then we obtain the following estimates:

$$\overline{Q_1} = \alpha\tau \left[W \frac{T_{low}}{T} + (W - \Delta W)\left(1 - \frac{T_{low}}{T}\right)\right] \quad (5)$$

$$\overline{Q_2} = \alpha \left[\delta W \frac{T_{low}{}^2}{2T} + (\delta W - \Delta W)(T - T_{low})\right] \quad (6)$$

It is seen that $\overline{Q_1}$ does not depend on δW, and $\overline{Q_2}$ depends on δW is linear, then there is a value δW, where the average in case of failure of any component of ICMS, to replace is unprofitable. This fact confirms that in case of failure of part of the ICMS should be possible to accurately restore the field to minimize the value δW. Thus, it is important to improve the recovery algorithms of the flux density of neutrons with partial loss of measurement information.

3. Recovering of missing height-sensor readings

Harmonic functions are used for the recovery of axial neutron flux density distribution, because they are eigenfunctions of one-dimensional nuclear reactor. The maximum number of harmonics is equal to the number of running sensors. However, these functions are not optimal for the reactor. The actual distribution of the neutron field may not be described by a superposition of these functions. In actual operation, for a neutron-physical and thermophysical feedbacks, displacement of the control rods, and others the distribution of the multiplying properties varies in time and space and is not uniform. The data used in the research was derived for the period from June 9 to August 17, 2011 of 4 measuring channels of unit 1 Tian Wan NPP (China) with VVER-1000 reactor. Using these data, we obtain a matrix of the correlation coefficients. Analysis of the matrix obtained on the basis of archives of power-generating units operating in the nominal regime showed that, as a rule, the matrices are not diagonal. This means that there exists a correlation between the fitting coefficients. In other words, the neutron flux density as a random function is represented in the form of a superposition of harmonic functions with correlated random coefficients. This fact confirms the bad choice of functions for approximation. There is the possibility to find the system of functions, which would have non-correlated coefficients in expansion. According to the theory of random functions [6], the convergence of an expansion of a random function in the series can always be improved by an identity transformation to a canonical expansion of the random function. The procedure for transforming to the canonical expansion is reminiscent of the Gram–Schmidt procedure for orthogonalizing vectors or the method of principal components [7]. The coefficients will be expressed in terms of the correlation moments of the random quantities. We obtain an approximate canonical expansion of the random function. This approach claims that new functions called 'natural' functions are linear combinations of initial harmonic functions (see figure 2). Therefore the solution of the approximation problem does not become more complex.

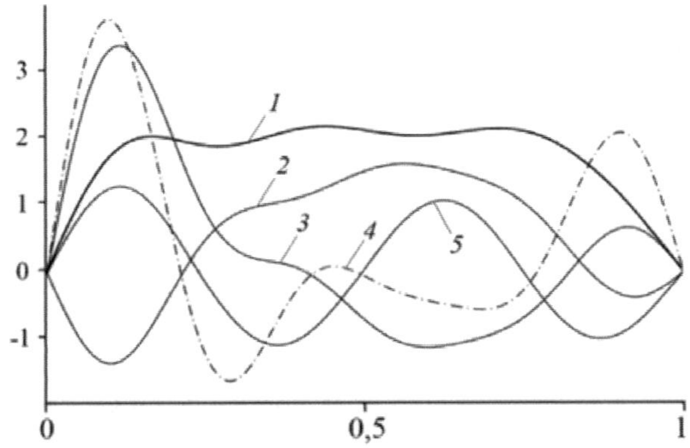

Figure 2. 'Natural' functions of reactor.

The following numerical experiment confirms the efficacy of a 'natural' fit to the axial distribution of the neutron flux. The sensors were prohibited fictitiously and the values of the sensor currents were reconstructed using harmonic functions and the 'natural' functions. The results of the experiment show that the correct choice of functions increases the accuracy of the fit. The advantage of the 'natural' functions is that each one contains several harmonics. For this reason, using fewer of them for the fit gives a smaller error than a large number of harmonic functions.

Obviously, the increase of the number of failed sensors affects the error [8]. However, VVER reactors have assemblies of seven SPNDs. The proposed technique enables to use accumulated data from an archive to restore the neutron field even after a serious loss of measurement information. Since the natural functions are found by processing archival information, their form must depend on the volume and character of the processed archival information. The simulation results [9] show that by using the archive data of 100 readings [10], lost reading may be recovered by special functions with an accuracy of about 2.5%. However, such a high accuracy is due to not only a good choice of functions but also the fact that the energy-release field in VVER type reactors operating in a stationary regime does not change strongly.

4. Conclusions

It is preferable to use 'natural' functions to reconstruct the height distribution of the energy release in the event of partial loss of measurement information. This choice of a basis makes it possible to reconstruct the field by using fewer functions on average with an order-of-magnitude smaller reconstruction error than with the use of harmonic functions. The reason is that information on the possible height distribution of the neutron flux density in the process of reactor operating is used to construct a new basis. Moreover, in essence, an approach to monitoring the energy release where its algorithm adapts during the operation of the reactor was proposed as an example of a solution to this problem.

References

[1] Dias A M, Silva F C 2016 Determination of the power density distribution in a PWR reactor based on neutron flux measurements at fixed reactor in-core detectors *Annals of Nuclear Energy* **94**

[2] Strindehag O 1971 *Self-powered neutron and gamma detectors for in-core measurements* (Nyköping: Aktiebolaget Atomenergi)

[3] Andruchechko S A, Afrov A M, Vasil'ev B Yu et al. 2010 *NPP with VVER-1000 reactor: from the physical principles of operation to the evolution of the design* (Moscow: Logos)

[4] Zagrebayev A M et al. 2011 *Mathematical modelling of a nuclear reactor under random perturbations of process parameters* (Moscow: NRNU MEPhI)

[5] Sinev N M 1987 *Economics of nuclear energy. Base of technology and economics of nuclear fuel production Economics of NPP* (Moscow: Energoatomizdat)

[6] Pugachev V S 1960 *Theory of random functions and its applications to control problems* (Moscow: Fizmatgiz)

[7] Aivazyan S A, Bukhshtaber V M, Enyukov I S, Meshalkin L D 1989 *Applied Statistics: Classification and Dimensional Reduction* (Moscow: Finansy i Statistika)

[8] Ovchinnikov F Ya, Semenov V V 1988 *Exploitation regimes o water-water power reactors* (Moscow: Energoatomizdat)

[9] Zagrebayev A M, Ovsyannikova N V, Ramazanov R N, 2015 Recovering of lost in-core energy field measuring data using archive data *Informatsionno-izmeritelnye i upravlyayushchie sistemy* **13 No. 8**

[10] Zagrebayev A M, Ovsyannikova N V, Ramazanov R N, Mil'to N V 2015 Reconstruction of missing height-sensor readings on the basis of archival neutron field monitoring data *Atomic Energy* **117 No. 3**

IOP Conf. Series: Journal of Physics: Conf. Series **781** (2017) 012029 doi:10.1088/1742-6596/781/1/012029

Generation of XS library for the reflector of VVER reactor core using Monte Carlo code Serpent

K I Usheva[1], S A Kuten[1], A A Khruschinsky[1] and L F Babichev[2]

[1]Institute for Nuclear Problems, Belarusian State University, Bobruiskaya 11, 220030 Minsk, Belarus
[2]Joint Institute for Power and Nuclear Research - Sosny, National Academy of Sciences of Belarus, Minsk, Belarus

Abstract. A physical model of the radial and axial reflector of VVER-1200-like reactor core has been developed. Five types of radial reflector with different material composition exist for the VVER reactor core and 1D and 2D models were developed for all of them. Axial top and bottom reflectors are described by the 1D model. A two-group XS library for diffusion code DYN3D has been generated for all types of reflectors by using Serpent 2 Monte Carlo code. Power distribution in the reactor core calculated in DYN3D is flattened in the core central region to more extent in the 2D model of the radial reflector than in its 1D model.

1. Introduction

Deterministic safety analysis of nuclear power plants is performed by using various reactor dynamics codes, such as DYN3D, BIPR, HEMERA, DIF3D, etc. Nodal reactor code DYN3D [1, 2], based on the solution of the transport equation in diffusion approximation is widely used for analysis of stationary and transient processes in the reactor core. A set of homogenized nuclear data (diffusion constants and group cross sections) is needed for each node to simulate neutron-physical processes in the reactor code. This data, called the XS library, can be prepared by using various spectral codes, such as Serpent [3], HELIOS, WIMS, TVS-M, some versions of the MCU and others. DYN3D deals with two-group diffusion coefficients and cross sections for fuel assemblies and the reflector.

A physical model of the reflector of VVER-1200-like reactor core has been developed for XS-generation library. Five types of radial and two types of axial reflectors with different material composition exist for the VVER reactor core [4]. A part of data available for VVER-1200 [5, 6] is used for calculation, the other data was taken from [7, 8] for VVER-1000.

2. Models of reflectors

The XS library of the reflector in the full core model is defined by the following characteristics, which are calculated in the two-group approximation:

- diffusion coefficient ($D_{1,2}$);
- group-removal cross section ($\Sigma_{rem1,2}$);
- group-transfer cross section (Σ_{S12});

The library for DYN3D [1] contains these two-group constants for each reflector type depending on the following parameters:

- coolant temperature (T_m);
- boron acid concentration in the coolant (C_b);

- coolant density (γ_m).

The dependence of the two-group diffusion coefficients and cross sections on these parameters in DYN3D is parameterized by the following form [1]:

$$\begin{aligned} \Sigma = \Sigma_0[1 + C_{R1}(\gamma_m - \gamma_{m0}) + C_{R2}(\gamma_m - \gamma_{m0})^2]* \\ *[1 + C_{C1}(C_b\gamma_m - C_{b0}\gamma_{m0}) + C_{C2}(C_b\gamma_m - C_{b0}\gamma_{m0})^2]* \\ *[1 + C_{Tm}(\frac{1}{(T_m)^{1/2}} - \frac{1}{(T_{m0})^{1/2}})] \end{aligned} \quad (1)$$

where Σ and Σ_0 are the actual and reference quantities of interest (cross sections or diffusion coefficients), C_{R1}, C_{R2}, C_{C1}, C_{C2}, C_{Tm} are parameterization coefficients. The quantities γ_m, C_b, T_m and γ_{m0}, C_{b0}, T_{m0} are parameters of the actual and reference states, respectively. The above-mentioned parameterization coefficients correspond to the DYN3D library in format Iwqs=2 [1].

Neutron production and energy production cross sections, and the data related to poisoning (yields of iodine, xenon and promethium, absorption cross section of xenon and samarium) as well as fuel temperature and burnup dependence are not included into this XS library since the reflector does not contain fissile materials.

Models for all types of radial reflector can be developed for the XS library generation in two different approaches: 1D and 2D models. The 2D model consists of hexagonal cells of radial reflector and one, two or three fuel assemblies.

A more simplified model for the radial reflector required less computing time in the 1D model composed of square cells and consisting of two fuel assemblies and homogeneous cells of the reflector and the surrounding region. Rectangular outer boundaries are used in both models. Black boundary conditions (BC) are used on the outer boundary in the 2D model, while the 1D model is calculated with black BC in one direction and reflective BC in another direction [9, 10] (see figure 3 and figure 4).

The radial reflector of the VVER reactor core can be modeled by one or more layers in radial direction. A three-layer structure of the reflector is shown in figure 1 for 60° symmetry sector of the core. It includes the corresponding part of the buffle, core barrel, coolant and the reactor vessel.

Taking into account 60° symmetry of the core and design features of the buffle, five types of radial reflector cells with different material composition for VVER reactor core can be constructed [4].

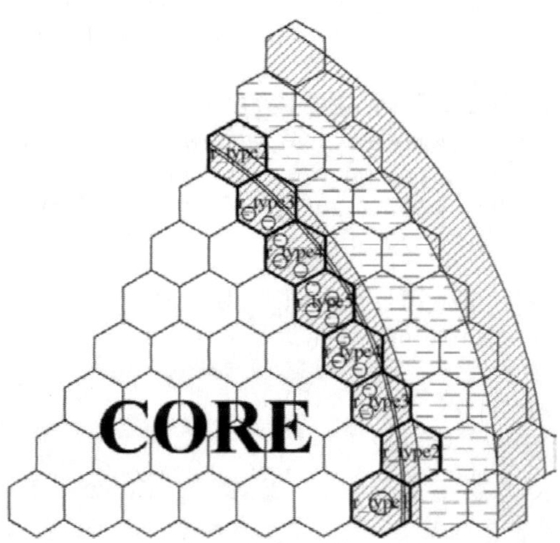

Figure 1. 60° sector of reflector for the VVER reactor core.

The correct boundary conditions for the neutron flux at the outer boundary of the reflector can be obtained by taking into account the surrounding (S) materials of the reflector region.

To cover all components of the surrounding region, two square layers of homogeneous surrounding are used in the 1D model (figure 3) of the radial reflector, while one hexagonal layer of heterogeneous surrounding is applied in the 2D model (figure 2).

The central cell of the reflector 2D model underlined in figure 2 is under consideration (for which the few-group cross section is calculated). Three FAs serve as a source of neutrons in r_type1 reflector while r_type2 needs only one FA and two FAs are used for other type of the radial reflector

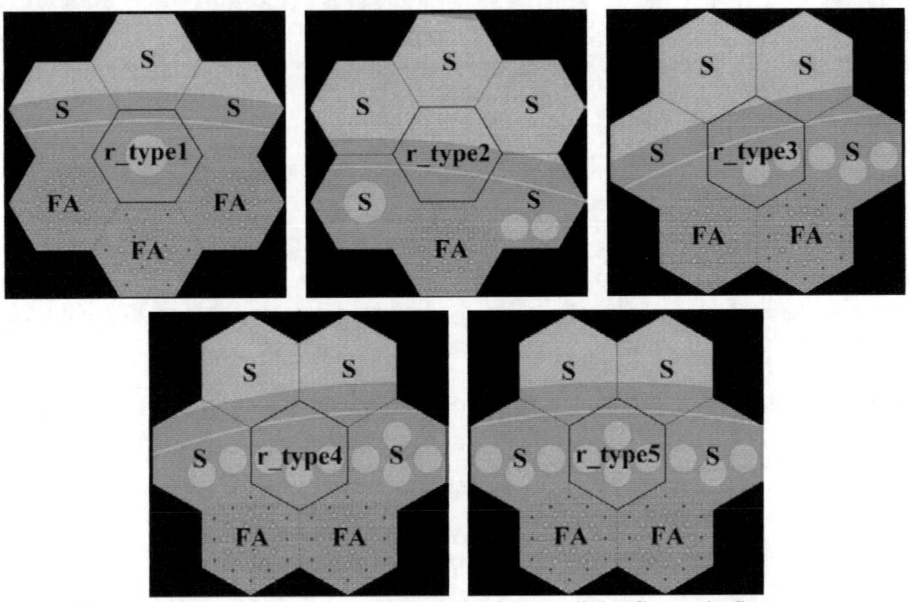

Figure 2. Five types of the 2D models of the radial reflector in Serpent.

Few-group constants for radial reflector are calculated by considering detailed geometry of all components (reflector, surrounding and fuel assembly). The above mentioned 2D model is rather accurate but needs considerable computing time.

To reduce the time of calculation, the 1D model for radial reflector was developed (see figure 3). Symmetry of the model reflects the specifics of boundary conditions (BC) for the system "reflector + neutron source (one or several FA)" available in Serpent. Each compartment of the system is presented as a square region. The boundary conditions in the 1D model are" black" in the direction of the reflector and reflective in the FA direction. The symbols RS_1 and RS_2 correspond to the first and second layers of the radial surroundings representing of homogenized regions with different water/steel ratio.

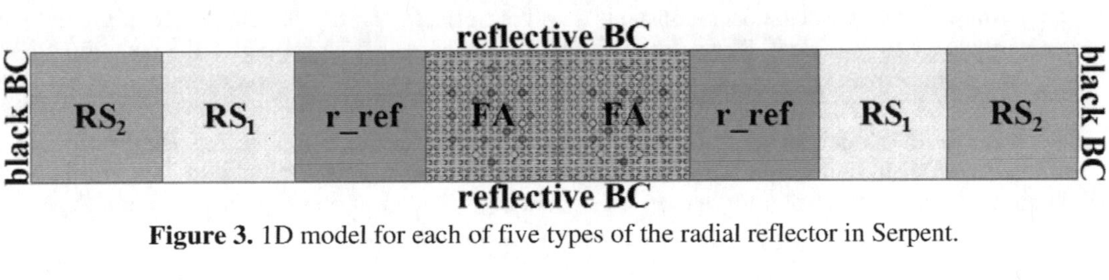

Figure 3. 1D model for each of five types of the radial reflector in Serpent.

The 1D model usually used for axial reflector is shown in figure 4. It is presented as an inhomogeneous three layer structure with different material composition in each layer [7]. The composition of each layer was calculated from the data of the FA design of VVER-1200 [6]. To simplify the 1D model for top and bottom axial reflectors the three layer structure was homogenized into one layer. The symbols AS's in both sides of figure 4 correspond to the layer of homogenized axial surroundings.

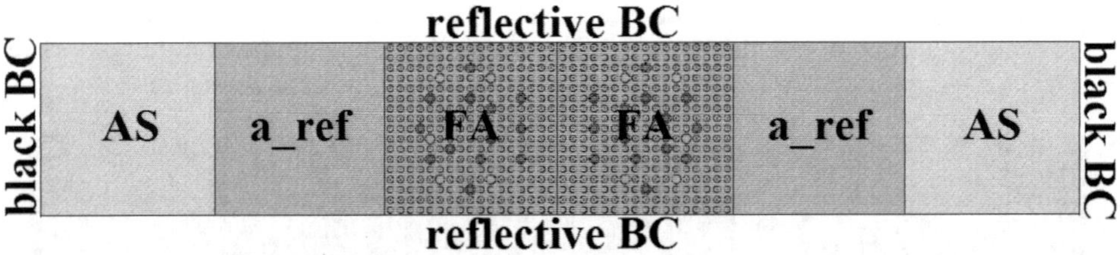

Figure 4. 1D models of the axial reflector in Serpent.

3. Generation of two-group XS library for reflector

The quantities of interest were determined by using Serpent 2 spectral code. The reference values of the parameters for calculation of two-group constants were the following: $T_f=T_m= 574$ K, $\gamma_m= 0.726098$ g/cm^3, $C_b= 0$ g/kg. The values of the coolant temperature, coolant density and boron acid concentration in the coolant used for calculation are presented in table 1 where different values of the parameters are in bold.

Table 1. Reactor core parameters used for calculation of two-group constants of the reflector.

parameter	T_{mod}, K	T_{fuel}, K	γ_{mod}, g/cm^3	C_b, g/kg
γ_{mod}	574	574	**0.5**	0.0
γ_{mod}	574	574	**0.6**	0.0
γ_{mod}	574	574	**0.8**	0.0
γ_{mod}	574	574	**0.9**	0.0
C_b	574	574	0.726098	**8.015**
C_b	574	574	0.726098	**16.03**
C_b	574	574	0.726098	**27.48**
C_b	574	574	0.726098	**40.075**
T_{mod}	**533**	533	0.726098	0.0
T_{mod}	**553**	553	0.726098	0.0
T_{mod}	**596**	596	0.726098	0.0
T_{mod}	**619**	619	0.726098	0.0

Two-group constants for five types of radial and two types of axial (top and bottom) reflector in the 2D and 1D models calculated for the reference parameters are presented in table 2.

Two-group parameter-dependent constants for the reflector were calculated according to the parameters values from table 1. The results of calculations are fitted according to the equation (1) to obtain the parameterization coefficients. As an example the calculated two-group constants depending on the boron acid concentration and results of their fitting are shown in figure 5 for the r_type5 radial reflector in the 1D model for the reference coolant density. The same for the top axial reflector is shown in figure 6. In both figures are shown the diffusion coefficients for the fast (D_1) and thermal (D_2) neutrons, group-removal cross sections (Σ_{rem1} and Σ_{rem2}) and group-transfer cross section (Σ_{S12}). In figure 5 and 6 arguments are $C_b\gamma_{m0}$, g^2/(kg·cm^3), $\gamma_{m0}= 0.726098$ g/cm^3.

Table 2. Reference values of two-group diffusion coefficients and cross sections of the radial and axial reflector.

	D_1, cm	D_2, cm	Σ_{rem1}, cm^{-1}	Σ_{rem2}, cm^{-1}	Σ_{S12}, cm^{-1}
Radial reflector					
2D model					
r_type1	1.05689	2.89315E-01	1.06416E-02	5.19650E-02	7.25384E-03
r_type2	1.02223	2.77912E-01	1.03660E-02	3.64028E-02	6.90553E-03
r_type3	1.04980	2.98462E-01	1.25117E-02	6.34929E-02	8.97670E-03
r_type4	1.06435	2.94033E-01	1.48532E-02	5.73539E-02	1.14607E-02
r_type5	1.07292	2.91831E-01	1.62414E-02	5.51828E-02	1.28671E-02
1D model					
r_type1	1.02548	3.33793E-01	1.29841E-02	9.98766E-02	8.89709E-03
r_type2	1.09480	3.18875E-01	2.09585E-02	8.52954E-02	1.74367E-02
r_type3	1.02912	3.32638E-01	1.38199E-02	9.87948E-02	9.75134E-03
r_type4	1.03598	3.30310E-01	1.55861E-02	9.54567E-02	1.16080E-02
r_type5	1.06443	3.24977E-01	1.83432E-02	9.05972E-02	1.45789E-02
Axial reflector					
1D model					
a_top	1.73119	2.99344E-01	3.59289E-02	1.17676E-02	3.51686E-02
a_bottom	1.33625	3.01945E-01	2.94695E-02	5.48096E-02	2.72188E-02

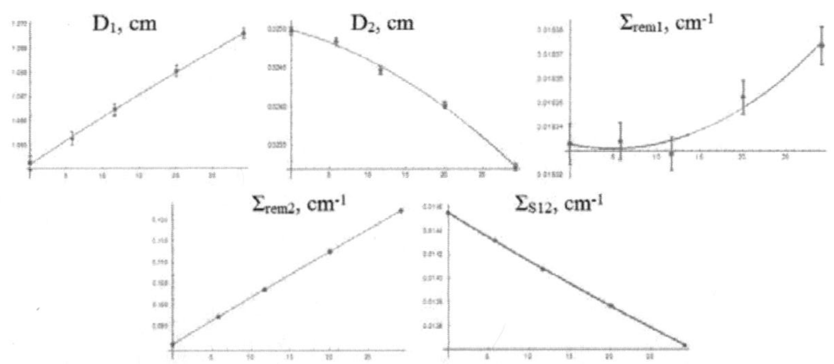

Figure 5. Dependence of two-group constants on the boron acid concentration and their fitting for the r_type5 radial reflector in the 1D model.

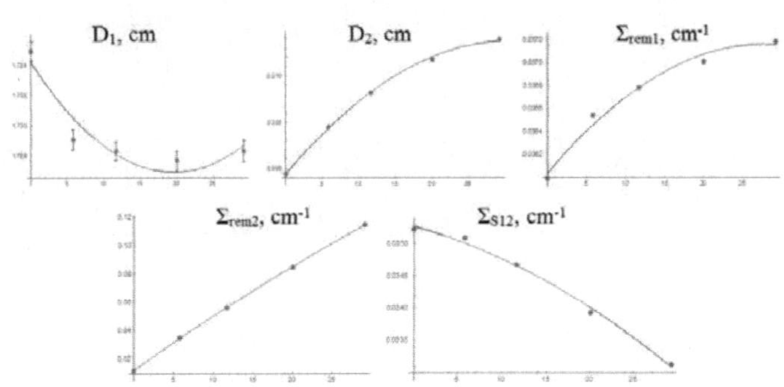

Figure 6. Dependence of two-group constants on the boron acid concentration and their fitting for the top axial reflector in the 1D model.

The similar plots were obtained for the XS dependence on the coolant temperature and coolant density for all types of the radial and axial reflectors. Fitting plots for all parameters of interest are in good agreement with the equation (1).

4. Calculation of normalized power distribution using DYN3D

A rough testing of the two-group constant library for the reflector was done in the reactor code DYN3D by using the fuel XS library in Iwqs=2 format. The normalized power distribution in the core with using the XS library with and without constants for the reflector is presented in figure 7.

The first and the second values in the hexagonal cells in figure 7 are the sequence number of the cell and the normalized power distribution in the core without reflector XSs, respectively. The third value is the difference between the normalized power distributions in the model without reflector and with reflector (1D radial and 1D axial) XSs. The fourth value is the difference between the normalized power distributions in the model without reflector and with reflector (2D radial and 1D axial) XSs.

The data shown in figure 7 indicate that the normalized power distribution in the reactor core is flattened in the central region of the core better by using the 2D model of the radial reflector than the 1D model.

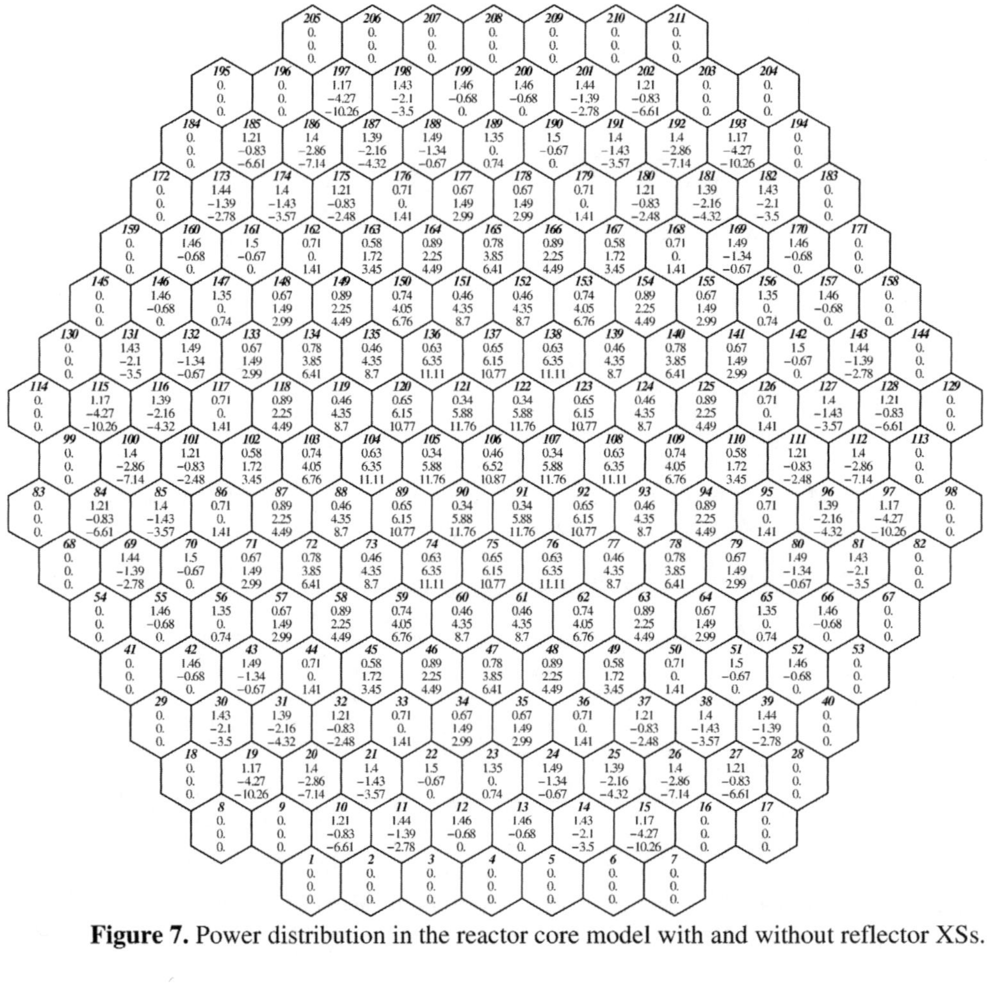

Figure 7. Power distribution in the reactor core model with and without reflector XSs.

5. Conclusion

A physical model of the radial and axial reflector of the VVER-1200-like reactor core has been developed. Five types of the radial reflector in the 1D and 2D models were developed. The axial top and bottom reflectors are described in the 1D model. The two-group XS library for the diffusion code DYN3D has been generated for all types of the reflectors by using Serpent 2 Monte Carlo code. A rough testing of the reflector library in DYN3D has shown that the normalized power distribution in the VVER-1200-like reactor core is flattened in the central region of the core better by using the 2D model of the radial reflector in comparison with the 1D model.

References

[1] Rohde U, Kliem S, Grundmann U, Baier S, Bilodid Y, Duerigen S, Fridman E, Gommlich A, Grahn A, Holt L, Kozmenkov Y and Mittag S 2016 The reactor dynamics code DYN3D – models, validation and applications *Progress in Nuclear Energy* **89** 170-190

[2] Kliem S, Bilodid Y, Fridman E, Baier S, Grahn A, Gommlich A and Nikitin E 2016 The reactor dynamics code DYN3D *Kertechnik* **81** 170-172

[3] Leppänen J *Serpent – a continuous-energy Monte Carlo reactor physics burnup calculation code. User's manual* (Finland: VTT Technical Research Centre) p 164

[4] Ovdiienko I, Ieremenko M, Kuchin A and Khalimonchuk V 2014 Development of cross-section library for DYN3D code *J. Yaderna ta Radyiatsyjna Bezpeka* **4(64)** 22–25

[5] Semchenkov Y and Styrin Y 2010 Advancing of VVER reactor core *BULATOM Int. Nuclear Forum on Nuclear Energy* (Bulgaria: Varna) p 31

[6] Dragunov Y, Ryzhov S, Vasilchenko I, Kobelev S and Vyalitzin V 2007 Core project for AES-2006 RP *Conf. Safety assurance of NPP with WWER* (Russia: Podolsk) p 11.

[7] Lotsch T, Khalimanchuk V, Kuchin A 2009 Proposal of a benchmark for core burnup calculation for a VVER-1000 reactor core *19th Symp. of AER on VVER Reactor Phys. and Reactor Safety* (Bulgaria: Varna) p 57

[8] Lotsch T, Khalimanchuk V, Kuchin A 2010 Corrections and additions to the proposal of a benchmark for core burnup calculation for a VVER-1000 reactor core *20th Symp. of AER on VVER Reactor Phys. and Reactor Safety* (Finland: Hanasaari, Espoo) p 57

[9] Fridman E and Leppänen J 2011 On the use of the Serpent Monte Carlo code for few-group cross section generation *Annals of Nuclear Energy* **38** 1399-1405

[10] Fridman E and Leppänen J 2012 Revised methods for few-group cross section generation in the Serpent Monte Carlo code *Proc. of PHYSOR-2012* (USA: Knoxville, Tennessee) p 12

Acknowledgments

The authors acknowledges support from the State program "Scientific support of nuclear power development in the Republic of Belarus for 2009 - 2010 and for the period up to 2020" and the sub-program "Nuclear energy, nuclear and radiation technologies".

Calculation evaluation of multiplying properties of LWR with thorium fuel

I V Shamanin[2], V M Grachev[1], V V Knyshev[2], S V Bedenko[2] and N G Novikova[2]

[1] National Research Nuclear University MEPhI, Moscow, Russian Federation
[2] National Research Tomsk Polytechnic University, Tomsk, Russian Federation

E-mail: shiva@tpu.ru

Abstract. The results of multiplying properties design research of the unit cell and LWR fuel assembly with the high temperature gas-cooled thorium reactor fuel pellet are presented in the work. The calculation evaluation showed the possibility of using thorium in LWR effectively. In this case the amount of fissile isotope is 2.45 times smaller in comparison with the standard loading of LWR. The research and numerical experiments were carried out using the verified accounting code of the program MCU5, modern libraries of evaluated nuclear data and multigroup approximations.

1. Introduction

In Russia at present the technologies connected with implementation of thorium nuclear cycle are innovative, so technically and economically they are not developed, consequently, they are connected with significant financial investments and risks. However, taking into consideration potential possibilities of thorium nuclear cycle, duration of the development and new nuclear technologies implementation phase there is need for works aimed at selection of the optimum alternative of using thorium in both operating and innovative constructions of reactors [1-4].

In works [4,5,11-15] 40 different constructions of low- and average power operating reactors and some innovative engineering solutions, which are being developed and licensed, were considered. Problems of using thorium for high temperature reactors (HTR), heavy water reactors (HWR) and light water reactors (LWR) are being researched. A very important conclusion that implementation of thorium nuclear cycle for fast neutron and light-water reactors is essential and timely has been made. In such reactors there are developed technologies which allow performing modernization of existing constructions of fuel assemblies and substitute traditional uranium fuel with thorium, providing necessary conditions.

BREST-300-OD is a well-developed construction of fast reactors with the power corresponding to low-power reactors in which thorium fuel can be used. This pilot and demonstration construction will become a new advanced model element of nuclear fuel cycle in Russia [1].

In work [3] behaviour of LWR of WWER-1000 type when active core is loaded with fuel assemblies with different types of thorium containing oxide fuel is analyzed. The calculation results showed working efficiency of using thorium in WWER-1000 of production design, as WWER-1000 operates in the open fuel cycle, the operation period and fuel burnup reach the peak value.

Constructions of high-temperature gas cooled nuclear reactors (HTGR) are supposed innovative solutions. In recent decades engineering and construction of modular HTGR of low and average power have been started or resumed [4,5,11-15].

In works [4, 5] calculation researches of low power HTGR were carried out. Physics of the reactor with fuel blocks and fuel pellets of different configurations were studied for the purpose of choosing the most optimum construction of the reactor core and its loading. Calculation results of multiplying properties and fuel burnup depth were shown for the chosen configuration. The research results allowed formulating conception of low power HTGR with the reactor core assembled from fuel blocks of the unified design. The number of fuel blocks and reactor core dimensions should be determined by the value of the reactor power which is required for solving the task-oriented problem, but fuel and fuel blocks will be produced in series (see fig.1).

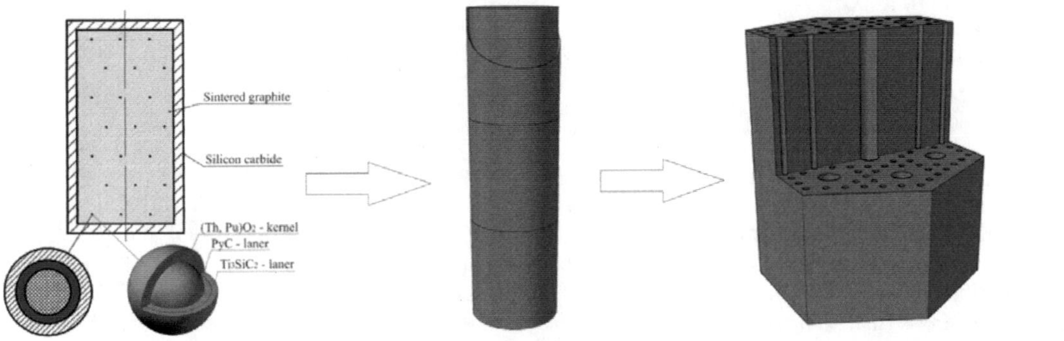

Figure 1. Scheme of fuel unit filling with fuel.

Economical assessment showed that if fuel pellet configuration suggested for HTGR is used in other reactor type, it will allow reducing production costs of the fuel and advancing efficiency of its application.

Therefore calculation evaluation of neutron-physical characteristics of LWR with thorium fuel pellet of HTGR was performed in the work.

2. Calculation model of LWR fuel assembly

Fuel elements for HTGR designed in the form of spheres, cylinders and prisms are made as carbon and graphite matrix composition containing microencapsulated fuel (coated particle fuel, microfuel). Microfuel is fuel kernel from fissionable material with some coating layers.

Various ceramic compositions (PyC, SiC, ZrC, TiC, etc.) and nanolaminates are applied as protective coatings very often [6-10]

The main function of such protective coatings is to retain different gaseous and solid fission products during the irradiation process.

Currently existing technological solutions are aimed at increasing the service life of microfuel by enlarging the number of layers, which in its turn results in the decrease of protective coatings damage rate.

But increase of the number of the layers does not solve the problem of deconfinement and fuel kernel migration at long-term irradiation and high temperature (see fig. 2, 3).

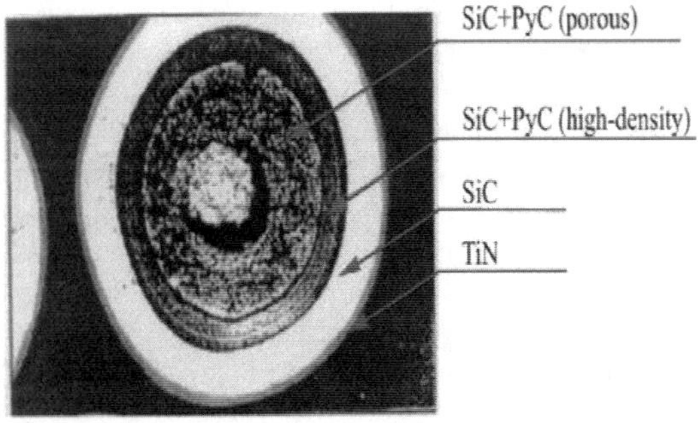

Figure 2. Results of microfuel irradiation with four-layer protective coatings on the basis of PyC-SiC.

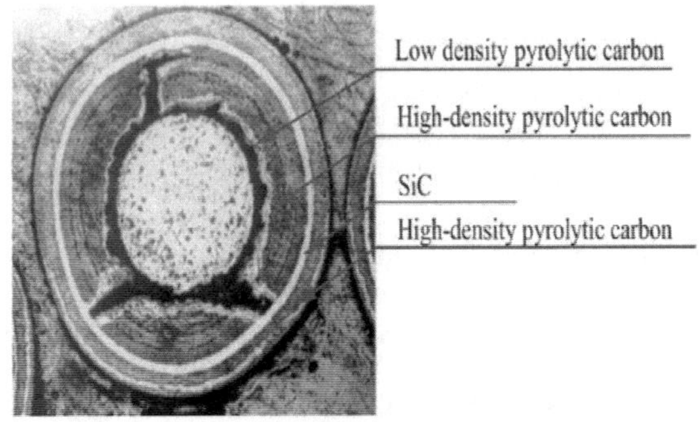

Figure 3. Results of microfuel irradiation with four-protective coatings on the basis of high density PyC-Si.

Thus it becomes evident that revolutionary materials and plating technologies are needed. At present "coated particles" technology is used for forming coatings for spherical fuel kernel [8]. This technology is also applied for cylindrical LWR fuel pellets of standard configuration [9]. However, this method has such disadvantages as low rate of coating deposition (~0.001 μm/s) and processing complexity of coating deposition from a wide range of materials.

Functional coating formation technology by the method of "semifluidized bed" had been developing in Germany for 20 years and the expenses were several billion deutschemarks [10].

This method is very technological as it provides the possibility to control and adjust quality and thickness of the coatings. The main disadvantages of this method are limited range of applied materials (C, SiC) and strict requirements to the continuity of thermodynamic parameters of the media. So surfacing metal coatings by the method of "semifluidized bed" does not seem possible at present.

The technology of coating deposition from different materials (including metallic) on the surface of spherical kernel with the diameter of (350−800) μm and cylindrical pellets with the diameter of (8−12) mm and height of (15) mm was developed at National Research Tomsk Polytechnic University. The possibility to modify properties of surface layers of construction materials is being researched [7].

The method has such an advantage as high rate of the coating process (≥0.003 μm/s) and it allows forming coatings from metallic materials, as well as from superstrength materials such as diamond [7].

The peculiarity of the material forming the coating in the given case is its structure. It is amorphous and is characterized by increased resistance to the effect of destroying neutron and γ-quantum flux.

Applied configuration of micro fuel and fuel pellet is presented in fig.1 [4, 5]. Low-density PyC-layer is used for location of gaseous fission products and is the first diffusion barrier which protects the second Ti_3SiC_2-layer from damage. Ti_3SiC_2 possesses excellent mechanical and thermal-physical characteristics and serves as the main power coating and diffusion barrier towards solid fission products. An additional safety barriers formed by a graphite matrix and a functional coating in the form of SiC-layer with the thickness of ~300 μm on the pellet surface. Thorium-plutonium fuel composition is used as fissile isotope. The isotopic composition of military Pu is presented in [16].

Service life irradiations of the fuel pellet of the given configuration show that radiation dimensional changes of the kernel (see Tab. 1), the coatings and the fuel pellet are insignificant, small in absolute magnitude, and structural characteristics are stable (id of the works is RFMEFI59114X0001, 2014). The service life of the fuel pellet towards metals impurities in the kernel is limited by the temperature compatibility of~ 1350 K; radiation stability is limited by the fast fluence of $\sim 0.6 \cdot 10^{26}$ m^{-2}; but the pellet is resistant for a short moment up to the fluence of$\sim 2 \cdot 10^{26}$ m^{-2}.

IOP Conf. Series: Journal of Physics: Conf. Series **781** (2017) 012030 doi:10.1088/1742-6596/781/1/012030

Table 1. Microfuel and fuel pellet specifications

Specifications	Value
Microfuel type	BISO
Fuel composition	$(Th,Pu)O_2$
Pu/Th composition	50%/50%
Pu isotopic composition (%)	military Pu
Microfule diameter (μm)	600
Fuel kernel diameter (μm)	350
PyC inner coating layer thickness (μm)	90
Ti_3SiC_2 outer coating layer thickness (μm)	35
Average density	
Fuel kernel (g/sm^3)	10.5
PyC (g/sm^3)	1.20
Ti_3SiC_2(g/sm^3)	4.16
Microfuel service life	
Fast fluence (м$^{-2}$)	$\sim 0.6 \cdot 10^{26}$
Temperature (K)	~1250
Fuel pellet	TRISO
Height (m)	0.0202
Diameter (m)	0.00757
SiC protective coating thickness (μm)	300
SiC average density (g/sm^3)	3.20

The fuel pellet and fuel assembly calculation model (see fig. 3,4) was created using program MCU5 (developed by Russian Research Center «Kurchatov Institute») A geometrical module of MCU5 allows simulating 3D systems with geometry of any complexity using the combinatorial approach, which is based on complex systems description by combinations of elementary bodies and surfaces [17].

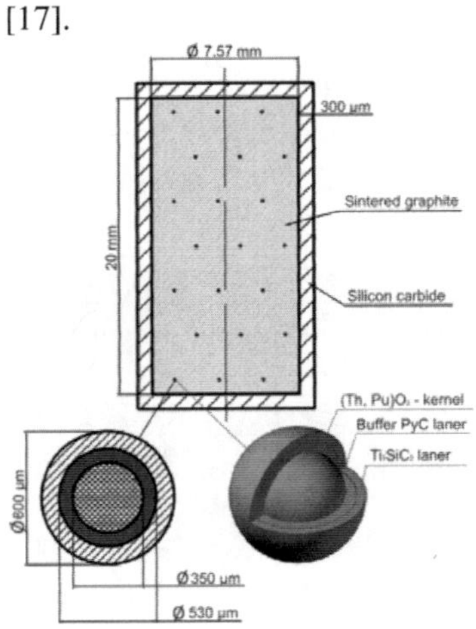

Figure 3. Microfuel and fuel pellet calculation model.

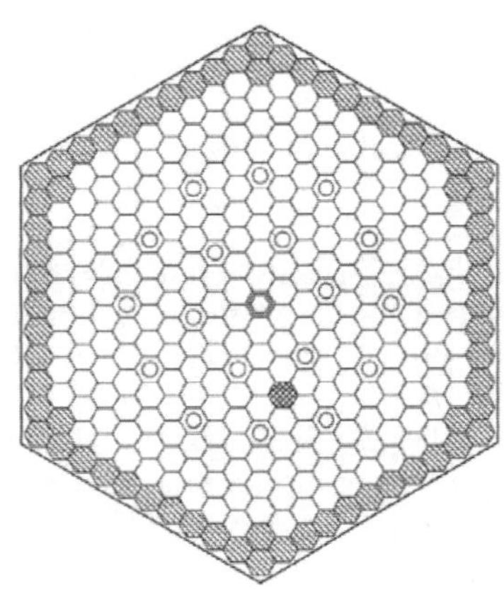

Figure 4. Fuel assembly calculation model.

2.1. Research methods.

Researches and numerical experiments were carried out using the calculation code of the MCU5 program, modern libraries of evaluated nuclear data (ENDF/B-VII.0, JEFF-3.1.1, JENDL-4.0, ROSFOND, BROND, BNAB, etc.) and multigroup approximations.

The MCU5 program is intended for precision process simulation of neutrons and photons transfer in any types of reactors by analogue and non-analogue Monte-Carlo methods on the basis of evaluated nuclear data, taking into consideration isotopic composition change of reactor materials.

For calculating the reactor life-time the calculation module BURNUP of the MCUP program is used. This module is meant for composition change calculation of fissile and absorbing reactor materials during its life-time. Also the module provides the possibility of calculation prediction of reactor materials nuclide composition and its multiplying properties depending on the fuel burnup. The calculation of reactor isotopic composition change during its life-time is performed at given dependence of average power density on time.

3. Results of neutron-physical researches

In fig. 5 and 6 the results of neutron-physical calculations of the unit cell and the LWR fuel assembly are presented. The calculation evaluation showed the possibility of efficient use of thorium in LWR. . Increased fuel burn-up depths are reached when military Pu is used (see tab.1). In this case the amount of fissile isotope in the fuel pellet with micro fuel is 2.45 times smaller than in the fuel pellet of the series LWR, and the fuel-element life-time is ~1500 effective days.

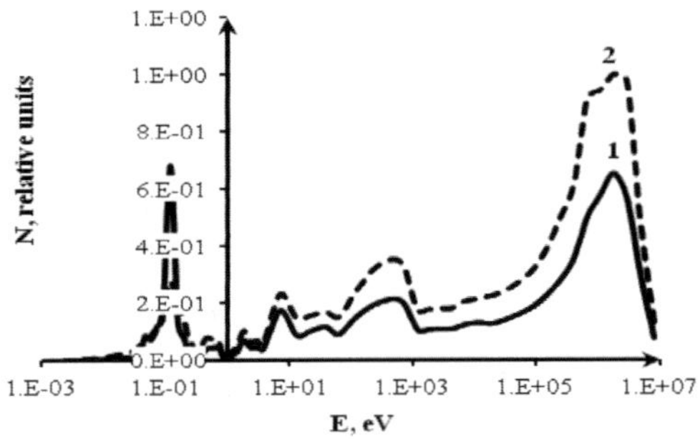

Figure 5. Neutrons spectrum in unit cell fuel portion:1 – fuel pellet with microfuel, 2 – fuel pellet of series LWR.

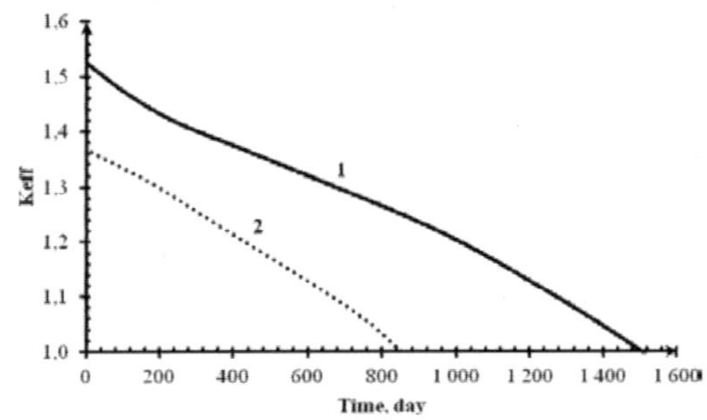

Figure 6. Multiplying properties of fuel assembly: 1 – fuel assembly with HTGR fuel pellet, 2 – fuel assembly with LWR fuel pellet.

4. Conclusion

The results of service life tests of HTGR thorium fuel pellet with the dispersed microfuel (BISO) were presented in the work. Also the results of calculation researches of multiplying properties of the unit cell and the LWR fuel assembly with HTGR fuel pellet were given. The calculation evaluation showed the possibility of using such replacement, because in this case the amount of fissile isotope is 2.45 times smaller compared with standard LWR loading. Increased fuel burn-up depths are reached when military Pu is used. The fuel-element life-time is ~1500 effective days, and the fuel burnup is ~150 GW day/tHM.

5. References

[1] http://www.i-russia.ru/nuclear/directions/36/ (accessed 20.03.2016).

[2] Baybakov D F, Godovykh A V, Martynov I S, Nesterov V N 2016 The dependence of the nuclide composition of the fuel core loaing from multiplying and breeding properties of the nuclear facility KLT-40S *Izv. Vyssh. Uchebn. Zaved. Yadernaya Energetika* **2** 99-111.

[3] Kulikov G G, Shmelev A N, Kryuchkov E F, et al 2002 Fizicheskie harakteristiki legkovodnogo jadernogo reaktora so sverhdlinnoj kampaniej torievogo topliva *Izv. Vyssh. Uchebn. Zaved. Yadernaya Energetika* **1** 18-26.

[4] Shamanin I V, Bedenko S V, Chertkov Yu B 2016 Thorium-loaded low-power reactor installation operated with super-long fuel residence time *Izv. Vyssh. Uchebn. Zaved. Yadernaya Energetika*, **2** 121-132.

[5] Shamanin I, Chertkov Y, Bedenko S, Gubaydulin I 2015 Gas-Cooled Thorium Reactor with Fuel Block of the Unified Design *Adv. in Mater. Science and Eng.* **2015** 1−8 doi:10.1155/2015/392721.

[6] Hernikov A S, Permyakov L N, Fedik I I 1999 Fuel elements on the basis of spherical fuel particles with protective coatings for high-safety reactors *Atomic Energy* **87(6)** 871-878.

[7] Linnik S A, Gaydaychuk A V 2013 Processes and parameters of diamond films deposition in AC glow discharge *Diamond and Related Materials* **32** 43-47.

[8] Brown P E, Fawcloth R L 1976 Metal fission product behavior in high temperature reactors UO_2 coated particle fuel *J. Nucl. Mater.* **59(1)** 29-41.

[9] Alkan Z, Kugeler K, Kaulbarsch R, Manter C. 2001 Silicon carbide encapsulated fuel pellets for lightwater reactors *Progress in Nuclear Energy* **38(34)** 411-414.

[10] Hase T, Suzuki H, Iseki T 1976 Formation process of β-SiC during reaction-sintering *Journal of Nuclear Materials* **59(1)** 42-48.

[11] Kiryushin A I, Kodochigov N G, Kuzavkov N G, et al 1997 Project of the GM-MHR high-temperature helium reactor with gas turbine *Nucl. Eng. Des.* **173** 119−129

[12] Black G , Black M A T., Solan D, et al., 2015 Carbon free energy development and the role of small modular reactors: A review and decision framework for deployment in developing countries *Renewable and Sustainable Energy Reviews* **43** 83−94.

[13] Hidayatullah H, Susyadi S, Subki M H 2015 Design and technology development for small modular reactors - Safety expectations, prospects and impediments of their deployment, *Progress in Nuclear Energy* **79** 127−135.

[14] Miletić M, Fukač R, Pioro I and Dragunov A 2014 Development of gas cooled reactors and experimental setup of high temperature helium loop for in-pile operation *Nucl. Eng. Des.* **276** 87−97.

[15] Gluhov L Ju, Kotkov S P, Kuznecov M S, Chursin S S 2015 Measurement of prompt neutrons production time of VIR-2M pulsed nuclear *Vopr. Atom. Nauki Tekhn. Ser. Fizika jadernyh reaktorov* **4** 58–63.

[16] Ridikas D, Bletzacker L, Deruelle O, et al. 2002 Comparative Analysis of ENDF, JEF and JENDL Data Libraries by Modelling High Temperature Reactors and Plutonium Based Fuel Cycles, *J. Nucl. Sci. Technol.* **39(2)** 1167–1170.

[17] Kalugin M A, Oleynik D S, Shkarovsky D A 2015 Overview of the MCU Monte Carlo Software Package *Annals of Nuclear Energy* **82** 54-62.

Modernization of the VVR-TS reactor core for the increasing of the radionuclides production

R V Fomin[1], V V Kolesov[1] and O Yu Kochnov[2]

[1] NRNU (MEPhI), Obninsk 249040, Kaluga reg., Russia
[2]The Branch of FSUE "Karpov Institute of Physical Chemistry", Obninsk 249033, Kaluga reg., Russia

E-mail: fominsparck@gmail.com, valeri-kolesov@yandex.ru, kochnov2000@mail.ru

Abstract. We have studied the possibility of increasing the ^{99}Mo and ^{131}I production through modernization of the VVR-TS reactor core. It has been found in particular that the introducing of beryllium reflector on the core periphery can significantly increase the starting reactivity margin. This, in turn, give us possibility to create an additional channel for increasing their production.

1. Introduction

The development of nuclear medicine, primarily associated with the diagnosis and treatment of cancer and are increasingly used in cardiac, neurological, endocrine and other diseases. In this regard, it is becoming increasingly important question of obtaining on nuclear power plant special isotopes. The need for such isotopes is increasing worldwide. The procedure for obtaining isotopes is very specific and time-consuming. It depends on many factors that require knowledge of the nuclear reactor features and neutron-physical processes occurring in the reactor core and target. In this case the main problem is to ensure the most efficient of use of reactor channels. From this depends directly on how the nomenclature of manufactured products, as well as production costs and, consequently, its availability to the end user.

The effectiveness of the production depends significantly on two main factors. Firstly is the operating conditions of a reactor. At any time the optimal conditions can be implemented in various experimental channels. Firstly the precise model of the reactor VVR-Ts was created to allow a high degree of accuracy to estimate the neutron-physical parameters of various channels, depending on the power and position of control rods and, thus, build the strategy to find the optimal location of the targets in the experimental channels of the reactor. Second, the effectiveness of production largely depend on the characteristics of the target and means of its placement in the experimental channels. Therefore, an important part of the research of the various special conditions that can be created for target, which may also allow to increase the yield of a particular nuclide.

The ^{131}I and ^{99}Mo production when targets irradiated by neutrons in the reactor core essentially depends on the duration of exposure, and its location [8,9]. Therefore, for good prediction of its production in the reactor is necessary to create an adequate model for the reactor campaign calculation. However, up to now neutron calculations for VVR-Ts reactor were carried out using engineering codes (HEXA code for VVR-Ts). The main objective of these codes is to ensure the proper operating of the reactor facility, rather than radionuclides production estimation [1,2]. To achieve this objective significantly more accurate calculation model was created, which allowed to solve this problem.

2. Modeling

The paper presents the results of calculations of neutron-physical characteristics of VVR-Ts reactor using the developed model of the reactor core and the results are compared with experimental data. As an example, Fig. 1 shows a comparison of the calculated reactivity margin during reactor campaign with real measurement.

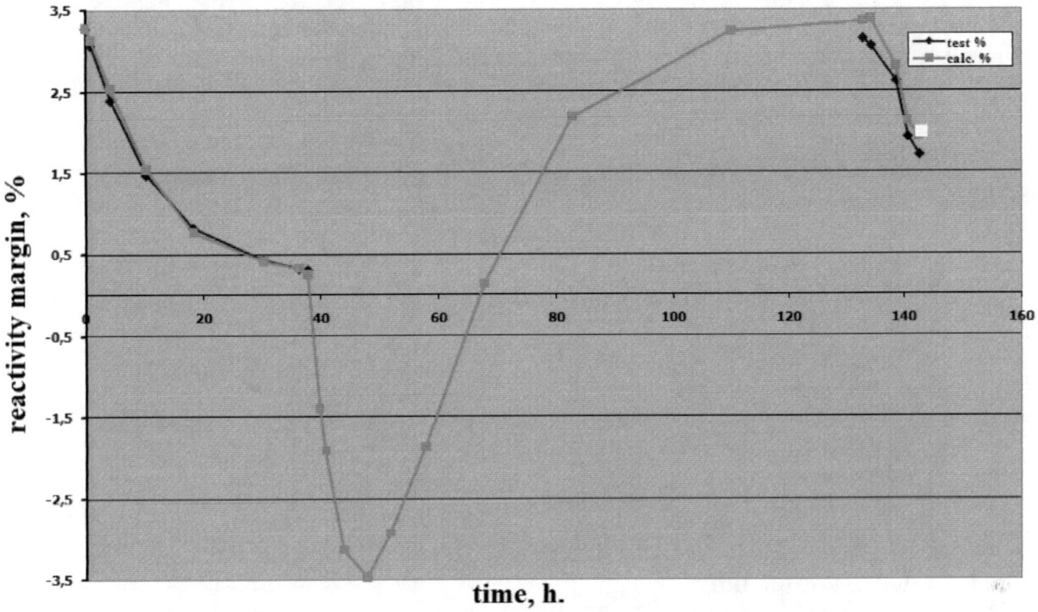

Figure 1. Comparison of the calculated reactivity margin with experiment.

The results of computational studies of the possibility of using beryllium reflectors to improve the neutron-physical characteristics of the VVR-Ts reactor, which is used at the Karpov Research Institute of Physical Chemistry to produce radionuclides, are presented. The calculations showed that in principle beryllium can be used by substituting it for some number of standard fuel assemblies. This made it possible to decrease the number of fuel assemblies from 70 in the standard core to 69 in a modified core and to increase the time between fuel overloads up to 14 days. The calculations also indicate that the neutron flux density in the experimental channels can be increased significantly, which is important for increasing the production of radionuclides [3].

The paper discusses also the possibility of increasing the ^{131}I and ^{99}Mo production for the uranium-containing and tellurium targets by placing beryllium blocks in the reactor core of the VVR-Ts reactor.[4,5] It was found that there was a reserve increase of these isotopes production. It can be obtained through increasing the neutron flux in the experimental channel and by mitigating the neutron spectrum.

The comparison of the ^{131}I and ^{99}Mo production for standard and modified targets was done for the standard core and for the core with a beryllium reflector. The calculations were performed for different variants of the beryllium reflector locations in the core. We considered four options:

- core without beryllium blocks (standard);
- two beryllium blocks are located on the perimeter of the core near the experimental channel;

- beryllium blocks are located around the perimeter of the core;
- beryllium blocks surround the experimental channel.

As an example, Fig. 2 shows all options mentioned above.

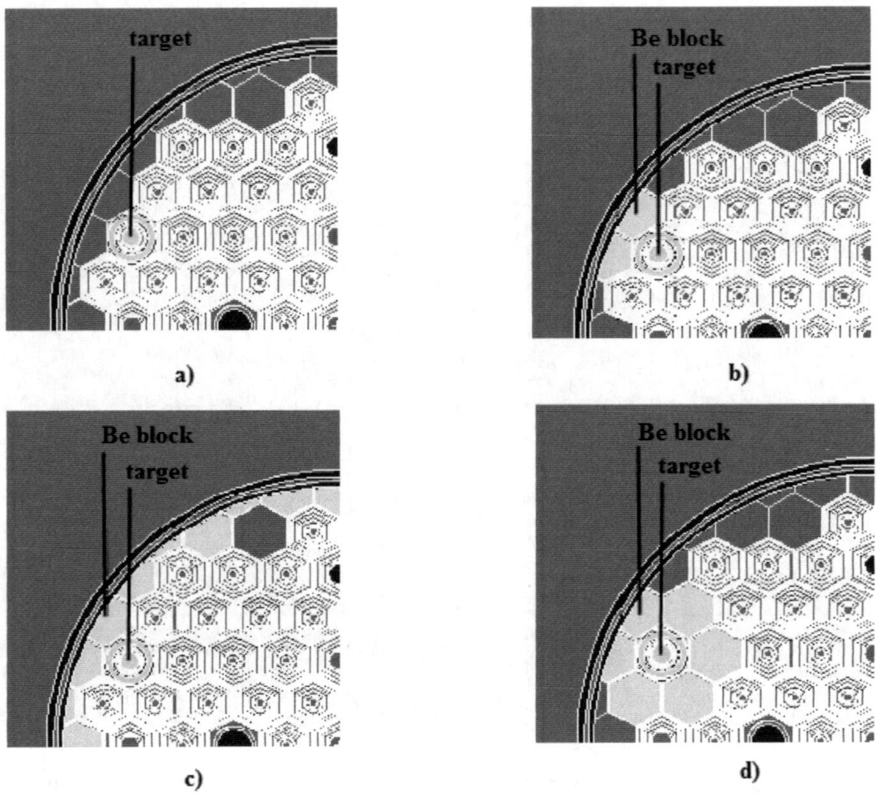

Figure 2. VVR-Ts core: a - core without beryllium blocks (standard); b - two beryllium blocks are located on the perimeter of the core near the experimental channel; c - beryllium blocks are located around the perimeter of the core; d - beryllium blocks surround the experimental channel.

The calculation results show that the use of beryllium inserts can significantly increase production ^{99}Mo. For example, option d, this increase was 7.5% for the initial target and more than 9% - for the modernized [7, 10]. Similar results were obtained for ^{131}I.

Changing the initial reactivity margin for the options discussed above is shown in Table 1.

Table 1. The change of the reactivity of the reactor core at the modernization of the VVR-Ts (variants a to d).

Option	a	b	c	d
Change of the reactivity, $\beta_{эфф}$	0	-0.3	-1.2	3.2

Thus, the option with the core surrounded by the beryllium reflector allowing us sigificantly increase initial reactivity margin. This result was the basis of proposals for the modernization of the core, which is scheduled for 2017-2018 years [6].

Also we investigated the possibility of using low-enriched uranium fuel in the reactor VVR-Ts. It was found that for steady state overload, number of overloaded low-enriched fuel assemblies are equal to 2-3 instead of 1-2 for highly enriched fuel. In this case neutron flux density in both cases were very similar by increasing the fuel load using new fuel assemblies. The duration of each campaign increased on average by 50%.

3. Conclusions

As a result of our studies, it was found that:

- the use of beryllium as a reflector in the reactor core allows us significantly improve its neutron-physical characteristics;
- when the low-enriched fuel is using in the reactor core, number of overloaded fuel assemblies are equal to 2-3 instead of 1-2 for high-enriched fuel and not leads to a considerable reduction in neutron flux density.

References

[1] Kolesov V.V., Kochnov O.Yu., Volkov Yu.V., Ukrainians V.F., Fomin R.V. Creating a precision model of the reactor VVR-Ts for further optimization of its design and operating time ^{99}Mo and other radionuclides. Izvestiya Wysshikh Uchebnykh Zawedeniy, Yadernaya Energetika, 2011, vol.4, p. 129-133.

[2] Kolesov V.V., Kochnov O.Yu., Volkov Yu.V., Ukrainians V.F., Fomin R.V. Creation and validation of the precision model of the WWR-C reactor for optimization of 99Mo productivity. Collection of works «XIII International conference NPP safety and personnel training», Obninsk, 2011, Volume 1,p 29-31.

[3] Kochnov O.Yu., Kolesov V.V., Fomin R.V. Improvement of the Neutron-Physical Characteristics of the VVR-Ts Reactor by Substituting a Beryllium Reflector for Some Fuel Assemblies in the Core. Atomic Energy, 2015, Volume 118, Issue 4, p 238-241

[4] Kochnov O.Yu., Kolesov V.V. Impact of beryllium reflector placed in the core of the reactor VVR-Ts on its neutron-physical characteristics. Izvestiya Wysshikh Uchebnykh Zawedeniy, Yadernaya Energetika, 2012, Volume 2, p. 50-52.

[5] Pozdeev V.V., Kochnov O.Y., Markina M.A., Grachev A.F. Determination of the Radiochemical Yield of Fission-Fragment 99Mo In VVR-Ts. Atomic Energy, 2016, Volume 120, Issue 3, p 170-179.

[6] Chusov I.A., Shelegov A.S., Kochnov O.Yu. Design features of water-cooled research reactors. Izvestiya Wysshikh Uchebnykh Zawedeniy, Yadernaya Energetika, 2016, Volume 3, p. 116-128.

[7] Kazantsev A.A., Sergeev V.V., Kochnov O.Yu. Calculation of the temperature in the container unit with a modified design for the production of 99Mo at the VVR-Ts research reactor facility (IVV.10M). Physics of Atomic Nuclei, 2015, Volume 78, Issue 12, p 1336-1344.

[8] Grachev A.F., Pozdeev V.V., Kochnov O.Yu., Zakharov N.V., Krasheninnikov A.I. Evaluation of characteristics of molybdenum-technetium generators prepared from regenerated uranium at the VVR-ts reactor, Radiochemistry, 2014, Volume 56, Issue 4, p 360-363.

[9] Pozdeev V.V., Kochnov O.Y., Krasheninnikov, A.I., Stepanov V.I., Grachev A.F. Production of medical 99Mo and molybdenum-technetium generators at the karpov research institute of physical chemistry. Atomic Energy, 2014, Volume 117, Issue 2, p 93-98.

[10] Kochnov O.Y., Pozdeev V.V., Krasheninnikov A.I., Zakharov, N.V. Production of fission 99Mo at the VVR-ts nuclear reactor in a closed-loop process with respect to U. Radiochemistry, 2012, Volume 54, Issue 2, p 173-177.

IOP Conf. Series: Journal of Physics: Conf. Series **781** (2017) 012032 doi:10.1088/1742-6596/781/1/012032

Improvement of Modeling HTGR Neutron Physics by Uncertainty Analysis with the Use of Cross-Section Covariance Information

V F Boyarinov[1], A V Grol[1], P A Fomichenko[1] and M Yu Ternovykh[2]

[1]National Research Centre "Kurchatov Institute",
Kurchatov Sq. 1, Moscow, Russia, 123182
[2]National Research Nuclear University MEPhI (Moscow Engineering Physics Institute), Kashirskoe highway, 31, Moscow, Russia

E-mail: Boyarinov_VF@nrcki.ru, Grol_AV@nrcki.ru

Abstract. This work is aimed at improvement of HTGR neutron physics design calculations by application of uncertainty analysis with the use of cross-section covariance information. Methodology and codes for preparation of multigroup libraries of covariance information for individual isotopes from the basic 44-group library of SCALE-6 code system were developed. A 69-group library of covariance information in a special format for main isotopes and elements typical for high temperature gas cooled reactors (HTGR) was generated. This library can be used for estimation of uncertainties, associated with nuclear data, in analysis of HTGR neutron physics with design codes. As an example, calculations of one-group cross-section uncertainties for fission and capture reactions for main isotopes of the MHTGR-350 benchmark, as well as uncertainties of the multiplication factor (k_∞) for the MHTGR-350 fuel compact cell model and fuel block model were performed. These uncertainties were estimated by the developed technology with the use of WIMS-D code and modules of SCALE-6 code system, namely, by TSUNAMI, KENO-VI and SAMS. Eight most important reactions on isotopes for MHTGR-350 benchmark were identified, namely: ^{10}B(capt), ^{238}U(n,γ), ν_5, ^{235}U(n,γ), ^{238}U(el), natC(el), ^{235}U(fiss)-^{235}U(n,γ), ^{235}U(fiss).

1. Introduction

This study is carried out in frames of the IAEA Coordinated Research Project "HTGR Reactor Physics, Thermal-hydraulics and Depletion Uncertainties Analysis" [1-6]. The main objective of the Project is elaboration of methodical basis for tracking of propagation of different uncertainties in modeling of neutron physics, thermal-hydraulics and fuel depletion, and development of corresponding computational tool. Special attention is paid to propagation of computational uncertainty in transition from one calculation stage to another. It is assumed that the whole chain of the uncertainty propagation from the basic data, technological uncertainties through different calculation stages will be tested on a number of benchmark exercises.

The main objective of our efforts is the analysis of uncertainties in modeling neutron physics of HTGR with prismatic fuel on the basis of MHTGR-350 benchmark [7] proposed by NEA OECD. One of the goals of this research is to study propagation of uncertainties contained in the libraries of the evaluated nuclear data to group wise microscopic cross-sections used for lattice physics calculations, and further to few-group macroscopic cross-sections used in codes for full scale reactor calculations.

Calculations are performed for two different variants of the specification of the MHTGR-350 benchmark, but obtained results showed that the conclusions drawn in both cases concerning the uncertainties are identical. Therefore, results submitted in this paper relate only to the second variant of the specification [8].

2. Preparation of the 69-group Library of Covariance Information for Main Isotopes

As a rule, WIMS-D is used in NRC "Kurchatov Institute" as a design code for HTGR lattice calculations. WIMS-D uses shielded 69-group cross-sections. Therefore, to track the propagation of the nuclear data uncertainties to the next calculation stage, as input one must have 69-group covariance information (group-wise standard deviations and covariance matrices).

An original code sequence was developed for preparation of 69-group library of covariance information from the basic 44-group binary library of SCALE-6 code. This technology uses the program modules **Cadillac** and **Cognac** [9] of SCALE-6 code package, **Angelo** code [10] and several program modules developed by authors of this work for the purposes of the Project. Generally, the functions of these modules are the following;

1) module Cadillac – extraction of covariance information for one isotope in binary form from the basic binary library of SCALE;
2) module Cognac – conversion of covariance information for one isotope from binary form to a formatted file (COVERX format [11]);
3) module **Cognac_Angelo** (NRC KI) – preparation of input files for Angelo code;
4) Angelo code – conversion of formatted 44-group covariance information to 69 groups;
5) module **Cov_matr-69** (NRC KI) – conversion of 69-group covariance information to a special format.

As a result, we get a separate file for each isotope in this library. This provides a possibility to easily extend the library content and change the number of groups.

Currently this library contains covariance information in the following form: group standard deviations, group correlation matrices, group covariance matrices. The library contains following 11 isotopes and elements: ^{235}U, ^{238}U, C-nat, ^{16}O, ^{4}He, Si-nat, ^{28}Si, ^{29}Si, ^{30}Si, ^{10}B, ^{11}B.

The use of this library together with design codes allows the automation of the entire process of calculation of uncertainties of neutron-physical calculation.

3. Calculation of One Group Cross-Sections Uncertainties for Main Isotopes and Multiplication Factor Uncertainty

3.1. Model problem

A model of a fuel compact cell of MHTGR-350 reactor (Figure 1) is used for calculation of the one-group cross-section uncertainties for fission and absorption reactions for main isotopes. Geometry of the spherical fuel particle is provided in figure 2 [1, 2].

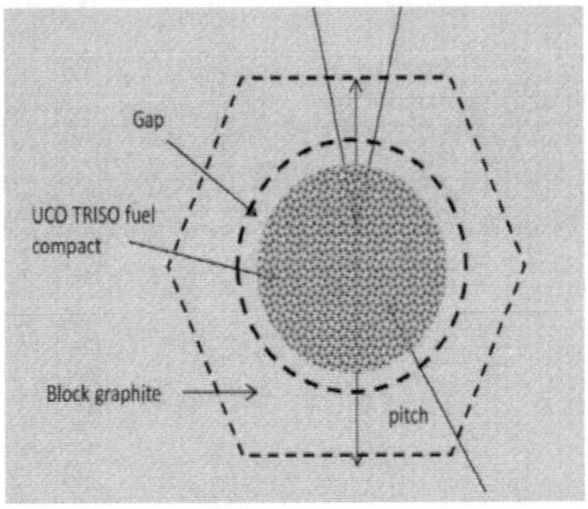

Figure 1. Fuel compact cell.

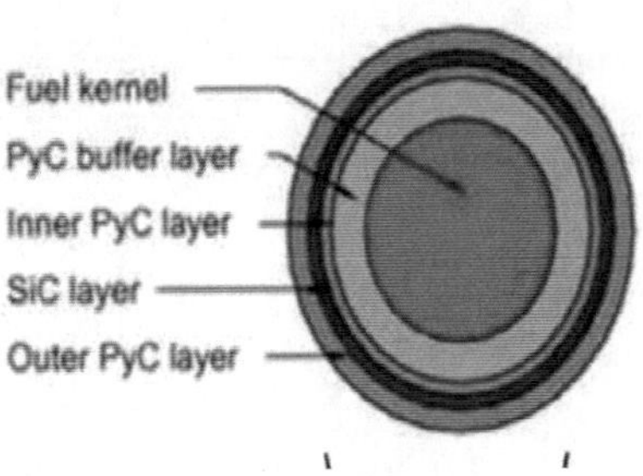

Figure 2. Spherical fuel particle.

The model is considered at the following temperatures states:

1) Cold Zero Power (CZP) – temperature of all materials in a fuel compact cell is 293 K;
2) Hot Full Power (HFP) - temperature of all materials in a fuel compact cell is 1200 K.

Calculation of the fuel compact cell model is performed in two variants:

1) with homogenization of UCO TRISO fuel particles within the zone of a fuel compact (Figure 1), helium gap zone, zone of the block graphite (I-1a);
2) with rigorous modeling of TRISO fuel particles structure (I-1b).

For additional information regarding the compact configuration see [8].

3.2. Code sequence developed for calculation of uncertainties

A special software technology (hereafter referred to as SUW technology) using the 69-group library of covariance information was developed for calculating one-group cross-section uncertainties for main isotopes presented in HTGR neutron physics problems. In application to the MHTGR-350 benchmark it consists of three modules: **MHTGR-350_Wd5, Reactions_Cross-sections**, and **Error_Sigl**.

Program module **MHTGR-350_Wd5** prepares an input file for WIMS-D and performs WIMS-D run. The real three-dimensional fuel compact cell is replaced by a two-dimensional cluster fuel compact cell with conservation of the mean chord in the fuel (with some correction factor) and mass of each material in the actual three-dimensional fuel compact cell and in three-dimensional cell built from two-dimensional cluster fuel compact cell with the same height. Calculation results of **MHTGR-350_Wd5** code are WIMS-D output files containing k_∞ value for the fuel compact cell and all information necessary for the calculation of many-group microscopic cross-sections of the main isotopes presented in the fuel compact cell.

Results of calculations of the multiplication factor are given in tables 1, 2. Values calculated by SERPENT code [12] taken from [13] are used as reference values. Calculations by SCALE-6.1.2 code package were performed by authors of this paper. In so doing, calculations named 'Scale-6.1, Keno-VI, 238 MG' for variant I-1a were carried out by TSUNAMI module, and for variant I-1b – by KENO-VI module with 'doublehet' option. Calculations named 'Scale-6.1, Keno-VI, CE' were carried out by KENO-VI module with library 'ce_v7'.

Table 1. k_∞ of MHTGR fuel compact cell model, I-1a.

Code	CZP		HFP	
	k_∞	Δk, %	k_∞	Δk, %
Serpent (ENDF-B-VII.0)	1.31698 ± 0.00019	–	1.18245± 0.00022	
WIMS-D5, WLUP (ENDF/ B-VI.8, CENDL-2.1)	1.33417	1.31	1.21093	2.41
Scale-6.1, Keno-VI, 238 MG	1.31161	-0.41	1.18302	0.05
Scale-6.1, Keno-VI, CE	1.31857	0.12	1.18911	0.5

Table 2. Value of k∞ for the MHTGR-350 fuel compact cell model, I-1b.

Code	CZP		HFP	
	k∞	**Δk, %**	**k∞**	**Δk, %**
Serpent (ENDF-B-VII.0)	1.37060 ± 0.00021	–	1.24461 ± 0.00023	¬
WIMS-D5, WLUP (ENDF/ B-VI.8, CENDL-2.1)	1.36180	-0.64	1.24474	0.01
Scale-6.1, Keno-VI, 238 MG	1.36595	-0.34	1.24443	-0.01

It should be noted the essential overestimation of the fuel compact cell k_∞ value calculated by the WIMS-D5 code in comparison with calculation by SERPENT and SCALE. This fact is a subject for further investigations.

The effect of homogenization of fuel particles and graphite matrix in a fuel compact is provided in table 3 and calculated by the following formula:

$$\delta k_\infty = \frac{k_\infty^{\hom} - k_\infty^{het}}{k_\infty^{het}} \tag{1}$$

Table 3. Effect of Homogenization.

Code	CZP, %	HFP, %
Serpent, ref.	-3.91	-4.99
WIMS-D5	-2.03	-2.72
Scale-6.1, Keno-VI, 238 MG	-3.98	-4.96

Module **Reactions_Cross-sections** uses the WIMS-D output file to calculate 69-group microscopic cross-sections for main isotopes in the fuel compact cell model and 69-group neutron fluxes in materials of the cell.

Module **Error_Sigl** calculates microscopic one-group fission and capture cross-sections for each isotope and their uncertainties in the form of dispersions and standard deviations.

Uncertainties of one-group microscopic cross-sections are calculated using formulas given in [2] which, in their turn, are taken from [14]. We rewrote them in a modified form, which is more correct and less complicated:

$$\Delta^2 = \sum_{g=1}^{G}\sum_{g'=1}^{G} \alpha_{x,g'} \operatorname{cov}(x,y)_{g',g} a_{y,g} = \alpha_x^T \operatorname{cov}(x,y)\alpha_y \tag{2}$$

where:

Δ^2 – uncertainty (dispersion) of one-group microscopic cross-sections of type "y" connected with uncertainty of cross-sections of type "x";

$\alpha_{x,g}$ – weighting factor based on data from "x" reaction in group g;

$\alpha_{y,g}$ – weighting factor based on data from "y" reaction in group g ;

$$\alpha_{x(y),g} = \frac{\sigma_{x(y),g}\Phi_g}{\bar{\sigma}_{x(y)}\Phi_{tot}} = \frac{\sigma_{x(y),g}\Phi_g}{\sum_{g=1}^{G}\sigma_{x(y),g}\Phi_g} \tag{3}$$

$\alpha_{x,y}$ – weighting group vector connected with corresponding reaction;

upper index "T" – transposed vector, i.e. a vector-row;

$\operatorname{cov}(x,y)_{g'g}$ – element of group relative covariance matric cov(x,y) for reactions types "x" and "y";

$\sigma_{x(y),g}$ – cross-section of corresponding reaction in the group g;

$\bar{\sigma}_{x(y)}$ - one-group cross-section of corresponding reaction;

Φ_g – neutron flux in the group g;

Φ_{tot} – sum of group fluxes.

3.3. Results of the calculations

The covariance information on reactions available in the SCALE-6 library for ^{235}U and ^{238}U is listed below in a symbolic form: σ_t - σ_t, σ_{in} - σ_{in}, $\sigma_{(n,2n)}$ - $\sigma_{(n,2n)}$, σ_f - σ_f, $\sigma_{(n,\gamma)}$ - $\sigma_{(n,\gamma)}$, ν_{tot}-ν_{tot}, σ_{el} - σ_{el}, σ_{el} - σ_t, σ_{el} - σ_{in}, σ_{el} - $\sigma_{(n,2n)}$, σ_{el} - $\sigma_{(n,f)}$, σ_{el} - $\sigma_{(n,\gamma)}$, $\sigma_{(n,\gamma)}$ - σ_t, σ_f - σ_t, σ_f - σ_c, χ - χ.

IOP Conf. Series: Journal of Physics: Conf. Series **781** (2017) 012032 doi:10.1088/1742-6596/781/1/012032

Covariance information used in further calculations is marked in bold, i.e. we took into account the matrix diagonal of reaction covariance information. We did not take into account the influence of the elastic scattering covariance information on uncertainties of fission and capture reactions and influence of the fission covariance information on the uncertainties of capture reactions.

Results of the calculations of one-group microscopic cross-sections, their uncertainties (dispersions) and standard deviations for ^{238}U and ^{235}U for two states of the fuel compact cell model for variant I-1a are provided in the table 4.

Table 4. Values and uncertainties of one-group cross-sections for isotopes ^{238}U and ^{235}U, variant I-1a.

	U-235		**U-238**	
	CZP	HFP	CZP	HFP
σ_f, barn	34.2201	31.0835	4.12395E-02	4.13055E-02
σ_c, barn	9.42600	8.84810	3.43927	4.15513
ν_{tot},	2.43871	2.43885	2.72996	2.72998
$(\Delta_f)^2$	9.29E-06	9.12E-06	2.94E-05	2.94E-05
$(\Delta_c)^2$	1.99E-04	2.12E-04	1.68E-04	1.76E-04
$(\Delta_\nu)^2$	2.15E-06	2.15E-06	3.73E-05	3.74E-05
Δ_f, %	0.3	0.3	0.5	0.5
Δ_c, %	1.4	1.5	1.3	1.3
Δ_ν, %	0.15	0.15	0.61	0.61

Results of the calculations of one-group microscopic cross-sections, their uncertainties (dispersions) and standard deviations for non-fissile isotopes for two states of the fuel compact cell for variant I-1a are provided in the table 5.

Table 5. Values and uncertainties of one-group cross-sections for non-fissile isotopes, variant I-1a.

	O-nat		**Si-nat**	
	CZP	HFP	CZP	HFP
σ_c, barn	9.55637E-04	9.56137E-04	1.20812E-02	1.12005E-02
$(\Delta_c)^2$	2.44E-01	2.45E-01	4.81E-04	4.53E-04
Δ_c, %	49.4	49.5	2.2	2.1
	C-nat, FCM		**C-nat, Block**	
	CZP	HFP	CZP	HFP
σ_c, barn	2.88391E-04	2.70435E-04	2.88063E-04	2.69184E-04
$(\Delta_c)^2$.	2.93E-03	3.25E-03	2.74E-03	3.28E-03
Δ_c, %	5.4	5.7	5.2	5.7

Results of the calculations of one-group microscopic cross-sections, their uncertainties (dispersions) and standard deviations for ^{238}U and ^{235}U for two states of the fuel compact cell model for variant I-1b are provided in table 6.

Table 6. Values and uncertainties of one-group cross-sections for isotopes ^{238}U and ^{235}U, variant I-1b

	U-235		**U-238**	
	CZP	HFP	CZP	HFP
σ_f, barn	33,4272	30.5008	4.10312E-02	4.09922E-02
σ_c, barn	9.12920	8.60040	3.14137	3.78015
ν_{tot},	2.43874	2.43891	2.73265	2.73267
$(\Delta_f)^2$	9.36E-06	9.22E-06	2.94E-05	2.94E-05
$(\Delta_c)^2$	1.96E-04	2.08E-04	1.67E-04	1.75E-04
$(\Delta_\nu)^2$	2.14E-06	2.14E-06	3.70E-05	3.70E-05
Δ_f, %	0.3	0.3	0.5	0.5
Δ_c, %	1.4	1.4	1.3	1.3
Δ_ν, %	0.15	0.15	0.61	0.61

Results of the calculations of one-group microscopic cross-sections, their uncertainties (dispersions) and standard deviations for non-fissile isotopes for two states of the fuel compact cell for variant I-1b are provided in table 7.

Table 7. Values and uncertainties of one-group cross-sections for non-fissile isotopes, variant I-1b.

	O-nat		**Si-nat**	
	CZP	HFP	CZP	HFP
σ_c, barn	9.76850E-04	9.75014E-04	1.23579E-02	1.14642E-02
$(\Delta_c)^2$	2.44E-01	2.45E-01	4.92E-04	4.64E-04
Δ_c, %	49.4	49.5	2.2	2.2
	C-nat, FCM		**C-nat, Block**	
	CZP	HFP	CZP	HFP
σ_c, barn	2.97546E-04	2.78865E-04	2.96010E-04	2.76416E-04
$(\Delta_c)^2$.	2.77E-03	3.06E-03	2.58E-03	2.86E-03
Δ_c, %	5.3	5.5	5.1	5.3
	C-nat, Fuel			
	CZP	HFP		
σ_c, barn	2.87152E-04	2.70533E-04		
$(\Delta_c)^2$.	3.08E-03	3.39E-03		
Δ_c, %	5.8	5.8		

Rather large capture cross-section uncertainty of O-nat attracts attention. In addition, it should be noted that we used covariance information for pair $\sigma_{(n,\gamma)}$- $\sigma_{(n,\gamma)}$ in calculating $(\Delta c)^2$ because covariance information for (n,α) and (n,p) reactions is not available in the library of SCALE-6 code.

3.4. Selection of the most important neutron reactions for the MHTGR-350 benchmark

Selection of the most important neutron reactions, i.e. reactions whose uncertainties give the largest contribution to the k_∞ uncertainty is carried out using one-group formula for k_∞ of the fuel compact cell model:

$$k_\infty=(\nu_5\sigma_f^5\rho_5\Phi_{fuel}+\nu_8\sigma_f^8\rho_8\Phi_{fuel})V_{fuel}/[(\sigma_f^5\rho_5\Phi_{fuel}+\sigma_f^8\rho_8\Phi_{fuel}+\sigma_c^5\rho_5\Phi_{fuel}+\sigma_c^8\rho_8\Phi_{fuel}+\sigma_c^{Onat}\rho_{Onat}\Phi_{fuel}$$
$$+\sigma_{c,\mathrm{fuel}}^{Cnat}\rho_{Cnat}^{\mathrm{Fuel}}\Phi_{\mathrm{fuel}})V_{fuel}+\sigma_{c,\mathrm{FCM}}^{Cnat}\rho_{Cnat}^{\mathrm{FCM}}\Phi_{\mathrm{FCM}}V_{\mathrm{FCM}}+\sigma_{\mathrm{c}}^{Sinat}\rho_{Sinat}^{\mathrm{Clad}}\Phi_{SiC}V_{\mathrm{Clad}}^{\mathrm{SiC}}+\sigma_{\mathrm{c,Block}}^{Cnat}\rho_{Cnat}^{\mathrm{Block}}\Phi_{\mathrm{Block}}V_{\mathrm{Block}}] \quad (4)$$

Let us introduce the notation for the 11 input parameters:

$$\alpha_i=\begin{Bmatrix}\nu_5,\nu_8,\sigma_f^5,\sigma_f^8,\sigma_c^5,\sigma_c^8,\sigma_c^{Onat},\\ \sigma_{\mathrm{c,Fuel}}^{Cnat},\sigma_{c,\mathrm{FCM}}^{Cnat},\sigma_{\mathrm{c}}^{Sinat},\sigma_{\mathrm{c,Block}}^{Cnat}\end{Bmatrix} \quad (5)$$

Dispersion of cell k_∞ connected with uncertainty of input parameter α_i with taking into account one-group covariance C_{ii}, can be written as:

$$\delta_{k_\infty,i}^2=S_{k_\infty,i}C_{ii}S_{k_\infty,i}, \quad (6)$$

Where $S_{k_\infty,i}$ - sensitivity coefficients of value k_∞ to the change of parameters α_i.

$$S_{k_\infty,i}=\frac{\alpha_i}{k_\infty}\frac{\partial k_\infty}{\partial\alpha_i} \quad (7)$$

C_{ii} are provided in the tables 6-7 as Δ^2.

Below there are the explicit expressions for the sensitivity coefficients $S_{k_\infty,i}$.

$$S_{k_\infty,1}=\frac{\nu_5\sigma_f^5\rho_5\Phi_{fuel}V_{fuel}}{k_\infty\gamma} \quad (8)$$

where:

$$\gamma=(\sigma_f^5\rho_5\Phi_{fuel}+\sigma_f^8\rho_8\Phi_{fuel}+\sigma_c^5\rho_5\Phi_{fuel}+\sigma_c^8\rho_8\Phi_{fuel}+\sigma_c^{Onat}\rho_{Onat}\Phi_{fuel}+\sigma_{c,\mathrm{fuel}}^{Cnat}\rho_{Cnat}^{\mathrm{Fuel}}\Phi_{\mathrm{fuel}})V_{fuel}$$
$$+\sigma_{c,\mathrm{FCM}}^{Cnat}\rho_{Cnat}^{\mathrm{FCM}}\Phi_{\mathrm{FCM}}V_{\mathrm{FCM}}+\sigma_{\mathrm{c}}^{Sinat}\rho_{Sinat}^{\mathrm{Clad}}\Phi_{SiC}V_{\mathrm{Clad}}^{\mathrm{SiC}}+\sigma_{\mathrm{c,Block}}^{Cnat}\rho_{Cnat}^{\mathrm{Block}}\Phi_{\mathrm{Block}}V_{\mathrm{Block}} \quad (9)$$

$$S_{k_\infty,2}=\frac{\nu_8\sigma_f^8\rho_8\Phi_{fuel}V_{fuel}}{k_\infty\gamma} \quad (10)$$

$$S_{k_\infty,3}=\frac{\sigma_f^5\rho_5\Phi_{fuel}V_{fuel}}{k_\infty}\frac{1}{\gamma}(\nu_5-k_\infty) \quad (11)$$

$$S_{k_\infty,4}=\frac{\sigma_f^8\rho_8\Phi_{fuel}V_{fuel}}{k_\infty}\frac{1}{\gamma}(\nu_8-k_\infty) \quad (12)$$

$$S_{k_\infty,i}=-(\alpha_i\cdot\omega_{i-4}\cdot\gamma^{-1})_i,\quad i=5,6,...,15 \quad (13)$$

$$\omega_i=\begin{Bmatrix}\rho_5\Phi_{fuel}V_{fuel},\rho_8\Phi_{fuel}V_{fuel},\rho_{Onat}\Phi_{fuel}V_{fuel},\\ \rho_{Cnat}^{\mathrm{Fuel}}\Phi_{\mathrm{fuel}}V_{fuel},\rho_{Cnat}^{\mathrm{FCM}}\Phi_{\mathrm{FCM}}V_{\mathrm{FCM}},\\ \rho_{Sinat}^{\mathrm{Clad}}\Phi_{SiC}V_{\mathrm{Clad}}^{\mathrm{SiC}},\rho_{Cnat}^{\mathrm{Block}}\Phi_{\mathrm{Block}}V_{\mathrm{Block}}\end{Bmatrix} \quad (14)$$

Formula for the total dispersion of k_∞ has the form:

$$\delta_{k_\infty}^2 = \sum_{i=1} \delta_{k_\infty,i}^2 \quad (15)$$

Comparative calculations of sensitivity coefficients of k_∞ of the fuel compact cell model to the change of input parameters α_i are provided in the tables 8, 9. Sensitivity coefficients are calculated using explicit formulas for variants I-1a, I-1b and by SCALE-6.1.2 code for variant I-1a for CZP and HFP states of the fuel compact cell. Note that SAMS module (SCALE-6) that calculates sensitivity and uncertainties does not work for double heterogeneity problems. Therefore, we do not provide results calculated for variant I-1b by SCALE-6.1.2.

Table 8. Sensitivity coefficients of k_∞ of the fuel compact cell to the change of input parameters α_i, CZP.

α_i	$S_{k_\infty,i}$		
	variantI-1b	variantI-1a	Scale-6.1, Keno-VI, variant I-1a
ν_5	**9.92E-01**	**9.93E-01**	**9.93E-01**
ν_8	7.35E-03	7.21E-03	6.64E-03
σ_f^5	***4.38E-01***	***4.50E-01***	***2.31E-01***
σ_f^8	3.69E-03	3.69E-03	3.46E-03
σ_c^5	**-1.51E-01**	**-1.50E-01**	**-2.04E-01**
σ_c^8	**-2.80E-01**	**-2.94E-01**	**-2.41E-01**
σ_c^{Onat}	-1.55E-04	-1.45E-04	-1.01E-04
$\sigma_{c,\mathrm{Fuel}}^{Cnat}$	-1.52E-05		
$\sigma_{c,\mathrm{FCM}}^{Cnat}$	-2.18E-03	-2.03E-03	-2.53E-03
σ_c^{Sinat}	-4.01E-03	-3.75E-03	-5.06E-03
$\sigma_{c,\mathrm{Block}}^{Cnat}$	-4.25E-03	-3.96E-03	-4.99E-03
$\sigma_{c,\mathrm{All}}^{Cnat}$	-6.45E-03	-5.99E-03	-7.52E-03
σ_{el}^8	–	–	**2.85E-02**
$\sigma_{el,FCM}^{Cnat}$	–	–	5.26E-02
$\sigma_{\mathrm{el,Block}}^{Cnat}$	–	–	1.68E-01
$\sigma_{\mathrm{el}}^{Cnat}$	–	–	**2.21E-01**

Table 9. Sensitivity coefficients of k_∞ of the fuel compact cell to the change of input parameters α_i, HFP.

α_i	$S_{k_\infty,i}$		
	variantI-1b	variantI-1a	Scale-6.1, Keno-VI, variant I-1a
ν_5	**9.92E-01**	**9.92E-01**	**9.93E-01**
ν_8	8.04E-03	7.94E-03	7.36E-03
σ_f^5	***4.86E-01***	***4.99E-01***	***2.57E-01***
σ_f^8	4.38E-03	4.42E-03	4.17E-03
σ_c^5	**-1.43E-01**	**-1.40E-01**	**-2.03E-01**
σ_c^8	**-3.38E-01**	**-3.54E-01**	**-2.93E-01**
σ_c^{Onat}	-1.55E-04	-1.45E-04	-1.02E-04
$\sigma_{c,\mathrm{Fuel}}^{Cnat}$	-1.43E-05		
$\sigma_{c,\mathrm{FCM}}^{Cnat}$	-2.04E-03	-1.90E-03	-2.54E-03
σ_c^{Sinat}	-3.72E-03	-3.47E-03	-5.08E-03
$\sigma_{c,\mathrm{Block}}^{Cnat}$	-3.97E-03	-3.70E-03	-4.97E-03
$\sigma_{c,\mathrm{All}}^{Cnat}$	-6.02E-03	-5.60E-03	-7.51E-03
σ_{el}^8	–	–	**3.08E-02**
$\sigma_{el,FCM}^{Cnat}$	–	–	5.64E-02
$\sigma_{\mathrm{el,Block}}^{Cnat}$	–	–	1.96E-01
$\sigma_{\mathrm{el}}^{Cnat}$	–	–	**2.52E-01**

Sensitivity coefficients marked in bold have the largest values. Besides, we marked in ***italic*** some values to draw attention that corresponding values of sensitivity coefficients calculated by explicit formulas are about two times larger than the values calculated by SCALE-6.1.2.

Comparative calculations of uncertainties (standard deviations $\delta_{k_\infty,i}$) of k_∞ of the fuel compact cell to the change of input parameters α_i are provided in the tables 10, 11. In so doing, uncertainties are calculated using the SUW technology for variants I-1a, I-1b and by SCALE-6.1.2 for variant I-1a for CZP and HFP states of the fuel compact cell model.

Table 10. Standard deviations of the k_∞ of the fuel compact cell model, CZP.

α_i	$\delta_{k_\infty,i}$, %		
	variantI-1b	variantI-1a	Scale-6.1, Keno-VI, variant I-1a
ν_5	**1.45E-01**	**1.46E-01**	**2.65E-01**
ν_8	4.47E-03	4.40E-03	7.93E-03
^{235}U(fiss)	***1.34E-01***	***1.37E-01***	*6.06E-02*
^{238}U(fiss)	2.00E-03	2.00E-03	1.82E-03
$^{235}U(n,\gamma)$= ^{235}U(capture)	**2.12E-01**	**2.11E-01**	**2.59E-01**
$^{238}U(n,\gamma)$= ^{238}U(capture)	**3.62E-01**	**3.81E-01**	**3.24E-01**
^{nat}O(capt.)	7.66E-03	7.17E-03	2.46E-04 2.45E-05(n,γ) 2.22E-04(n,α)
^{nat}C(capt)	2.62E-02	2.35E-02	2.19E-02 1.82E-02(n,γ) 1.22E-02(n,α)
^{nat}Si(capt)	8.89E-03	8.23E-03	2.01E-02 2.00E-02(n,γ) 2.44E-03(n,p) 1.15E-03 (n,α)
^{238}U(el)	–		**1.06E-01**
^{nat}C(el)	–		**1.11E-01**
^{235}U(fiss)-$^{235}U(n,\gamma)$	–		**9.70E-02**
δ_{k_∞}, %	4.65E-01[a] –	4.80E-01[a] –	4.97E-01[a] 0.5276[b]

a – total k_∞ uncertainty, caused by uncertainty of the input parameters, listed in the second column (9 parameters).

b - total k_∞ uncertainty, caused by the uncertainty of all input parameters (reactions) taken into account in calculation performed by SCALE-6 code (83 pairs of reactions).

Table 11. Standard deviations of the multiplication factor of the fuel compact cell model, HFP.

α_i	$\delta_{k_\infty,i}$, %		
	variantI-1b	variantI-1a	Scale-6.1, Keno-VI, variant I-1a
ν_5	**1.45E-01**	**1.45E-01**	**2.62E-01**
ν_8	4.89E-03	4.86E-03	8.79E-03
^{235}U(fiss)	***1.47E-01***	***1.51E-01***	*6.65E-02*
^{238}U(fiss)	2.37E-03	2.40E-03	2.20E-03
^{235}U(n,γ)= ^{235}U(capture)	**2.06E-01**	**2.04E-01**	**2.56E-01**
^{238}U(n,γ)= ^{238}U(capture)	**4.47E-01**	**4.70E-01**	**4.00E-01**
^{nat}O(capt.)	7.67E-03	7.18E-03	2.24E-04 2.45E-03(n,γ) 2.23E-02(n,α)
^{nat}C(capt)	2.41E-02	2.40E-02	2.19E-02 1.81E-02(n,γ) 1.23E-02(n,α)
^{nat}Si(capt)	8.02E-03	7.39E-03	2.04E-02 2.02E-02(n,γ) 2.45E-03(n,p) 1.15E-03(n,α)
^{238}U(el)	–		**1.23E-01**
^{nat}C(el)	–		**1.28E-01**
^{235}U(fiss)-^{235}U(n,γ)	–		**1.01E-01**
δ_{k_∞}, %	5.34E-01[a] –	5.54E-01[a] –	5.47E-01[a] 0.5796[b]

a – total k_∞ uncertainty, caused by the uncertainty of the input parameters, recorded in the second column (9 parameters).
b - total k_∞ uncertainty, caused by the uncertainty of all input parameters (reactions), taken into account in calculation performed by SCALE-6 code (83 pairs of reactions).

It is should be noted, that names in the first columns of the tables 10, 11 are shown in an abbreviated form. Every reaction name means in fact a couple of identical reactions, for example ^{238}U(n,γ) means ^{238}U(n,γ)- ^{238}U(n,γ).

Analysis of the results, provided in the tables 10, 11, allows us to suggest the following seven parameters as parameters giving the largest contributions to the uncertainty of the multiplication factor:

- ^{238}U(n,γ);
- ν_5;
- ^{235}U(n,γ) ;
- ^{238}U(el) ;
- ^{nat}C(el) ;
- ^{235}U(fiss)–^{235}U(n,γ) ;
- ^{235}U(fiss).

Note that we included parameter ^{235}U(fiss) as a candidate for these parameters although in all calculations by SCALE code this parameter is not the one of the most important. We did this because all calculations by SCALE code were carried out only for variant 1a, not taking into account the double heterogeneity. In the same time, in calculations by SUW technology this parameter is surely one of the most important ones both for variant 1a and for variant 1b. Besides, it should be noted that parameters ^{238}U(el), ^{nat}C(el), ^{235}U(fiss)–^{235}U(n,γ) give close values of contribution to the uncertainty of multiplication factor (k_∞) of the fuel compact cell model.

4. Lattice Physics: Calculation of the Few-Groups Macroscopic Cross-Sections

4.1. Fuel block of the MHTGR-350 benchmark

Fuel block of the MHTGR-350 (Figure 3) in Hot Full Power state [8] is used to calculate few-groups macroscopic cross-sections. Fresh fuel block includes six burnable poison compact in the corners of the block. The problem requires accounting the double-heterogeneity effects, i.e. self-shielding that occurs both within the fuel and the burnable poison. Burnable poison compacts consist of a several thousands of coated particles from B_4C with buffer (porous graphite) and PyC layers (dense graphite).

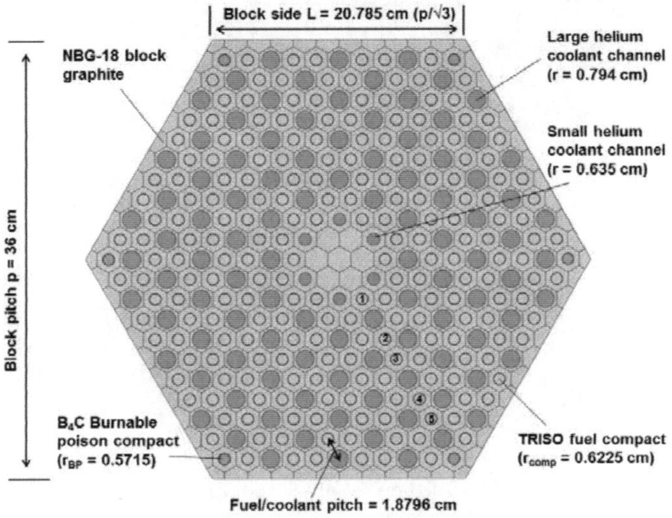

Figure 3. Model of the fuel block.

For additional information regarding the fuel block configuration, please refer to the [8, 13]

4.2. Simplified Model for Fuel Block

For calculation of the fuel block of the MHTGR-350 benchmark we used our experience obtained at elaboration of simplified models for fuel block for GT-MHR [15]. A simplified model for the fuel block of the MHTGR-350 is presented in Figure 4. The basic assumptions for elaboration of this simplified model are the following:

- Three dimensional cell of the fuel compact is replaced by a two dimensional cluster cell of the fuel compact with 52 cylindrical rods.
- Diameter of fuel rods is chosen from condition of equality of mean chord in fuel with some coefficient.
- A cell of the fuel compact is surrounded by materials of the fuel block (graphite and helium), accounted per one cell of the fuel compact.
- There are eight poison rods in surrounding.

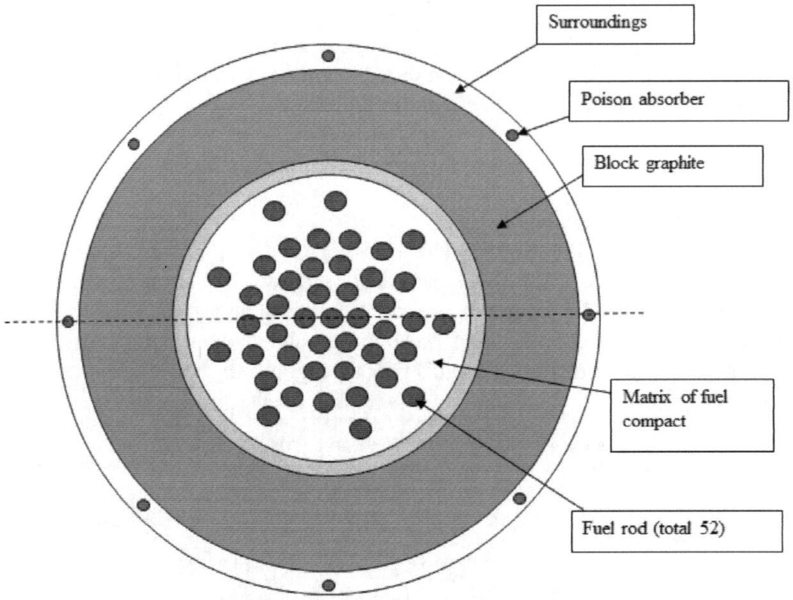

Figure 4. Simplified model for Fuel Block for WIMS-D code.

4.3. Calculation of k_∞ of a simplified model of the fuel block

Calculation of the k∞ of the simplified model of the fuel block is performed by WIMS-D5 code (Figure 4). Table 12 shows the k_∞ of the full model of the fuel block presented in Figure 3 calculated by SERPENT code with two presentations of fuel particles in fuel compacts: location in random order and in the regular lattice [13].

Table 12. Multiplication factor values of the fuel block model.

Code	k_∞	Δk, %
Serpent - random (ENDF-B-VII.0), reference	1.05679	–
Serpent - regular lattice (ENDF-B-VII.0), reference	1.06120	0.42
WIMS-D5, WLUP (ENDF/B-VI.8, CENDL-2.1)	1.05928	0.24

4.4. Values and uncertainties of one-group microscopic cross-sections

Table 13 shows the calculation results of one-group microscopic fission and radiation capture (n,γ) cross-sections and the number of secondary neutrons per fission, as well as their uncertainties in the form of relative dispersion and standard deviation in % for isotopes ^{235}U, ^{238}U for the simplified fuel block model.

IOP Conf. Series: Journal of Physics: Conf. Series **781** (2017) 012032 doi:10.1088/1742-6596/781/1/012032

Table 13. Values and uncertainties of one-group microscopic cross-sections for isotopes ^{235}U, ^{238}U for the simplified fuel.

	U-235	U-238
σ_f, barn	36.5367	4.14150E-02
σ_c, barn	9.9340	3.89793
ν_{tot},	2.43882	2.73507
$(\Delta_f)^2$	9.69E-06	2.93E-05
$(\Delta_c)^2$	1.72E-04	1.69E-04
$(\Delta_\nu)^2$	2.21E-06	3.25E-05
Δ_f, %	0.31	0.54
Δ_c, %	1.31	1.30
Δ_ν, %	0.15	0.57

Table 14 shows the resulting microscopic one-group capture (including reactions (n,γ), (n,p), (n,α)) cross-sections, as well as their uncertainties in the form of relative dispersion and standard deviation in percent for non-fissile isotopes for the simplified fuel block model.

Note that uncertainties of the absorption cross sections of oxygen and boron isotopes are very large.

Table 14. Values and uncertainties of one-group microscopic cross-sections for non-fissile isotopes for the simplified model.

	O-nat	Si-nat
σ_c, barn	1.01083E-03	1.31777E-02
$(\Delta_c)^2$	2.44E-01	5.05E-04
Δ_c, %	49.39	2.25
	C-nat, FCM	**C-nat, Block**
σ_c, barn	3.11093E-04	3.08934E-04
$(\Delta_c)^2$	2.44E-03	2.21E-03
Δ_c, %	4.94	4.70
	C-nat, Fuel	**C-nat, OKP**
σ_c, barn	3.15565E-04	3.08305E-04
$(\Delta_c)^2$	2.83E-03	2.19E-03
Δ_c, %	5.32	4.68
	B-10	**B-11**
σ_c, barn	253.469	3.79880E-04
$(\Delta_c)^2$	1.92E-02	3.43E-01
Δ_c, %	13.85	58.53
	C-nat, BP	
σ_c, barn	3.12322E-04	
$(\Delta_c)^2$	2.57E-03	
Δ_c, %	5.07	

Rather large capture cross-section uncertainty of O-nat and B-10 attracts attention.

4.5. Selection of the most important reactions on isotopes of the fuel block model of the MHTGR-350 benchmark

Selection of the most important reactions on isotopes, i.e. reactions whose uncertainties give the largest contribution to the k∞ uncertainty is carried out using one-group formula for multiplication factor k_∞ of the simplified fuel block model:

$$k_\infty = (\nu_5\sigma_f^5\rho_5\Phi_{fuel} + \nu_8\sigma_f^8\rho_8\Phi_{fuel})V_{fuel} / [(\sigma_f^5\rho_5\Phi_{fuel} + \sigma_f^8\rho_8\Phi_{fuel} + \sigma_c^5\rho_5\Phi_{fuel} + \sigma_c^8\rho_8\Phi_{fuel} + \sigma_c^{Onat}\rho_{Onat}\Phi_{fuel} + \sigma_{c,\text{fuel}}^{Cnat}\rho_{Cnat}^{\text{Fuel}}\Phi_{\text{fuel}})V_{fuel} + \sigma_{c,\text{FCM}}^{Cnat}\rho_{Cnat}^{\text{FCM}}\Phi_{\text{FCM}}V_{\text{FCM}} + \sigma_{\text{c}}^{Sinat}\rho_{Sinat}^{\text{Clad}}\Phi_{SiC}V_{\text{Clad}}^{\text{SiC}} + \sigma_{\text{c,Block}}^{Cnat}\rho_{Cnat}^{\text{Block}}\Phi_{\text{Block}}V_{\text{Block}} + \sigma_{\text{c,Okr}}^{Cnat}\rho_{Cnat}^{\text{Okr}}\Phi_{\text{Okr}}V_{\text{Okr}} + \left(\sigma_{c,\text{BP}}^{Cnat}\rho_{Cnat}^{\text{BP}} + \sigma_{\text{c}}^{B10}\rho_{B10}^{\text{BP}} + \sigma_{\text{c}}^{B11}\rho_{B11}^{\text{BP}}\right)\Phi_{BP}V_{\text{BP}}] \quad (16)$$

k_∞ are calculated by the formula (14) with using one-group cross-sections given in Tables 13, 14 is 1.059297, which is very close to corresponding value from table 13.

Let us introduce the notation for the formula (16) input parameters:

$$\alpha_i = \begin{Bmatrix} \nu_5, \nu_8, \sigma_f^5, \sigma_f^8, \sigma_c^5, \sigma_c^8, \sigma_c^{Onat}, \sigma_{\text{c,Fuel}}^{Cnat}, \sigma_{\text{c,FCM}}^{Cnat}, \\ \sigma_{\text{c}}^{Sinat}, \sigma_{\text{c,Block}}^{Cnat}, \rho_{Cnat}^{\text{Okr}}, \sigma_{c,\text{BP}}^{Cnat}, \sigma_{\text{c}}^{B10}, \sigma_{\text{c}}^{B11} \end{Bmatrix} \quad (17)$$

Dispersion of the k_∞ of the fuel block model connected with the uncertainties of the input parameters α_i taking into account one-group covariance C_{ii} (table 13, 14) is calculated by the formula (6). Sensitivity coefficients of the k_∞ to the change of α_i are calculated by the formula (7).

Below there are the explicit expressions for the sensitivity coefficients $S_{k_\infty,i}$.

$$S_{k_\infty,1} = \frac{\nu_5\sigma_f^5\rho_5\Phi_{fuel}V_{fuel}}{k_\infty\gamma} \quad (18)$$

where:

$$\gamma = (\sigma_f^5\rho_5\Phi_{fuel} + \sigma_f^8\rho_8\Phi_{fuel} + \sigma_c^5\rho_5\Phi_{fuel} + \sigma_c^8\rho_8\Phi_{fuel} + \sigma_{\text{c}}^{Onat}\rho_{Onat}\Phi_{fuel} + \sigma_{\text{c,fuel}}^{Cnat}\rho_{Cnat}^{\text{Fuel}}\Phi_{\text{fuel}})V_{fuel} + \sigma_{\text{c,FCM}}^{Cnat}\rho_{Cnat}^{\text{FCM}}\Phi_{\text{FCM}}V_{\text{FCM}} + \sigma_{\text{c}}^{Sinat}\rho_{Sinat}^{\text{Clad}}\Phi_{SiC}V_{\text{Clad}}^{\text{SiC}} + \sigma_{\text{c,Block}}^{Cnat}\rho_{Cnat}^{\text{Block}}\Phi_{\text{Block}}V_{\text{Block}} + \sigma_{\text{c,Okr}}^{Cnat}\rho_{Cnat}^{\text{Okr}}\Phi_{\text{Okr}}V_{\text{Okr}} + \left(\sigma_{c,\text{BP}}^{Cnat}\rho_{Cnat}^{\text{BP}} + \sigma_{\text{c}}^{B10}\rho_{B10}^{\text{BP}} + \sigma_{\text{c}}^{B11}\rho_{B11}^{\text{BP}}\right)\Phi_{BP}V_{\text{BP}}. \quad (19)$$

$S_{k_\infty,i}$, i=2..4, are calculated by the formulas (10)-(12).

$$S_{k_\infty,i} = -(\alpha_i \cdot \omega_{i-4} \cdot \gamma^{-1})_i, \quad i = 5,6,...,15 \quad (20)$$

$$\omega_i = \begin{Bmatrix} \rho_5\Phi_{fuel}V_{fuel}, \rho_8\Phi_{fuel}V_{fuel}, \rho_{Onat}\Phi_{fuel}V_{fuel}, \\ \rho_{Cnat}^{\text{Fuel}}\Phi_{\text{fuel}}V_{fuel}, \rho_{Cnat}^{\text{FCM}}\Phi_{\text{FCM}}V_{\text{FCM}}, \rho_{Sinat}^{\text{Clad}}\Phi_{SiC}V_{\text{Clad}}^{\text{SiC}}, \\ \rho_{Cnat}^{\text{Block}}\Phi_{\text{Block}}V_{\text{Block}}, \rho_{Cnat}^{\text{Okr}}\Phi_{\text{Okr}}V_{\text{Okr}}, \rho_{Cnat}^{\text{BP}}\Phi_{BP}V_{\text{BP}}, \\ \rho_{B10}^{\text{BP}}\Phi_{BP}V_{\text{BP}}, \rho_{B11}^{\text{BP}}\Phi_{BP}V_{\text{BP}} \end{Bmatrix} \quad (21)$$

Total dispersion is calculated by the formula (15).

Table 15 provides calculations of sensitivity coefficients of k_∞ of the simplified fuel block model to the change of input parameters α_i. Sensitivity coefficients are calculated using explicit formulas.

Table15. Sensitivity coefficients of k_∞ of the simplified fuel block model to the change of input parameters α_i.

α_i	$S_{k_\infty,i}$
ν_5	9.89E-01
ν_8	6.76E-03
σ_f^5	5.59E-01
σ_f^8	4.14E-03
σ_c^5	-1.17E-01
σ_c^8	-2.47E-01
σ_c^{Onat}	-1.14E-04
$\sigma_{c,Fuel}^{Cnat}$	-1.18E-05
$\sigma_{c,FCM}^{Cnat}$	-1.65E-03
σ_c^{Sinat}	-3.10E-03
$\sigma_{c,Block}^{Cnat}$	-3.21E-03
$\sigma_{c,Okr}^{Cnat}$	-2.29E-03
$\sigma_{c,BP}^{Cnat}$	-3.01E-05
$\sigma_{c,All}^{Cnat}$	-7.20E-03
σ_c^{B10}	-1.94E-01
σ_c^{B11}	-1.17E-06

Table 16 provides uncertainties (standard deviation) of k_∞ of the simplified fuel block model to the change of input parameters α_i. Uncertainties are calculated using SUW technology.

Analysis of the results, provided in the tables 10, 11, 16, allows us to suggest the following eight parameters whose uncertainties give the largest contribution to the k_∞ uncertainty of the fuel block model:

- ^{10}B(capt)
- ^{238}U(n,γ);
- ν_5;
- ^{235}U(n,γ) ;
- ^{238}U(el);
- ^{nat}C(el);
- ^{235}U(fiss)–^{235}U(n,γ);
- ^{235}U(fiss).

It is should be noted that ^{10}B(capt) (((n,α)+ (n,γ)) reaction contribution to the k_∞ uncertainty of simplified fuel block model is almost ten times more than the other reactions.

Table 16. Standard deviations of the k_∞ of the simplified fuel block model to the change of input parameters α_i.

α_i	$\delta_{k_\infty,i}$, %
ν_5	**1.48E-01**
ν_8	3.87E-03
^{235}U(fiss)	**1.75E-01**
^{238}U(fiss)	2.25E-03
^{235}U(n,γ)= ^{235}U(capture)	**1.54E-01**
^{238}U(n,γ)= ^{238}U(capture)	**3.22E-01**
^{nat}O(capt.)	5.64E-03
^{nat}C(capt)	1.98E-02
^{nat}Si(capt)	6.83E-03
^{10}B(capt)	**2.64**
^{11}B(capt)	6.75E-05
δ_{k_∞}, %	2.68

5. Conclusions

The main achievements of this work are the following:

- Development of an original 69-group library of covariance information in a special format for main isotopes typical for HTGR. At the moment library contains covariance information for the following isotopes and elements: ^{235}U. ^{238}U. C-nat, ^{16}O, ^{4}He, Si-nat, ^{28}Si, ^{29}Si, ^{30}Si, ^{10}B, ^{11}B. The library consists of individual files for each isotope or element, and can be easily extended in the future. The library is planned to be used for estimation of nuclear data uncertainty in calculating HTGR parameters in design calculations.
- Results of uncertainty calculations for one-group fission and capture cross-sections for main isotopes of the MHTGR-350 benchmark and uncertainties of the multiplication factor of the fuel compact model. One-group uncertainties are calculated using the developed SUW code technology and modules of the SCALE-6 code package (TSUNAMI, KENO-IV, SAMS).
- Results of calculations of sensitivity coefficients and standard deviations that allowed to distinguish eight nuclear reactions whose uncertainties give the largest contribution to the multiplication factor uncertainty of the fuel block model:
  - ^{10}B(capt);
  - ^{238}U(n,γ);
  - ν_5;
  - ^{235}U(n,γ) ;
  - ^{238}U(el);
  - ^{nat}C(el);
  - ^{235}U(fiss)–^{235}U(n,γ);
  - ^{235}U(fiss).

It is shown that the main contribution to the multiplication factor uncertainty of the fuel block model makes the reaction ^{10}B(capt) (((**n**,α)+ (**n**,γ)).

Acknowledgments
This study is carried out with partial support of IAEA in frames of Coordinated Research Project "HTGR Reactor Physics, Thermal-hydraulics and Depletion Uncertainties Analysis". Authors are grateful to IAEA for possibility to take part in this IAEA Project.

References

[1] Reitsma F, Strydom G, Tyobeka B, Ivanov K 2012 The IAEA coordinated research program on HTGR reactor physics, thermal-hydraulics and depletion uncertainty analysis: description of the benchmark test cases and phases *HTR-2012 Conference,* Tokyo, Japan, October 28 – November 1

[2] Strydom G, Reitsma F, Ivanov K 2013 *HTGR Reactor Physics, Thermal-hydraulics and Depletion Uncertainty Analysis, Prismatic HTR Benchmark, Draft, Rev. 1, IAEA HTGR UAM CRP – PMR 350 MW Benchmark Specification,* Vienna, Austria

[3] Strydom G, Reitsma F, Ivanov K *2014 HTGR Reactor Physics, Thermal-hydraulics and Depletion Uncertainty Analysis, Prismatic HTR Benchmark Definition: Phase I, Draft, Rev. 2, IAEA HTGR UAM CRP – PMR 350 MW Benchmark Specification,* Vienna, Austria

[4] Strydom G, Bostelmann F, Reitsma F, Ivanov K 2014 *IAEA Coordinated Research Project on HTGR Reactor Physics, Thermal-hydraulics and Depletion Uncertainty Analysis. Prismatic HTGR benchmark Definition: Phase I. Draft Rev. 4, IAEA,* 26 November

[5] Bostelmann F, Strydom G 2014 *PMR 350 MW Benchmark Specifications. Phase I – Exercise I-1a/b. INL, Nuclear Science & Engineering Division. 2nd RCM Meeting: HTGR UAM,* December 2-5, Vienna, Austria

[6] Reitsma F, Strydom G, Bostelmann F, Ivanov K 2014 *The IAEA Coordinated Research Program on HTGR Uncertainty Analysis: Phase I Status and Initial Results. Paper HTR 2014-51106*

[7] Ortensi J *et al* 2013 *Prismatic Coupled Neutronics/Thermal Fluids Transient Benchmark of the MHTGR-350 MW Core Design: Benchmark Definition, OECD Nuclear Energy Agency NEA/NSC/DOC DRAFT (03/01/2013)*

[8] Strydom G, Bostelmann F 2015 *IAEA Coordinated Research Project on HTGR Physics, Thermal-hydraulics and Depletion Uncertainty Analysis: Prismatic HTGR Benchmark Definition: Phase I. INL/ EXT-15-34868, Revision 1,* September

[9] Greene N M, Dunn M E 2009 User's guide for AMPX utility modules *ORNL/TM-2005/39*, Version 6, Vol. **III**, Sect.M15

[10] Kodeli 2003 Manual for ANGELO2 and LAMBDA codes *NEA-1264/05 package*

[11] Rearden B T, Jessee M A 2009 TSUNAMI utility modules *ORNL/TM-2005/39*, Version 6, Vol. **III**, Sect.M18

[12] Leppänen J 2007 *Development of a new Monte Carlo reactor physics code*, PhD Thesis, Helsinki University of Technology

[13] Strydom G, Bostelmann F, Su Jong Yoon 2015 *Results for Phase I of the IAEA Coordinated Research Project on HTGR Uncertainties. INL/EXT-14-32944, Revision 2,* September

[14] Ivanov K, Avramova M, Kamerow S, Kodeli, Sartori E, Ivanov E, Cabellos O 2012 *Benchmark for Uncertainty Analysis in Modelling (UAM) for Design, Operation and Safety Analysis of LWRs, Volume I: Specification and Support Data for the NeutronicsCases (Phase I), Version 2.0, OECD Nuclear Energy Agency NEA/NSC/DOC,* March

[15] Boyarinov V F 2003 *Investigation of Some Models and Approximations Applied at Calculation of GT-MHR Fuel Assemblies / Proceeding of International Conference M&C,* Gatlinburg, Tennessee, USA, April 6-11

Experimental analysis of liquid-metal reactor scram rod kinematic characteristics

F D Konovalenko[1] **and S I Kondrashov**[2]

[1] Engineer, Ph. D student, JSC «NIKIET», Moscow, Russia
[2] Engineer, JSC «NIKIET», Moscow, Russia

E-mail: f.konovalenko@nikiet.ru

Abstract. This article represents the results of computational and experimental research of liquid-metal research reactor control rod kinematics. In this research liquid-metal coolant (sodium) was simulated by water. Investigation of control rod scram-mode movement duration and investigation of velocity of movable parts near the bump of damper are the purposes of this research. Also mathematic simulation of control rod movement in scram mode was performed. Computational results for some modes of water circulation comply with experimental results well. Results of this work will be used for tests of scram rod drive of above-named research reactor. It will significantly simplify the scram rod drive testing stand construction.

1. Introduction

In this work the results of experimental evaluation and mathematical simulation of kinematics of scram rod of research liquid-metal nuclear reactor are presented. In this research reactor control and scram rods are moved in guiding pipes. Liquid-metal coolant flows through these pipes in bottom-upwards direction. The scram rod is moved by rod drive. In case of scram signal or zeroing of rod drive scram rod moves into the core by gravity or by scram springs force.

The definition of scram rod model movement time in case of opposite liquid flow and discovering the hydraulic damper characteristics were the purposes of this work. Received data will be used for scram rod drive testing. In this testing scram rod mass simulator will be moved in slack water. Therefore testing stand construction can be simplified.

Also the mathematical simulation of scram rod movement and comparison simulation and experimental results were provided.

Depending on construction, technical characteristics and work conditions rod drives can include different damping devices [1]. On the one hand, damper must provide smooth scram rod slowdown to prevent it breakdown. On the other hand, scram rod movement time is important parameter, which influence on nuclear safety. Hydraulic damper was chosen for this research nuclear reactor. The main advantage of hydraulic damper is no rebound after the scram rod slowdown. Similar researches (without damper characteristics discovering) were provided in India [2].

2. Testing object

The object of experimental research was a scram rod model and a hydraulic damper model. Scram rod model was mounted in guiding pipe and can be moved axially over a distance of (650 ± 5) mm. Geometric parameters of scram rod model, guiding pipe and hydraulic damper model consist with geometric parameters of real scram rod, pipe and hydraulic damper. Scram rod model mass and real scram rod mass are equal.

During the experiments the liquid-metal coolant (sodium) was substituted by water. Scram rod movement time depends on hydraulic drag force from the flow. Hydraulic drag force is proportional to pressure fall. Water friction coefficient is more than sodium friction coefficient for about 14 % in case of same water and sodium velocities. Hence, we may suppose, that friction pressure fall is independent of Reynolds criteria [3].

Pressure fall at local hydraulic resistances depends on geometrical characteristics, fluid density and fluid velocity. Therefore pressure falls on scram rod for water and sodium will be equal under the following conditions:

$$\frac{\rho_{H2O} \cdot v_{H2O}^2}{2} = \frac{\rho_{Na} \cdot v_{Na}^2}{2} \quad (1)$$

here ρ_{H2O} – water density, kg/m^3;

ρ_{Na} – sodium density, kg/m^3[4];

v_{H2O} – water velocity, m/s;

v_{Na} – sodium velocity, m/s.

The necessary water velocity for correct sodium velocity simulation can be found from this equation.

Testing stand consists of flowing circuit and testing devices: pipe, scram rod model and hydraulic damper. In different testing modes water flow varies from 0 to 12 m^3/h.

Initially hydraulic damper plunger leans on hydraulic damper shank. Scram rod model is moved upwards over a distance (650 ± 5) mm and is fixed in the top position. Scram rod model falling is provided by gravity. For measuring scram rod model coordinate and falling time is used wire displacement sensor WDS-3000-SR-I.

The wire of the displacement sensor is fixed to the scram rod model. The wire is coiled on the drum of the sensor. Rotation angle of the drum is transduced into current. Output signal of the sensor is given to oscilloscope through the current-to-voltage converter. Scram rod model displacement and displacement time can be measured by the oscillogramm. The scheme of the testing stand is presented on figure 1.

3. Testing methods

In [1, 5] the following method is described: the pressure fall on the scram rod is measured and the scram rod velocity is calculated. In our tests scram rod coordinate is measured directly, and scram rod velocity could be calculated as the first time derivative of the scram rod coordinate.

At the tests common mass of scram rod model and moving parts of the stand was equal to real masses (36±1) kg. Scram springs force has been ignored: it was assumed, that this force will be compensated by the friction force at the moving parts of the scram rod drive.

Oscillogramms of scram rod model falls were saved to graphics files, and then the scram rod model falling time and scram rod model velocity at the bottom mechanical stop were determined.

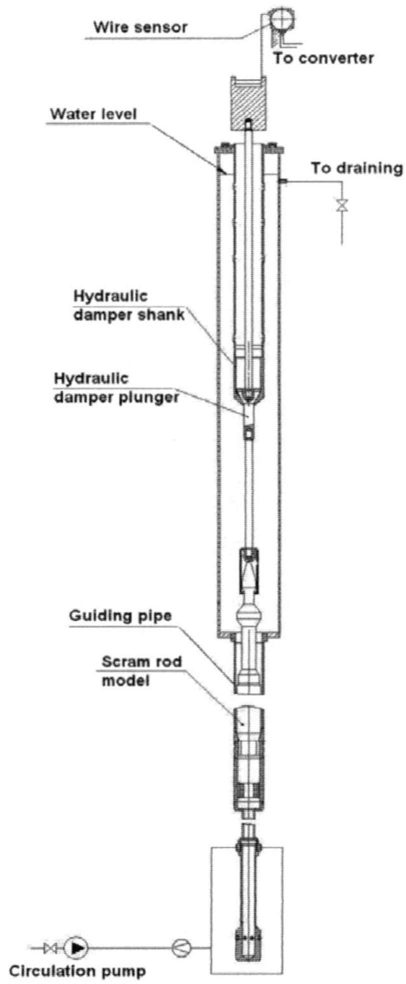

Figure 1. Scheme of the testing stand.

Experiments were provided with different damper plunger construction: with or without vertical slashes for liquid flow (figure 2).

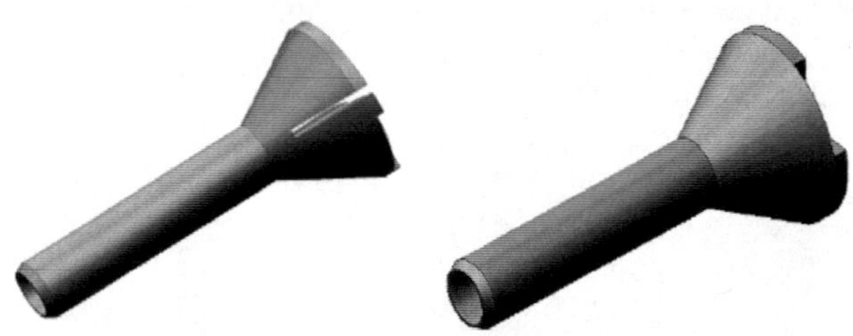

Figure 2. Different damper plunger constructions.

4. Experiment results

As the result of the experiments the scram rod model falling time for different values of water flow was determined. Also the scram rod model velocity at the bottom can be determined by the oscillogramms (figure 3).

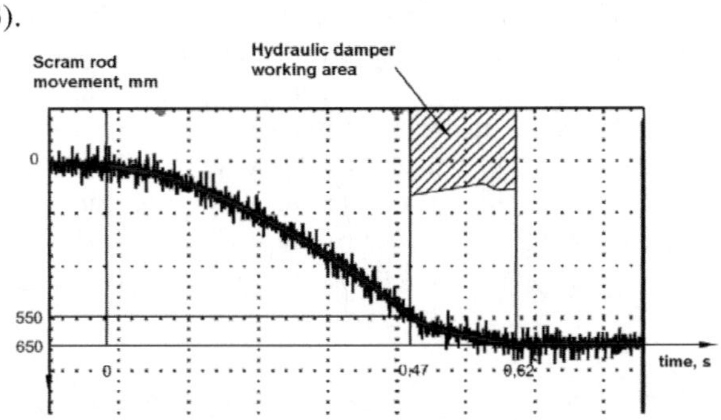

Figure 3. Example of oscilogramm (water flow – 2 m^3/h, plunger without slashes).

Common graphics are presented on figure 4.

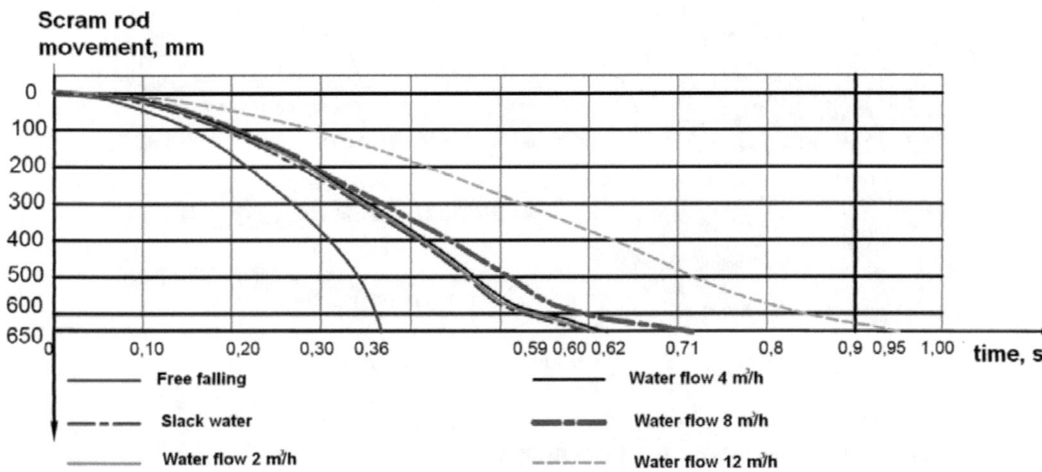

Figure 4. Scram rod model displacement for different values of water flow.

The test results analysis lead to the following conclusions:

- while water flow increasing the scram rod falling time is also increasing because of the hydraulic resistance force increases (but not more than 1 s);
- hydraulic damper without plunger slashes provides more smooth slowdown than the other one;
- the scram rod model displacement time is practically independent of the water flow value for water flow values lower than 4 m^3/h.

The given results for models with real geometric parameters allow to determine the acceptable value of scram rod model falling time at scram rod drive tests in the slack water.

5. Analysis results

The approximate correlations for simple-formed bodies movement in viscous fluid calculations are presented in [1, 5]. These correlations are inapplicable because of complex form of the hydraulic damper plunger and short scram rod displacement time. Therefore the scram rod movement can be calculated numerically. To simplify the mathematical model the following method was applied: the pressure falls on scram rod and on hydraulic damper plunger were calculated separately, and the common pressure fall was calculated as the sum of partial pressure falls.

For example in work [6] pressure falls were calculated analytically, and the viscous friction was neglected. But in work [6] moving parts had simple form (cylinder). Therefore methods presented in [6] are unacceptable in our case. For example, the results of plunger movement calculation with the constant plunger velocity hypothesis are presented in work [7].

In our work the simulation was provided with SALOME-6.6 (grid generation and results representation) and Code-Saturne 4.0 [8] (flow calculation). The calculation was carried out for different plunger constructions: with and without slashes (figure 5). K-ε turbulence model [9] was used for the flow calculation. And the pressure falls were received as the result of the calculation.

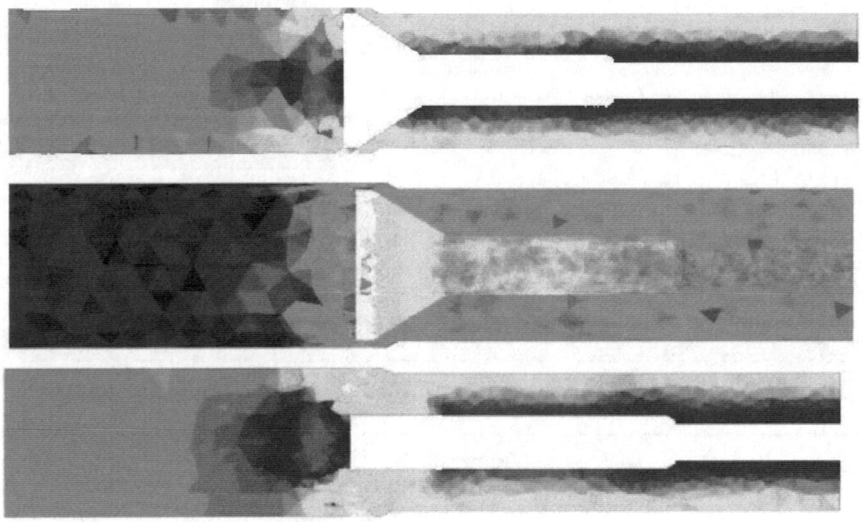

Figure 5. SALOME calculation result representation.

For the simulation the following assumptions were accepted:

- fluid properties in the testing area are constant;
- the hydraulic resistance coefficient is constant for each part of plunger way (before and after hydraulic damper plunger pinch), therefore the pressure fall is proportional to velocity squared;
- some construction simplifications were setted.

The following calculation method was adopted: the pressure fall was calculated numerically for fluid velocity 1 m/s, then the resistance force was calculated for each value of fluid velocity, using the hydraulic resistance coefficient constance hypothesis. The resistance force was used in equation (2) to determine the scram rod speed and coordinate for each iteration.

$$\mathbf{x}''=\left(\mathbf{m}\cdot\mathbf{g}-\mathbf{F_a}-\frac{\mathbf{P}\cdot((\mathbf{x}'+\mathbf{u})/\mathbf{u_0})^2}{\mathbf{S}}\right)/\mathbf{m} \qquad (2)$$

here x, x', x'' – scram rod coordinate, m; velocity, m/s; acceleration, m/s^2, respectively;

m – moving parts mass, kg;

F_a – Archimedes force, N;

P – calculated pressure fall for fluid velocity u_0,Pa;

u_0 – fluid velocity for simulation, 1 m/s;

S_{lob} – plunger face area, m^2.

The received results have good agreement with the experimental results for low values (lower than

4 m^3/h) of water flow (figure 6).

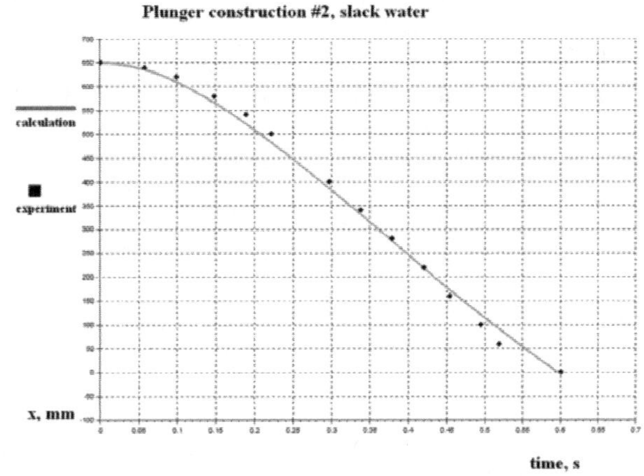

Figure 6. The comparison of experimental and calculation results (slack water).

For the water flow values more than 4 m^3/h the value of the scram rod movement velocity is overestimated. This fact can be explained with model geometrical construction simplifications.

6. Conclusions

The experimental and calculation investigation of liquid-metal research reactor scram rod kinematics were provided [10, 11]. The practical utility of this work consists in probability to simplify construction of scram rod drive testing stand: it is possible to test the drive in slack water using the results of the present work.

The results of the mathematical simulation have good agreement with the experimental results for low values of water flow. Hence, such calculation method can be used for estimation of hydraulic dampers kinematic characteristics only, but it is enough to solve the design problems.

References

[1] Emelyanov I Y, Voskoboinikov V V, Maslenok B A 1978 *Principle of Design of System Actuators of Reactor Control and Safety System* (Moscow: Publishing house «Atomizdat»).

[2] V Rajan Babu, G Thanigaiyarasu, P Chellapandi 2014 Mathematical Modeling of Performance of Safety Rod and Its Drive Mechanism in Sodium Cooled Fast Reactor During Scram Action *Nuclear Engineering and Design* **№ 278.** P. 601…617.

[3] Kirillov P L, Yuriev Y S, Bobkov V P 1990 *Guide of Thermohydraulic Calculations: Nuclear Reactors, Coolers, Steam Generators* (Moscow: Publishing house «Energoatomizdat»).

[4] Borishanskiy V M, Kutateladze S S, Novikov I I, Fedynskiy O S 1976 *Liquid-metal Coolants* (Moscow: Publishing house «Atomizdat»).

[5] Ionaytis R R, Stobeckiy V N 1972 *Nuclear Reactor Control System Hydraulics* (Moscow: Publishing house «Atomizdat»).

[6] Sujuan Jiao, Yu Wang, Lei Zhang, Hongxing Hua 2010 Shock Wave Characteristics of a Hydraulic Damper for Shock Test Machine *Mechanical systems and signal processing* **№ 24.** P. 1570…1578.

[7] Branislav Titurus, Jonathan du Bois, Nick Lieven, Robert Hansford 2010 A Method for the Identification of Hydraulic Damper Characteristics From Steady Velocity Inputs *Mechanical systems and signal processing* **№ 24.** P. 2868…2887.

[8] Code_Saturne version 4.0.0 practical user's guide. EDF R&D, 2015. 202 p.

[9] Kalishevskiy L L, Nikitushkina G P, Streltsov V Y 2000 *Calculation Methods of Turbulent*

Boundary Layer (Moscow, BMSTU publishing house).

[10] Dragunov U G, Lopatkin A V, Lukasevich I B, Romanova N V, Tretyakov I T, Svyatkin M N, Kochetkov L A 2013 Experimental capabilities of MBIR reactor *VANT* **№33**, The The research reactors P. 36-40.

[11] Dragunov U G, Lopatkin A V, Lukasevich I B, Romanova N V, Tretyakov I T, Multipurpose Fast Research Reactor (МБИР) as an Innovative Tool for Development of Nuclear Energy Technologies *Atomic energy* 2012 **№113-1** P. 25-28.

Application of ATHLET/DYN3D coupled codes system for fast liquid metal cooled reactor steady state simulation

V Ivanov[1], A Samokhin[1], I Danicheva[1], N Khrennikov[1], J Bouscuet[2], K Velkov[2] and I Pasichnyk[2]

[1]Scientific and Engineering Centre for Nuclear and Radiation Safety (SEC NRS), Malaya Krasnoselskayast, 2/8, bld. 5, 107140, Moscow, Russian Federation
[2] Gesellschaft für Anlagen- und Reaktorsicherheit (GRS), Forschungszentrum, Boltzmannstraße 14, 85748 Garching bei München, Germany

E-mail: asamokhin@secnrs.ru

Abstract. In this paper the approaches used for developing of the BN-800 reactor test model and for validation of coupled neutron-physic and thermohydraulic calculations are described. Coupled codes ATHLET 3.0 (code for thermohydraulic calculations of reactor transients) and DYN3D (3-dimensional code of neutron kinetics) are used for calculations. The main calculation results of reactor steady state condition are provided. 3-D model used for neutron calculations was developed for start reactor BN-800 load. The homogeneous approach is used for description of reactor assemblies. Along with main simplifications, the main reactor BN-800 core zones are described (LEZ, MEZ, HEZ, MOX, blankets). The 3D neutron physics calculations were provided with 28-group library, which is based on estimated nuclear data ENDF/B-7.0. Neutron SCALE code was used for preparation of group constants. Nodalization hydraulic model has boundary conditions by coolant mass-flow rate for core inlet part, by pressure and enthalpy for core outlet part, which can be chosen depending on reactor state. Core inlet and outlet temperatures were chosen according to reactor nominal state. The coolant mass flow rate profiling through the core is based on reactor power distribution. The test thermohydraulic calculations made with using of developed model showed acceptable results in coolant mass flow rate distribution through the reactor core and in axial temperature and pressure distribution. The developed model will be upgraded in future for different transient analysis in metal-cooled fast reactors of BN type including reactivity transients (control rods withdrawal, stop of the main circulation pump, etc.).

1. Introduction

The next generation of fast reactors is under development in Europe. Implementation of advanced calculation tools for transient simulation is important to provide safety analysis of such systems. To improve computational accuracy, a coupled multi-physics approach is used. Such approach allows performing static and dynamic analyses integrating all relevant physical phenomena: neutronics, thermal-hydraulics and structural mechanics. Current tendency of performing realistic steady-state and transient simulations is to couple best estimate thermo-hydraulics codes with 3D neutronic, thermo-mechanic, burnup calculation and other physical codes. Such approach appears to be more consistent especially for Beyond Design Basis Accidents.

In this work the first results obtained with coupled version of 3D diffusion code DYN3D and ATHLET for fast sodium cooled reactor transient simulations are represented. DYN3D is a reactor physic calculation code for Steady States and Transients developed at Helmholtz-Zentrum Dresden-Rossendorf in Germany [2]. It uses the multigroup neutron diffusion approximation for hexagonal or quadratic fuel assemblies. Coupled version was successfully used and verified for PWR and WWER reactors transient calculations, including Kalinin-3 benchmark [3].

ATHLET is an advanced best-estimate code which has been initially developed for the simulation of design basis and beyond design basis accidents (without core degradation) for light water reactors [1]. Then the ATHLET 3.0 version was upgraded with properties of metal coolants such as sodium, lead-bismuth, lead and helium. But it is still in validation phase. These updates allow made possible to use ATHLET for metal cooled fast reactors calculations.

Previously an attempt to perform comprehensive validation of ATHLET and DYN3D for sodium cooled reactors was made. Preliminary verifications of DYN3D multigroup diffusion code were performed on the example of the BN-type international benchmarks [4]. ATHLET developer team has performed standard thermo-hydraulic tests for sodium coolant [1]. As far as verification showed plausible results for both ATHLET and DYN3D stand-alone versions it is possible to perform first test coupled calculations with ATHLET/DYN3D codes.

BN-800 reactor was chosen for simulation. It is 2100 MWt sodium cooled power reactor designed in Russia by OKBM. BN-800 represents the next generation of BN-reactor types and has been in operation at the Beloyarskaya NPP site since 2015. Test core model was developed in accordance to BN-800 type reactor design and based on public open source materials. The resulting test model for coupled calculations uses simplified approach where only core with upper and lower plenums are modeled. Such approach allows simulating main BN-800 core design features and it is an important step towards coupled version verification as well as first step in development of BN-800 model for safety analysis. Steady-state calculations were performed as part of the testing.

2. Modeling approach for coupled calculations

The next approach was used for the first sodium reactor simulations with coupled ATHLET/DYN3D version: the designed model is limited by core, lower and upper plenum; fixed boundary conditions are used for inlet flow and temperature, and for outlet coolant pressure and enthalpy. The resulting so called “open core” proved to be good as prior step before the full scale model development (core, loops, technological channels) and is applicable for the analysis of transient behavior in the core.

The BN-800 core layout [5] which was planned for the start of operation ("fresh" fuel in the core) was used for the simulation. The simplified approach was used for the modeling: every assembly is homogenized, core and blanket areas have the same axial dimensions, only one type of assemblies is assigned to every fuel zone. Simplified axial assemblies description is based on [6]. This description was used both for neutronic and thermo-hydraulic modeling.

For the testing model only two thermal-hydraulic feedbacks are considered: the fuel temperature and the coolant density. For that case parameterized cross section library was prepared for 4 fuel temperatures and 4 coolant temperatures. Linear approximation was used to calculate reactivity feedbacks during coupled calculations.

Homogeneous macroscopic cross sections are required for each assembly for DYN3D calculation. They were prepared by the lattice code NEWT from SCALE 6.1 package. The work [4] gives an overview of the applied studies and methods for generation of parameterized few-group macroscopic cross sections with SCALE code, adapted to the test transient simulation.

The described approach allows performing test steady-state calculations. The calculated thermo-hydraulic parameters can be used to validate coupled version results by comparing with BN-800 design features.

2.1. Neutron model

The model consists of only the core region (including radial, upper and lower reflectors). The model consists of low, medium, high enrichment sub-cores and MOX-sub-core. Each sub-core is assigned as one thermal-hydraulic channel group.

The core (figure 1) consists of three uranium dioxide fuel regions i.e., inner (low enrichment zone, LEZ), middle (middle enrichment zone, MEZ), and outer (high enrichment zone, HEZ) core regions which have 136, 94 and 139 fuel assemblies, respectively. Uranium zone is surrounded with layer of MOX zone. The MOX core zone is bounded by radial blanket zone (RB) followed by one row of radial reflector (RF) assemblies. To ensure correct axial neutron leakage, top and bottom assembly's technological parts were homogenized to be used as axial shielding.

The fuel is uranium dioxide with the enrichment of 17, 21, 26 % for inner, middle and outer cores, respectively. The MOX fuel is considered as a mixture of natural uranium with 17% plutonium-239. Radial and axial blankets are composed of natural uranium. For the core structural material austenitic (fuel cladding) and ferritic/martensitic (assembly cladding and structure) steels are used [6].

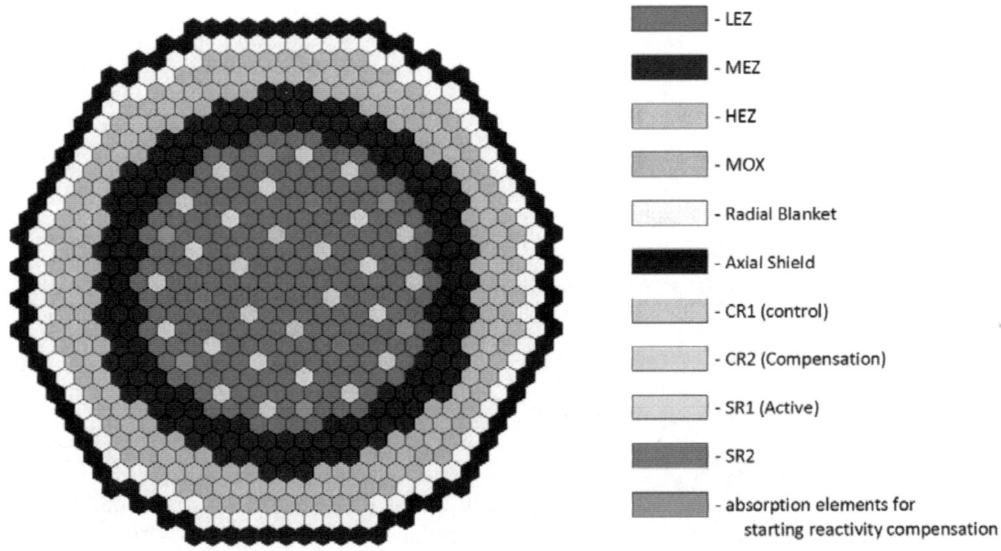

Figure 1. Core layout

2.2. Thermo-hydraulic model

The core hydraulic nodalization scheme consists of 6 hydraulic zones: one zone for each fuel zone (HEZ, LEZ, MEZ and MOX), one zone for radial blanket and one zone for channels with control rods. The assemblies of each hydraulic zone have difference in cross-section area of its inlet part. This difference is necessary for correct coolant mass-flow rate core profiling. The corresponding value of hydraulic cross-sectional area was chosen on the reactor radial power distribution. To achieve suitable mass flow in every hydraulic zone the normalized radial power distribution obtained from standalone DYN3D calculation (with critical positioning of control rods) was used. With accordance to power distribution, hydraulic cross-sectional area distribution was selected.

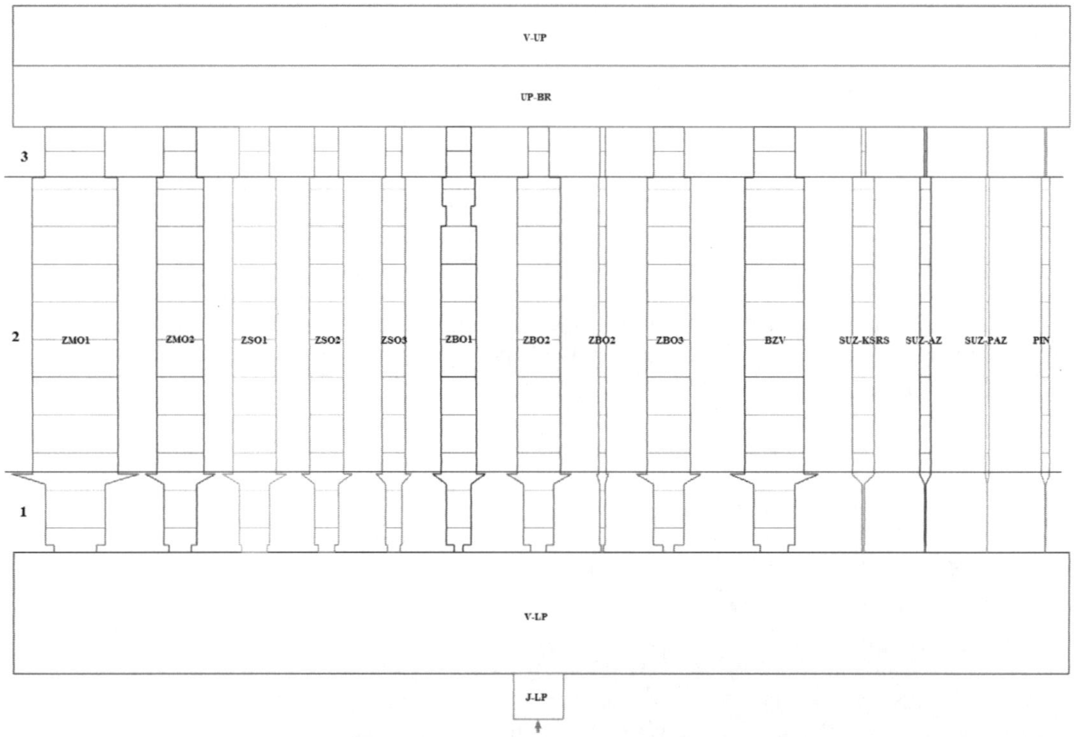

ZMO1 –ZMO2 – LEZ assemblies, ZSO1 – ZSO3 - MEZ assemblies, ZBO1 - ZBO3 – HEZ assemblies, BZV – blanket assemblies, SUZ-KSRS, SUZ-AZ, SUZ-PAZ, PIN – control rods, J-LP – core mass flow rate boundary condition, V-LP – lower plenum, UP-BR – upper plenum, V-UP – pressure and enthalpy boundary condition.

1 – inner part of assemblies, 2 – middle part of assemblies, 3 – outer part of assemblies

Figure 2. Core hydraulic nodalization scheme

3. Steady state calculation results

As first step, reaching of stabilized reactor parameters at nominal power (steady state conditions) is required to get the correct neutronic and thermo-hydraulic starting parameters before transient calculations. The calculation process by A/D coupled code system consists of three phases: «zero transient» - the model initialization, «steady state» - the stabilization of reactor parameters and «transient» - the calculation of a transient. In the present model, the stabilization of the reactor parameters was achieved after 60 seconds. The transient calculations can be started after this stabilization period, but no any transients were considered in this article. The main reactor parameters [6] had the following values after stabilization: maximum fuel temperature - 1260°C; maximum fuel cladding temperature - 480°C; total power - 2100 MWt and inlet/outlet core coolant temperature - 354/547°C, respectively. Graphics of reactor total power, mean outlet core coolant temperature and maximum fuel temperature and fuel rods cladding temperature during the withdrawal are presented in Figures 3-6.

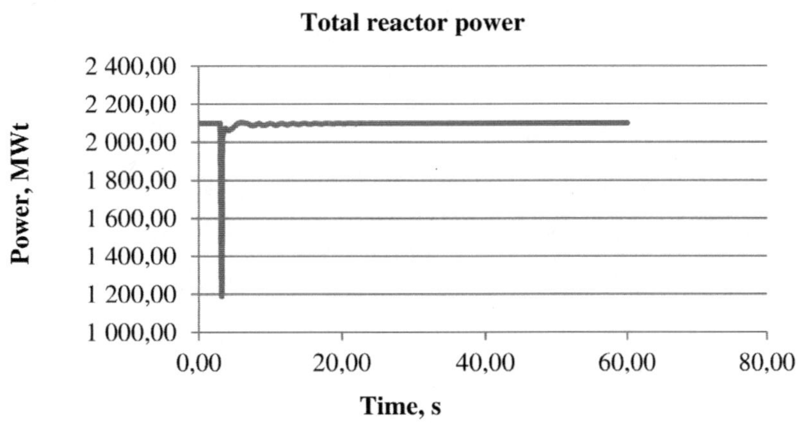

Figure 3. History of the total reactor thermal power.

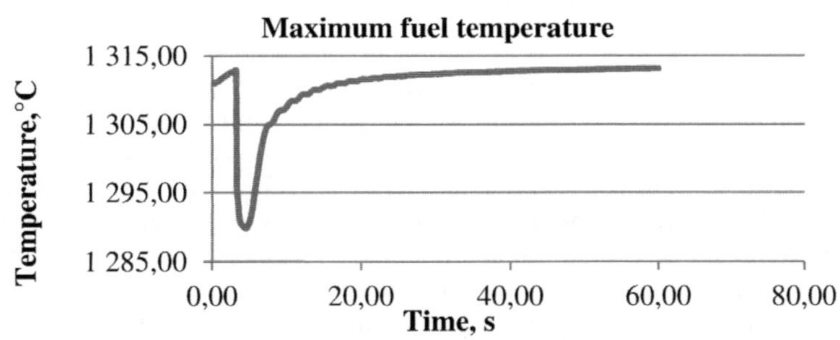

Figure 4. History of the reactor maximum fuel temperature.

So, as it follows from Fig.3-6, the calculation by A/D coupled codes provides physically plausible results. That means that data transfer between coupled codes is correctly. So, it can be concluded that A/D coupled codes and developed model can be applied for calculation of different transients on fast metal cooled reactors.

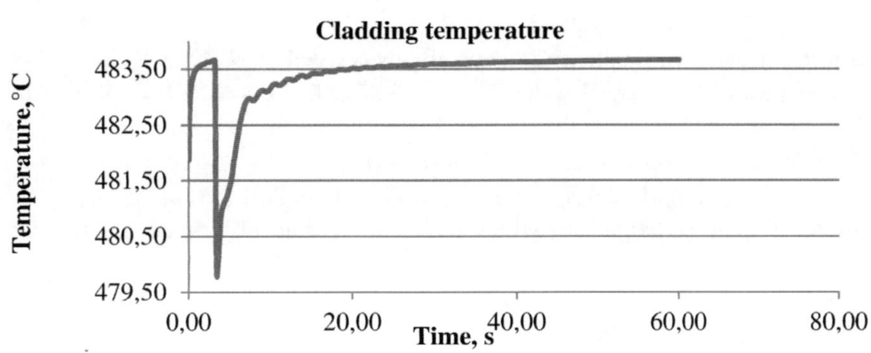

Figure 5. History of the reactor maximum cladding temperature.

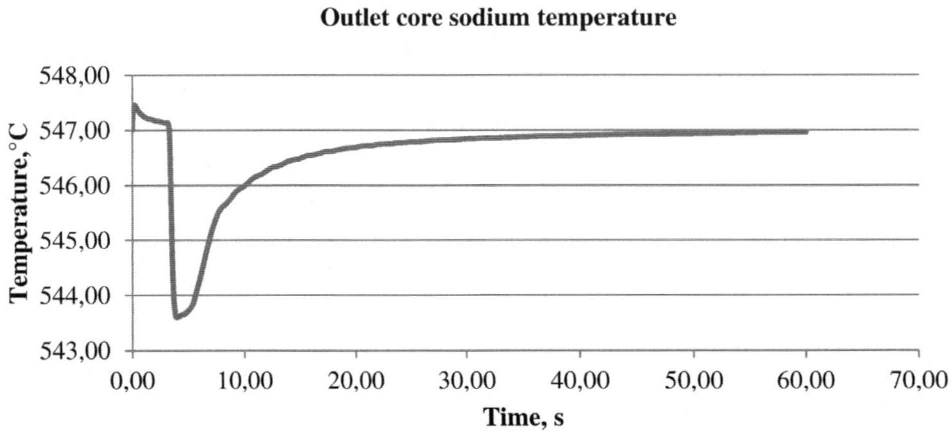

Figure 6. History of the core outlet sodium temperature.

4. Conclusion

The paper describes the developed model of BN-800 type core and the simulation of steady state reactor parameters at nominal power performed by using of the coupled code system A/D.

The steady state calculations showed that data transfer between coupled codes is correct, obtained results are acceptable and developed model can be used as basis for future investigations.

This work is the first step toward liquid metal cooled reactor independent safety analysis with A/D coupled codes. Further developments of the applied methodology will be continued and will include the model updating to the whole BN-800 reactor description for calculation of different transients, basis and beyond basis accidents It is also planned to model the thermal expansion of different core construction parts and to research it's influence on the reactivity.

References

[1] ATHLET Mod 3.0 Cycle A. User's Manual, (2012). GRS - P - 1/ Vol. 1 Rev. 6 – GRS, Garching bei München, Germany

[2] Rohdeet U. al., (2016). The reactor dynamics code DYN3D - models, validation and applications, Progress in Nuclear Energy 89, 170-190

[3] Kozmenkov Y., Kliem S., Rohde U., (2015). Validation and verification of the coupled neutron kinetic/thermal hydraulic system code DYN3D/ATHLET. Ann. Nucl. Energy 84, 153–165.

[4] Bousquet J., Ivanov V., Danitcheva I., Khrennikov N., Velkov K., Klim S., Samokhin A. (2015). Assessing Reactor Physics Codes SCALE, SERPENT and DYN3D to Simulate Fast Reactors on the Example of the BFS-62-3ABenchmark. In: IPPE Nyutronik conference, Obninsk 12-16 October 2015. –p. 20. (In Russian);

[5] Vasiliev B., Farakshin M., Belov S., Kuznetsov A. (2016). Prospects of development of BN-800 reactor core. In: 10th International Scientific and Technical Conference, Safety Efficiency and Economy of Nuclear Power Industry, Moscow 25-27 May 2016

[6] I. Kuznetsov, V. Poplavski, (2012). Safety Aspects of NPP's with Fast Reactors, ISBN 978-5-86656-257-2, Moscow (In Russian)

IOP Conf. Series: Journal of Physics: Conf. Series **781** (2017) 012035 doi:10.1088/1742-6596/781/1/012035

Nuclear Safety of RBMK Storage Pool under Seismic Impact

A Fedosov

Head of laboratory, NRC "Kurchatov institute", Moscow, Russia

E-mail: amfedosov@mail.ru

Abstract. Nuclear safety of RBMK storage pool of spent fuel during of the maximum design earthquake is evaluated. The lower ends of the fuel assemblies are not fixed and they can deviate from the vertical position. The seismic action may be one of the reasons for such deviations. 3D model of fuel assemblies movements caused by seismic impact is used. The simulation of the dynamics of a fuel assemblies group under seismic impacts allows to find the dangerous configuration of closest approach of the fuel assemblies. Three-dimensional neutron program STEPAN calculates the Keff of the most dangerous systems. The maximum design earthquake is the design basis accident. In this case according to the regulatory documents the fuel is considered with zero burn-up. Nuclear safety of RBMK storage pool under considered conditions is provided.

1.Introduction

The RBMK spent fuel pool is located directly in the reactor hall. The pool consists of two concrete compartments. Each compartment is covered with corrosion-resistant steel plates and is 10.3 m long, 4.2 m wide and of 17.3 m deep. The compartment is filled with water up to 16.6 m and closed with caps. Fuel assemblies are hanged on the 2 m long cantilever beams. The distance between the centers of the beams is 0.25 m. There is a canyon in the middle of the compartment between the ends of the beams. The canyon is used to transport fuel assemblies under water to their place of storage.
In the initial project it was planned to put the fuel assemblies into steel pipes. But to increase the capacity of the pool 2.2 times the fuel assemblies are placed now without the pipes by a triangular lattice. The lower ends of the fuel assemblies are not fixed and they can deviate from the vertical position. The seismic action may be one of the reasons for such deviations.

According to the regulations nuclear safety during storage and transportation of nuclear fuel is provided when the neutron multiplication factor Keff is less than 0.95 in normal operation, in case of violations of normal operations and in case of design basis accidents. The Keff≤0.95 criterion is not required for beyond design basis accidents. It is sufficient to show that a self-sustaining fission chain reaction is impossible. For beyond design basis accidents regulatory documents require a realistic (non-conservative) analysis. This means that the actual fuel burn-up should be considered in the calculations.

The maximum design earthquake is the design basis accident. In this case the burn-up credit can be used if a special device for the measurement of fuel burn-up is available. Otherwise, all of the fuel in the pool should be considered as fresh (with zero burn-up). The conservative static analysis has shown that if a lot of assemblies with fresh fuel deviate from the vertical and their lower ends are in contact, the criterion Keff≤0.95 may be violated. That brought up the task of modeling of the actual behavior of a fuel assemblies group during an earthquake.

2. Fuel assemblies motion simulation

The problem can be divided into two parts. The first part is the simulation of the dynamics of a fuel assemblies group under seismic impacts with regard to their collisions and water resistance. This simulation allows to find the dangerous configuration of closest approach of the fuel assemblies (FA). In the second phase the three-dimensional neutron program calculates the Keff of the most dangerous systems.

Seismic action moves the pool with cantilever beams. Since the FA fixed only at the top, they start to oscillate like a pendulum relative to the walls and bottom of the pool due to the inertia.

When the horizontal oscillations occur, the lower part of the water at the depth of more than ¾ of the pool width is considered to be rigidly connected with the walls, i.e. it moves the same way as the pool itself [1]. Above the level of ¾ of the pool width there is a complex movement of water, including the possibility of wave oscillations of liquid free surface. Water that moves together with the pool walls resists the movement of the FA. Fuel assemblies are fixed to the beams by the thin hangers. Resistance of water to hangers can be neglected. To simplify the task the bending of the fuel assemblies is not considered.

Figure 1 shows a scheme of the fuel assembly with the hanger. The hanger consists of two parts 14 and 36 mm diameter.

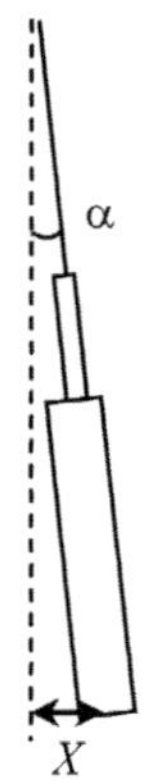

Figure 1. Scheme of fuel assembly.

As a result of the seismic action assembly deviates from the vertical by an angle α, which leads to displacement of the lower part of FA by a distance X. Water resistance (force) for dz element of moving fuel assembly at a height z is determined by the formula:

$$F_c(z) = \frac{1}{2} C_x(z) \rho u^2(z) D dz + \frac{\mu}{L} a(z) dz \tag{1}$$

where $C_x(z)$ - drag coefficient; ρ – water density; μ - added mass of water; $u(z)$, and $a(z)$ – velocity and acceleration at the height z; D, L - diameter and length of the fuel assembly or hanger. The drag coefficient is determined according to the formula $C_x = cRe^{-b} k$ [2], where c - constant depending on the geometry; Re - Reynolds number; b - exponent, k - correction factor that takes into account the non-uniformity of movement and the mutual influence of the fuel assembly.

From dynamic equation considering the moments of all the forces, one can get an equation that describes the behavior of a harmonic oscillator with a nonlinear damping and external disturbing force f (force of inertia):

$$\ddot{X} + \beta \dot{X}^{2-b} + \omega_0^2 X = -f(t) \tag{2}$$

where X - deviation of lower part of FA from of the vertical axis; β - coefficient determined by the geometry of the assembly and the viscosity of water; ω_0 - the natural frequency of the assembly with added mass of water.

The movement of the lower part of each assembly, composed of the oscillation in X and Y directions, is described by equation (2). All assemblies move in the same way before the collision with the walls or other assemblies. Distance between them is almost constant. The collision of the FA lower parts with the wall and its reflection cause a chain reaction of mutual collisions. As a result regular lattice is broken. In some places the lower parts of fuel assemblies form a compact group with a reduced lattice pitch that leads to an increase of the neutron multiplication factor.

3. Evaluation of the neutron multiplication factor

STEPAN-БВ code is used for the analysis of nuclear safety. This code is a modification of a well-known STEPAN code [3] that is certified for calculations of RBMK. Changing lattice pitch in the horizontal direction for the layers with different heights allows to simulate the FA deviation from the vertical axis in emergency situations. Two-group system of constants has been prepared for the three-dimensional calculations. The constants depend on fuel burn-up, lattice pitch, water density and temperature. Calculation model was verified by comparison with the precision calculations by the Monte Carlo MCNP code [4]. Keff calculation example for different types of lattice is shown in the table 1.

Table 1. Neutron multiplication factor in the RBMK pool (fresh 2.6% enriched fuel).

Lattice type	MCNP	STEPAN-БВ	Δ(STEPAN – MCNP)
Square 8x8 cm	1.2069 ±0.0003	1.2105	0.0036
Triangle 12,5 cm	0.7974 ±0.0003	0.8018	0.0044
Rectangle 8x12,5 cm	0.9954 ±0.0003	1.0055	0.0101

4. The results of the maximum design earthquake simulation

FA dynamics in a fragment of the pool of the 1st unit of Kursk NPP caused by the maximum design earthquake of 7 points was considered. Accelerogram (acceleration g) during an earthquake in a horizontal direction is shown in figure 2.

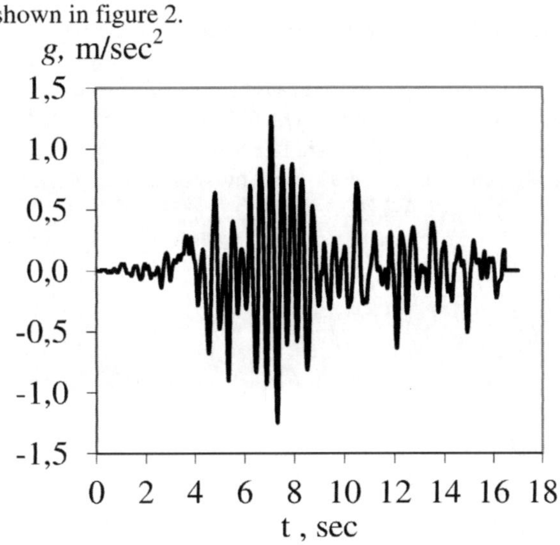

Figure 2. Accelerogram of the 7 points earthquake.

Pool fragment 20x14 FA is shown in figure 3. The pool walls are to the left and at the bottom of the scheme. Canyon is at the top of scheme. The figure 3 shows the moment of closest approach of the FA lower parts.

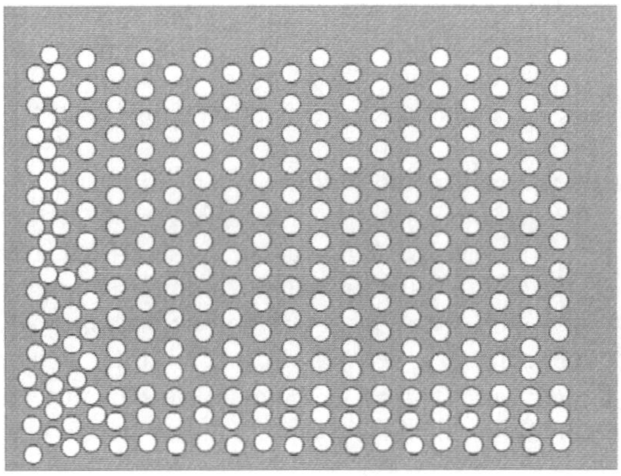

Figure 3. Horizontal section of pool fragment (lower parts of FA).

In the initial state before the earthquake Keff=0.790 for 2.6% enrichment fuel with a zero burn-up. For the configuration shown in figure 3 Keff=0. 937. Nuclear safety criterion is ensured. Safety can be improved by leaving the left row empty (free of fuel assemblies). In this case Keff=0.823.

5. Conclusion

Presented calculations have shown that nuclear safety of storage pool RBMK in the maximum design earthquake of 7 points is provided.

References

[1] Birbraer A N 1988 Calculation of Structures for Seismic Stability (St. Petersburg: Nauka) (in Russian).

[2] Batchelor G K 1970 An Introduction to Fluid Dynamics (Cambridge at the University Press)

[3] Babaytsev M N, Fedosov A M, Glembotsky A V, Krayushkin A V, Kubarev A V, Romanenko V S 1993 The STEPAN Code for RBMK Reactor Calculation Preprint IAE-5660/5

[4] MCNP – A General Monte Carlo N-Particle Transport Code Version 4B 1997 Los Alamos National Laboratory Report LA-12625-M

Recent neutronics developments for reactor safety studies with SIMMER code at KIT

A Rineiski, M Marchetti, L Andriolo and F Gabrielli

Institut für Kern- und Energietechnik (IKET), Karlsruher Institut für Technologie (KIT), Hermann-von-Helmholtz-Platz 1, 76344 Eggenstein-Leopoldshafen, Germany

Andrei.Rineiski@kit.edu

Abstract. The SIMMER family of codes is applied for safety studies of sodium fast reactors and reactors of other types. Both neutronics and fluid-dynamics parts of SIMMER are under development. In the paper new neutronics capabilities are presented. In particular developments for neutron transport solvers and a new technique for taking into account thermal expansion effects are described. These new capabilities facilitate 3D simulations and improve accuracy of modelling for the initiation transient phase during a hypothetical severe accident.

1. Introduction

The SIMMER family of codes is developed by JAEA (Japan) in cooperation with CEA (France), KIT (Germany) and other partners [1]. SIMMER-III is a 2D code version and SIMMER-IV is a 3D one. SIMMER includes neutronics, fluid-dynamics and structure parts. They are employed together to simulate reactor behavior. In particular sodium-cooled fast reactors (SFRs), lead-cooled, LBE-cooled, He-cooled fast reactors, experimental water-cooled reactors, accelerator driven systems and molten salt reactors were studied with SIMMER at KIT. SFR analyses at KIT are mainly done for 2D (RZ) models representing the reactor vessel. 3D (XYZ) simulations are time-consuming, but possible for parts of the vessel that include the core and its surroundings.

Transient simulations of hypothetical severe accidents in SFRs may include several phases, starting from the initiation phase, during which the fuel subassembly can-walls remain intact, that prevents the radial material movement in the core after coolant boiling, clad and fuel failure. The initiation phase may end after a power excursion, the amplitude of which depends on the reactor design and accident initiator. Then a transition to massive core melting may occur. The melting enables radial movements of molten/broken materials that may lead to subsequent re-criticalities and appreciable mechanical energy releases.

The neutronics part of SIMMER gets material and temperature distributions from the other parts and computes the power shape and amplitude employed in the other parts. The set of neutronics subroutines includes a subset that generates macroscopic cross-sections from SIMMER nuclear data libraries, a subset that performs reactivity and power amplitude computations and a subset that calculates time-dependent neutron flux and power shapes.

2. SIMMER cross-section libraries and validation of neutronics models at nominal conditions

Several cross-section libraries in the extended CCCC format [2] are employed in SIMMER studies at KIT. The libraries include data in ISOTXS format, in particular group-average cross-sections and group-to group scattering cross-sections, their angular Legendre expansion coefficients, and data in BRKOXS format, in particular Bondarenko f-factors. An 11-group library is usually applied for SFR production calculations, this library being produced directly from evaluated nuclear data files by employing a weighting function related to a SFR neutron spectrum. For treating water-cooled systems, a 40-group library is used. It includes cross-sections for 10 neutron groups below 1 eV and data on neutron up-scattering. Also a 72-group library is applied for fast reactor calculations for investigating effects related to the limited number of groups in the 11-group library and different nuclear data. The 40-group and 72-group libraries are obtained from a "master" 560 group library [3]. It was shown that for simple geometries, the 560-group library provides results very close to those obtained with point-wise cross-sections while employing a Monte-Carlo code [4].

Before transient simulations, the power shapes and major reactivity effects, including coolant density/void and Doppler effects, computed by SIMMER at steady-state are compared to those computed with reference neutronics codes, such as ERANOS [5]. In most cases the deviations between the compared values are acceptable in view of existing uncertainties, in particular due to nuclear data and modeling assumptions.

Note that XYZ geometry models are applied currently in SIMMER for 3D simulations, in plane each HEX being represented by two rectangles with the same area as HEX together. For the fluid-dynamics part, the node interface areas are provided by the user in addition to other geometry data, while in the neutronics part, these areas between neighboring nodes of the same axial location are approximate. For an experimental benchmark on the EBR-II reactor with metal fuel, a very good agreement between XYZ and HEX-Z models was demonstrated [6] after applying VARIANT [7], a code that includes both XYZ and HEX-Z options. VARIANT is a 3D ERANOS neutron transport solver. Effects of using of XYZ geometry for neutron transport simulations in fast reactors with hexagonal subassemblies will be studied also in the future.

3. Improved quasi-static scheme in SIMMER

In SIMMER, the improved quasi-static scheme [8], IQS, is employed. The neutron flux is represented by a product of (1) a neutron flux amplitude function depending on time, and (2) a neutron flux shape function, depending on the time, spatial, energy (group) and angular variables. The shape function is assumed to be normalized, while establishing the IQS equations. The amplitude is computed by solving the point-kinetics equation. The equation for the shape is obtained from the neutron transport equation by substituting the neutron flux by the product of the amplitude and shape functions. Therefore the shape calculations can fully or partially compensate inaccuracies in amplitude computation, in particular when these inaccuracies are due to approximate weighting functions used for reactivity computations. The adjoint flux at steady-state is usually used as the weighting function.

The time-dependent equation for the shape function is solved by employing a fully implicit first order scheme for discretization in time. This approach transforms the time-dependent equation into a sequence of steady-state-like problems with external-source-like terms related to the values of the shape function at the beginning of time step and to delayed neutron precursors. In case of using the same time step for shape and amplitude computations, IQS is not much more efficient than the "direct" fully implicit method for solving the time-dependent transport equation for the neutron flux. When time-consuming shape calculations are performed less frequently as compared to fast-running amplitude calculations, the IQS method is faster than the direct one.

The point-kinetics parameters, such as reactivity and effective delayed neutron fractions, are re-calculated during the transient by employing the time-dependent shape function. These calculations are done for so called reactivity steps, which are smaller than shape steps, but larger than the amplitude ones. The user may provide time-dependent "external reactivity" values, which are added to the computed transient reactivity for adjusting its value. One of the reasons is the existence of some phenomena affecting the reactivity, but not yet considered in SIMMER, such as e.g. thermal vessel expansion. If "external" reactivity values are used, corresponding corrections are applied also to the neutron production term in the equation for the shape function. For molten-salt reactors, the scheme was extended, the effects related to the movement of precursors being treated as an integrally negative neutron source term [4].

4. PARTISN-based neutron transport solver

The current SIMMER version includes a neutron solver based on DANTSYS, a static neutronics code [9]. The set of DANTSYS subroutines was introduced into SIMMER and extended - to solve time-dependent equations for the shape function - at KIT in the past. Also the treatment of delayed neutrons was included. Unlike DANTSYS, PARTISN [10] includes an option for transient calculations. It is based on the semi-implicit scheme. The latter scheme is efficient for small time steps, but less suited for IQS if large shape steps are used. Therefore the time-discretization scheme in PARTISN was extended by implementing a theta-scheme, that is semi-implicit for theta=0.5 and fully implicit for theta=1, the theta parameter being provided by the user. A delayed neutron treatment was first introduced [11] via coupling of PARTISN with the KIN3D code [12]. KIN3D is a kinetics extension for VARIANT. The extended code was validated by simulations of fast transients in zero-power subcritical experiments performed earlier and also by comparisons with other codes [11]. Then the extended PARTISN code was coupled to SIMMER framework as an external executable, and it was demonstrated that SIMMER/PARTISN could be used to provide similar results as SIMMER/DANTSYS. [13]. Unlike DANTSYS, PARTISN is a parallelized code. A high efficiency of parallelization was demonstrated for 3D cases for a large number of neutronics nodes, the computation time per shape step being reduced by about a factor of 50 while employing 64 processors. For smaller number of nodes, in particular for 2D problems, smaller gains were observed employing a smaller number of processors, up to 8, while with more processors the computation time was not reduced. Further studies on parallelization are planned.

5. VARIANT-based neutron transport solver

An effort was also done to extend VARIANT capabilities and make possible its application as a neutron transport solver in SIMMER. Unlike DANTSYS and PARTISN, which are both based on the Sn method, VARIANT is based on the Variational Nodal Method and cannot be employed for modeling neutron transport in vacuum sub-regions. Though a complete vacuum does not occur in the systems currently considered with SIMMER, regions with a very low material density may appear after coolant elimination due to its boiling or a fission gas release from failed pins. For such regions, the VARIANT calculations may be less fast and accurate. A spatial kinetics extension for VARIANT was developed in the past [12]. More recently, options for application of large spatial nodes were studied at KIT. A technique for treating heterogeneous nodes was developed [14, 15]. It allows application of nodes with different cross-sections for axial sub-nodes; e.g. a node with two sub-nodes, with and without a control rod absorber, can be treated. Using of larger nodes reduces the node number and accelerates calculations. More effort is needed to improve the VARIANT-performance, in particular to implement parallelization options.

6. Treatment of core thermal expansion effects in SIMMER

A conventional scheme for performing transient analyses for SFRs in Europe and Japan includes two steps: (1) SAS4A [16] or a similar code, based on the channel approach, is used for taking into account fuel irradiation effects and is employed for initiation phase simulation and (2) SIMMER is used for later transient phases. A special coupling code is used to transfer the values obtained at the end of the first step to SIMMER. The coupling introduces appreciable uncertainties because of quite different models used in different codes. The scheme applicability to the most recent SFR designs, in particular to those containing an axial heterogeneous fertile layer, is limited. Therefore an effort is under way to perform once-through analyses with SIMMER starting from nominal conditions, while SAS4A or special fuel performance codes should be used for irradiation simulations.

In view of the two-step scheme mainly applied in the past, SIMMER did not include before recently a model for taking into account reactivity effects due to core thermal expansion. These effects are important mainly for the initiation phase. For calculations with SIMMER starting from nominal conditions, these effects should be taken into account.

In the recently implemented in SIMMER-III scheme [17, 18], the reactivity effects due to core thermal expansion can be taken into account. The contribution to the effect from each node is computed as the sum of two ones: (1) due to homogeneous expansion, i.e. same for all materials, including fuel, structure, coolant, etc., and (2) due to a relative variation of the coolant density compared to solid material densities. The second contribution is calculated on the basis of the assumption that the fuel mass, steel mass and coolant density do not vary due to expansion in any node. The SIMMER framework is already established for computing the second contributions.

For taking into account the first contribution, an "external reactivity"-like approach is used. The reactivity and neutron production term in the equation for the shape are modified, but the mesh is kept constant in fluid-dynamics and neutron transport calculations. The "external reactivity" due to expansion is computed in two steps. A technique based on the equivalence method [19] is applied in these computations. According to this method, instead of an isotropic system expansion determined by a factor f, one can consider an equivalent - with respect to the reactivity and other neutronics values - non-expanded system with densities divided by f.

First, the equivalence principle is applied node-wise, i.e. with different factors for each node (that is an approximation), so that one obtains an "equivalent" system with initial dimensions in plane, but with "equivalent" densities and axial dimensions. Second, additional transformations are done to bring also the axial dimensions to the original mesh, while making further modifications with material and related densities. Several options have been developed for the second step. One of the options is to establish equivalence coefficients between axial dimension and density variations globally or ring-wise by performing k-eff calculations at the beginning of the transient with modified axial dimensions.

Alternatively, an "equivalent" system with initial axial dimensions can be obtained first and then additional transformations are done to take into account the radial mesh variations.

Thus the task is finally reduced to modifications of material/nuclear densities employed in the calculations and computing the reactivity effects induced by these modifications. In this way rather complex variations, e.g. related to bowing of fuel assemblies could be treated. After a structure mechanics code is coupled to SIMMER, the full potential of the incorporated scheme can be investigated. Currently effects due to expansions in plane leading to a cylindrical or conical core shape can be computed. Axially expanded dimensions are computed independently for different radial locations.

7. Conclusions

The SIMMER code is used and developed in Japan and Europe for simulating hypothetical core disruptive accidents in sodium cooled fast reactors and other systems. The code neutronics part has been extended recently to accelerate 3D calculations via employing a new neutron transport solver based on the PARTISN code. Another neutron transport code, VARIANT, is extended to treat axially heterogeneous nodes that facilitates its application as neutron solver in SIMMER as a smaller number of nodes can be treated. A new technique has been implemented in SIMMER for taking into account core thermal expansion effects. The extensions improve the SIMMER performance and applicability range.

References

[1] Sa. Kondo, K. Morita, Y. Tobita and N. Shirakawa, SIMMER-III: An Advanced Computer

Program for LMFBR Severe Accident Analysis, Proc. Int. Conference on Design and Safety of Advanced Nuclear Power Plants (ANP'92), Tokyo, Japan, Oct. 25-29, No. 40-5 (1992), see also W. Maschek, A. Rineiski, T. Suzuki, et al., SIMMER-III and SIMMER-IV Safety Code Development for Reactors with Transmutation Capability, M&C 2005, Avignon, France (September 12-15, 2005)

[2] R. D. O'Dell, Standard Interface Files and Procedures for Reactor Physics Codes, Version IV, LA-6941-MS (1977)

[3] A. Rineiski, V. Sinitsa, W. Maschek, C^4P, a Multigroup Nuclear CCCC Data Processing System for Reactor Safety and Scenario Studies, German Annual Meeting on Nuclear Technology JK 2005, Nürnberg, Germany (May 10 - 12, 2005)

[4] A. Rineiski, V. Sinitsa, W. Maschek, S. Wang, Kinetics and cross-section developments for analyses of reactor transmutation concepts with SIMMER, M&C 2005, Avignon, France (September 12-15, 2005)

[5] G. Rimpault, D. Plisson, J. Tommasi, et al., The ERANOS Code and Data System for Fast Reactor Neutronic Analyses. PHYSOR 2002, Seoul, Korea (October 2002)

[6] B. Vezzoni, M. Marchetti, F. Gabrielli, A. Rineiski, IAEA EBR-II Neutronics Benchmark: Impact of Modeling Options on KIT Results, PHYSOR 2016, Sun Valley, Idaho, USA (May 1–5, 2016)

[7] G. Palmiotti, E. E. Lewis, and C. B. Carrico. VARIANT: VARIational Anisotropic Nodal Transport for Multidimensional Cartesian and Hexagonal Geometry Calculation. ANL-95/40, ANL, 9700 South Cass Avenue, Argonne Illinois 60439 (October 1995).

[8] K. O. Ott, D. A. Meneley, Accuracy of quasi-static treatment of spatial reactor kinetics, Nuclear Science and Engineering, 36 (1969), p. 402

[9] R. E. Alcouffe, R. S. Baker, F. W. Brinkley, et al., DANTSYS: A Diffusion Accelerated Neutral Particle Transport Code System, LA-12969-M (June 1995).

[10] R. E. Alcouffe, R. S. Baker, J. A. Dahl et al., PARTISN: A Time-Dependent, Parallel Neutral Particle Transport Code System, LA-UR-08-07258 (Revised Nov. 2008)

[11] F. Gabrielli, A. Rineiski, W. Maschek, M. Marchetti, Reactor transient analyses with KIN3D/PARTISN, M&C 2013, Sun Valley, Idaho (May 5-9, 2013).

[12] A. Rineiski, KIN3D: Module de Cinetique Spatiale et de Perturbations pour TGV2, Report SPRC/LEPh 97-203, CEA (1997).

[13] M. Marchetti, F. Gabrielli, A. Rineiski, and W. Maschek, The SIMMER/PARTISN Capability for Severe Accident Analyses, PHYSOR 2014, Kyoto, Japan (September 28th - October 3rd, 2014)

[14] A. Rineiski, Intra-Nodal Reactivity Calculations Based on the Variational Nodal Method, M&C 2001, Salt Lake City, Utah (September 2001).

[15] M. Marchetti and A. Rineiski, Variational Nodal Method with Heterogeneous Nodes: Application to Reactor Analyses. Transaction of the American Nuclear Society, 107:535–536, San Diego, California (November 11-15 2012).

[16] J. E. Cahalan and T. Wei, Modeling Developments for the SAS4A and SASSYS Computer Codes, Proc. International Fast Reactor Safety Meeting, Snowbird, Utah (August 12-16, 1990)

[17] L. Andriolo, A. Rineiski, B. Vezzoni, et al., An Innovative Methodology for Evaluating Core Thermal Expansion Feedbacks in Transient Analyses, ICAPP'15,Nice, France (May 2015).

[18] L. Andriolo, Impact of innovative sphere-pac fuels on safety performances of sodium cooled fast reactors. Fluids mechanics [physics.class-ph]. Université Grenoble Alpes, 2015. English.<NNT : 2015GREAI067>.<tel-01257910>

[19] S. B. Shikhov, Perturbation theory formulas for the effect of the dimensions on the critical mass in a fast reactor. The Soviet Journal of Atomic Energy, 1959, 6(2), p. 90–94.

The search of the best mode of the reserve power supply consumption during the nuclear reactor's emergency shutdown procedures in case of force majeure circumstances

A M Zagrebaev and A V Trifonenkov

National Research Nuclear University MEPhI (Moscow Engineering Physics Institute), 31 Kashirskoe shosse, 115409 Moscow, Russia

E-mail: AVTrifonenkov@mephi.ru

Abstract. This article deals with the problem of the control mode choice for a power supply system in case of force majeure circumstances. It is not known precisely, when a force majeure incident occurs, but the threatened period is given, when the incident is expected. It is supposed, that force majeure circumstances force nuclear reactor shutdown at the moment of threat coming. In this article the power supply system is considered, which consists of a nuclear reactor and a reserve power supply, for example, a hydroelectric pumped storage power station. The reserve power supply has limited capacity and it doesn't undergo the threatened incident. The problem of the search of the best reserve supply time-distribution in case of force majeure circumstances is stated. The search is performed according to minimization of power loss and damage to the infrastructure. The software has been developed, which performs automatic numerical search of the approximate optimal control modes for the reserve power supply.

1. Introduction

Nuclear reactor today is one of the most widespread types of energy sources. Nuclear energy is continuously spreading all over the world, approving to be convenient for a variety of purposes.

In present, the growth of nuclear power industry allows nuclear power plants to be built in areas of unstable climatic conditions or with an increased risk of seismic activity, as well as in isolated regions, which are difficult of access and don't have a developed power supply system infrastructure, so a nuclear reactor may be the only source of energy there [1].

Such power plants may undergo force majeure circumstances, which cause abrupt shutdown of nuclear reactors and their consequent idleness. The consumers of energy produced by the reactor may suffer damages induced by an unexpected shutdown and outage of the chief source of power [1, 2].

In this article a power supply system is considered, which consists of a nuclear reactor and reserve power supply unit. The reserve power supply unit is meant to have huge capacity and energy emission rates, comparable to the same values for a nuclear reactor. For example, such conditions may be satisfied by hydroelectric pumped storage power stations [3, 4, 5]. It is supposed, that the reserve power supply capacity is limited and does not undergo the incident. In the system analysed, the reserve supply can only be released in case of force majeure circumstances.

If such power supply system is isolated, then there is no other way of obtaining electricity or thermal energy for the infrastructure, but from the nuclear reactor. Any lack of energy production can't be compensated. In regions, which are difficult of access, the resumption of power production after unplanned reactor shutdown may occur not earlier than the reactor is restored or an external source of power is connected; both scenarios may take very long time.

Thus, in isolated systems any interruption in power production may cause considerable damage on the infrastructure. This is why methods of nuclear reactor control should be considered, which may help to bypass an unexpected shutdown or at least minimize the damage, caused by the reactor outage.

The time when force majeure situation occurs can't be predicted precisely. A menacing period is considered, during which the incident is expected, given the function of probability distribution. Force-majeure circumstances considered force nuclear reactor shutdown at the moment of the incident coming. After the reactor is shut down, there is a period of time, when the consequences of the accident are eliminated, and the reactor can't be started up. After that, the reactor power may be maximized again.

In this article the problem of the search for the best way of the reserve power supply time-distribution is considered, with aim of the minimization of damage, caused by the nuclear reactor power losses due to emergency control being applied and unexpected shutdown being carried out.

2. Mathematical model of the nuclear reactor

The nuclear reactor and the reserve power supply unit are considered separately, with two different control sets. The control process for both elements of the system starts at the time, when the menacing period has been declared.

The control process for the nuclear reactor has to comply with the limitation of xenon-135 concentration during the reactor operation [6, 7]. To simulate the situation, one-group point model of a nuclear reactor [8] was used in the research. The process of xenon poisoning is described by the system of differential equations [9]. The reactor is supposed to be in steady state at the initial time.

$$\begin{cases} \frac{di(t)}{dt} = \sigma_X \varphi(t) - \lambda_I i(t) \\ \frac{dx(t)}{dt} = \lambda_I i(t) - \lambda_X x(t) - \sigma_X \varphi(t) x(t) \end{cases} \quad (1)$$

where t – time, h,

$i(t)$, $x(t)$ – iodine-135 and xenon-135 concentrations, normalized by the xenon-135 concentration for the "infinite" neutron flux, relative unit,

$i(t) = \frac{\sigma_X}{\gamma_X \Sigma_f} I(t)$ – iodine-135 concentration, relative unit,

$x(t) = \frac{\sigma_X}{\gamma_X \Sigma_f} X(t)$ – xenon-135 concentration, relative unit,

γ_X – xenon-135 yield, relative unit,

Σ_f – macroscopic fission cross-section, sm^{-1},

λ_I, λ_X – decay constants for iodine-135 and xenon-135, s^{-1},

σ_X – microscopic xenon-135 absorption cross-section, sm^2,

φ – neutron flux density, $sm^{-2}s^{-1}$.

Then, the system, which describes the change of state of the reactor after shutdown, appears to be (1) with zero neutron flux ($\varphi = 0$):

$$\begin{cases} \frac{di(t)}{dt} = -\lambda_I i(t) \\ \frac{dx(t)}{dt} = \lambda_I i(t) - \lambda_X x(t) \end{cases} \quad (2)$$

The control of the reactor before the menace comes usually causes power production loss due to local power decreases or high iodine concentration, which leads to outage after shutdown. During time intervals, when the reactor power is less then maximum, the reserve power supply comes into use to compensate the lack of the reactor energy production. It is difficult to search all variety of reactor and reserve supply controls simultaneously, so it was decided to fix the set of possible reactor controls.

In this research, the nuclear reactor control has been fixed with respect to the search of optimal reserve supply consumption mode. Also, the control to be applied does not depend on the moment when menace comes – the power of the reactor at this moment is just set to zero. The control fixed has been defined by the following.

In case of force majeure circumstances, when the time of menace coming is unknown, it is consistent, that the reactor should be shut down as fast as possible. But instant shutdown at the time when menacing period was declared may cause strong poisoning of the reactor with xenon-135. So, some control should be chosen, which shuts down the reactor and satisfies the limitation on xenon-135 concentration.

The control, which complies with the above, is the solution of optimal performance problem. Such control does the optimum time transition of the state of a nuclear reactor to the xenon-safe state, in which the reactor shutdown won't cause xenon limitation violation. On the other hand, in case when shutdown occurs earlier, than the reactor has reached xenon-safe state, xenon poisoning duration is equal or less than in case of instant shutdown. Also, such optimal performance control does not have to end with the reactor shutdown: if menace has not come yet, the reactor may produce energy according to the mode, which leaves the concentrations of iodine-135 and xenon-135 in safe bounds. The control is formed according to principles given in [10]. The form of the optimal performance control is presented in fig. 1:

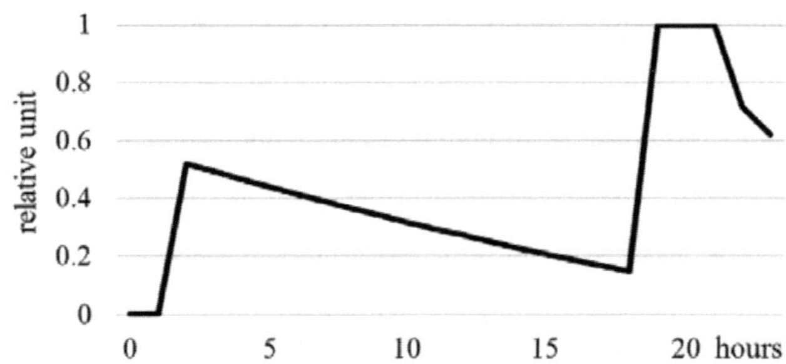

Figure 1. The form of the optimal performance control

Besides the optimal performance control has been chosen, two more reactor control alternatives have been considered to compare them to each other. These two controls are: instant stop at the moment of menacing period start and last-minute stop, which instantly stops the reactor at the time of menace coming.

3. Mathematical model of the reserve power supply

As the reserve power supply capacity is limited, the distribution of the supply in time should be carried out by some criterion, defining the effect of power compensation in time.

Let Q_R be the energy capacity of the reserve power supply. The capacity is sequenced in time by fractions, which sizes are defined by the chosen criterion. So, the aim of the search is to find such fractions of which fit the limitation conditions. For example, the timeline of the optimal performance control may be split into 4 time intervals. Then, the conditions of the search are:

$$\varepsilon_1, \varepsilon_2, \varepsilon_3, \varepsilon_4 : 0 \le \varepsilon_i \le 1, 0 \le \sum_{i=1}^{4} \varepsilon_i \le 1 \quad (3)$$

According to the ε_i fractions, the control function $u_R(t)$ for the reserve power supply is formed. The control function is proportional to the power of the supply. It is formed on principle of smoothing the total power graph within a time interval to the uniform state.

The lack of total power at the moment t is proportional to the following: $\Delta u(t) = u_{max} - u(t) - u_R(t)$.

The cost functional is defined by some parameter, which depends on the state of the system and the control function. For example, the power generation loss before the reactor shutdown, as well as the reactivity margin or the concentration of xenon-135 in the reactor core at the time of the shutdown, etc. may be used as such parameters.

Penalty function was introduced, which defines the value of damage, caused by the lack of power supply system energy production due to decreased power of the reactor. The penalty is defined by the difference between designed reactor capacity and current total power, produced by the power supply system.

So, in this article, the optimization criterion is represented by the functional to be minimized. The functional describes total energy production loss:

$$\Delta Q(u_{\mathrm{R}}) = \int_{t_0}^{t_1} C(t)\Delta u(t)\,\mathrm{d}t \tag{4}$$

where $C(t)$ – penalty function, for example: $C(t) = \mathrm{e}^{\Delta u(t)}$.

The form of penalty function reflects the situation, when the lack of power production causes more damage, than

The force-majeure incident during the emergency period may occur at any moment of time τ, which is random. The probability is defined by the probability distribution $f(\tau)$. Given $f(\tau)$ and boundaries of the emergency period $[t_0; T]$, mathematical expectation of total power production loss is calculated as follows:

$$M[\Delta Q] = \int_{t_0}^{T} f(\tau) \int_{t_0}^{\tau} C(t)\Delta u(t)\,\mathrm{d}t\,\mathrm{d}\tau = \int_{t_0}^{T} f(\tau)\Delta Q(\tau, u_{\mathrm{R}})\,\mathrm{d}\tau \tag{5}$$

In this article $f(\tau)$ given is the uniform probability distribution. In this case, practically, the average value of losses is calculated:

$$\overline{\Delta Q} = \frac{1}{T - t_0} \int_{t_0}^{T} \Delta Q(\tau, u_{\mathrm{R}})\,\mathrm{d}\tau \tag{6}$$

The software has been developed to perform automatic numerical search of reserve power supply optimal time-distribution approximations.

4. Simulation and results

The algorithm implemented accepts the steady state parameters of the reactor, the xenon-135 concentration limitation value, the emergency period duration and the capacity of the emergency power supply as input parameters, and calculates numerically the control modes for the nuclear reactor and reserve power supply unit, average power supply system losses and relative loss values.

The system power loss for three different controls applied has been calculated. These calculations have been performed for different values of the limitation on xenon-135 concentration (see fig. 2).

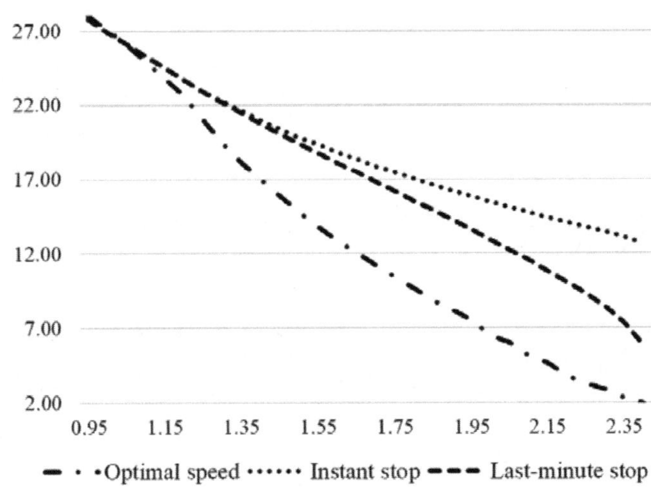

Figure 2. The dependence of power production losses for different types of reactor control from the value of xenon-135 limitation

In the figure, it can be noticed, that the best result has been derived for the optimal speed (optimal performance) reactor control. It can be explained: instant shutdown of the reactor leads to long-time xenon poisoning, which means practically full-power loss for the whole time until real menace comes;

as for the last-minute shutdown, the control is more profitable due to maximum power production for the whole time until menace comes, but after the reactor is stopped, xenon poisoning may end even later, than in previous case.

The minimal losses almost for each value of xenon limitation are shown in graph for the optimal performance control. This means, that power production losses reduction with the reserve power supply should be performed for this type of nuclear reactor control.

The time-distribution of the reserve supply is retrieved in the following form:

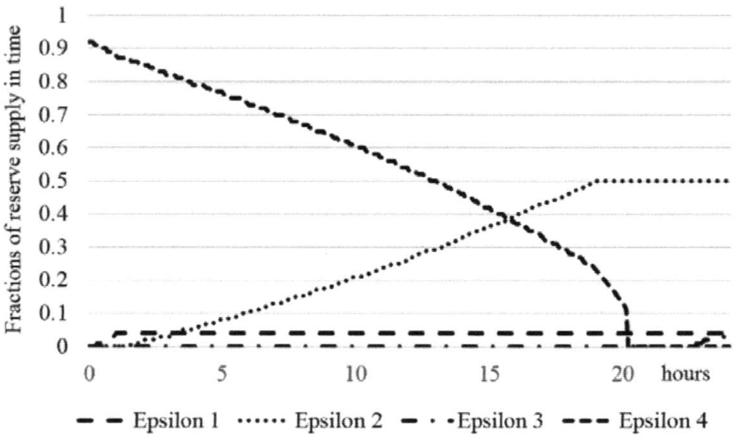

Figure 3. The distribution of reserve power supply in time by fractions, depending on the moment of menace coming

Actually, optimal modes strongly depend on the specific system chosen, because, for example, practical power supply capacity may vary during day and night workload changes and penalty functions are defined by the infrastructure features.

The algorithms implemented and results derived may be used for planning control modes for power supply systems during menacing period.

References

[1] Zagrebayev A M, Ovsyannikova N V and Sadchikov S M 2012 Optimization of a mode of change of a nuclear reactor capacity during the threatened period under force-majeur circumstances *Modern problems of science and education* **4** 1–7

[2] Zagrebayev A M, Ovsyannikova N V and Sadchikov S M 2012 The average loss of energy production at a random stop of a reactor with a limited operational reserve of reactivity *Natural and technical sciences* **58** 418–22

[3] Edwards J, Bindra H, Sabharwall P 2016 Exergy analysis of thermal energy storage options with nuclear power plants **96** 104–11

[4] Chi-Jen Yang 2016 *Storing Energy. Chapter 2 – Pumped Hydroelectric Storage* 25–38

[5] Barbour E, Wilson G, Radcliffe J, Ding Y and Li Y 2016 A review of pumped hydro energy storage development in significant international electricity markets *Renewable and Sustainable Energy Reviews* **61** 421–32

[6] Dollezhal' N A and Emel'yanov I Ya 1980 *Channel Nuclear Power Reactor* (Moscow: Atomizdat)

[7] Ovchinnikov F Ya, Golybev L I, Dobrynin V D, Klochkov D I, Semenov V V and Tsybenko V I 1979 *Operating conditions of water-moderated-water-cooled power reactors* (Moscow: Atomizdat)

[8] Zweifel P 1997 *Reactor Physics* (Moscow: Atomizdat)

[9] Rudik A P 1979 *Optimization of reactor physical characteristics* (Moscow: Atomizdat)

[10] Rudik A P 1971 *Nuclear reactors and Pontryagin's maximum principle* (Moscow: Atomizdat)

Test case specifications for coupled neutronics-thermal hydraulics calculation of Gas-cooled Fast Reactor

F Osuský[1], R Bahdanovich[2], G Farkas[1], J Haščík[1] and G V Tikhomirov[2]

[1] Institute of Nuclear and Physical Engineering, Faculty of Electrical Engineering and Information Technology, Slovak University of technology in Bratislava, Ilkovičova 3, 81219 Bratislava, Slovakia
[2] Department of Theoretical and Experimental Physics of Nuclear Reactors, National Research Nuclear University MEPhI (Moscow Engineering Physics Institute), Kashirskoe Shosse 31, 115409 Moscow, Russia

E-mail: filip.osusky@stuba.sk

Abstract. The paper is focused on development of the coupled neutronics-thermal hydraulics model for the Gas-cooled Fast Reactor. It is necessary to carefully investigate coupled calculations of new concepts to avoid recriticality scenarios, as it is not possible to ensure sub-critical state for a fast reactor core under core disruptive accident conditions. Above mentioned calculations are also very suitable for development of new passive or inherent safety systems that can mitigate the occurrence of the recriticality scenarios. In the paper, the most promising fuel material compositions together with a geometry model are described for the Gas-cooled fast reactor. Seven fuel pin and fuel assembly geometry is proposed as a test case for coupled calculation with three different enrichments of fissile material in the form of Pu-UC. The reflective boundary condition is used in radial directions of the test case and vacuum boundary condition is used in axial directions. During these condition, the nuclear system is in super-critical state and to achieve a stable state (which is numerical representation of operational conditions) it is necessary to decrease the reactivity of the system. The iteration scheme is proposed, where SCALE code system is used for collapsing of a macroscopic cross-section into few group representation as input for coupled code NESTLE.

1. Introduction

The future prognosis of depletion of fossil fuel and limitation of CO_2 release into the atmosphere cause a new interest in nuclear energy as the only energy source with high capacity without CO_2 production. Together with increasing pressure from public to reduce the amount of long-lived radionuclides and enhance safety of nuclear power plants, the new concepts of Generation IV reactors have been proposed. The proposed designs should enter to the operation within the first half of the 21st century. One of these concepts refers as Gas-cooled Fast Reactor (hereinafter GFR) and objectives of the design are to improve all aspect on nuclear power generation such as: 1) safety and reliability, 2) economics, 3) sustainability, 4) availability, 5) proliferation and physical protection. There is a large amount of reactor-grade plutonium available, and thus breeding of fissile material is no longer the primary target for the fast reactors in the short term. This enables a shift from high power density fast reactors to designs that are self-sustaining and put more emphasis on the safety, especially passive and inherent safety systems [1,2]. The purpose of the paper is to provide the test case model of GFR assembly for the comprehensive analysis and for the cross verification of the coupled codes.

2. GFR2400 design

Research of the GFR2400 is directed towards fulfilling the ambitious long term goals of Generation IV concepts. This reactor is the newest on the evolutionary path of fully ceramic GFRs featuring ceramic fuel. Structural materials in the form of silicon carbide fiber are allowing high temperatures and good mechanical stability across wide range of temperature gradient. An important innovation of the current design is the application of refractory metallic liners in the form of $W_{14}Re$ layer to enhance the fission product retention of the cladding, resulting in a significant neutronic penalty during normal operation. However, this concept is advantageous under transient conditions and involves spectrum softening. Helium is used as an efficient primary coolant. Since it introduces practically no moderation the GFR's neutron spectrum is one of the hardest among the fast reactors, making it ideal for recycling all actinides, including minor actinides (MAs). Helium is inert and transparent, eliminating most problems related to coolant interaction with structural materials and making online visual inspection of the core possible. Also void reactivity effect is low due to neutronic transparency of coolant. The core outlet temperature is not limited by the coolant characteristics, making it attractive for potential heat applications [3,4].

As mentioned in [5] the European experience in gas cooled reactor technology was unparalleled with more than a thousand reactor years of gas thermal reactor operating experience and a number of in-depth design studies has been developed for the gas cooled fast rectors.

The evolution of fuel designs includes designs of coated particle fuel with or without a binding matrix, silicon carbide blocks with dispersed microparticle fuel inside, the idea of silicon carbide plates with fuel pellets arranged in a honeycomb structure, finally arriving to the current design of a hexagonal lattice of cylindrical fuel rods consisting of a column of fuel pellets inside the composite silicon carbide cladding (Figure 1) [3].

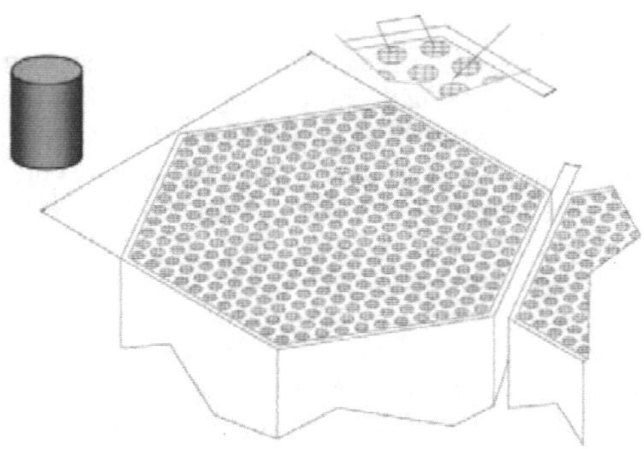

Figure 1. Hexagonal lattice of cylindrical fuel type representation [6].

The starting concept of the fuel assembly in the GoFastR project was the plate-type concept consisting of the carbide fuel with silicon carbide fiber reinforced silicon carbide (SiC_f/SiC) cladding. The goal was to achieve a more realistic and feasible, fully ceramic model which would satisfy all the ambitious GFR requirements. The power generation rate of the Gas-cooled Fast Reactor is 2400 MW_{th} thermal power and helium is used as a coolant at a high pressure of 7 MPa to ensure adequate heat transfer. The coolant volume fraction in the core is high in order to have a small pressure drop that allows decay heat removal by natural circulation under pressurized conditions (the large coolant volume fraction of roughly 43 $V/V_{tot}\%$). To deal with depressurized accidents three dedicated decay heat removal loops (each capable of removing 100% of the decay heat) and six gas reservoirs are available. During normal operation, the power is transferred to the secondary side via three 800 MW power conversion systems equipped with heat exchangers (IHX) and blowers. The secondary coolant is a helium nitrogen mixture and the overall plant efficiency is expected to be around 45% in an indirect Brayton cycle. The design parameters of GFR2400 are specified in Table 1 [3,7].

The core consists of 516 fuel assemblies in a hexagonal geometry. Two different zones are used with different content of Pu: an inner core with lower and an outer core with higher plutonium content (both zones are almost equal size). Reactivity control equipment is represented by 18 control system devices (CSDs) and 13 diverse safety devices (DSDs). The absorber rods in both groups are made up by boron carbide (B_4C, with boron enriched in ^{10}B to 90%). Wrapper structural material is a special steel alloy (AIM1) and cladding is made from silicon carbide fiber. The axial reflectors and the 480 radial reflectors are from Zr_3Si_2. For these core elements no detailed layout was made, only their expected volumetric composition was determined [3,7]. The core is show in Figure 2.

According to [8] the technical feasibility of Gas-cooled Fast Reactor is critical. The main issues are arising within the reactor safety and operation. This technology requires complex, innovative and expensive security systems to ensure continued pressurisation of the reactor pressure vessel (RPV) and core cooling in accidental situations.

Table 1. Basic design parameters of GFR2400 [3].

Parameter	Value	Parameter	Value
Thermal power [MW]	2400	Primary coolant	He
Primary pressure [MPa]	7	Pressure drop in core [MPa]	0.143
Mass flow rate [kgs^{-1}]	1213	Bypass flow rate [kgs^{-1}]	60
Core inlet temp. [°C]	400	Core outlet temp. [°C]	780
Secondary coolant	20%He, 80%N_2	Secondary pressure [MPa]	6.5
IHX inlet temp. [°C]	346	IHX outlet temp. [°C]	750

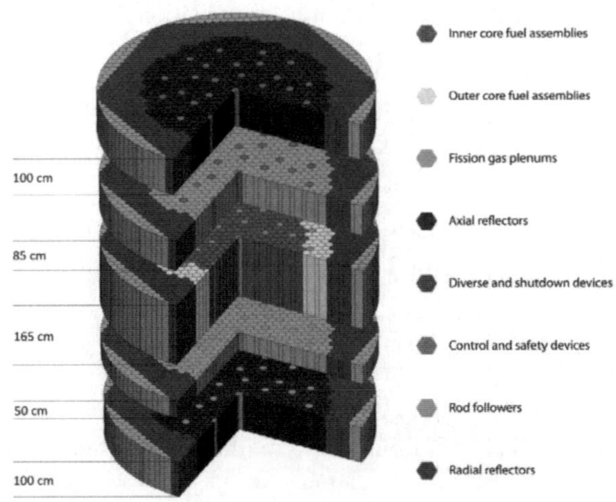

Figure 2. Core configuration of GFR2400 [3].

3. Test case

3.1. Geometry and material composition

Hexagonal lattice of the cylindrical fuel rods consisting of the column of the fuel pellets is the proposed design as a test case. The fuel pin with geometric dimensions is shown in Figure 3. As mentioned in section 2, $W_{14}Re$ layer enhances the fission product retention and significantly influences neutronic properties of the system. Outer surface of the fuel pin is covered by silicon carbide fiber which is allowing high temperatures and good mechanical stability. First proposed geometry as the test case is illustrated in Figure 4 which consists of 7 identical fuel pins with height

165 cm and lattice pitch 11.57 mm. The helium coolant flows between the fuel pins and is pressurized to 7 MPa. Three cases are proposed for investigation in Table 2. Case 1 consists of carbide fuel type and achieves the average value of PuC enrichment within whole core. Case 2 represents inner core fuel assembly material properties with lower enrichment of PuC and case 3 outer fuel assembly with higher enrichment (see Figure 2). Silicon carbide is used for cladding with coating in the form of the $W_{14}Re$ layer and silicon carbide fiber for each case. Flow rate of helium is 273 kgh^{-1} for 7 pins. Core inlet temperature 400 °C and core outlet temperature should approach to 780 °C according to the power generation equal to 0.15 MW_{th} for 7 pins.

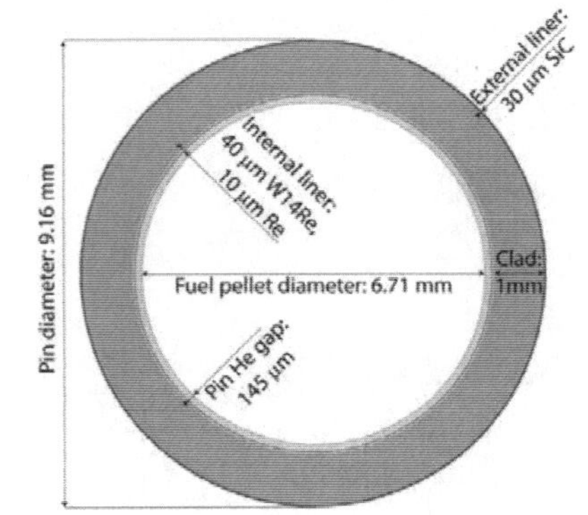

Figure 3. Cross sectional view of the fuel pin of GFR2400 [3].

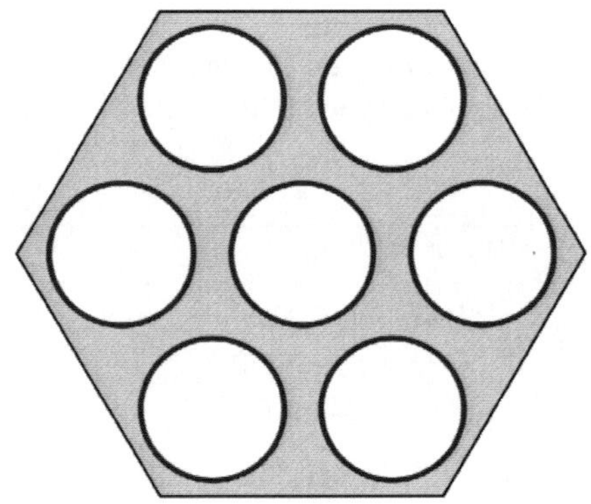

Figure 4. Cluster of 7 pins (Lattice pitch 11.57 mm, height 1650 mm).

Table 2. Fuel material composition [3].

Case 1		Case 2		Case 3	
Density [gcm^{-3}]	10.9001984	Density [gcm^{-3}]	10.9006112	Density [gcm^{-3}]	10.899764
$\%_V$ in PuC	15.84	$\%_V$ in PuC	14.12	$\%_V$ in PuC	17.65
$\%_V$ in UC	84.16	$\%_V$ in UC	85.88	$\%_V$ in UC	82.35
$\%_W$ Pu	14.54422677	$\%_W$ Pu	13.32679852	$\%_W$ Pu	16.65979296
$\%_W$ Am-241	0.105392948	$\%_W$ Am-241	0.093945206	$\%_W$ Am-241	0.117440635
$\%_W$ U-235	80.14544	$\%_W$ U-235	81.7803	$\%_W$ U-235	78.42491
$\%_W$ C-nat	4.798422563	$\%_W$ C-nat	4.798957869	$\%_W$ C-nat	4.797859203

Next proposed geometry is whole fuel assembly (Figure 5 and Figure 6). Inner flow rate of helium for whole assembly is 8462.91 kgh^{-1} and outer flow within inter-assembly gap is 418.6 kgh^{-1}. Inlet temperature is also 400 °C and power generation rate reaches 4.651 MW_{th}.

3.1.1. k_{inf} results

The calculation of k_{inf} was performed for all fuel types. The calculation was performed by SCALE6 KENO VI code [9] system using the multi-group (MG) approach (the used library is ENDF/B-VII.0-238) and the results are shown in Table 3.

Table 3. Results of k_{inf} for all fuel types.

Test case	k_{inf}	deviation
Case 1	1.173878	± 0.000052
Case 2	1.103342	± 0.000048
Case 3	1.243441	± 0.000051

For zero incoming current boundary condition, all cases achieve very deep sub-critical state. Therefore, it is necessary to stabilize nuclear system with reflective boundary condition. Obviously, the boron concentration within the helium coolant cannot be used. The first approach is to surround test case with depleted pins to decrease the reactivity of the system (Figure 7). NESTLE code system [10] is used for coupled calculations. NESTLE is able to treat whole assemblies, thus to mitigate the influence of increasing diameter on thermo-hydraulics calculation, it is necessary to surround the test case with the depleted fuel assemblies (Figure 8). All these approaches can be applied also for the test case with whole fuel assembly. Additionally, it is possible to calculate k_{inf} of the whole fuel assembly with included absorber in the form of B_4C in the chosen representative pin position (red pins in Figure 6). The absorber part influences the neutronic properties of the system, but has negligible effect on thermos-hydraulics.

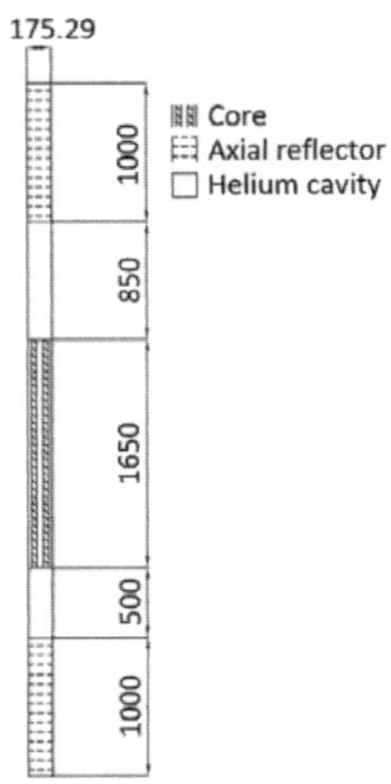

Figure 5. Axial cut of fuel assembly (dimensions in mm).

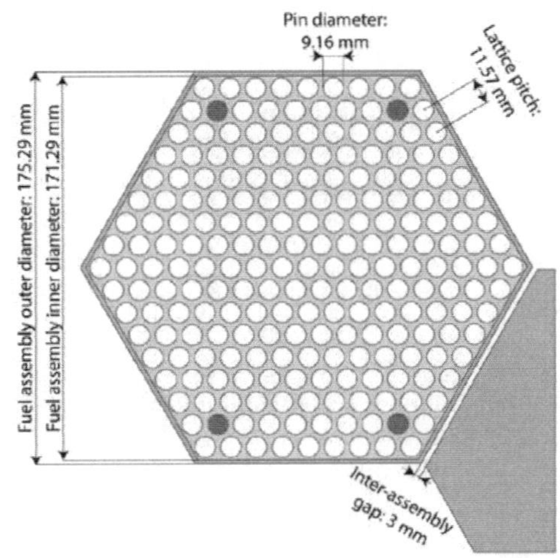

Figure 6. Cross sectional view of the fuel assembly of GFR2400 [3].

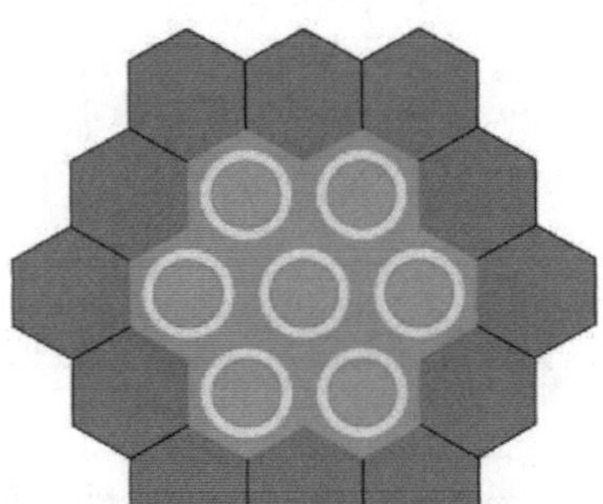

Figure 7. Depleted fuel pins (blue colour) on the edge of system.

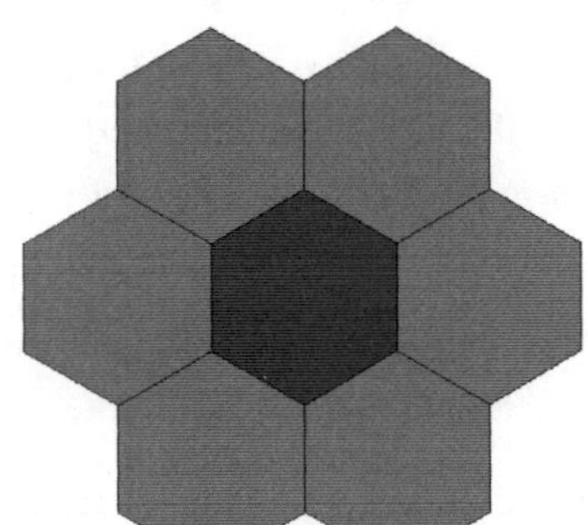

Figure 8. Depleted fuel assemblies (red colour) on the edge of system around the test case of 7 fuel pins (blue).

3.2. Iteration scheme

Iteration scheme is presented in Figure 9. NESTLE code system is able to solve either the nodal or Finite Difference Method representation of 4 or 2-group neutron diffusion equation. Therefore, it is necessary to prepare 4 or 2-group macroscopic cross-section library. The sub-module TRITON within the SCALE code system is used for this purpose. TRITON can be used to provide automated, problem-dependent cross-section processing followed by multigroup transport calculations for one-, two-, and three-dimensional (1D, 2D, and 3D) configurations [9,10].

The convergence of the k_{eff} is highly dependent from thermo-hydraulic feedback, therefore the criterion for convergence in NESTLE is set to k_{eff} eigenvalue or to representation of k_{eff} error criteria. After successful convergence of the coupled calculation, the script checks if stable state is achieved ($k_{eff} \approx 1.0$). If not, the changes in the peripheral depleted or absorber area have to be made and the calculation is performed again until the stable state is not achieved. It is also possible to add small control rods in the peripheral area to achieve the stable state.

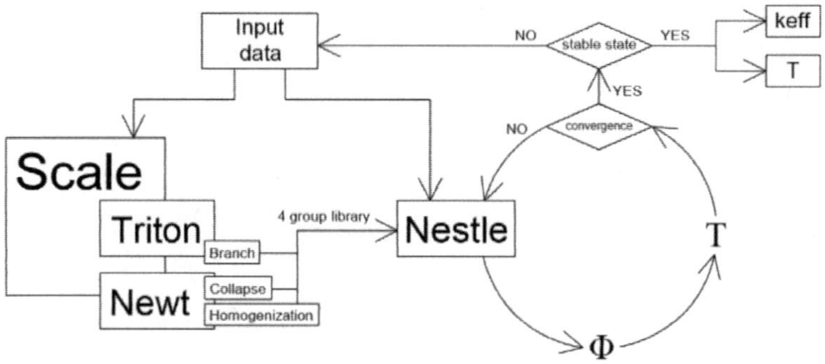

Figure 9. Iteration scheme.

3.2.1. SCALE characteristic equations

The multigroup Monte Carlo procedures are used in SCALE. Typical integro-differential form of the Boltzmann transport equation is solved [9]:

$$\begin{aligned}&\frac{1}{v}\frac{\partial}{\partial t}\phi(\bar{r},E,\overline{\Omega},t)+\overline{\Omega}\cdot\nabla\phi(\bar{r},E,\overline{\Omega},t)+\Sigma_t(\bar{r},E)\phi(\bar{r},E,\overline{\Omega},t)\\&=S(\bar{r},E,\overline{\Omega},t)+\iint dE'd\overline{\Omega}'\Sigma_s(\bar{r},E'\to E,\overline{\Omega}'\to\overline{\Omega})\phi(\bar{r},E',\overline{\Omega}',t)\end{aligned} \quad (3)$$

where $(\bar{r},E,\overline{\Omega},t)$ denotes the general seven-dimensional phase space, $\bar{r}$ position variable, E the particle's kinetic energy, v the particle's speed corresponding to its kinetic energy E, $\overline{\Omega}$ represents unit vector which describes the particle's direction of motion, t time variable, $\phi(\bar{r},E,\overline{\Omega},t)$ stands for the time-dependent angular flux, $\Sigma_t(\bar{r},E)$ represents the total cross-section at the space point $\bar{r}$ for particles of energy E, $\Sigma_s(\bar{r},E'\to E,\overline{\Omega}'\to\overline{\Omega})$ is the differential scattering cross-section and $S(\bar{r},E,\overline{\Omega},t)$ stands for source particles emitted per unit volume and time.

3.2.2. NESTLE characteristic equations

As mentioned above, the NESTLE code originates from the multi-problem solution capability, abbreviating Nodal Eigenvalue, Steady-state, Transient, Le core Evaluator. The eigenvalue problem allows criticality searches to be completed, and the external fixed-source steady-state problem can search a specified power level. Transient problems model delayed neutrons via precursor groups. Several core properties can be input as time dependent. The non-linear iterative strategy associated with the Nodal Expansion Method (NEM) is employed. An advantage of the non-linear iterative strategy is that NESTLE can be utilized to solve either the nodal or Finite Difference Method (FDM) representation of the few-group neutron diffusion equation. NESTLE solves the standard NEM formulation for the solution of the three-dimensional, Cartesian geometry, multi-group, eigenvalue neutron diffusion equation [10]:

$$-\bar{\nabla}\cdot D_g\bar{\nabla}\phi_g+\Sigma_{tg}\phi_g=\sum_{g'=1}^{G}\Sigma_{sgg'}\phi_{g'}+\frac{\chi_g}{k}\sum_{g'=1}^{G}\nu_{g'}\Sigma_{fg'}\phi_{g'} \tag{4}$$

where D_g is diffusion coefficient [cm], ϕ_g represents neutron flux [cm^{-2}sec^{-1}], Σ_{tg} stands for total macroscopic cross section [cm^{-1}], $\Sigma_{sgg'}$ for group-to-group scattering cross section [cm^{-1}], χ_g represents fission neutrons yield, *k* is so called multiplication factor (or critical eigenvalue), ν_g stands for average number of neutrons created per fission and Σ_{fg} represents macroscopic fission cross section [cm^{-1}].

Macroscopic cross section is written in form:

$$\hat{\Sigma}_{xg}=a_{1_{xg}}+\sum_{n=1}^{2}a_{(n+1)_{xg}}(\Delta\rho_c)^n+a_{4_{xg}}\Delta T_c+a_{5_{xg}}\Delta\left(T_{F_{eff}}\right)^{1/2}+\sum_{n=1}^{3}a_{(n+5)_{xg}}(\Delta N_{sp})^n \tag{5}$$

where $\hat{\Sigma}_{xg}$ represents a macroscopic cross-section for reaction type x and energy group g without transient fission products corrected to local conditions, $a_{j_{xg}}$ is an expansion coefficients, $\Delta\rho_c=\rho_c-\rho_c^{(0)}$ is a change in coolant density [gcm^{-3}] from reference condition, $\Delta T_c=T_c-T_c^{(0)}$ stands for change in coolant temperature [°F] from the reference condition, $\Delta\left(T_{F_{eff}}\right)^{1/2}=\left(T_{F_{eff}}\right)^{1/2}-\left(T_{F_{eff}}^{(0)}\right)^{1/2}$ is a change in square root of effective fuel temperature [°F] from the reference condition and $\Delta N_{sp}=N_{sp}-N_{sp}^{(0)}$ represents a change in soluble poison number density [cm^{-3}10^{-24}] from the reference condition.

The effective fuel temperature can be evaluated by equation:

$$T_{F_{eff}}=T_c+W_c\left[W_pT_F+(1-W_p)T_{F_{Srf}}-T_c\right] \tag{6}$$

where W_p is a pellet weighting factor, which accounts for resonance flux depression in the interior of the pellet, W_c represents a core statistical weighting factor, that compensates for the lack of detail in the spatial description of the core, T_F stands for a volume average fuel pellet temperature and $T_{F_{Srf}}$ is a surface average fuel pellet temperature. Thermal-hydraulic feedback is modelled employing a Homogenous Equilibrium Mixture (HEM) model, allowing two-phase flow to be treated. The hydrodynamic model models single and two phase coolant flow up in closed coolant channels. However, only the continuity and energy equations for the coolant are solved, implying a constant pressure treatment. The constant pressure assumption removes the need to consider the momentum equation. The one-dimensional, mass continuity equation along a specified channel for a radial node *ij* is [10]:

$$A_C^{ij}(z)\frac{\partial\rho_C^{ij}(z,t)}{\partial t}=-\frac{\partial}{\partial z}G_C^{ij}(z,t)A_C^{ij}(z) \tag{7}$$

and the energy conservation equation assuming constant pressure is given by:

$$\begin{aligned}A_C^{ij}(z)\frac{\partial}{\partial t}\left(\rho_C^{ij}(z,t)U_C^{ij}(z,t)\right)&=-\frac{\partial}{\partial z}\left(G_C^{ij}(z,t)A_C^{ij}(z)U_C^{ij}(z,t)\right)-\\ P\frac{\partial}{\partial z}\left(\frac{G_C^{ij}(z,t)A_C^{ij}(z)}{\rho_C^{ij}(z,t)}\right)&+q_S^{ij}(z,t)S_F^{ij}+q_C^{ij}(z,t)A_C^{ij}(z)\end{aligned} \tag{8}$$

where ρ_C^{ij} represents coolant density, G_C^{ij} stands for coolant mass velocity, U_C^{ij} is coolant internal energy, q_S^{ij} is volumetric power density from heat deposited directly in the coolant, A_C^{ij} represents total cross-sectional area for the coolant flow within the node, S_F^{ij} is total fuel rod surface area per unit axial length within the node and P stands for coolant pressure [10].

4. Results

Calculated preliminary results of the test case 1 are shown in Figure 10 and Figure 11. For simplification, the geometry was changed to the square fuel assembly consisting of 9 pins, taking into the account the same volume fractions of materials. 2-group macroscopic cross-section library was prepared by SCALE code system (the used library was ENDF/B-VII.0-238) as the input for NESTE coupled code. The deviance of axial power distribution and fuel temperature from study [11] is caused by the value of k_{eff} = 1.243582 together with vacuum boundary condition at the top and the bottom of the fuel assembly. In the study [11], the authors calculated the whole fuel assembly with axial reflectors, that caused increase of the power at the top and the bottom of the assembly. Also the gathering of data from figure in [11] results in the approximate error in power distribution chart. Another results are calculated by FlowVision CFD simulation [12], where the input power distribution is calculated by neutronic code MCU (in this case, the precise geometry was modelled).

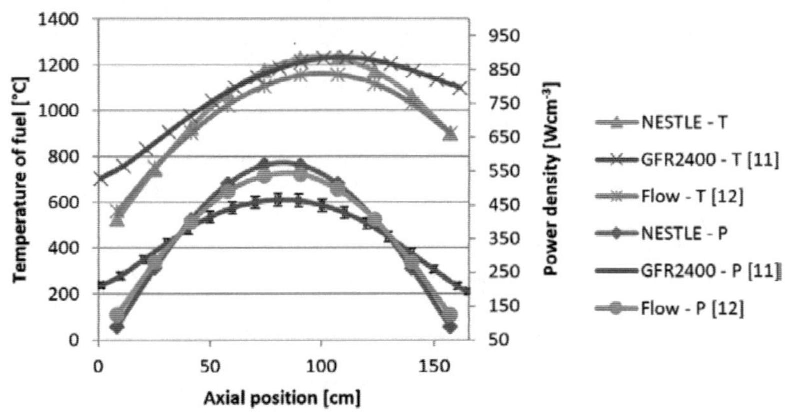

Figure 10. Power density and fuel temperature vs. axial position.

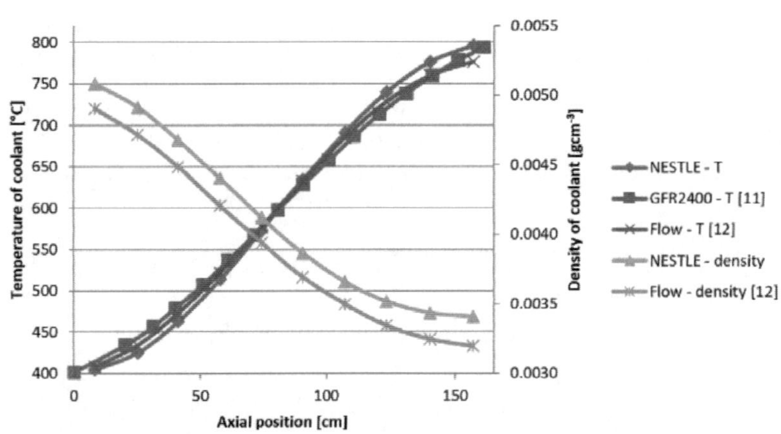

Figure 11. Coolant density and coolant temperature vs. axial position.

5. Conclusion

The overview of the GFR2400 concept is proposed in the paper together with the possible test cases including different enrichment and geometries. Also the procedures how to decrease excessed reactivity are described by incorporation of the depleted peripheral areas to the system or by replacement of appropriate number of the fuel pins with absorber. The preliminary results are showing good agreement with results of GFR2400 study [11] and FlowVision calculation [12]. However, there is necessary to improve the input data for coupled calculation and to achieve the stable state with $k_{eff} \approx$ 1.0 in the future. The cross checking of the results with other calculation procedures will be performed with other European institutions.

Acknowledgments
This work was financially supported by ENEN-RU II project with grant agreement number 605149.

References

[1] Rooijen W F G 2009 Gas-cooled fast reactor: A historical overview and future outlook *Science and Technology of Nuclear Installations* **2009** pp 1–12 (doi:10.1155/2009/965757)
[2] Abram T and Ion S 2008 Generation-IV nuclear power: A review of the state of the science *Energy Policy* **36** pp 4323–4330
[3] Perkó Z, Pelloni S, Mikityuk K, Křepel J, Szieberth M, Gaëtan G, Vrban B, Lüley J, Čerba Š, Halász M, Fehér S, Reiss T, Kloosterman J L, Stainsby R and Poette Ch 2015 Core neutronics characterization of the GFR2400 gas cooled fast reactor *Progress in Nuclear Energy* **83** pp 460–481
[4] Saadi M K, Abbaspour A and Pazirandeh A 2012 Startup of ''CANDLE'' burnup in a gas-cooled fast reactor using monte carlo method *Annals of Nuclear Energy* **50** pp 44–49
[5] Stainsby R, Peers K, Mitchell C, Poette Ch, Mikityuk K and Sommers J 2011 Gas cooled fast reactor research in Europe *Nuclear Engineering and Design* **241** pp 3481–3489
[6] Technische Universität München 2009 *Gas-Cooled Fast Reactor* Lehrstuhl für Nukleartechnik, Garching, Boltzmannstr. 15 85747 p 16
[7] Girardin G, Rimpault G, Morin F, Bosq J C, Coddington P, Mikityuk K and Chawla R 2008 Development and characterization of the control assembly system for the large 2400 MWth generation IV gas-cooled fast reactor *Annals of Nuclear Energy* **35** pp 2206–2218
[8] Locatelli G, Mancini M and Todeschini N 2013 Generation IV nuclear reactors: current status and future prospects *Energy Policy* **61** pp 1503–1520
[9] Oak Ridge National Laboratory 2011 *SCALE: A Comperhensive Modeling and Simulation Suite for Nuclear Safety Analysis and Design* (ORNL/TM-2005/39)
[10] Electric Power Research Center North Carolina State University 2003 *NESTLE Version 5.2.1 Few-Group Neutron Diffusion Equation Solver Utilizing The Nodal Expansion Method for Eigenvalue, Adjoint, Fixed-Source Steady-State and Transient Problems* (NC 27695-7909)
[11] European Gas Cooled Fast Reactor Collaborative Project 2012 *GFR2400 core neutronics and thermal-hydraulics characterization* (Work Package 1.1: GFR Conceptual Design)
[12] CAPVIDIA 2015 *FlowVision: Installation and Setup guide* (Version 3.09.04)

Non-invasive studies of multiphase flow in process equipment. Positron emission particle tracking technique

B V Balakin[1,2], T C H Adamsen[3,4], Y-F Chang[2], P Kosinski[2] and A C Hoffmann[2]

[1] Bergen University College, Department of Mechanical and Marine Engineering, Norway
[2] University of Bergen, Department of Physics and Technology, Norway
[3] University of Bergen, Department of Chemistry, Norway
[4] Haukeland University Hospital, Department of Nuclear Medicine, Norway

E-mail: boris.balakin@hib.no

Abstract. Positron emission particle tracking (PEPT) is a novel experimental technique for non-invasive inspection of industrial fluid/particle flows. The method is based on the dynamic positioning of a positron-emitting, flowing object (particle) performed through the sensing of annihilation events and subsequent numerical treatment to determine the particle position. The present paper shows an integrated overview of PEPT studies which were carried out using a new PET scanner in the Bergen University Hospital to study multiphase flows in different geometric configurations.

1. Introduction

Positron emission particle tracking (PEPT) is a modification of the well-known positron emission tomography (PET) method widely employed in nuclear medicine. Both techniques combine sequential detection of back-to-back gamma-rays resulting from a positron/electron annihilation event with advanced mathematical post-processing of a sensing log. A typical PET system therefore requires the positioning of a positron-emitting isotope within a set of gamma-detectors, e.g. scintillators, distributed in space in a way allowing detection of gamma-rays from a single annihilation event. The detectors are integrated in a PET-scanner unit where their output is collected and pre-analysed by advanced hardware in a data-processing controller. The classical PET technique is primarily focused on the registration of annihilation events from spatial regions where a β^+-emitter is distributed and the subsequent image reconstruction, while PEPT defines a point source of β^+ performing cross-triangulation directly from a sensing log of individual "lines of response" (LORs). A LOR is a line drawn between two sensors of the scanner, which detect photons emitted from the same annihilation event, i.e. a double-detection must be performed within a very narrow temporal gap. In principle, the point of intersection of multiple LORs is recognized in PEPT as the spatial position of the point radioactivity source, however, in practise a number of factors lead to some spatial scatter in the LORs [1]. Due to relatively high energy of gamma-rays (511 keV), they are able to penetrate a wide range of industrial surfaces opaque to visible light, PEPT is therefore suitable for non-invasive control of industrial equipment where

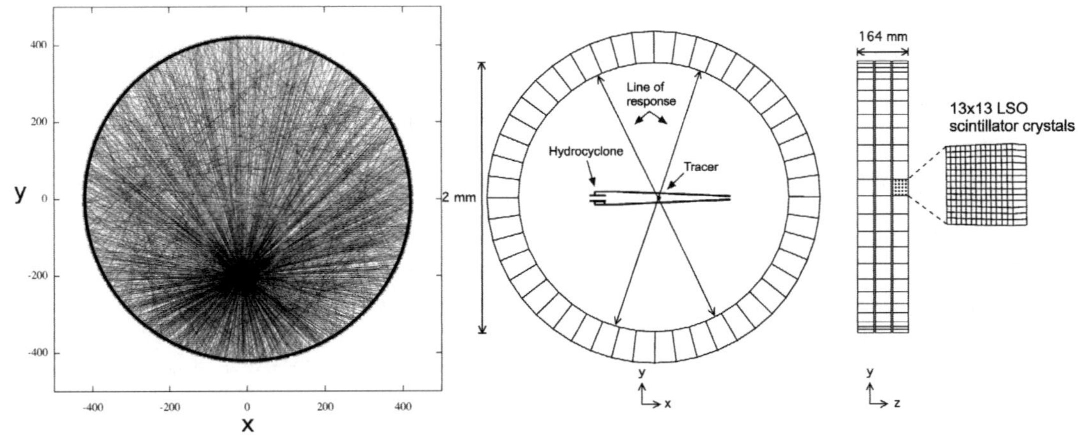

Figure 1. Left: LORs from a tracer particle registered by the camera in 1 ms in a 2D cross-section of camera. Right: the hydrocyclone located in the PET scanner.

direct visual access to the flow is impossible [2]. Comparing basic principles of PET and PEPT it can be concluded that the later operates with a precise and compact dataset which results in far better spatial and temporal resolution for the position of a point source.

The first records on PEPT experiments come from the University of Birmingham [3] where the technique was used to elucidate flow in rotating drums [4], stirred vessels [5], turbines [6], casting [7], paste flows [8] and solid-liquid flow systems [9]. The technique was subsequently adopted by different research groups reporting studies of fluidized beds [10], granulators [11] and other process equipment. The in-situ PEPT-study of an industrial multiphase system by PEPT was reported by Ingram et al. [12]. The present paper provides a detailed overview of the PEPT experiments performed at the PET center of Haukeland University Hospital (Norway) to explore pneumatic conveying lines and hydrocyclones at the conditions similar to industrial.

2. Materials and methods

The experiments were conducted with the use of Siemens TruePoint medical PET scanner with the energetic detection interval 425-650 keV. The scanner, presented schematically in Fig. 1 (right), comprises three 855.2-mm detecting rings. Each ring is equipped with 48 sensor blocks combined of 13×13 lutetium oxyorthosilicate (LSO) scintillator crystals with the size $4 \times 4 \times 20$ mm. The blocks are separated in radial and axial directions by one crystal gap so the axial depth of the detection zone is 164 mm. The co-incidence time of a single-LOR detection is set to 4.5 ns [2], so that if two photons are detected within this window, they are assumed to emanate from one annihilation event. An example of the LORs from 1 ms reconstructed from a stationary emitter of positrons in presented in Fig. 1 (left). It can be seen in the figure that, in addition to the afore-mentioned scatter, there is a certain number of "false" LORs appear due to a random co-incidence, multiple co-incidence and Compton interaction [2].

Multiphase flow was studied in the two types of processing systems: a tangential-inlet 414-mm length hydrocyclone with a diameter of 40 mm (Fig. 3(left)) and a vertical 50-cm cylindrical PVC-pipeline section for pneumatic conveying with the diameter of 45.2 mm combined with two $50 \times 88.5^{\circ}$ bends and a replaceable flow straightener of the honeycomb/mixing-plate type (Fig. 4). The carrier phase in case of the hydrocyclone was water, in some cases with dissolved NaCl, resulting in a density of 1100 kg/m^3 intended to match that of the tracer particle and therefore to study the flow of the fluid in the hydrocyclone; and air at normal conditions in the case of the pneumatic conveying study. The test equipment was connected to flow loops comprising flow agitators (centrifugal pump and gas compressor) generating continuous flow through the test sections of the loop in the field of view of the camera, dispersed phase injectors, flow meters and a system of by-pass lines. The details of the flow loops design may be found elsewhere [2, 13].

The dispersed phase, and, in the case of the neutral-density tracer particle, the continuous phase, was represented by the tracer particles emitting positrons. Amberlyst anion-exchange styrene divinylbenzene beads with the mean size 430±56 μm and average density 1070 kg/m^3, shown in Fig. 2, were used. The beads were labelled with $_{18}$F ($t_{\frac{1}{2}}$=110 min) as follows: at first a sample of 0.5 ml of water was exposed by protons in cyclotron establishing $_{18}$O(p,n)$_{18}$F nuclear reaction, the beads are then immersed into the resulting radioactive solution and agitated there for 10 minutes until $_{18}$F is adsorbed by the bead material via the ion-exchange mechanism; the bead, with the individual activity up to 2000 μCi, is then removed from the solution, dried with a filter paper and injected into the flow.

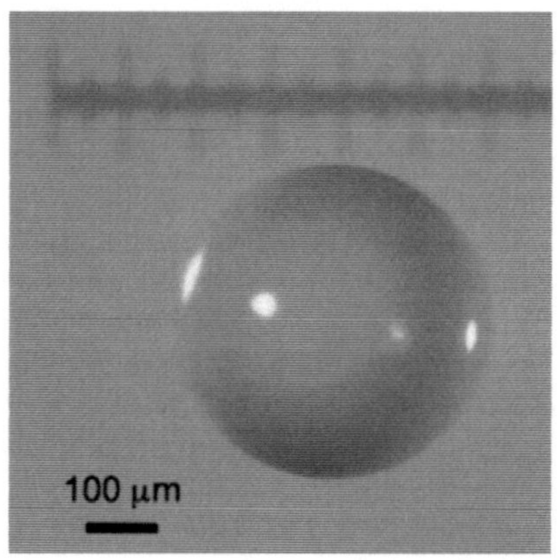

Figure 2. Amberlyst A21 anion exchange resin bead

The position of the tracer in the flow of the carrier phase is detected using the algorithm originating from Hoffmann et al. [1]. The algorithm initially detects all the LORs and the two-dimensional (x-y) co-ordinates of their intersections are determined from the sensor positions obtained in the 64-bit list-mode log files from the PET scanner. The average of all the intersection coordinates is computed and outlying intersections eliminated, which first are those intersections generated by "false" LORs. The rest of the intersection coordinates are averaged again and the procedure of eliminating outliers and narrowing the spatial averaging window is repeated until only intersections remain that are within a final spatial window with a diameter below 20 mm [2]. The final position of the tracer is found by averaging the intersections located in this window. Optimization of the size of the final window is done by verifying that this size results in the lowest standard deviation of the positions of a stationary tracer. The coordinate of the particle in the third dimension is obtained by averaging of the intersection between the remaining LORs from the last dataset. The computational costs may further be reduced assuming the first window to be located in the vicinity of the tracer position at the previous time step. The first window is centered on the position minizing the sum of the squared distances to all of the LORs, an analytical expression for this position was derived.

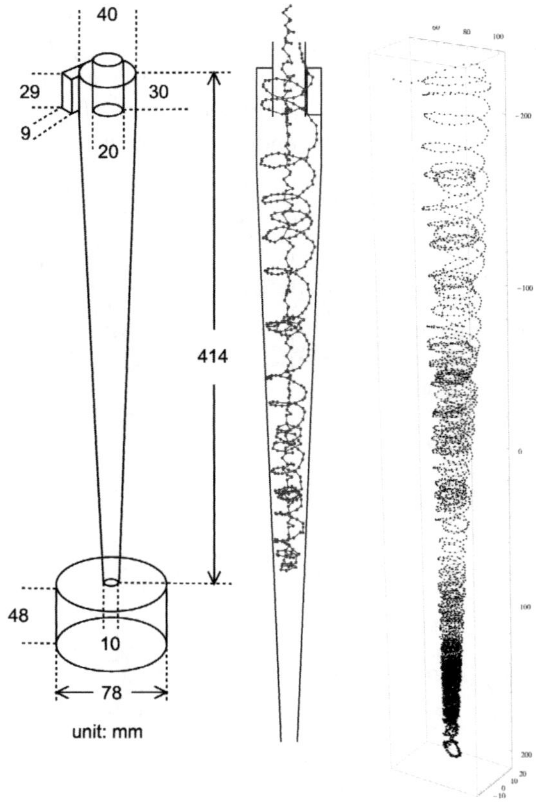

Figure 3. Left: dimensions of the hydrocyclone. Middle: Track of a neutral-density particle after 505 ms with very low residual activity, inlet flow velocity 3.6 m/s. Right: track of a tracer particle larger than the cyclone cut-size in water with a position recorded every 500 μs.

3. Results and discussion

The trajectory of a neutral-density tracer in the hydrocyclone at Re=158 400 is presented in Fig. 3 (middle), where the points represent the particle position with an acquisition step of 1.0 ms. As is seen in the figure, the particle enters the cyclone and, instead of flowing smoothly to the underflow close to the wall of the cyclone, as would be expected for a solid particle larger than the cut-size of the separator, it quickly starts moving radially in and out due to the turbulence in the flow. Low in the cyclone it reverses the direction of its axial velocity, moving into the inner, upwardly flowing, vortex and moves out through the liquid outlet in the cyclone roof following the liquid flow. The solid particle in water, to the right in the figure, represents the behaviour of a particle larger than the cut-size of the cyclone. This track is of the normal high quality since the activity remains on the particle and it is feasible to determine a position every 500 μs. One would expect such a particle to move smoothly to the particle outlet in the bottom of the cyclone. However, it can be seen to take several excursions into the inner vortex of upwardly directed flow on the way. This is unexpected, and is an effect that can only be observed with Lagrangian particle tracking. Moving further downwards the conical section of the cyclone, the particle experiences significant deviations from the swirling motion, probably due to the presence of the "end of the vortex" at this position, where the vortex end is attached to the wall and precesses resulting in the generation of a complex fluid flow pattern and a less-than-optimal downward particle transport. Under this position, a swirling motion is restored, albeit very much weaker, induced by the precessing of the vortex. In the experiments [2], the density of the tracer particle was matched by adding salt to the fluid in order to study the flow of the fluid in the cyclone. This proved very challenging, however, since the activity leached from the partticle very quickly, leading to tracking under difficult conditions, and the development of post-processing methods to reduce the resulting high scatter [2]. The second experimental system, presented in Fig. 4

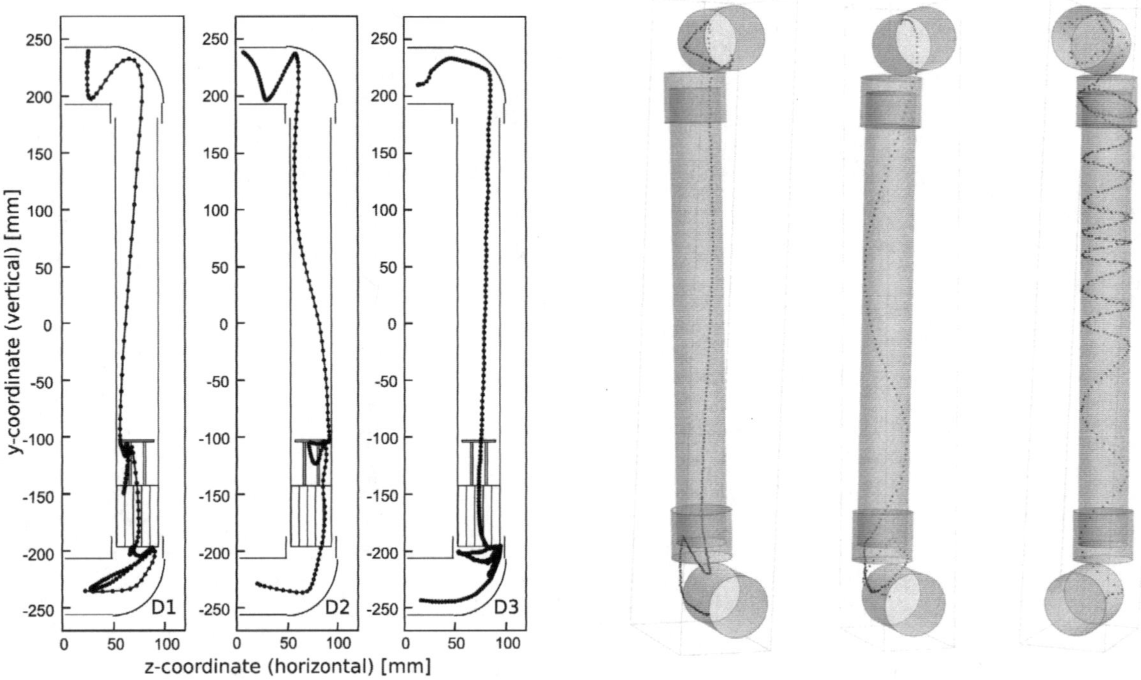

Figure 4. Particle tracks obtained in the pneumatic system. Left: interaction of the tracer with the flow straightener in 2D projection at 15.7, 10.6, 4.4 m/s (left→right). Right: 3D tracks in an obstacle-free section at 5.0, 16.9 and 17.1 m/s (left→right).

demonstrates the particle behaviour in a pneumatic conveying system with (left) and without (right) a flow straightener installed at Reynolds numbers in the interval 10500–41500. First we focus on the system with flow straightener where tracks are presented for mean flow velocities of 15.7,10.6 and 4.4 m/s. As it can be observed in the figure, the tracer is approaching the test section moving nearly horizontally and experiences multiple reflections from the straightener elements with magnitudes proportional to the flow velocity. The reflections, which take place with a coefficient of restitution of 0.81 [14], give rise to reversed motion of the particle. The fluid and particle phases are only weakly coupled (the particle Stokes number is in the range 48–187). The particulate phase flow after the straightener becomes aligned over the rest of the test section and experiences rotational motion caused by the formation of assymmetrical Dean vortices (one vortex becomes dominant) formed in the carrier phase after the second bend.

It is further possible to observe from the 3-D images to the right in Fig. 4 that a spiral-like motion of the tracer is induced in some cases in the test section when the straightener is removed. As in the cyclonic case in the conical section, the frequency of the swirling motion increases with the mean flow velocity. This might be expected, taking into account possible interference of Dean vortices at higher Re. Although the particles experience minor reflections from the elements forming the bend, they do not affect the swirling motion. The particle tracks are rather insensitive to gravity at the Reynolds numbers selected, while the influence of electrostatic forces potentially induced on the tracer due to friction with plastic walls could be considered as possible. It is finally reasonable to conclude that the flow straightener conditions the upward flow better than the empty elbow.

4. Conclusions

The present paper illustrates the behaviour of multiphase flow in two types of process equipment: the hydrocyclone and the pneumatic system with obstacles, captured by means of positron emission particle tracking. This paper and ref. [2] demonstrate that the technique is able to visualise the behaviour of both the carrier phase, making the tracer neutrally buoyant in the carrier fluid, or the flow of the inertial dispersed phase where other forces than the drag become important. The temporal resolution of the technique is 1 ms or better while the spatial resolution is determined by the inherent scatter, the activity on the tracer particle and the sophistication of the algorithm used to position the particle. The technique is non-invasive and the algorithm analysing the PET scanner logs takes less than 60 min on an ordinary PC. This method is therefore suitable for the benchmarking of numerical models [15]. PEPT is applicable to most types of the process equipment which is confirmed by other studies performed by our group for the dispersion in Hartmann tube [14] and the liquid-solid flow in a recirculated flow loop [15].

5. Acknowledgments

The authors would like to thank Jan Selbu and Øyvind Steimler at Haukeland University Hospital for their support during the experiments.

References

[1] Hoffmann A C, Dechsiri van de Wiel C F, Dehling H G PET investigation of a fluidized particle: spatial and temporal resolution and short term motion 2005 *Measurement Science and Technology* **16** 851–858
[2] Chang Y F, Ilea C G, Aasen Ø L, Hoffmann A C Particle flow in a hydrocyclone investigated by positron emission particle tracking 2011 *Chemical Engineering Science* **66** 4203 – 4211
[3] Parker D J, Hawkesworth M R, Broadbent C, Fowles P, Fryer T D, McNeil P A Industrial positron-based imaging: Principles and applications 1994 *Nuclear Instruments and Methods in Physics Research Section A: Accelerators, Spectrometers, Detectors and Associated Equipment* **348** 583–592
[4] Ding Y L, Forster R, Seville J P K, Parker D J Segregation of granular flow in the transverse plane of a rolling mode rotating drum 2002 *International Journal of Multiphase Flow* **28** 635 – 663
[5] Barigou M, Fan X, Parker D, Nienow A, Guida A Positron emission particle tracking in a mechanically agitated solid-liquid suspension of coarse particles 2009 *Chemical Engineering Research and Design* **87** 421–429
[6] Fishwick R P, Winterbottom J M, Parker D J, Fan X F, Stitt E H Hydrodynamic measurements of up-and down-pumping pitched-blade turbines in gassed, agitated, vessels using positron emission particle tracking 2005 *Industrial and Engineering Chemistry Research* **44** 6371–6380
[7] Griffiths W D, Beshay Y, Parker D J, Fan X The determination of inclusion movement in steel castings by positron emission particle tracking (PEPT) 2008 *Journal of Materials Science* **21** 6853–6856
[8] Wildman R D, Blackburn S, Benton D M, McNeil P A, Parcker D J Investigation of paste flow using positron emission particle tracking 1999 *Powder Technology* **103** 220–229
[9] Fairhurst P, Barigou M, Fryer P, Pain J-P, Parker D Using positron emission particle tracking (PEPT) to study nearly neutrally buoyant particles in high solid fraction pipe flow 2001 *International Journal of Multiphase Flow* **27** 1881 – 1901
[10] Chan C W, Seville J, Fan X, Baeyens J Solid particle motion in a standpipe as observed by positron emission particle tracking 2009 *Powder Technology* **194** 58 – 66
[11] Schaafsma S, Marx T, Hoffmann A Investigation of the particle flowpattern and segregation in tapered fluidized bed granulators 2006 *Chemical Engineering Science* **61** 4467 – 4475
[12] Ingram A, Hausard M, Fan X, Parker D J, Seville J P K, Finn N, Kilvington R, Evans M Portable positron

emission particle tracking (PEPT) for industrial use 2007 *Proceedings of 12th International Conference on Fluidization* 13–17

[13] Middha P, Balakin B V, Leirvaag L, Hoffmann A C, Kosinski P, PEPT - a novel tool for investigation of pneumatic conveying 2013 *Powder Technology* **237** 87 – 96

[14] Berg A E Investigation of dust dispersion in a modified Hartmann tube. Master Thesis (under consideration) *the University of Bergen* 2016

[15] Ghaffari M, Chang Y-F, Balakin B, Hoffmann A C CFD modelling of PEPT results of particle motion trajectories in a pipe over an obstacle 2012 *AIP Conference proceedings* **1479** 193–196

IOP Conf. Series: Journal of Physics: Conf. Series **781** (2017) 012040 doi:10.1088/1742-6596/781/1/012040

Nuclide radioactive decay data uncertainties library

D S Barabanova[1] **and G M Zherdev**[2]

[1] FSUE RFNC All-Russian Research Institute of Experimental Physics, Mira Ave., 37, Sarov, Nizhny Novgorod region 607188, Russia
[2] JSC «State Scientific Center of the Russian Federation - Institute for Physics and Power Engineering n.a. A I Leypunsky», Bondarenko Sq., 1, Obninsk, Kaluga Region 249033, Russia

E-mail: dassha20081@mail.ru, jerdev@ippe.ru

Abstract. The results of the developing the library of uncertainties of radioactive decay data in the ABBN data library format are described. Different evaluations of uncertainties were compared and their effects on the results of calculations of residual energy release were determined using the test problems and experiment. Tables were generated in the ABBN format with the data obtained on the basis of libraries in ENDF-6 format. 3821 isotopes from the ENDF/B-7 data library, 3852 isotopes from the JEFF-3.11 data library and 1264 isotopes from the JENDL-4.0 data library were processed. It was revealed that the differences in the evaluations accepted in different decay data libraries are not so big, although they sometimes exceed the uncertainties assigned to the data in the ENDF/B-7 and JEFF-3.11 libraries, which as a rule, they agree with each other. On the basis of developed method it is supposed to create a library of data uncertainties for radioactive decay within the constant data system in FSUE RFNC-VNIIEF with its further connection with CRYSTAL module.

1. Introduction

In 2012 the CRYSTAL software module [1] was developed in the Institute of Theoretical and Computational Physics of All-Russian Scientific Research Institute of Experimental Physics. It is designed for solving computational problems of nuclear fuel burning up and residual energy release in both thermal and fast reactor facilities. CRYSTAL is used for isotopic kinetics calculations in CONCORD software package [2]. A specific characteristic of CRYSTAL module is that it includes production of a complete set of nuclides (more than 1300) in fuel material and can be used in both multigroup and Monte-Carlo computations.

Figure 1 demonstrates the task complexity exemplified in the scheme of ^{86}Kr transmutation which shows its production as a function of precursor and descendant nuclei. To find isotope concentrations it is necessary to solve a stiff linear set of differential equations using the Rosenbrock-Wanner method.

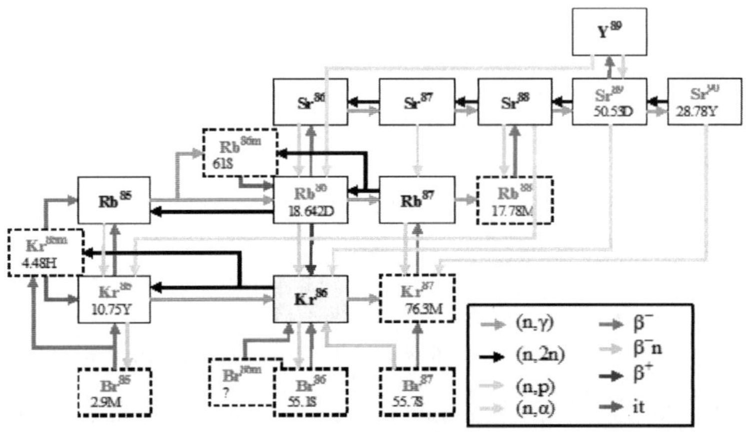

Figure 1. Implicit bonds character between ^{86}Kr descendants nuclei and ^{86}Kr precursor nuclei

Nuclear data uncertainties estimation used in calculations is an important factor in numerical simulation of isotopic composition change. This work describes the method to develop a library of the radioactive decay nuclear data uncertainties. The ABBN radioactive decay constant data system [3, 4] controlled and developed for many years by JSC «SSC RF – IPPE» in the city of Obninsk (earlier SSC RF – IPPE) is considered as a possible base for the CRYSTAL module. The library is widely used in the Russian Federation. This work contains methods of collecting radioactive decay data in the ABBN format.

2. Processing evaluated radionuclide radioactive decay data files

Russian (ABBN), American (ENDF/B-7), European (JEFF-3.11) and Japanese (JENDL-4.0) libraries were used in the work. Foreign libraries data are presented and stored in the internationally accepted ENDF-6 format [5, 6].

Fortran-based DECAPRO Code (Decay Processing) [7], which extracts radioactive decay data from MF=8 file of the library in the ENDF-6 format and generates decay data tables with its uncertainties in the ABBN format, was used for evaluated nuclear data file processing.

The same format as the one used for the data themselves was suggested for storing the uncertainties of the decay data, only with a different value of the MT identifier determining in this format the type of the data contained in the tables. It was suggested for tables of uncertainties to use the MT identifier with value of 81(MT=81).

ENDF/B-7, JEFF-3.11 and JENDL-4.0 libraries were processed using this program (3821 isotopes from the ENDF/B-7 data library, 3852 isotopes from the JEFF-3.11 data library and 1264 isotopes from the JENDL-4.0 data library), and tables in the ABBN format were generated for these evaluations.

3. Application of the data on the uncertainties of energy release

The data from libraries were analyzed earlier in [7]; energy release calculation uncertainties were evaluated for the test model of a reactor in comparison with models of different authors within the SKALA system [8, 9].

Then additional calculations of residual energy release and its uncertainties were performed for the same fast reactor model and for «Flash» experiment, the idea of which is in the short-term irradiation of samples followed by the measurement of energy release.

Figure 2 shows that uncertainty is two orders of magnitude smaller than energy release and almost doesn't change. As shown in the Figure 3, for exposure time 1 – 10^4 seconds the uncertainty is small, and after that it increases and suffers sharp changes. It is presumably connected with fission products accumulation, on which less data is available.

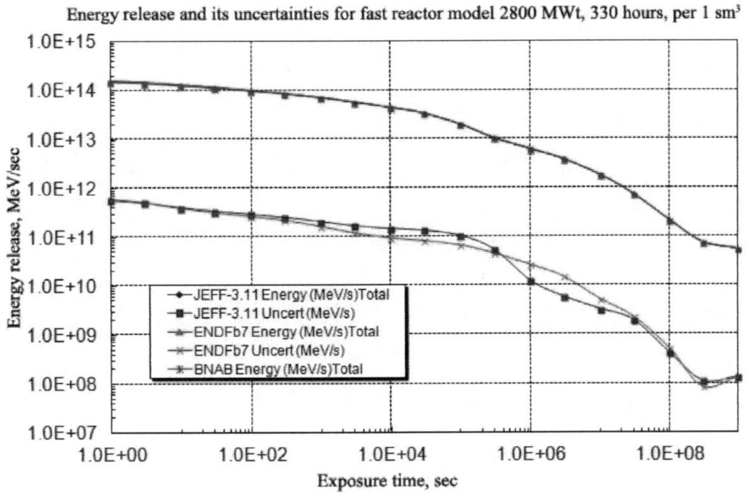

Figure 2. Dependence of energy release and absolute uncertainty on exposure time in different evaluations.

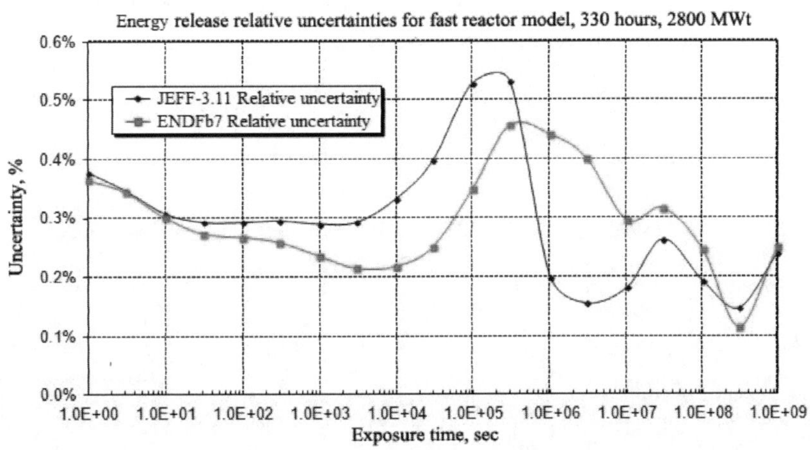

Figure 3. Dependence of relative uncertainty on exposure time in different evaluations.

Besides, different energy releases evaluations were compared to the ABBN data. Figure 4 illustrates that the ENDF/B7 and JEFF-3.11 data results differ by less than 1%, while difference from the ABBN can exceed 3%. Apparently, this can be attributed to the absence of data on short-lived isotopes in the earlier published ABBN library.

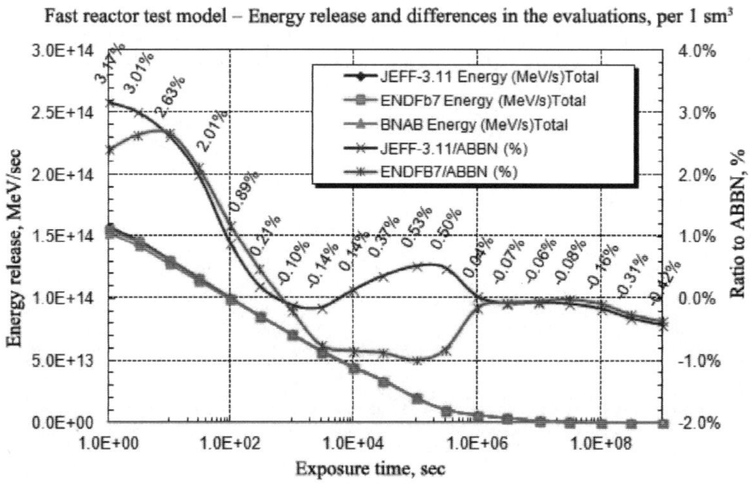

Figure 4. Dependence of energy releases and differences in the evaluations on exposure time.

«Flash» experiment [10] data are shown below. Dickens (ORNL), Karlsruhe, Yayoi (Tokyo University), Lowell (University of Massachusetts, Lowell) named in Figure 5 are experimental results, and JNDC (JAEA), AFPA (NRNU MEPhI), ORIGEN (ORNL), CARE-E (JSC «SSC RF – IPPE») are calculation results. The results of this work calculation are SKALA JEF. The plots in Figure 5 show that the calculation results in our work insignificantly differ from the results of other authors and agree with the experiment within uncertainties for times over 10 seconds both for U-235 (left plot) and Pu-239 (right plot).

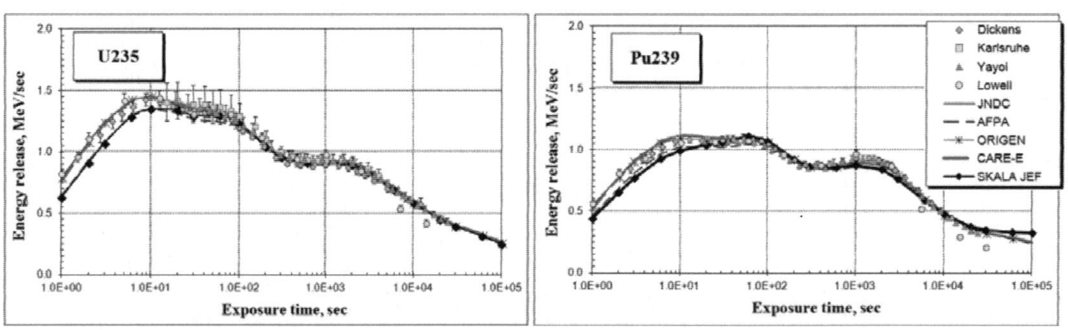

Figure 5. Comparison of the residual energy release value in benchmark experiments with the calculated values for the exposure time from 1 to 10^5 seconds for U-235 and Pu-239 («Flash» experiment).

4. Conclusions

- We have mastered the technology of evaluating radioactive nuclides decay data uncertainties in the ABBN format.
- Energy release and its uncertainties for test models within the SKALA system (JSC «SSC RF – IPPE») were calculated; calculation results were analyzed and compared with different authors and experiments.

- On the basis of developed method it is supposed to create a radioactive decay data uncertainties library within the FSUE RFNC-VNIIEF constant support system and that will be further linked with the CRYSTAL module.

References

[1] Mzhachih S V *et al* 2014 Estimation of isotopic composition of reactor facilities when modeling burning up and residual nuclear fuel energy release processes using CONCORD software package *Nuclear science and technology issues. Ser. Mathematic simulation of physical processes* **1** 49-64

[2] Alekseev A V *et al* 2013 CONCORD software package for nuclear reactors cells calculations on supercomputers *Proc. XIV Int. Conf. Supercomputing and mathematical modeling* (Sarov: FSUE RFNC - VNIIEF) 15-27

[3] Zabrodskaja S V *et al* 2000 Decay data library and fission product release in the constant data system *Nuclear science and technology issues. Ser. Nuclear constants* **2** 71

[4] Manturov G N *et al* 1996 BNAB-93 group constants system. Part 1: Nuclear constants for calculation of neutron and photon radiation fields *Nuclear science and technology issues. Ser. Nuclear constants* **1** 59

[5] Trkov A *et al* Nov 2011 ENDF-6 Formats Manual, Data Formats and Procedures for the Evaluated Nuclear Data Files ENDF/B-VI and ENDF/B-VII ed M Herman (New York: NNDC Brookhaven National Laboratory Report)

[6] ENDF-6 Formats Manual. Russian translation by Zabrodskaja S V 2005; available at http://www.ippe.ru/podr/abbn/libr/endf.php

[7] Barabanova D S and Zherdev G M 2015 Radioactive decay data uncertainties library of isotopes for ABBN constant system *Higher education institution news, Nuclear power industry* **1** 150-6

[8] Zherdev G M September 12-15 2005 SKALA – The computing system for estimation of nuclear and radiation safety on CD-ROM - sect. 315 *Proc. Int. Conf. «M&C 2005»* (Avignon: France Nuclear Energy Society)

[9] Vlaskin G N *et al* 2002 Intensity estimation of radioactive radiation sources (SOURCE program) *Nuclear science and technology issues. Ser. Nuclear reactors physics* **4** 39-60

[10] JNDC Nuclear Data Library of Fission Products 2nd Version 1990, available at http://jolissrch-inter.tokai-sc.jaea.go.jp/pdfdata/JAERI-1320.pdf

IOP Conf. Series: Journal of Physics: Conf. Series **781** (2017) 012041 doi:10.1088/1742-6596/781/1/012041

Simulation of motions of the plasma in a fusion reactor for obtaining of future energy

Askhat Zhumabekov

Teacher Education College, JAMK University of Applied Sciences, 35 Rajakatu street, 40200 Jyvaskyla, Finland

Zhumabekov_Askhat@sm.nis.edu.kz

Abstract. According to the most conservative estimates, by the middle of the XXI century in the world energy consumption will double. This will be a consequence of the global economic development, population growth and other geopolitical and economic factors. Energy consumption in the world is growing much faster than its production and industrial use of new advanced technologies in the energy sector, for objective reasons, will not begin until 2030. This paper discusses how to obtain and develop nuclear energy on the experience of the National Nuclear Center. Implemented model for the problem of plasma confinement, and also presents the main achievements of modern construction and Megaproject National Nuclear Center in Kurchatov, the Republic of Kazakhstan. Spend a social survey in the East Kazakhstan region on the theme: "Prospects for the development of nuclear energy in Kazakhstan" and the citizens' opinion. Narration new priorities for May 22, 2015 in Ust-Kamenogorsk in the industrial park "Altai" based on the competition of innovation projects green technology in the international exhibition "OSKEMEN EXPO – 2015", with the participation of the regional authorities of the Republic of Kazakhstan, representatives of JSC NC "Astana Expo" and delegations from Japan, Russia, Canada, USA, South Korea.

1. Introduction

Objectives:

1. Study and investigate principle and features of Kazakhstan thermonuclear material Tokamak reactor KTM under the ITER project.
2. Prepare the computer model by using the 3D Max of the motions of the plasma in Kazakhstani thermonuclear material "Tokamak reactor KTM".

Tasks:

1. Explore devices from different reactor types to compare them with the installation of ITER and Kazakhstan thermonuclear material "Tokamak reactor KTM".
2. Carry out the social survey on the prospects of nuclear energy development in Kazakhstan.
3. To study the process of energy production in nuclear reactors on fast neutrons (ITER).

Hypothesis:

If we prepare a computer model of Kazakhstan thermonuclear material Tokamak reactor KTM, we can demonstrate the motions of the plasma process, as well as offer solutions with its path of instability, tendency to buckle, thickening and formation of constrictions.

Relevance:

Energy consumption in the world is growing much faster than its production and industrial use of new advanced technologies in the energy sector, for objective reasons, will not begin until 2030. All the more acute the problem of shortage arises fossil fuels. The possibilities of building new hydropower plants are also very limited. Do not forget about the fight against the greenhouse effect, imposes restrictions on the burning of oil, gas and coal in thermal power plants. The solution may be an active development of nuclear energy. Every year in Kazakhstan the problem of shortage of energy resources is becoming increasingly important and concerns every citizen of our republic. Nuclear power margin consists of several components that will develop the economy of Kazakhstan, and also bring many advantages. This theme will help us to analyze the acute problems facing the nuclear power industry in Kazakhstan and learn about the main advantages of the installation ITER and its superiority over other nuclear reactors.

Novelty:

The novelty of this project lies in the fact that at the moment, nuclear power is not used in Kazakhstan, despite the fact that the reserves (by the IAEA) of uranium in the country equal to 900 thousand tons. The main deposits are in the south of Kazakhstan (South Kazakhstan region and Kyzylorda region) in the west of Mangystau, in northern Kazakhstan (Semizbay deposit).

Now the issue of building a new nuclear power plant of 600 MW in Aktau. The country has maintained around 5 research reactors.

Methods:

1. Study, generalization, analysis (online resources, and literature).
2. Compare (comparison of various types of nuclear reactors with Kazakhstan ustanovkoyITER and thermonuclear material reactor - tokamak KTM).
3. Survey (conducting social survey on the prospects of nuclear energy development in Kazakhstan).
4. Modeling (creating a computer model of Kazakhstan thermonuclear material Tokamak reactor KTM).

Research Steps:

- The research expedition to the National Nuclear Center of Kurchatov, town to study experimental Tokamak KTM.
- Creating a computer model of the tokamak in 3DMax program.
- Conduct social survey on "Development of nuclear energy in Kazakhstan".
- Identification of the main problems for the instability of the plasma column.

All the more acute the problem of shortage arises fossil fuels. The possibilities of building new hydropower plants are also very limited. Do not forget about the fight against the greenhouse effect, imposes restrictions on the burning of oil, gas and coal in thermal power plants.

But one of the most important and significant events in the preparation of our project took place at the seminar "Radiation in everyday life", which was held December 22, 2014 the Association "Nuclear Society of Kazakhstan", headed by the Minister of Energy of the Republic of Kazakhstan V. Shkolnik. The seminar was organized by the Executive Director of the Nuclear Society of Kazakhstan N.A. Zhdanova and Vice-President of the International Antinuclear Motion "Nevada-Semipalatinsk" S.U. Kartoyev (figure 1). To speak at the workshop with the students, we prepared more than one day, because the invited veterans of labor, scientists of the National Nuclear Center and the public: non-governmental organizations, national cultural centers and the media. Our students spoke about how the study was conducted, which put forward the hypothesis, confirmed in a study, and what does not. This research led to a very lively interest among organizers, veterans, order bearers and scientists of the National Nuclear Center of the republic. At this meeting they discussed issues such as the construction of nuclear power plants, the state of the radiation background at this point, and more. The students were able to answer all the questions and talk about the future work plan.

Figure 1. Seminar "Radiation in everyday life".

2. The research expedition to the National Nuclear Centre of the Republic of Kazakhstan

The uniqueness of the experimental base NNC, the results and experience gained from the fuel behavior studies, materials and construction of nuclear reactors in the simulation of transient and accident operation modes cause certain interest among specialists from different countries of the world (fig. 2). The most significant in recent years is the research on two projects: experimental studies in support of the safety of light water reactors (COTELS project) and experimental studies to support the safety of fast neutron reactors (EAGLE project). Towards the creation of the first nuclear fusion reactor the world scientific community is faced with a number of complex issues, both scientific and technological plan. It is currently expected that the first installation, which will help to achieve the use of fusion energy for commercial use is the installation of ITER (International Thermonuclear Experimental Reactor). ITER - project of International Thermonuclear Experimental Reactor. The objective of ITER is to demonstrate the possibility of commercial use of nuclear fusion reactor and the solution of physical and technological problems that may be encountered along the way. The reactor design is completely finished and the place chosen for its construction - the research center Cadarache (French Cadarache.) in the south of France, 60 km from Marseille. Preparation of the construction site in Cadarache in the south of France began in January 2007. Today, there are two promising project of fusion reactors in the world: the tokamak (toroidal chamber with magnets coils), and the stellarator. In both setups the plasma is confined by the magnetic field, but it has the shape of a tokamak toroidal cord through which electrical current is passed, while the magnetic field is induced Stellarator external coils. In fusion reactors, there are reactions of synthesis of heavy elements of lung (helium from hydrogen isotopes - deuterium and tritium), in contrast to conventional reactors, which initiates the process of decay of heavy nuclei into lighter.

Figure 2. The research expedition to the National Nuclear Center of the Republic of Kazakhstan.

Theoretical tokamak scheme was proposed in 1951 by Soviet physicists Andrei Sakharov and Igor Tamm, and in 1954 the first plant was built in the USSR. However, scientists have not been able to for a long time to maintain the plasma in a steady state, and by the mid 1960s, the world was convinced that controlled thermonuclear fusion is not possible on the basis of the tokamak. But three years later on the T-3 at the Institute of Kurchatov Atomic Energy under the guidance of Leo Artsimovich managed to heat the plasma to temperatures of more than five million degrees Celsius and hold it for a short while; Scientists from the UK attending the experiment on their equipment recorded a temperature of about ten million degrees. After that, the real boom began tokamak in the world, so that there are about 300 units, the largest of which are in Europe, Japan, the United States and Russia have been built. The research was visualized various instabilities of the plasma column, which tends to buckle, the formation of constrictions and bulges using program Adobe Flash 8 (fig. 3).

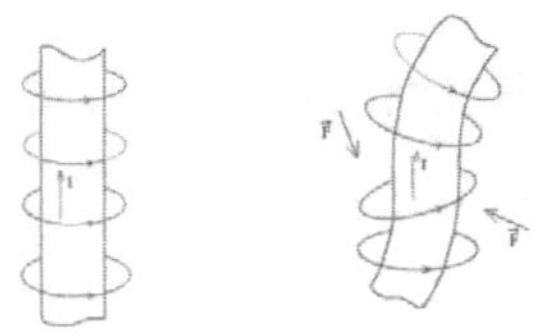

Figure 3. The instability of the plasma pinch.

2.1. A trial run of KTM tokamak in Kurchatov
September 5, 2010 conducted a test launch of KTM Tokamak. Work on the preparation and conduct of the trial run of KTM Tokamak made in the implementation of the budget program 045 "Creating Kazakhstan material tokamak KTM". The purpose of the trial run - getting the working gas breakdown in the vacuum chamber of KTM Tokamak and formation of the plasma column with current 10-30 kA.

2.1.1. Implementation of the project.
These images are an example for the realization of our project. As a basis we took the Kazakhstan material tokamak KTM under the ITER project, a joint Russian-Kazakh project of material testing tokamak KTM, National Nuclear Center of the Republic of Kazakhstan (fig. 4).

Figure 4. Several stages of work KTM Tokamak reactor in the Republic of Kazakhstan.

References

[1] Djumagulova K N and Zhukeshov A M and Ramazanov T S 2007 *The interaction of plasma flows with structural materials CF units* (Almaty: LLP Print-S)
[2] Nurabaeva H W 2006 *Neutron physics* (Kokshetau: Publishing Center of Kokshetau University)
[3] Galeev A A and Sudan R 1984 *Fundamentals of Physics Plasma* (Moscow: Energoatomisdat)
[4] Thomson J J 1996 *Conduction of Electricity through Gases* (Cambridge University Press, New York) p 109
[5] Goldston R J and Rutherford P H 1995 *Introduction to Plasma Physics* (Bristol: Institute of Physics) ISBN: 0-7503-0325-5
[6] Ackleh A S and Allen E J and Kearfott R B and Seshaiyer P 2009 *Classical and Modern Numerical Analysis: Theory, Methods and Practice* (Hoboken: CRC Press) ISBN: 1420091573
[7] Chen F F 1984 *Introduction to Plasma Physics and Controlled Fusion* vol 1 (Boston, MA: Springer US) ISBN: 9781475755954
[8] Averback R S and Diaz de la Rubia T 1998 *Solid State Physics* vol 51 (New York: Academic) pp 281–402
[9] Allen M P and Tildesley D J 1989 *Computer Simulation of Liquids* (Oxford: Oxford University Press)
[10] Roth J 1981 *Sputtering by Particle Bombardment* vol 1 (Berlin: Springer) pp 91–146

The use of the neutronic calculation code CORNER for evaluating the protection of fast neutron reactor and CNFC equipment

M E Shekhanova

National Research Nuclear University MEPhI (31, Kashirskoe shosse, Moscow, 115409, Russia)
Nuclear Safety Institute of the Russia Academy of Sciences (52, B.Tulskaya st., Moscow, 115119, Russia)

E-mail: marie.altavista@gmail.com

Abstract. In this paper we propose a method of using neutronic calculation code CORNER to the analysis of experiments on the protection of fast neutron reactor and CNFC equipment. An example of Winfrith Graphite Benchmark experiment calculation using this approach is presented. This task can be considered as one step in the general theme of the safety analysis of FR with liquid metal coolant, their fuel cycles and related equipment. CORNER implement a solution of the kinetic equation with a source in the three-dimensional hexagonal geometry based on Sn-method. The purpose of this paper is a demonstration of the application of CORNER's possibilities for the analysis of the actual reactor problems.

1. Approximations

Neutron transport equation in multi-group approximation in some spatial domain G for the energy group g, $g = 1,2,...,N_n$ has the form:

$$(\vec{\Omega} \cdot \nabla \Psi^g(\vec{r},\vec{\Omega})) + \Sigma_T^g(\vec{r}) \cdot \Psi^g(\vec{r},\vec{\Omega}) = R_n^g(\vec{r},\vec{\Omega}) \quad (1)$$

where the right-hand-side of the equation looks like:

$$R_n^g(\vec{r},\vec{\Omega}) = \sum_{h=1}^{g} \int_{4\pi} \mathrm{P}_n^{h \to g}(\vec{r},\vec{\Omega}\cdot\vec{\Omega}')\Psi^h(\vec{r},\vec{\Omega}')d\vec{\Omega}' + \chi_p^g \sum_{h=1}^{N_n} \nu\Sigma_f^h(\vec{r})\Phi^h(\vec{r}) + S^g(\vec{r}) \quad (2)$$

$$\Phi^g(\vec{r}) = \frac{1}{4\pi} \int_{4\pi} \Psi^g(\vec{r},\vec{\Omega})d\vec{\Omega} \quad (3)$$

At the boundary Γ of spatial domain G values of angular flow set to zero $\Psi^g(\vec{r},\vec{\Omega})$ for directions $\vec{\Omega}$ inside this area:

$$\Psi^g(\vec{r},\vec{\Omega})\Big|_{\Gamma,\vec{\Omega}\cdot\vec{n}<0} = 0 \quad (4)$$

The system of equations (1) - (4) describes a stationary distribution of neutrons taking into account the fission process as well as the specified internal sources [1].

Calculation task of K_{eff} (homogeneous problem) is described by the system of equations (1)–(4) with zero boundary conditions and with zero internal sources $S^g(\vec{r})$ and with a multiplier $1/K_{eff}$ in front of the second term in the right-hand side of (2).

In solving these problems to calculate the collision integral in the right side of the transfer equation is necessary to set the angular quadrature (Sn-approximation method of discrete ordinates) and Pm-approximation of the scattering function, which corresponds to the solution of the transport equation in SnPm-approximation.

2. Description of the code

Computer code CORNER is based on SN discrete ordinates method and PM approximation of scattering cross section. It allows to solve two types of stationary problems of neutrons and gamma rays transport in the three-dimensional hexagonal (HEX-Z) geometry: the problem on Keff (homogeneous) and the problem with the source (heterogeneous). To approximate the spatial dependence implemented WDD (Weighted Diamond Difference) and nodal scheme were realized.

CORNER software tool is developed in Fortran and has a modular structure. The main modules are: module of preparation of neutron constants in ANISN format; geometric module; module of preparation of angular quadrature; module of input data containing the parameters of the approximation and the control parameters; module of neutronic calculation and the calculation results processing unit.

Energy dependence is represented by multi-group approximation. Discretization of angular variable is carried out by introducing the angular quadrature. Using of LQn and Pn-Tn quadrature kits also provides the opportunity to quadrature user sets. Use an iterative process solutions, including external iteration on the source of division and inner iterations from the scattering source. The output of the iterative process, both in terms of accuracy, and the number of iterations is provided.

3. Calculation of the test problem

An experimental graphite protection composition irradiated at the installation ASPIS, shown in Fig.1, is considered for test calculation [2].

The initial model is described using a hexagonal geometry. The size of "turn-key", equal to 5 cm, chosen from consideration of the conservation area of the cross-section of individual elements of the protective composition. As a result, cartogram in the plane contains 4624 cells. Nonuniform grid with 26 layers is constructed to analyze the high-altitude partitioning (along the Z axis). Example of partitioning is shown in Fig.2. Energy dependence is represented by 299-group approach, using group constants training program CONSYST with ABBN- RF library. Calculation of the code CORNER carried out in S4 approximation method of discrete ordinates, scattering anisotropy is taken into account in P1 approximation [3]. Estimated time is ~ 50 min. As a result of the calculation the rates of the reactions S32(n, p)P32, A127(n, a)Na24, In115(n, n ')In115m and Rh103(n, n')Rh103m were obtained, studied and compared with experimental ones.

An example of the comparison of CORNER result with the results of other calculation codes ones is presented in Fig.3.

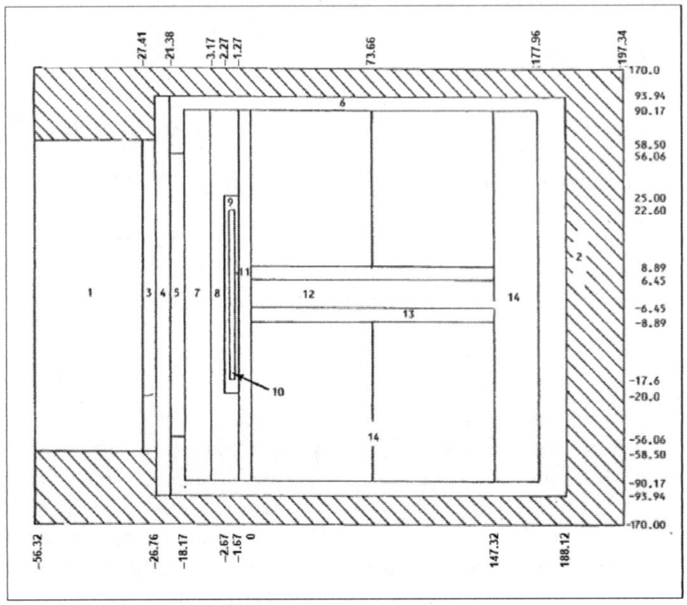

Figure 1. Scheme of protection composition in the installation ASPIS for the graphite experiment. (1-14 – numbers of zones).

```
9
26
  1     3    4      5      7     8     9       10       9     11    12
0.   29.   30.   35.   38.   53.   53.5   53.75   54.25   54.5   56.  61.
```

Figure 2. Example of partitioning in the input file "geom".

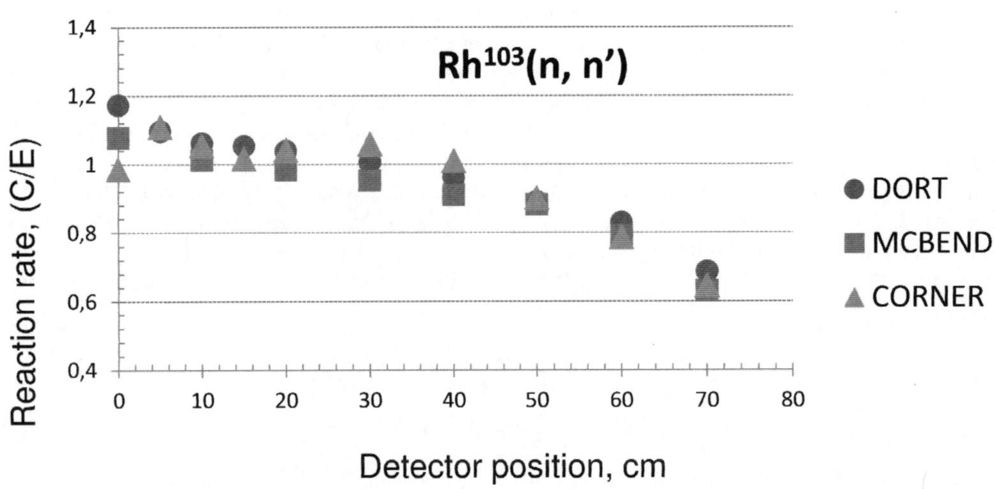

Figure 3. Comparing of the calculated result with the results of other codes on example of rate of Rh103(n, n') reaction.

IOP Conf. Series: Journal of Physics: Conf. Series **781** (2017) 012042 doi:10.1088/1742-6596/781/1/012042

4. Conclusion

A method of applying the neutronic calculation code CORNER for evaluating experiment with graphite protection was developed. Similar calculations for the protection of fast-neutron reactor and CNFC equipment are planned during further development of this work. For example, the proposed method makes it possible to evaluate the protection of facilities and equipment of process stages of closing the nuclear fuel cycle (processing module, fabrication module, module of radioactive waste management).

References

[1] Bell D., Glesston S. "The theory of nuclear reactors", Moscow, Atomizdat, 1974.

[2] M.D. Carter, P.C. Miller, A. Packwood "The ASPIS Graphite Benchmark Experiment. Part 1 Experimental Data and Preliminary Results", *NEACRP-A-630*, 1984.

[3] V.P.Bereznev, E.F.Seleznev, D.S.Asatryan "The CORNER neutronic calculation code", *Izvestiya VUZov. Nuclear power engineering*, Moscow, 2015.

[4] Alan F. Avery, AEA-RS-5628, private communication.

[5] G. A. Wright, A. Avery, M. J. Grimstone, H. F. Locke, S. Newbon Benchmarking of the JEFF2.2 Data Library for Shielding Applications, Proceedings, *8th International Conference on Radiation Shielding,* April 24-28, 1994, Arlington, Texas, USA., vol.2, p.816.

[6] Seleznev E.F., Belov A.A., Manturov G.N., Peregudov A.A. and other Uncertainty Analysis for fuel Flux Calculations of Fast Reactors with External Fuel Cycle. *International Conference on Fast Reactors and Related Fuel Cycles: Safe Technologies and Sustainable Scenarios (FR13).* Paris, France, 4-7.03.2013, IAEA

[7] Korn G., Korn T. Handbook of mathematics for scientists and engineers. – Moscow, Nauka, 1970. - P. 575-576.

[8] Abagyan A.A., Seleznev E.F., Ivanov V.B. "The possibility of long-term development of nuclear energy on the basis of modern technologies", *Atomic Energy,* 2004 vol.97, issue 4, p.252-260.

[9] Seleznev E.F., Asatryan D.S., Belov A.A. et al. "Neutron-physical codes". *International scientific-technical conference "Innovative nuclear energy projects and technologies"*, Moscow, 27-29 November 2012, p.79.

[10] Zabrodskaya S.V., Ignatyuk A.V., Koshcheev V.N. And others. ROSFOND - Russian The National Library of evaluated neutron data. *VAST,* Ser. Nuclear constant 2007 vol.1-2, p.3-21.

IOP Conf. Series: Journal of Physics: Conf. Series **781** (2017) 012043 doi:10.1088/1742-6596/781/1/012043

The choice of core unit cells boundaries in surface harmonics method by the test tasks solving example

A Elshin, A Pakhomov and V Riuchin

[1]Institute of Nuclear Power Engineering (branch) of Peter the Great Saint-Petersburg Polytechnic University, Sosnovy Bor 188544, Leningrad Region, Russian Federation.

E-mail: avelshin@gmail.com, rjuchinvadim@yandex.ru, andy19867@yandex.ru

Abstract. The surface harmonics method (SHM) is one of the approaches to neutronic reactor computation reasoning and developing method of homogenization. The symmetry of cells in the method of homogenization and in the SHM is still being used while equations deriving. The cells can be asymmetric in reality. The surface harmonics method allows us to obtain finite-difference equations for a heterogeneous reactor and for the reactor core with asymmetric cells. This allows one to check SHM recommendations on choosing boundaries of the unit cells of the reactor. The examples of test tasks with asymmetric cells are shown in this paper for substantiation of the unit cells boundaries choice for the reactor core.

1. Introduction

The symmetry of cells is used in derivation of the finite-difference equations for a heterogeneous reactor by surface harmonics method, in particular for ranking the trial functions depending on their importance. If neutron flux densities are expected to be equal at different faces of the cells in modulus from the symmetry of cells, the equations will become simpler. In reality, the cells can appear asymmetric (and they become asymmetric because of burnout process). The regularity of a mesh may also be disrupted in the reactor core for example in the gap between the PWR assembles.

Therefore, in the paper [1] finite-difference equations for a heterogeneous reactor are derived without cells symmetry assumption. The trial functions were computed as a cell "response" to the single flux inflow (odd moment) of neutrons from a g-group from any cell edge.

The obtained finite-difference equations for a heterogeneous reactor (with irregular mesh and asymmetric cells) allow us to check the SHM thesis that the unit cells boundaries should be choose in moderator in order to describe the distribution of neutrons at the cell boundaries using less angular moments. The obtained equations possibilities for calculating the neutron distribution in the asymmetric cells system are used to demonstrate the boundaries choice influence on the calculation results in this paper. For simplicity one-dimensional one-group test tasks were used.

2. One-group one-dimensional test tasks

Test tasks are taken from [2]. In this paper test tasks were used in the method of homogenization to investigate the influence of different ways of diffusion coefficient calculating on the computing accuracy.

In the test tasks the problem was solved to get the eigenvalue for a one-dimensional reactor that is made up of different number of plates (5-75) with different sizes and properties (Figure 1) with the given one-group cross sections for fast sodium reactor.

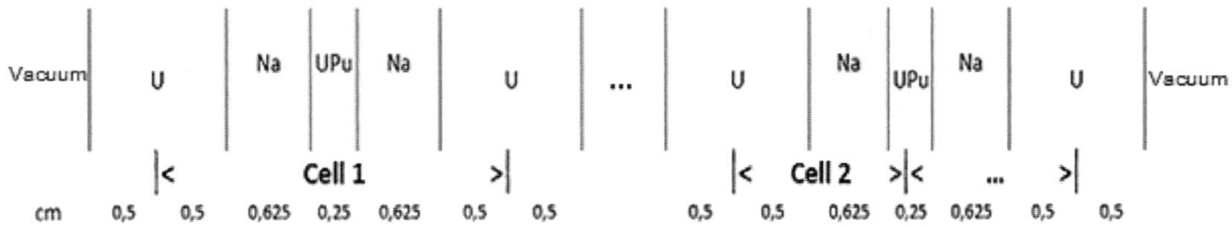

	U	Na	UPu	
Σt	0,18155	0,04529	0,25265	cm^{-1}
νΣf	0,00998	0	0,08604	cm^{-1}
Σs	0,11789	0,02677	0,18500	cm^{-1}

Figure 1. Geometry and materials properties in the test task.

The reference solution in the paper [2] was computed using the discrete ordinates method. In this paper the spherical harmonics method was used to obtain the reference solution. The spherical harmonics method was applied analytically for planar geometry with a minimum of VBA language coding for numerical results. In order to implement the vacuum boundary conditions in the SHM in the paper [3] the connection matrix has been obtained for odd and even angular moments of the neutron distribution at the edge with vacuum. The reference solution was computed to the P_{29}-approximation. This approximation would be sufficient for the convergence of the flux density in space (Figure 2).

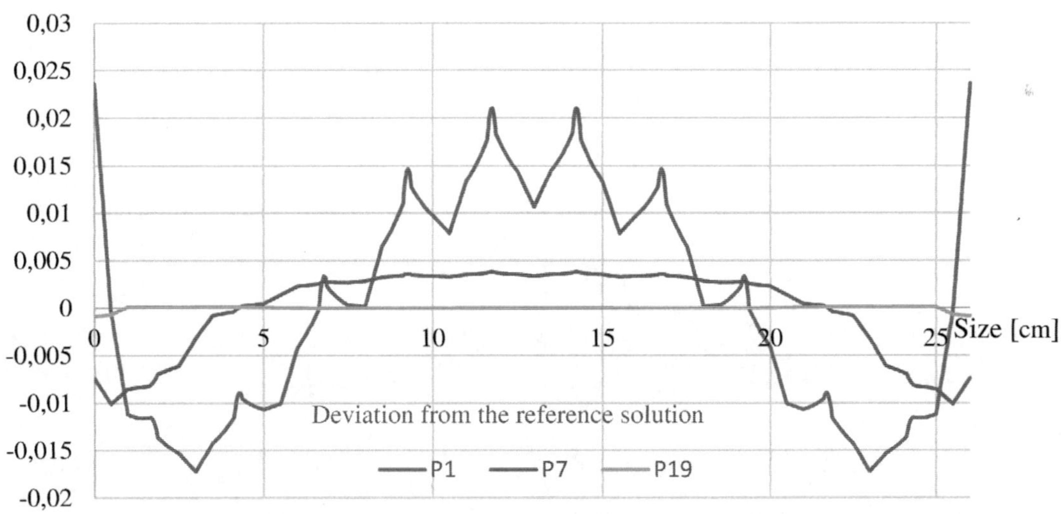

Figure 2. Calculation with unit normalization of neutron flux for the task with 11 U-Pu plates (deviation from the reference solution for P_1-, P_7-, P_{19} approximations).

The convergence of neutron flux angular distribution is not apparently achieved especially in sodium (Figure 3).

While receiving the numerical results it turned out more preferable to compute the trial functions with the same boundary conditions even in asymmetric cells and the flux inflow schemes (in the general case, of the odd angular moments of the neutron distribution) to be left in the ordered symmetry. In this case even angular moments on the right (the first) and left (the second) faces of one-dimensional asymmetric cell will be different. With this computation of trial functions the succession of finite-difference equations is obvious. It is also very important to have close affinity between the equations and finite-difference approximation of the diffusion equation.

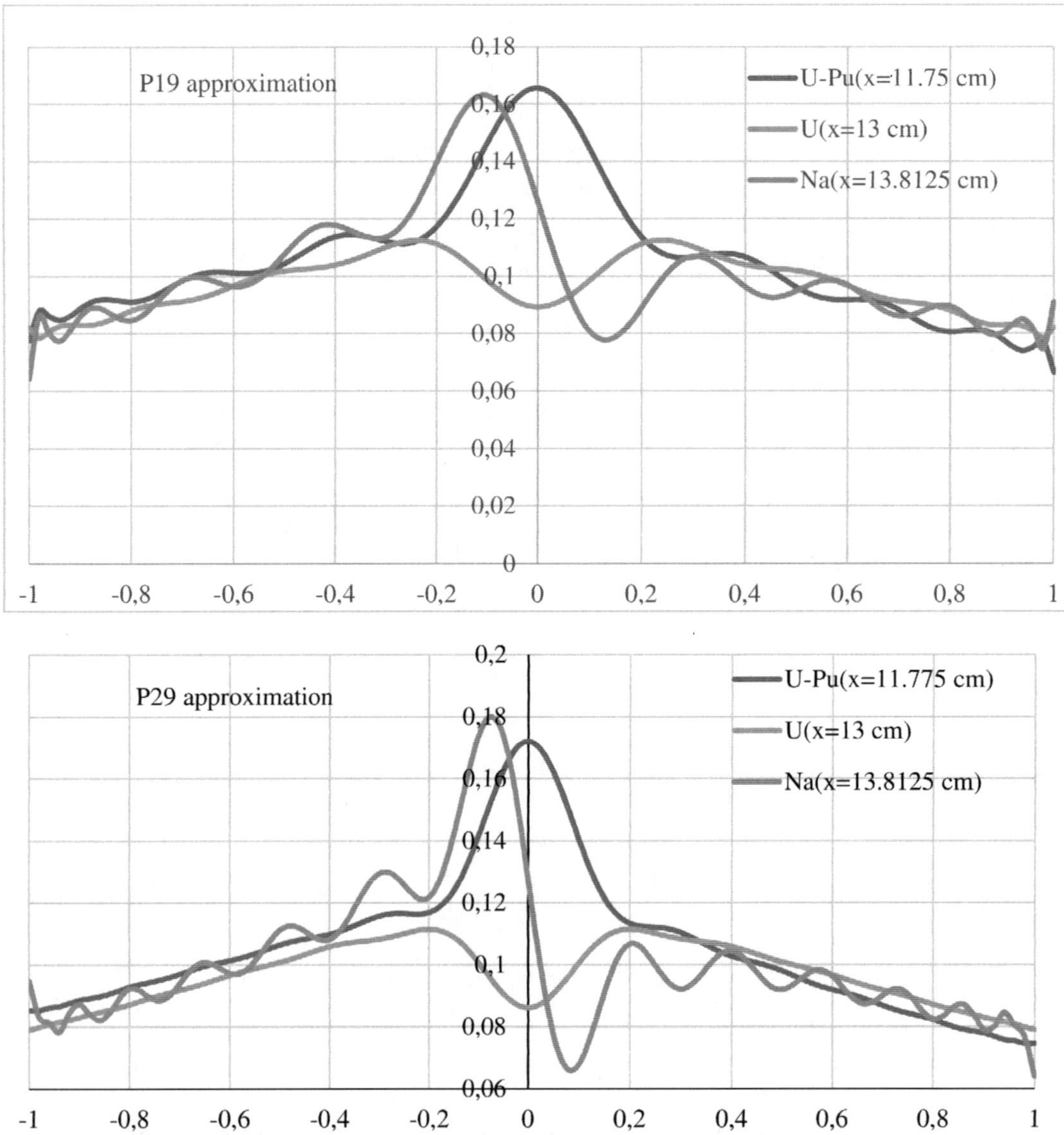

Figure 3. Convergence of neutron flux angular distribution for U-Pu, U, Na.

It is not so important for the purpose of this paper, since it is enough that the trial functions inside the cell are computed to the same P_{29}-approximation as in the reference solution.

The resulting system of finite-difference equations in the case of asymmetric cells looks in this way:

IOP Conf. Series: Journal of Physics: Conf. Series **781** (2017) 012043 doi:10.1088/1742-6596/781/1/012043

$$(\boldsymbol{\psi}_{k-1}^{1}+\boldsymbol{\psi}_{k}^{2})^{-1}(\boldsymbol{\Phi}_{k-1}^{1}-\mathbf{T}_{k}\boldsymbol{\Phi}_{k}^{1})+(\boldsymbol{\psi}_{k}^{1}+\boldsymbol{\psi}_{k+1}^{2})^{-1}(\mathbf{T}_{k+1}\boldsymbol{\Phi}_{k+1}^{1}-\boldsymbol{\Phi}_{k}^{1})-\boldsymbol{\Sigma}_{k}\boldsymbol{\Phi}_{k}^{1}=0 . \quad (1)$$

In the system of equations (1) lower index is the one of the cell and top index is the one of the cell face (visible from (1) that the values of even angular moments of "quasisymmetrical" $\hat{\varphi}$ - functions and the "quasiasymmetrical" $\boldsymbol{\psi}$ - functions are different at different faces). $\boldsymbol{\Phi}_{k}^{1}$ is the vector close to the one of even moments of the neutron distribution at the first (right) face of the cell.

Since one-group case is considered in the test the matrix - coefficients of the equation (1) are composed of even neutron distribution moments at the face of the cell for different trial functions (with non-zero odd moments at the cell face). Matrix - $\mathbf{T}_{k}=(\hat{\varphi}_{k}^{2}-\boldsymbol{\psi}_{k}^{2})(\hat{\varphi}_{k}^{1}-\boldsymbol{\psi}_{k}^{1})^{-1}$ would be unit for symmetric cells. The role of the diffusion coefficient (coefficient matrix in the lattice) in the equation (1) has $\mathbf{D}_{k}=\boldsymbol{\psi}_{k}^{-1}/2$, the role of the neutron absorption cross section (and neutron generation) has $\boldsymbol{\Sigma}_{k}=2(\varphi_{k}^{1}-\boldsymbol{\psi}_{k}^{1})^{-1}$.

3. The results of test tasks solving

The method of surface harmonics advises us to choose cell boundaries in the places where the angular distribution of neutrons can be characterized by the small number of angular moments. The middles of zones should meet this condition. So the unit cell in the test task calculation was chosen with boundaries in the middle of uranium layer («Cell1» Figure 1). The cells at the edges of the system are half of the uranium layer. All cells in this case are symmetric. All sizes of cells are included into the trial functions, so the mesh irregularity is not important for the SHM. To investigate the effect of cells asymmetry on the calculation of eigenvalues (the neutron multiplication factor) and neutron flux density the face of the cell was shifted from the uranium layer center to some distance (both the left edge and the right one and the cell becomes asymmetric). In this case, uranium cells with different sizes remain at the edges of the system. The calculation results with 11 uranium layers are shown on the Figure 4.

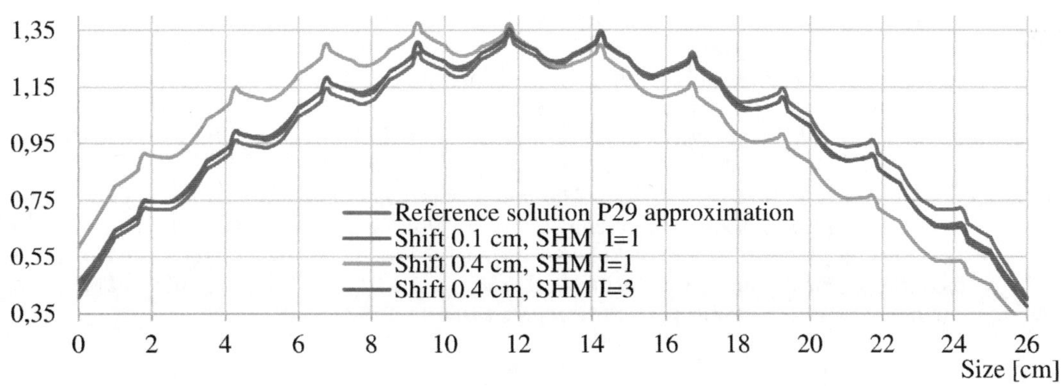

Figure 4. Neutron flux density at the different boundary choice of unit cells.

The obtained results show that are the closer the chosen unit cells boundaries to the zone middle, the closer the obtained solution is to the reference one. It is interesting to note, that in the lowest approximations of SHM with "bad" choice of boundaries an asymmetric solution in symmetric task is possible (however, it is impossible to solve the problem with asymmetric cells using homogenization method). It is the result of the fact that neutron distribution inside a cell depends on the angular distribution of the neutrons flowing into the cell.

When choosing the unit cell boundaries in the middles of zones the problem can be solved arbitrarily selecting cells, for example with the cells which boundaries are located in the middle of plutonium and uranium layers (asymmetric cells, «Cell 2» on Figure 1). Results for system with these cells are shown on the Figure 5.

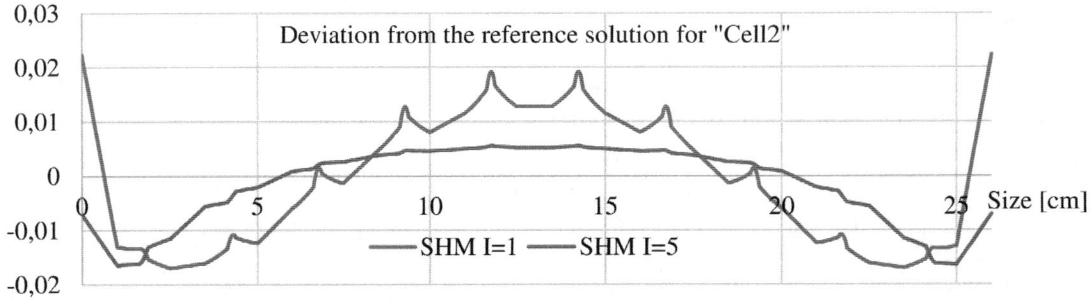

Figure 5. Neutron flux density in the task with asymmetric cells. Comparing with the reference solution.

We have symmetric solution.

Conclusion

The test problem computations have shown the validity of the SHM recommendations on choosing boundaries of the core unit cells of a heterogeneous reactor. The importance of correct choice of the unit cell boundaries in the SHM is demonstrated in the SHM application to the asymmetric cells problems. The surface harmonic method is shown to be also applicable with asymmetric cells while following the SHM recommendations on selecting boundaries of the unit cells.

References

[1] A V Elshin *Finite difference equation for the distribution of neutrons and their value in three-dimensional heterogeneous reactor with unstructured mesh* Problems of atomic science and technology Series Physics of nuclear reactors. In printing, 2016

[2] T J Trahan and E W Larsen 2012 *An Asymptotic Homogenized Neutron Diffusion Approximation. II Numerical Comparison.* Proceeding of Physor2012 – Advances in Reactor Physics – Linking Research, Industry, and Education, Knoxville, Tennessee, USA

[3] A V Elshin and N V Tenishcheva 2014 *Marshak conditions at boundary with a vacuum in the method of surface harmonics.* Scientific Forum with international participation «Science Week at the Peter the Great Saint-Petersburg Polytechnic University» : materials of scientific-practical conference. Institute of Energy and transport systems of St. Petersburg State University. P. 1. Publisher - Polytechnic Institute, P. 143–145

IOP Conf. Series: Journal of Physics: Conf. Series **781** (2017) 012044 doi:10.1088/1742-6596/781/1/012044

Multi-scale model for point defects behaviour in uranium mononitride

S Starikov, A Kuksin and D Smirnova

Joint Institute for High Temperatures of Russian Academy of Sciences
Nuclear Safety Institute of Russian Academy of Sciences

starikov@ihed.ras.ru

Abstract. A multiscale approach was used to study the properties of point defects in uranium mononitride (UN). In this work we used combination of several methods: *ab initio* calculations; molecular dynamics simulations with a new interatomic potential; thermodynamic model. Density functional theory (DFT) calculations are used for fitting of the parameters of the angular-dependent interatomic potential, as well as for evaluation of the defects formation and migration energies. Molecular dynamics (MD) simulations are applied to analyze what migration/formation mechanisms are activated at finite temperatures and to calculate diffusion coefficients of point defects. The thermodynamic model for description of concentrations and diffusivities for point defects in non-stoichiometric UN_{1+x} is proposed.

1. Introduction
Uranium mononitride (UN) has been studied extensively due to the huge interest in application of this material as a basis of nuclear fuel. Due to the high thermal conductivity, high melting temperature and high density UN is promising fuel for fourth-generation reactors, such as lead-cooled fast reactor. In spite of the importance of this type of fuel, many physical features of UN have not been studied well.

The present work is devoted to determination of concentrations and diffusivities of point defects in the UN, which is useful for microscopic description of nuclear fuel behavior during burn-out. For instance, the data on concentrations and diffusion coefficients of the defects are important for analysis of the dissociation and sintering of the nitride fuel [1, 2] or for investigation of fission products evolution [3-5].

2. Multi-scale model
In this work we used a multiscale approach for study of point defects behaviour. This research combines three models describing various scales: (1) density functional theory calculations; (2) molecular dynamics; (3) thermodynamic model. To connect DFT and MD we create the novel interatomic potential, which accurately describes the structure and thermodynamic properties of uranium mononitride. In this combination of computational approaches one method discovers important features, which must be taken into account when the other one is applied. Stochastic nature of MD (detailed discussion can be found in Ref.[6]) provides convenient way for analysis of the diffusion processes activated at finite temperatures. The developed interatomic potential becomes a bridge between DFT and MD approaches. In this manner, this study describes two basic levels (quantum and atomistic modeling) of the multiscale approach. In addition, we use the thermodynamic model for description of point defects concentrations in non-stoichiometric UN.

IOP Conf. Series: Journal of Physics: Conf. Series **781** (2017) 012044 doi:10.1088/1742-6596/781/1/012044

2.1. Density functional theory

Main details of DFT calculation were described in our work [7]. All quantum calculations were performed using the DFT code VASP 5.2 [8]. Characteristics of the point defects have been studied by DFT computations and classical MD simulations. It is noteworthy that the DFT calculations are used in two different ways. On the one hand, results of the DFT calculations provide a database necessary for the development of an interatomic potential for the atomistic simulation. On the other hand, DFT is used for direct calculations of properties of the defects. At the same time, the simulation based on application of created interatomic potential helps to find important atomic configurations for the DFT computations. Atomic configurations obtained from the energy minimization performed with the MD usually do not correspond exactly to the energy minima according to DFT and should be subjected to additional relaxation within the appropriate procedure in DFT. However, preliminary calculation/minimization with MD significantly simplifies search of the stable defects states. In such a manner, these two methods (DFT and atomistic simulations with the potential) supplement each other.

2.2. Novel interatomic potential

In this work we develop an interatomic potential for UN in the form of the Angular-Dependent Potential (ADP). Details of development and verification were described in our work [7]. The force-matching method [9] is used for the development of the potential, as implemented in the Potfit code [10]. This method provides a way to construct physically justified interatomic potentials from the fitting database that does not contain experimental data. The idea is to adjust the interatomic potential functions to optimally reproduce per-atom forces (together with total energies and stresses) computed at the *ab initio* level for a fine-tuned set of reference structures.

The verification shows that the potential reproduces defect energies in agreement with the DFT estimations [7]. In addition, ADP correctly describes the thermodynamic properties.

2.3. Atomistic simulation

Characteristics of point defects have been studied by DFT computations and classical MD simulations. Most of the details were described in [7]. We used created Angular-Dependent Potential to describe atomic system in the molecular dynamics.

It is shown that the uranium antisite defects (U_N) play an important role in defects behavior. Uranium atom from the interstitial site (I_U) passes to the place of nitrogen atom in a regular lattice position. In this case, formation of U_N occurs with displacing nitrogen atom to the interstitial defect (I_N). Antisite defects are common for compounds formed from the components of similar chemical nature, such as SiC or metallic alloys. In case of uranium mononitride, U_N is quite favorable defect due to relatively small displacements produced in the uranium sublattice. Table 1 shows the basic formation energies of point defects and energy barrier for formation of U_N and I_N from I_U .

Table 1. The formation energies of composition-conserving defects and the energy barrier for formation of the pair of defects $U_N + I_N$ (obtained from DFT and MD). The energies are given in eV.

	MD	DFT
Uranium Frenkel pair, E_{FP}^{U}	6.2	7.1
Nitrogen Frenkel pair, E_{FP}^{N}	4.0	4.9
Schottky defect, E_{SD}	4.6	4.7
$I_U \rightarrow U_N + I_N$, E_{As}	0.7	0.6

It should be noted that the energies from DFT calculations are more accurate than the values that were calculated in MD. However, the difference between the energies calculated by various methods is small, and MD agrees well with quantum calculations.

In this work we calculate the diffusivity for various point defects (i.e. diffusion coefficients of atoms *via* defect diffusion). The calculation was realized for the simulation box containing approximately 2000 atoms and a one point defect. The figure 1 shows results of molecular dynamics simulation in the Arrhenius form. Details of such calculation method were described in [7, 11]. The diffusivity of atoms (U or N) at the presence of given defect is determined by the displacements of all atoms in a system with time:

$$D_{def}^{Y} = \frac{\sum_{i=1}^{M} \delta r_i^2}{6t}, \quad (1)$$

where δr is displacement of the *i*-th atom, *t* is the computation time, index *Y* may be U or N, index *def* may be I (interstitial) or V (vacancy), *M* is the number of atoms of the same type (uranium or nitrogen). Such calculated coefficient is related to the diffusion coefficient of point defect: $D_{def}^{Y} = \alpha \cdot D_{def}$, where D_{def} is diffusion coefficient of point defect and α is a constant about 1 (see [12] for details). Unlike the uranium dioxide [12], the diffusivities of atoms in both sublattices of UN are quite close to each other. The existence of two regions in the dependence $D_I^U(T)$ was discussed in [7].

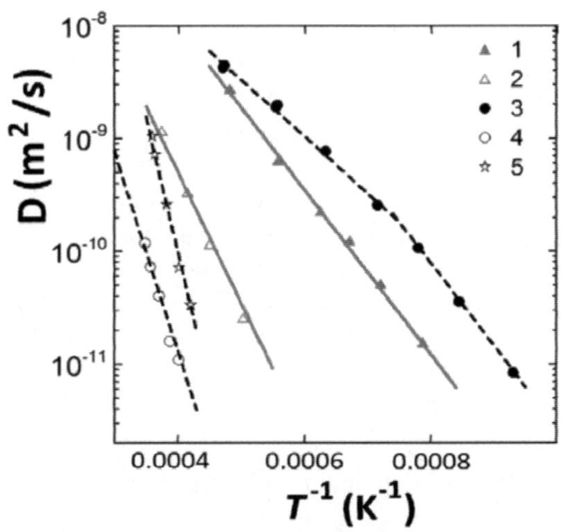

Figure 1. Temperature dependence of diffusion coefficients of nitrogen *via* defects: 1 – interstitial D_I^N; 2 – vacancy D_V^N. Diffusion coefficients of uranium *via* defects: 3 – interstitial D_I^U; 4 – vacancy D_V^U; 5 – uranium in the antisite state D_{As}^U.

2.4. Thermodynamic model

The thermodynamic model contains several equations for point defects concentrations that are described by thermodynamics equilibrium:

$$[I_U][V_U] = \exp\left(S_{FP}^U / k\right) \cdot \exp\left(-E_{FP}^U / kT\right), \quad (2)$$

$$[I_N][V_N] = \exp\left(S_{FP}^N / k\right) \cdot \exp\left(-E_{FP}^N / kT\right), \quad (3)$$

$$[V_N][V_U] = \exp\left(S_{SD} / k\right) \cdot \exp\left(-E_{SD} / kT\right), \quad (4)$$

$$\frac{[U_N][I_N]}{[I_U]} = \exp\left(S_{As} / k\right) \cdot \exp\left(-E_{As} / kT\right), \quad (5)$$

$$x = [I_N] - [I_U] + [V_U] - [V_N] - 2 \cdot [U_N], \quad (6)$$

IOP Conf. Series: Journal of Physics: Conf. Series **781** (2017) 012044 doi:10.1088/1742-6596/781/1/012044

where [def] is defects concentration, k is Boltzmann constant, S_{def} is vibration formation entropy of the defect and x is deviation from stoichiometry in UN_{1+x} . Symbols I and V denote interstitial defect and vacancy. Such system of equations may be resolved in direct way for the given x. The values of S_{def} may be estimated as $6k$ [13]. The figure 2 illustrates the solution of equations (2)-(6) at various temperatures T.

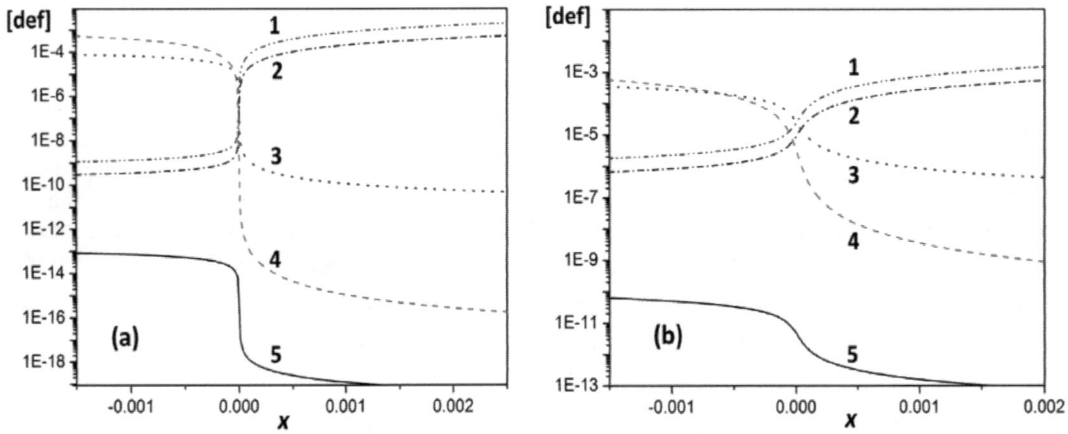

Figure 2. The dependencies of defects concentrations on x: $1 - V_U$; $2 - I_N$; $3 - V_N$; $4 - U_N$; $5 - I_U$. The results are given for two different temperatures: (a) T = 1500K; (b) T = 2000K.

There is a boundary of physically justified solution for the equations (2)-(6). The uranium antisite defects may be considered as nanosized nuclei of pure uranium. At negative value of x, the concentration $[U_N]$ may be about 0.001 that can be interpreted as the formation of uranium. On the other hand, at positive value of x, large concentrations $[V_U]$ and $[I_N]$ correspond to formation of U_2N_3. It can be assumed that in case when any defect concentration [def] achieves value about 10^{-3} , the formation of pure uranium (at negative x) or U_2N_3 (at positive x) takes place. The existence of the narrow area of UN homogeneity agrees well with the experimental data. However, more accurate model must take into account thermodynamic properties of all phases. The question about accurate description of phase transitions and phase decomposition requires further investigation and will be considered in future works.

For verification of our model we calculated coefficient of uranium self-diffusion (D_S^U) and nitrogen self-diffusion (D_S^N) at various temperatures and at x = 0. The calculated defects concentrations and diffusivities of atoms *via* defects were used together to get estimation of self-diffusion coefficients:

$$D_S^U = [I_U] \cdot D_I^U + [V_U] \cdot D_V^U + [U_N] \cdot D_{As}^U \tag{7}$$

$$D_S^N = [I_N] \cdot D_I^N + [V_N] \cdot D_V^N + [U_N] \cdot D_{As}^N \tag{8}$$

The results of simulation are given in figure 3. We compared our results with self-diffusion coefficients of nitrogen from the experiment [14] and with self-diffusion coefficients of plutonium in uranium-plutonium mononitride from the experiment [15]. The difference between calculated and measured values of D_S^N may be caused by difference in conditions. We use x = 0, but more correct condition must be connected with the nitrogen pressure P_{N2} (for instance, nitrogen pressure is equal to 1 atm.) The nitrogen pressure may be estimated using equation $P_{N2} = A \cdot [I_N]^2$, where A is a constant.

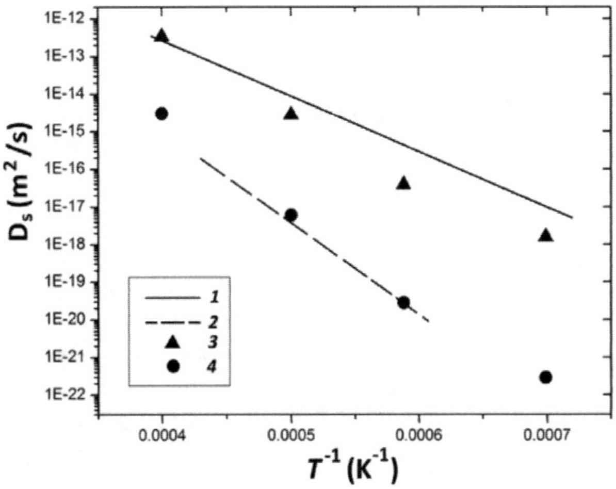

Figure 3. Temperature dependence of self-diffusion coefficients of nitrogen and actinide: 1 – D_S^N from experiment [14]; 2 – D_S^{Pu} from experiment [15]; 3 – D_S^N (calculated in this work); 4 – D_S^U (calculated in this work)

3. Conclusion

It is shown that the U-antisite defects play an important role at negative deviation from stoichiometry in UN_{1+x} (in U-rich area). During migration, the interstitial uranium is able to knock-out nitrogen atom, and this act leads to formation of U in antisite and N interstitial. This effect results in dependence of the diffusivity of U-interstitials on the concentration of defects in the N sublattice. Another peculiarity of UN is the large athermal concentration of U-vacancies in the N-rich UN_{1+x}. This is due to close formation energies of nitrogen Frenkel pairs and Schottky defects.

References

[1] Baranov V G, Lunev A V, Mikhalchik V V, Tenishev A V and Shornikov D P 2016 *Conf. Series: Materials Science and Engineering* **130** 012022
[2] Lunev A V, Mikhalchik V V, Tenishev A V and Baranov V G 2016 *Journal of Nuclear Materials* **475** 266–273
[3] Claisse A, Klipfel M, Lindbom N, Freyss M, Olsson P 2016 *Journal of Nuclear Materials* **478** 119–124
[4] Lopes D A, Claisse A, Olsson P 2016 *Journal of Nuclear Materials* **478** 112–118
[5] Veshchunov M S, Boldyrev A V, Kuznetsov A V, Ozrin V D, Seryi M S, Shestak V E, Tarasov V I, Norman G E, Kuksin A Y, Pisarev V V, Smirnova D D, Starikov S V, Stegailov V V, Yanilkin A V 2015 *Nuclear Engineering and Design* **295** 116-126
[6] Norman G E and Stegailov V V 2013 *Mathematical Models and Computer Simulations* **5** 305.
[7] Kuksin A Y, Starikov S V, Smirnova D E, Tseplyaev V I 2016 *Journal of Alloys and Compounds* **658** 385-394.
[8] Kresse G and Furthmuller J 1996 *Phys. Rev. B* **54** 11169
[9] Ercolessi F and Adams J B 1998 *Europhys. Lett.* **26** 583.
[10] Brommer P and Gahler F 2007 *Model. Simul. Mater. Sci. Eng.* **15** 295.
[11] Smirnova D E, Kuksin A Y, Starikov S V 2015 *Journal of Nuclear Materials* **458** 304-311.
[12] Kuksin A Y and Smirnova D E 2014 *Physics of the Solid State* **56** 1214-1223.
[13] Mehrer H "Diffusion in Solids" 2007
[14] Holt J B and Almassy M Y 1969 *J. Am. Ceram. Soc.* **52** 631
[15] Bradbury M H and Matzke H 1978 *Journal of Nuclear Materials* **75** 68-76

4. Acknowledgements

The authors are grateful to E.A. Kotomin, D. Gryaznov, P.A. Korzhavyi and A.V. Yanilkin for suggestions and discussions. This study is supported by Rosatom State Nuclear Energy Corporation (State contract No. N.4x44.90.13.1084 on March 22, 2013) and the President RF Grant MK-7688.2015.8.

Test problem for thermal-hydraulics and neutronic coupled calculation fore ALFREAD reactor core

A Filip[1], G Darie[1], I S Saldikov[2], A D Smirnov[2] and G V Tikhomirov[2]

[1] Department of Nuclear Engineering, Power Engineering Faculty, University Politehnica of Bucharest, Independentei street, 313, Bucuresti, Romania, 060042
[2] Department of Theoretical and Experimental Physics of Nuclear Reactors, National Research Nuclear University MEPhI (Moscow Engineering Physics Institute), Kashirskoe highway, 31, Moscow, Russia, 115409

Corresponding author's e-mail address: ISSaldikov@mephi.ru

Abstract. The beginning of a new era of nuclear reactor requires technological advances and also multiples studies. The European Liquid metal cooled Fast breeder Reactor is one of the designs for the generation IV nuclear reactor, selected by ENEA. A pioneer of its time, ELFR needs a demonstrator in order to prove the feasibility of this project and to acquire more data and experience in operating a LFR. For this reason the ALFRED project was started and it is expected to be under operation by the year 2030. This paper has the objective of analyzing the neutronic and thermohydraulics of the ALFRED core by the means of a coupled scheme. The selected code for neutronic simulation is MCNP and the selected code for thermohydraulics is ANSYS.

1. Introduction

The main goal of the ALFRED project is to play the role of a demonstrator for the European concept of a LFR, thus proving the safety and reliability – in all operating conditions – of the simple engineering solutions adopted (e.g., the extension of the FA up to the cover gas), while reducing uncertainties in design, construction and operation to the largest possible extent. Since ALFRED and ELFR are characterized by different thermal powers, of course not all their core parameters can be maintained equal. Therefore, the issue arose in the choice, for the sake of the demonstration, of the parameters that are necessarily or conveniently to be kept and those which are not. In particular the fuel enrichment and the breeding ratio are different in the two cores and therefore the closure of the fuel cycle does not represent a goal for the ALFRED core. On the other hand, the materials are the same in the two cores, except for the cladding materials: the expected time required for the full qualification of advanced claddings and coatings in lead (foreseen in the ELFR) is not compatible with the foreseen roadmap for ALFRED. [1]

2. ALFRED

ALFRED is a small-size (300 MWth) pool-type LFR and its primary system current configuration is depicted in figure 1 [2].

Journal of Physics: Conference Series **781** (2016) 012045 doi:10.1088/1742-6596/781/1/012045

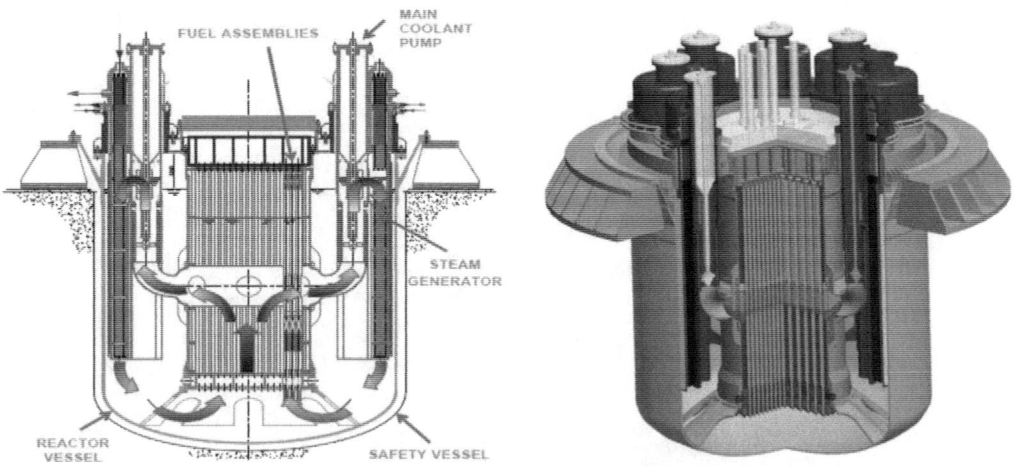

Figure 1. ALFRED core layout [2].

The 171 FAs are subdivided into two radial zones with different plutonium fractions guaranteeing an effective power flattening, and surrounded by two rows of dummy elements serving as a reflector. Two different and independent control rod systems have been foreseen, namely, Control Rods (CRs) and Safety Rods (SRs), which are assigned regulation/compensation and scram functions assuring the required reliability for cycle reactivity swing control and safe shut-down. [4]

Table 1. ALFRED main parameters [3]

Thermal power	300 MW_{th}
Coolant mass flow rate	25984 kg/s
Total number of FAs	171
Pins per FA	127
Coolant inlet temperature	400 ^{0}C
Coolant outlet temperature	480 ^{0}C
Mean assembly power	1.75 MW
Mean rod power	13.814 kW
Mean wall heat flux	0.7 W/m^{2}
Lead flow average FA	144.1 kg/s
Bypass flow average FA	2.76 kg/s
Clad maximum temperature	550 ^{0}C

2.1. The Core

The core is made of hexagonal wrapped fuel assemblies. MOX fuel is used in the form of tubular beads. The Active height is large facilitating the natural circulation. The power of the active core is 300 MWt. The reactor consists of 171 fuel assemblies, 12 control rods and 4 shot down rods surrounded by 108 dummy elements (ZrO2-Y2O3), protecting the interior vessel (figure 2) [4]. Each fuel assembly is 8 meters in length and consists of 127 fuel rods, fixed to the base grating. The springs in the upper part prevents the floating of the assemblies caused by rising lead flow and reduces the effects of expansions. The top ends of the assemblies exceed the free surface of molten lead, entering the plenum gas, which facilitates inspection activities and refueling. In this way, the refueling can be done without the intrusion of refueling equipment [3].

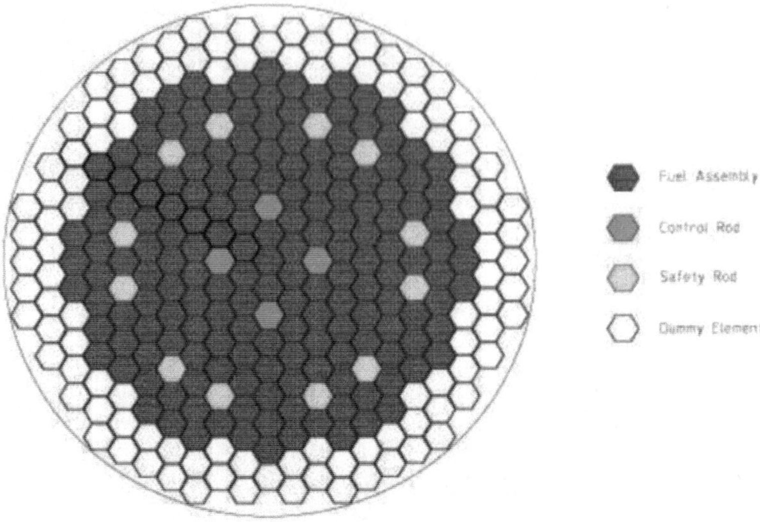

Figure 2. ALFRED core section [4].

2.2. Fuel assembly

Each FA is constructed by 127 fuel rods. The main elements are (a) the spike, which fits into the lower diagrid to ensure the correct positioning of the FA and which is provided with a high number of orifices along the entire circumference to allow for the lead inlet, thus minimizing the risk for sudden significant flow reduction due to occlusion; (b) the funnel, which facilitates the lead outlet toward the steam generators through a "nose" able to host neutron and temperature instrumentation and sniffers for failed cladding detection; (c) a long stem allowing the upper head to emerge from the coolant for an easy refueling in full visibility and without moving parts in lead, and for hosting a ballast which counterbalances the buoyancy during refueling (when the upper diagrid is removed). Along the extension of the shrouds, at different heights, the hexcan is bumped to make pads against core compaction in case of an earthquake (figure 3) [3].

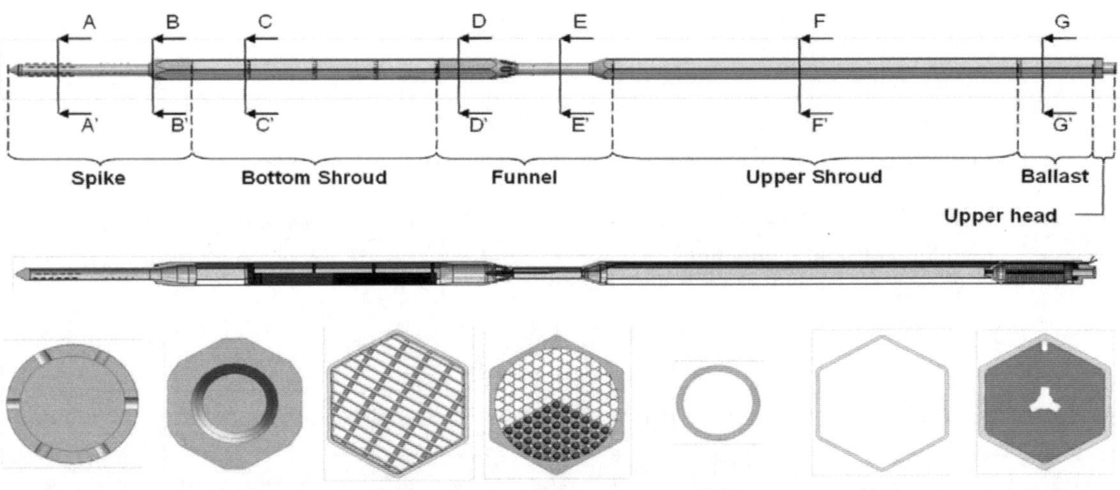

Figure 3. ALFRED Fuel assembly [3].

The presence of a limited number of control and safety rods (12 control rods and 4 safety rods) and the absence of experimental equipment above the core, make feasible the typical refueling operation, characterized by FAs loading from the above of the core(typical solution adopted in SFRs and LWRs).

During normal operating conditions, when the primary pumps are active, the FAs are held in their position thanks to the core upper plate.

2.3. ALFRED fuel

The reactor core is composed of wrapped hexagonal Fuel Assemblies (FAs), each one containing 127 fuel pins arranged in a triangular lattice. The 171 FAs are subdivided into two radial zones (57 inner and 114 outer) with different plutonium enrichments, and surrounded by two rows of dummy elements serving as reflector. In particular, the fuel considered for ALFRED consists of annular U–Pu Mixed Oxide (MOX) pellets. As far as the cladding is concerned, the well known Ti-15-15 steel has been selected because it was already qualified and licensed in fast reactors (Phenix, Superphenix). In Tables 3 and 4, the main specifications of the ALFRED reactor and fuel pin parameters are presented. The fuel is composed of MOX pellets with 95% theoretical density and an Oxygen-to-Metal ratio (O/M) of 1.97. The fuel isotopic composition refers to a typical reactor grade Pu from the reprocessing of Pressurized Water Reactor (PWR) spent fuel (4.5% initial enrichment in 235U, 45 MWd t−1 of burn-up, 15 years of cooling and storage [3].

Table 3. ALFRED pin design parameters [3].

Fuel type	MOX
Enrichment as Pu/(Pu+U)[wt.%] (inner zone)	21.7
Enrichment as Pu/(Pu+U)[wt.%] (outer zone)	27.80
Fuel density [%] of theoretical density	95
O/M	1.97
Cladding	Ti-15-15
Fill gas	He
Pre pressurization [MPa]	0.1
Upper plenum volume [mm^3]	30,000
Upper plenum length [mm]	120
Active length [mm]	600
Lower plenum length [mm]	550
Cladding outer diameter [mm]	10,5
Cladding inner diameter [mm]	9,3
Fuel pallet outer diameter [mm]	9
Fuel pallet inner diameter [mm]	2
Pin pitch [mm]	13,86

Table 4. Isotopic specification for U and Pu [3].

Plutonium		Uranium	
Isotope	Fraction (wt.%)	Isotope	Fraction (wt.%)
^{238}Pu	2.332	^{234}U	0.003
^{239}Pu	56.873	^{235}U	0.404
^{240}Pu	26.997	^{236}U	0.010
^{241}Pu	6.105	^{238}U	99.583
^{242}Pu	7.693		

2.4. Cladding material

The cladding material proposed for the ALFRED reactor is the austenitic stainless steel Ti-15-15 (Table 5). This class of steels has been adopted as cladding material in fast reactors in the early phase of the development of this technology. After several years of reactor management (Superphenix) and irradiation campaigns (EBRII; Phenix), austenitic chromium-nickel (Cr-Ni) stainless steels have proven the best performance in reactor, showing excellent creep and good swelling resistance. In order to reduce the high swelling propensity caused by irradiation, several titanium-stabilized steels has been proposed, exploiting the capability of TiC precipitates to increase the swelling incubation time.

Moreover, the presence of silicon and boron can change the swelling and the creep features, respectively. Indeed, the concentration of each alloyed element has a paramount effect on the creep, on the swelling behavior and on the ruptures train [5].

Table 5. Ti-15-15 Composition.

Steel	Cr[wt.%]	Ni[wt.%]	Mo[wt.%]	Mn[wt.%]	Si[wt.%]	Ti[wt.%]	C[wt.%]	B[ppm]
AIM 1	15.0	15.0	1.50	1.50	0.90	0.40	0.09	60

2.5. Coolant (Lead)

Pure lead is an interesting coolant in fast reactors due to its high mass number and its low absorption cross-section, which are fundamental properties for a fast reactor coolant. It has good thermal characteristics for heat removal. With respect to sodium and lead-bismuth eutectic, it has high melting temperature (327°C, 600 K) and an high boiling point (1749°C, 2022 K) allowing a wide range of operation. [5]

It is steel-aggressive for the core materials both for erosion and corrosion issues (therefore, a flow velocity of less than 2 m s-1 is envisaged, and an oxygen control and/or a coating for cladding are required, respectively) and it is very toxic and polluting, requiring particular attention from production to disposal or recycling. Nevertheless, the stainless steel of which the cladding is made requires further protection, and different coating layers are under development. Among the proposed solutions, the most promising are the FeCrAlY alloy, realized through the so-called GESA (Gepulste ElektronenStrahlAnlage) treatment and the Al2O3 coating realized through Pulse Laser Deposition (PLD). For ALFRED FA, the modeling solution that can be adopted is substitute an external layer of the cladding (40 μm) by a coating which is not affected by corrosion and is featured by a thermal conductivity of 16 W m-1 K-1. Mechanical and irradiation effects on the coating are presently not taken into account.

3. MCNP-ANSYS coupled scheme

3.1. MCNP

MCNP is a general-purpose Monte Carlo N-Particle code that can be used for neutron, photon, electron, or coupled neutron/photon/electron transport. Specific areas of application include, but are not limited to, radiation protection and dissymmetry, radiation shielding, radiography, medical physics, nuclear criticality safety, Detector Design and analysis, nuclear oil well logging, Accelerator target design, Fission and fusion reactor design, decontamination and decommissioning. The code treats an arbitrary three-dimensional configuration of materials in geometric cells bounded by first- and second-degree surfaces and fourth-degree elliptical tori.[6]

3.2. ANSYS

ANSYS CFX is an element-based finite-volume method with second-order discretization schemes in space and time. It uses a coupled algebraic multigrid algorithm to solve the linear systems arising from discretization. The discretization schemes and the multigrid solver are scalable parallelized. ANSYS CFX works with unstructured hybrid grids consisting of tetrahedral, hexahedral, prism, and pyramid elements.

3.3. MCNP-ANSYS

In order to perform a neutronic and thermohydraulic analysis of ALFRED reactor core two computational codes are needed to simulate the steady state functioning of the reactor.

For this paper, MCNP and ANSYS code were chosen for creating a coupling scheme (Fig 4) [8]. Although MCNP library does not include materials such as Ti-15-15, it is possible to create is according to weight fraction of isotope composition. In creating the input file necessary for neutronic simulation of the ALFRED core, the precise geometry of a fuel rod is necessary, provided by Table 6 [7]. Data such as density, an estimated temperature of the fuel and material composition is required for a better resolution. According to the used coupled scheme, after the first iteration density of the fuel and temperature will be replace with the new data obtained. After the first run of MCNP, power distribution data will be extracted from the output file and used in creating the input file for ANSYS.

The second part of the coupled scheme is represented by the termohydraulic code ANSYS. It requires the full geometry of the fuel assembly (or a simplified design) considering the same geometrical data used in creating the MCNP input file. Thermo physical properties of the materials are also necessary. Material composition, geometry, viscosity, inlet temperature of coolant are requiring in ANSYS. The power distribution is resulted from MCNP simulation and is used as an input date for the termohydraulic code. The resulting data for density and temperature of the coolant and are used to recreate the input file for MCNP code. The iteration continues until power distribution and temperature converges.

Table 6. MCNP and ANSYS Geometry [7].

Inner gap diameter	Fuel pallet outer diameter	Gap thickness	Cladding thickness	Cladding outer diameter	Height
2 [mm]	9 [mm]	0.15 [mm]	0.6[mm]	10.5 [mm]	6000 [mm]

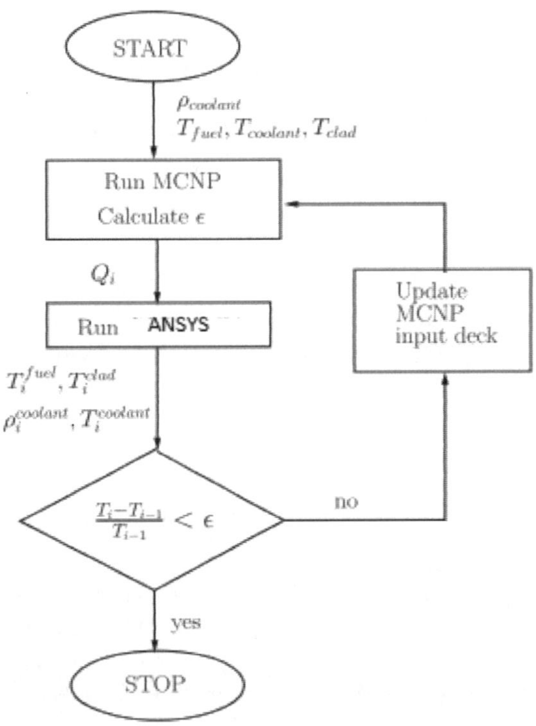

Figure 4. MCNP – ANSYS coupled scheme [8].

3.4. Thermophysical correlation of lead

Density: The density of liquid lead can be related to temperature with a simple linear correlation. Even if purity is not well specified, agreement within data is very strong, with uncertainty under 1%.

$$\rho=11367-1.1944T \text{ [kg*m}^{-3}\text{] [5]} \quad (1)$$

Specific heat capacity at constant pressure: the correlation is suggested as a reasonable choice between available data.

$$c_p= 175.1-4.961*10^{-2}T+1.985*10^{-5}T^2-2.099*10^{-9}T^3-1.524*10^{6}T^{-2} \text{ [J kg}^{-1}\text{ K}^{-1}\text{] [5]} \quad (2)$$

Thermal Conductivity: The data regarding the thermal conductivity of liquid lead are very dispersed. The proposed correlation is linear and fits well recent data.

$$\lambda=9.2+0.011T\ [Wm^{-1}K^{-1}] \quad (3)$$

3.5. Thermo physical properties of Ti-15-15

Density: It is calculated as a function of temperature using the linear expansion correlation.

$$\rho(T)=\rho_0*\left(\frac{1}{1+\zeta(T)}\right)^3\ [kgm^{-3}] \quad (4)$$

Specific heat: Specific heat measurements have been carried out in the range 20÷1000°C (293÷1273 K). The correlation has been selected and implemented.

$$c_p=431+0.177T+0.0000872T^2\ [jkg^{-1}K^{-1}] \quad (5)$$

Thermal conductivity: The thermal conductivity of the cladding material is a fundamental property because it determines the temperature gradient across the cladding, controlling thermal stress and temperature levels (which determine all the other mechanical properties).

$$\lambda=13.95+0.01163T\ [Wm^{-1}K^{-1}] \quad (6)$$

4. Expected results

Through this work it is expected to create a flexible coupled scheme between MCNP code and ANSYS code with the purpose of validation and verification of deterministic calculation. The coupled scheme will provide important data for ALFRED reactor core. Axial power distribution, axial fuel temperature and also fuel temperature can be studied. The iteration process continues until convergence.

Acknowledgment

The authors appreciate the support and effort of National Research Nuclear University MEPhI. This work was accomplished with the financial support of ENEN by the means of ENEN RU II Project number 605149. This work was supported by Competitiveness Program of National Research Nuclear University MEPhI.

References

[1] Application of the TRANSURANUS code for the fuel pin design process of the ALFRED reactor - Lelio Luzzi, Antonio Cammia, Valentino Di Marcellob, Stefano Lorenzia,Davide Pizzocria, Paul Van Uffelenba.
[2] Petri-net based modeling approach for ALFRED reactor operation and control system design - Roberto Ponciroli , Antonio Cammi, Stefano Lorenzi, Lelio Luzzi.
[3] The core design of ALFRED, a demonstrator for the European lead-cooled reactors - Grassoa, C. Petrovicha, D. Mattiolia, C. Artiolia, P. Sciorab, D. Gugiuc, G. Bandinia,E. Bubelisd, K. Mikityukea.
[4] The lead fast reactor: demonstrator (Alfred) and ELFR design - A. Alemberti, M. Frogheri, L. Mansani.
[5] Modeling and analysis of fuel pin behavior for innovative lead cooled FBR – A. Del Nevo, M. Tarantino.
[6] A flexible coupling scheme for monte carlo and thermal-hydraulics codes - J. Eduard Hoogenboom, Aleksandar Ivanov, Victor Sanchez, Cheikh Diop.
[7] Monte Carlo modeling of lead-cooled fast reactor in adiabatic equilibrium state - Przemysław Stanisz, Mikołaj Oettingen, Jerzy Cetnar.
[8] Experience in Neutronic/Thermal-hydraulic Coupling in Ciemat - Miriam Vazquez Francisco Martín, Fuertes Aleksandar Ivanov.

Preparation macroconstants to simulate the core of VVER-1000 reactor

V Y Seleznev

Reactor development and design department, INPE NRNC MEPhI, Obninsk, 249040, Russian Federation.

E-mail: vad9232@yandex.ru

Abstract. Dynamic model is used in simulators of VVER-1000 reactor for training of operating staff and students. As a code for the simulation of neutron-physical characteristics is used DYNCO code that allows you to perform calculations of stationary, transient and emergency processes in real time to a different geometry of the reactor lattices [1]. To perform calculations using this code, you need to prepare macroconstants for each FA. One way of getting macroconstants is to use the WIMS code, which is based on the use of its own 69-group macroconstants library. This paper presents the results of calculations of FA obtained by the WIMS code for VVER-1000 reactor with different parameters of fuel and coolant, as well as the method of selection of energy groups for further calculation macroconstants.

1. Initial data for FA

1.1. FA types of VVER-1000 reactor

Figure 1 shows the fuel assemblies types of VVER-1000 [2] reactor used in the calculations:

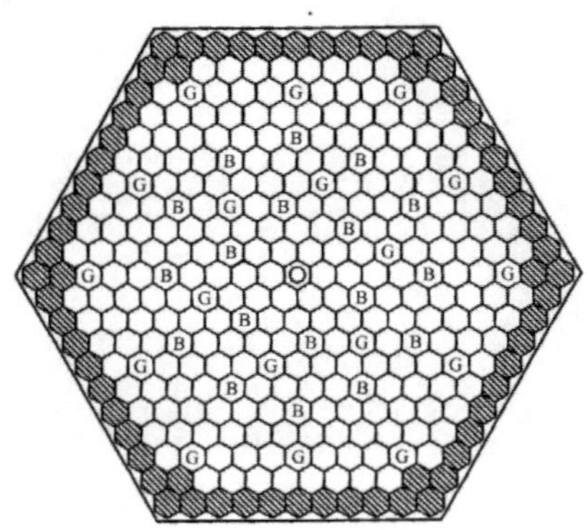

Figure 1. VVER-1000 FA type «1»: - 4.4% uranium enrichment (UO_2); - 4% uranium enrichment (UO_2); (G) - fuel with burnable absorber with 5% uranium enrichment (Gd_2O_3); (B) - channels for CPS (B_4C).

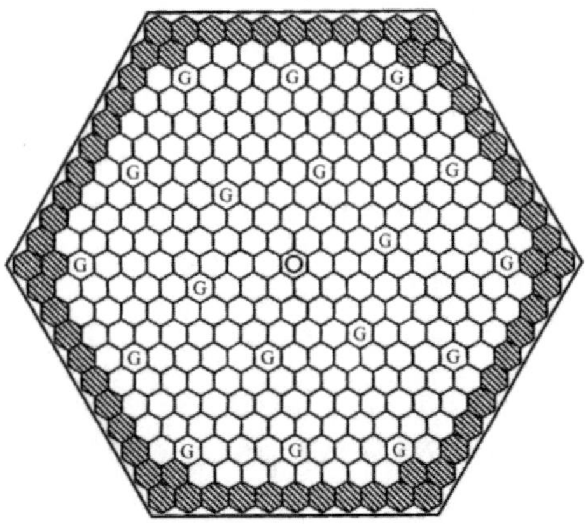

Figure 2. VVER-1000 FA type «2»: - 4.4% uranium enrichment (UO_2); - 4% uranium enrichment (UO_2); (G) - fuel with burnable absorber with 5% uranium enrichment (Gd_2O_3); (B) - channels for CPS (B_4C).

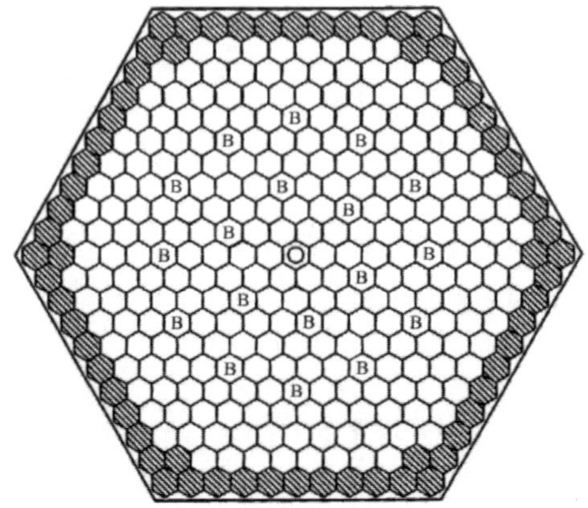

Figure 3. VVER-1000 FA type «3»: - 4.4% uranium enrichment (UO_2); - 4% uranium enrichment (UO_2); (B) - channels for CPS (B_4C).

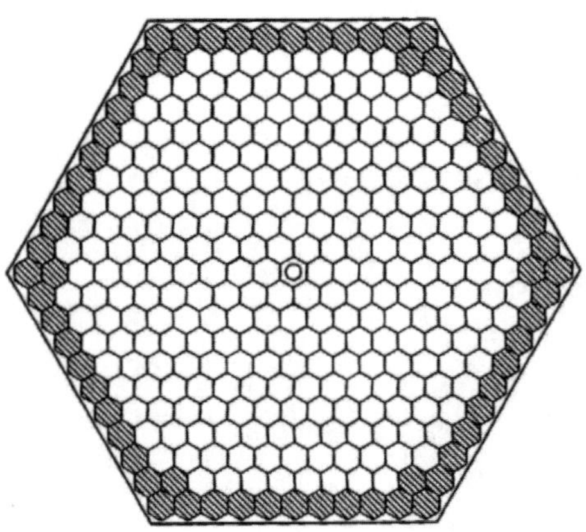

Figure 4. VVER-1000 FA type «4»: - 4.4% uranium enrichment (UO_2); - 4% uranium enrichment (UO_2).

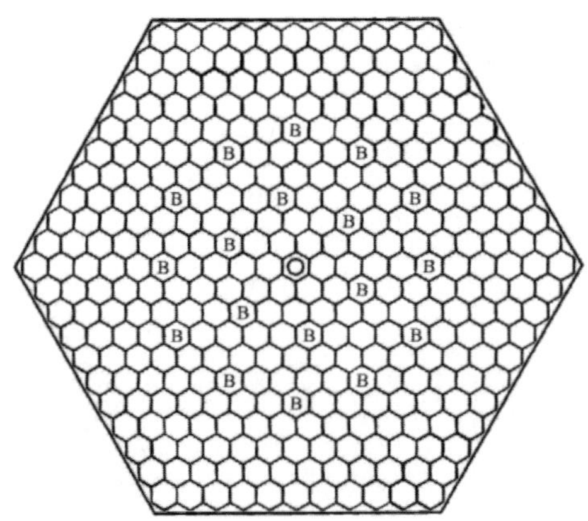

Figure 5. VVER-1000 FA type «5»: - 4.4% uranium enrichment (UO_2); (B) - channels for CPS (B_4C).

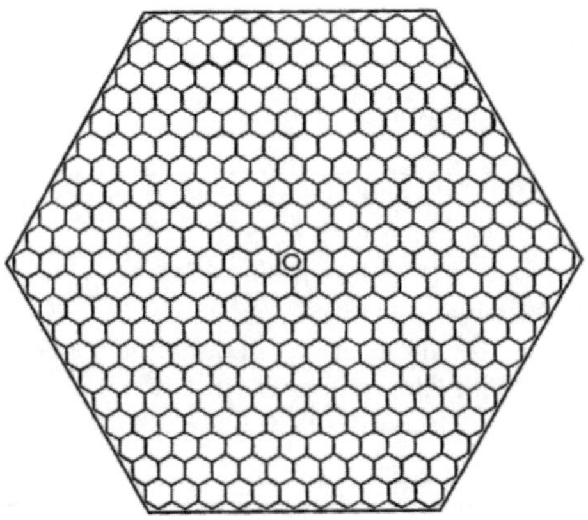

Figure 6. VVER-1000 FA type «6»: - 4.4% uranium enrichment (UO_2).

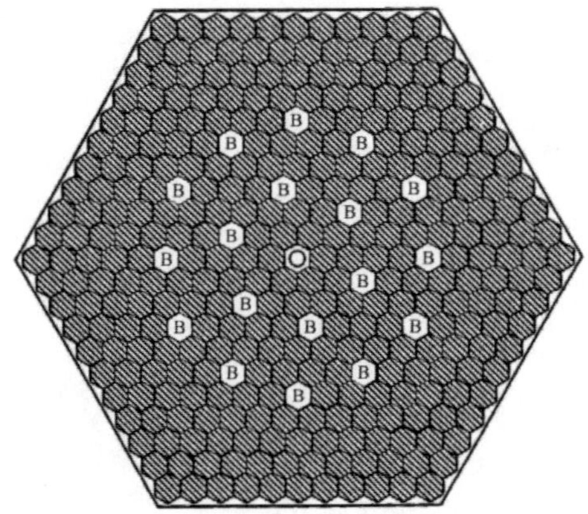

Figure 7. VVER-1000 FA type «7»: - 4% uranium enrichment (UO_2); - channels for CPS (B_4C).

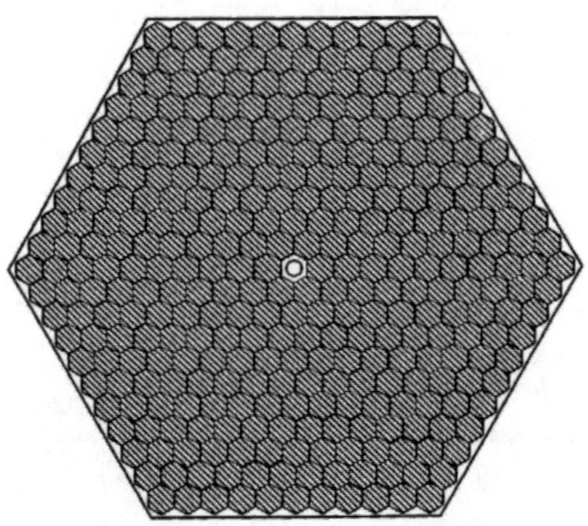

Figure 8. VVER-1000 FA type «8»: - 4% uranium enrichment (UO_2).

Table 1 describes the VVER-1000 FA types used in the subsequent calculations.

1.2. Selecting groups

For the selection was chosen the range of group from 15 to 69 [6], as it contains the resonance and thermal energies neutrons. For calculation was used each second group range. Group at number 69 is a boundary group for field of thermal neutrons, so it is selected for the default calculation. Selection was carried out by selecting a group in which a result of calculation k_∞ was maximized in groups ranging from 15 to 68, further calculation began with the start of the range, but instead a selected group (with maximum k_∞), involved the following order. Thus, was selected the groups with numbers 19, 25, 27, 47, 55, 59, 69. All calculations were performed at a fuel temperature of T = 900K. Data selection presented in Table 2.

Table 1. FA types.

FA type	Figure	Condition
1	a	4 and 4.4% uranium enrichment, fuel with burnable absorber with 5% uranium enrichment, with CPS
2	b	4 and 4.4% uranium enrichment, fuel with burnable absorber with 5% uranium enrichment, without CPS
3	c	4 and 4.4% uranium enrichment, without fuel with burnable absorber, with CPS
4	d	4 and 4.4% uranium enrichment, without fuel with burnable absorber, without CPS
5	e	4% uranium enrichment, without fuel with burnable absorber, with CPS
6	f	4% uranium enrichment, without fuel with burnable absorber, without CPS
7	g	4.4% uranium enrichment, without fuel with burnable absorber, with CPS
8	h	4.4% uranium enrichment, without fuel with burnable absorber, without CPS

The data in Table 2 illustrate the procedure of selecting the group with the maximum value of k∞. According to Table 2 were plotted k∞ on the specific group of energy and calculation. Changes in k∞ parameter is shown in Figure 9.

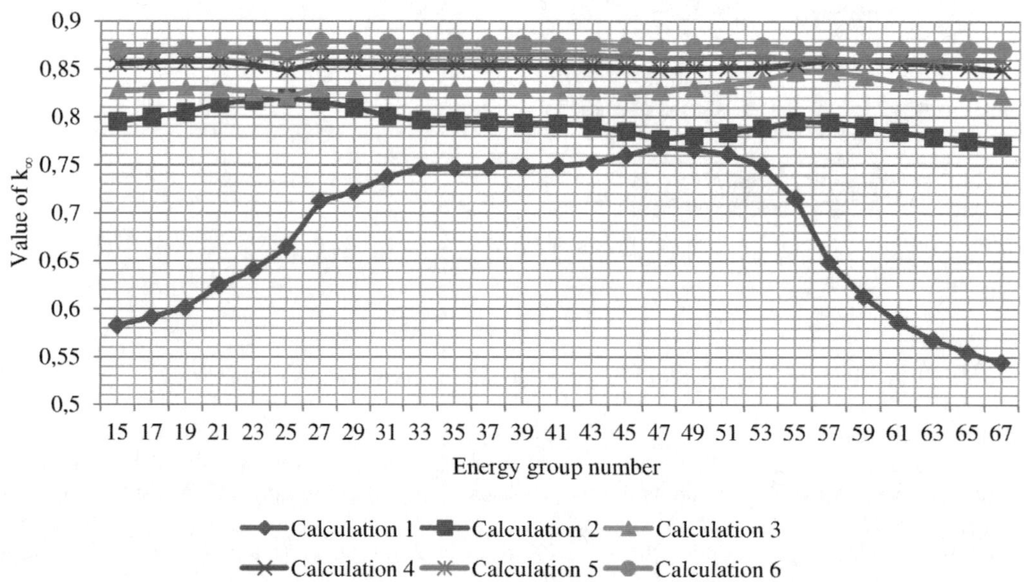

Figure 9. Changes in k∞ parameter, depending on energy group selected for calculation.

IOP Conf. Series: Journal of Physics: Conf. Series **781** (2017) 012046 doi:10.1088/1742-6596/781/1/012046

Table 2. The result value of parameter k_∞ depending on selected group.

Number of calculation	Number of group	Value of parameter k_∞					
		2 groups 69 +	3 groups 69, 47+	4 groups 69, 47, 25+	5 groups 69, 47, 25, 55+	6 groups 69, 47, 25, 55, 59+	7 groups 69, 47, 25, 55, 59, 19
1	15	0.5828	0.7952	0.8275	0.8556	0.8671	0.8698
2	17	0.5912	0.7997	0.8286	0.8568	0.8683	0.8697
3	19	0.6013	0.8048	0.8297	0.8579	**0.8695**	0.8708
4	21	0.6244	0.8138	0.8294	0.8577	0.8692	0.8716
5	23	0.6405	0.8170	0.8261	0.8542	0.8657	0.8715
6	25	0.6637	**0.8196**	0.8210	0.8488	0.8602	0.8710
7	27	0.7123	0.8158	0.8298	0.8565	0.8679	**0.8788**
8	29	0.7220	0.8097	0.8297	0.8562	0.8676	0.8785
9	31	0.7376	0.8010	0.8294	0.8554	0.8667	0.8776
10	33	0.7460	0.7963	0.8288	0.8546	0.8659	0.8768
11	35	0.7466	0.7954	0.8286	0.8544	0.8657	0.8766
12	37	0.7473	0.7945	0.8284	0.8542	0.8655	0.8764
13	39	0.7482	0.7936	0.8282	0.8540	0.8653	0.8761
14	41	0.7492	0.7927	0.8280	0.8537	0.8651	0.8760
15	43	0.7518	0.7905	0.8275	0.8533	0.8646	0.8755
16	45	0.7601	0.7845	0.8262	0.8519	0.8632	0.8740
17	47	**0.7684**	0.7769	0.8273	0.8499	0.8612	0.8720
18	49	0.7661	0.7804	0.8306	0.8506	0.8621	0.8730
19	51	0.7614	0.7836	0.8338	0.8516	0.8629	0.8737
20	53	0.7495	0.7883	0.8390	0.8518	0.8632	0.8740
21	55	0.7149	0.7955	**0.8474**	0.8557	0.8614	0.8723
22	57	0.6482	0.7946	0.8473	0.8584	0.8612	0.8720
23	59	0.6128	0.7897	0.8424	**0.8588**	0.8595	0.8703
24	61	0.5863	0.7842	0.8365	0.8568	0.8599	0.8707
25	63	0.5678	0.7791	0.8311	0.8543	0.8600	0.8706
26	65	0.5545	0.7747	0.8264	0.8517	0.8597	0.8705
27	67	0.5443	0.7708	0.8221	0.8491	0.8593	0.8701

Thus, in Figure 9 clearly shows that as a result of the calculations were able to establish the amount of energy groups that equal to seven, three of which belong to the group of resonance neutrons, and four - the group of thermal. The boundaries of established energy groups, and the corresponding them energies are given in Table 3.

Table 3. The boundaries of established energy groups, and the corresponding them energies

	Number of group	Energy, eV
Resonance	19	1.4251E+3
	25	2.7700E+1
	27	9.877E+0
Thermal	47	0.500E+0
	55	0.180E+1
	59	6.700E-2
	69	5.000E-3

To check the consistency of the results obtained by the WIMS complex for 69 and 7 energy groups, the calculation of the $k\infty$ was held at various parameters of coolant (temperature and density at constant pressure). As the coolant water is used, pressure was equal to 15.7 MPa [7, 8]. The calculation results are shown in Table 4.

Table 4. The result value of parameter k_∞ at various parameters of coolant and fuel.

Parameters of coolant	Fuel temperature, K	69 groups, k_∞	7 groups, k_∞	Deviation, %
T=583 K, ρ=0.7056 kg/m^3	900	0.8871	0.8788	0.94
	600	0.8956	0.8873	0.94
	300	0.9066	0.8981	0.95
T=310 K, ρ=1 kg/m^3	900	0.9017	0.8930	0.97
	600	0.9105	0.9018	0.96
	300	0.9219	0.9130	0.97
T=617 K, ρ=0.6 kg/m^3	900	0.8856	0.8773	0.95
	600	0.8941	0.8857	0.95
	300	0.9050	0.8965	0.95

From Table 4 it is seen that for any parameters of the coolant and fuel value of the deviation less than 0.97%, it means that the models and the original constants library are selected correctly and can then be used to calculate macroconstants.

1.3. Preliminary calculation of assemblies

The calculation of k_∞ for a different types of VVER-1000 FA was held in this section [9, 10]. Types of FA are shown in Figures 1-8. The calculation was performed for 69 and 7 energy groups of neutrons. Conditions and results of the calculations are shown in Table 5.

Table 5. Calculation conditions and obtained results.

FA type	T=900 K			T=300 K		
	k_∞ for 69 groups	k_∞ for 7 groups	Deviation, %	k_∞ for 69 groups	k_∞ for 7 groups	Deviation, %
1	0.8871	0.8788	0.94	0.9066	0.8982	0.94
2	1.1499	1.1419	0.70	1.1650	1.1569	0.70
3	1.0206	1.0222	0.16	1.0333	1.0348	0.15
4	1.3504	1.3548	0.33	1.3683	1.3726	0.31
5	1.0018	1.0036	0.18	1.0143	1.016	0.17
6	1.3371	1.3415	0.33	1.3548	1.3591	0.32
7	1.0237	1.0254	0.17	1.0336	1.0351	0.15
8	1.3530	1.3573	0.32	1.3686	1.3728	0.31

The data in Table 5 shows that for calculations performed for 7 and 69 energy groups for different types of FA deviation is not more than 0.94%, it means that the models and the original constants library are selected correctly and can then be used to calculate macroconstants.

References

[1] A V Levchenko, I N Leonov and S V Kovalchuk 2011 Verification of the three-dimensional dynamic DYNCO code on the basis of international test cases for pressurized water reactors (Obninsk: Simulation systems Ltd.) p 8

[2] S A Andrushechko, A M Afrov, B Yu Vasiliev, V N Generalov, K B Kosourov,

Yu M Semchenkov and V F Ukraincev 2010 *Nuclear power plants with VVER-1000. From the physical bases of operation to the evolution of the project (Moscow: Logos)* pp 197-199

[3] V A Sarychev, A N Dudkin and A A Uvarov 1990 Codes for calculation of cells and reactors: laboratory practical work on the course: «Calculation methods of nuclear reactors» (Obninsk: INPE NRNC MEPhI) pp 26-27

[4] C I Costescu, J R Deen and W L Woodruff 1995 WIMS-D4M user manual rev.0 (Illinois: University of Illinois) p 86

[5] V S Okunev and I S Lisitsyn 2011 Neutron physics calculation lattice nuclear reactor based on gas-kinetic transport theory (Moscow: Bauman Moscow State Technical University (BMSTU)) p 145

[6] A Trkov, D Lopez Aldama and F Leszczynski 2003 WIMS-D Library Update Int. (Vienna: Atomic Energy Agency) p 308

[7] G Ya Rumyantsev and G I Toshinskiy 2006 *Physics and dynamics of nuclear reactors: training aid* (Obninsk: INPE NRNC MEPhI) p 210

[8] G N Kolpakov and O V Selivannikova 2009 *Construction rods, channels and active zones of nuclear power reactors: training aid* (Tomsk: National Research Tomsk Polytechnic University) p 118

[9] Yu A Kazanskij и Ya V Slekenichs 2009 *The kinetics of nuclear reactors. The coefficients of reactivity. Introduction to dynamics: training aid for the course of lectures "Physical theory of nuclear reactors: the kinetics and dynamics of reactors"* (Moscow: National Research Nuclear University MEPhI) p 300

[10] G G Bartolomej, G A Bat, V D Bajdakov и M S Altuhov 1989 *Fundamentals of the theory and methods of calculation of nuclear power reactors: training aid* (Moscow: Energoatomizdat) p 512

Journal of Physics: Conference Series **781** (2016) 012047 doi:10.1088/1742-6596/781/1/012047

Specifications for a coupled neutronics thermal-hydraulics SFR test case

A Tassone[1], A D Smirnov[2] and G V Tikhomirov[2]

[1] Department of Astronautical, Electrical and Energy Engineering Nuclear Section, Sapienza Universit di Roma, Corso Vittorio Emanuele II 244, Roma, Italy, 00186

[2] Department of Theoretical and Experimental Physics of Nuclear Reactors, National Research Nuclear University MEPhI (Moscow Engineering Physics Institute), Kashirskoe Shosse, 31, Moscow, Russia, 115409

Abstract. Coupling neutronics/thermal-hydraulics calculations for the design of nuclear reactors are a growing trend in the scientific community. This approach allows to properly represent the mutual feedbacks between the neutronic distribution and the thermal-hydraulics properties of the materials composing the reactor, details which are often lost when separate analysis are performed. In this work, a test case for a generation IV sodium-cooled fast reactor (SFR), based on the ASTRID concept developed by CEA, is proposed. Two sub-assemblies (SA) characterized by different fuel enrichment and layout are considered. Specifications for the test case are provided including geometrical data, material compositions, thermo-physical properties and coupling scheme details. Serpent and ANSYS-CFX are used as reference in the description of suitable inputs for the performing of the benchmark, but the use of other code combinations for the purpose of validation of the results is encouraged. The expected outcome of the test case are the axial distribution of volumetric power generation term (q'''), density and temperature for the fuel, the cladding and the coolant.

1. Introduction

Multi-physics is the field of computer-aided engineering who is interested in the simulation of multiple physical models. Although no complete multi-physics platform has yet reached the maturity for a general release, the increasing requirements of computational accuracy for design and safety analysis, especially in the nuclear industry, have caused a growing interest in the scientific community toward this approach.

Several neutronics and thermal-hydraulics codes have been coupled in the past to perform analysis of both thermal and fast reactor cores [1][2][3][4]. In particular, it has been demonstrated the possibility of employing coupled neutronics and thermal-hydraulics codes in the study of sodium cooled fast reactors (SFR), which are one of the six types of reactors actually being developed in the framework of the OECD Generation IV roadmap [5][6][7][8].

2. ASTRID core design

The Advanced Sodium Technological Reactor for Industrial Demonstration (ASTRID) is a 1500 MW_th pool-type sodium cooled fast reactor developed since 2010 by the French public government-funded research organisation CEA and its linked partners. Expected to be built after 2030, ASTRID is a industrial demonstrator for the SFR technology and can be regarded as

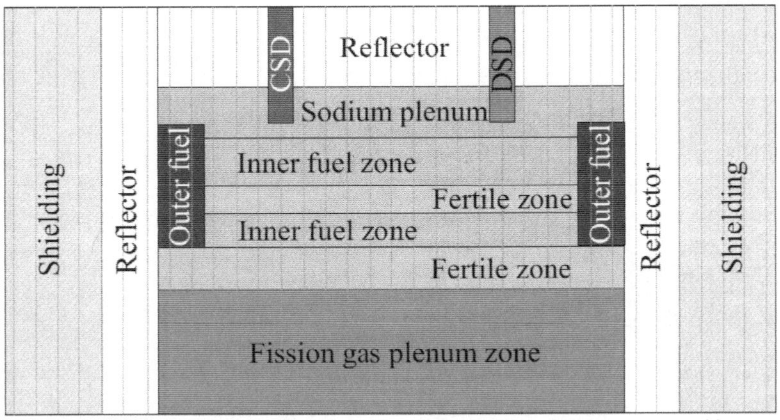

Figure 1. ASTRID core map, axial cross-section [11].

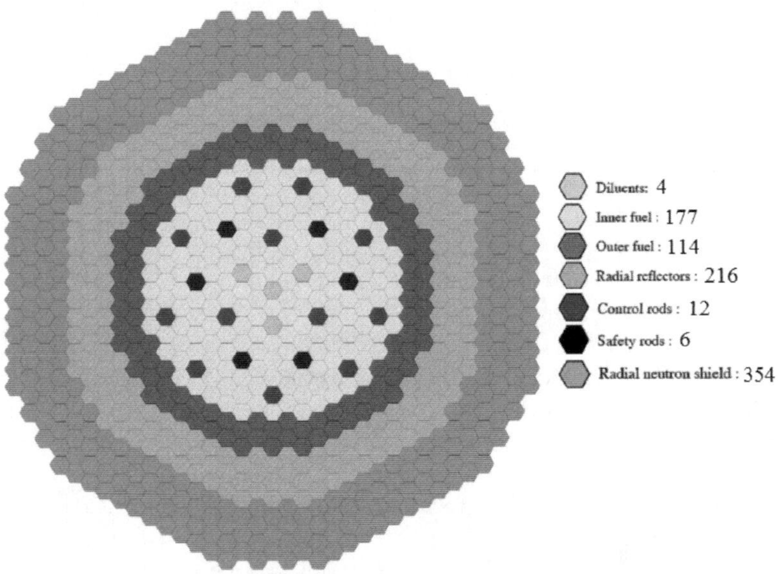

Figure 2. ASTRID core map, top view [12].

the next iteration on the French SFR reactor fleet [9][10]. To improve the safety performance of the reactor, an axially and radially heterogeneous core was designed to reduce the sodium void reactivity coefficient. ASTRID core maps are provided in Figure 1 and 2.

The reference fuel for the CFV core is a uranium-plutonium mixed oxide (MOX), with an enrichment in plutonium varying from the 20% of the outer zone to the 23.5% of the inner zone. The fuel is contained inside 291 hexagonal sub-assemblies, of which 177 are found in the inner region and 114 in the outer region. Each sub-assembly is composed of 217 fuel pins, arranged in 8 rows around the central pin, and delimited by the wrapper can. The pin are wrapped by a wire and enclosed in a cladding, a gap is present between the fuel and the cladding to house the gaseous fission products. The AIM1 austenitic stainless steel is employed as cladding, whereas the material chosen for the wrapper is the EM10 alloy.

Table 1. Pin radial dimensions and SA parameters [10][13][14].

Pin radial dimensions			SA parameters		
Pellet diameter	9.065	mm	Number of pins	217	
Gap thickness	0.075	mm	SA pin rows	8	
Clad inner diameter	9.14	mm	Inter-SAs gap	4.73	mm
Clad thickness	0.56	mm	Last pin–wrapper pitch (g)	10.44	mm
Clad outer diameter (D_p)	9.7	mm	Space between flats (e)	193.78	mm
Pin-to-Pin pitch (P)	10.8	mm	SA pitch	198.41	mm
Spacing wire diameter (D_w)	1.1	mm	q'_{max} [W cm^{-1}]	484	
Helical pitch of wire	180	mm	$\overline{q}'''$ [W cm^{-3}]	234	

Table 2. Pin axial dimensions [11].

INNER FUEL ZONE			OUTER FUEL ZONE		
Upper fissile layer	35.228	cm	Fissile fuel	90.685	cm
Middle breeder plate	20.152	cm			
Lower fissile layer	25.228	cm			
Axial blanket	30.228	cm	Axial blanket	30.228	cm
Inner zone pin height	110.836	cm	Outer zone pin height	120.913	cm

3. Benchmark specifications

Since the ASTRID design details are still under development an ASTRID-like geometry is employed. Two different test case are proposed, one for the inner and one for the outer zone sub-assembly. The radial dimensions of the fuel pin for the two cases are identical (see Table 1). The complete geometrical data for the axial direction are available in Table 2. For the purpose of the neutronics simulation the presence of the wrapping wire can be neglected, as well as the the sodium and gas plena, the structural support of the pin and of the fuel assemblies.

The fissile fuel is a mixed uranium-plutonium oxide (MOX) with an enrichment in plutonium of 23.5% for the inner zone and of 20% for the outer zone, whereas the fertile sections of the pin are composed by natural uranium oxide. The test case fuel is constituted by the isotopes U^{235}, U^{238}, Pu^{238}, Pu^{239}, Pu^{240}, Pu^{241} and Pu^{242}. The oxygen is assumed as composed completely by O^{16} and in perfect stoichiometric ratio with both the uranium and plutonium. Fresh fuel is modelled and the isotopic distribution variation it is neglected. Data about the fuel composition for both the fissile and fertile sections are available in Table 3.

The pin-cladding gap is filled by gaseous helium at 4 MPa, which it is composed exclusively by He^4. Moreover, the coolant is considered as natural and perfectly pure sodium. The composition of the AIM1 cladding and the EM10 wrapper is assumed to be equal to the average data in the reference [16]. Thermo-physical properties about the fuel, the coolant, the cladding and helium can be found in the references [17], [18] and [19].

4. Coupling scheme

The approaches that can be followed to couple a neutronics code and a thermal-hydraulics one are divided in two categories: a loose coupling, where the calculations are performed separately and informations between the models is exchanged only at specified points, and an internal coupling, where to models are integrated in the same platform and communicate continuosly.

The loose coupling can considered equivalent to the solving of a set of differential equations through the operator splitting method. The main advantage of this approach is that it is possible

Table 3. Fissile and fertile fuel mass and atomic fractions [15].

	MOX outer zone		MOX inner zone		UO_2	
Isotope	WF	AF	WF	AF	WF	AF
U-235	0.507%	0.194 %	0.485%	0.186%	0.220%	0.084 %
U-238	69.910%	26.440 %	66.837%	25.283%	87.929%	33.241 %
Pu-238	0.496%	0.188 %	0.579%	0.219%		
Pu-239	9.132%	3.439 %	10.696%	4.029%		
Pu-240	4.770%	1.789 %	5.610%	2.105%		
Pu-241	1.880%	0.702 %	2.220%	0.829%		
Pu-242	1.454%	0.541 %	1.724%	0.642%		
O-16	11.851%	66.707 %	11.849%	66.707%	11.850%	66.674 %

to employ verified and validate code for the solution of each model, without the necessity to modify the source code. It is the state-of-the-art for coupled calculations and it is implemented for this test case [7].

The codes used as reference for this test case are Serpent, for the neutronics part, and ANSYS-CFX, for the thermal-hydraulics part.

Serpent is a three-dimensional continuous-energy Monte Carlo reactor physics burnup calculation code developed by the VTT Technical Research Center of Finland since 2004 and first distributed in 2009. Serpent allows the description of any two- or three-dimensional fuel or reactor configuration. Integrated in the code are cross section libraries based on JEF-2.2, JEFF-3.1, JEFF-3.1.1, ENDF/B-VI.8 and ENDFB/B-VII for 432 nuclides at 6 temperatures between 300 and 1800 K. [20]

ANSYS-CFX is a general purpose Computational Fluid Dynamics three-dimensional code. The program is able to simulate a wide range of physical phenomena and flows. A tailored programming language (CEL) is available to extend the capability of the code, i.e. adding new materials to the software libraries and modifying the solver equations. [21].

The Serpent output supplies the distribution of the volumetric power generation (q''') inside the fuel, which is employed in the writing of the CFX input file. In turn, the CFX results provide the distribution of the temperature and the density of the fuel, the cladding and the coolant which are integrated as input for the next iteration of Serpent. Local power, k_{eff} and the temperature distribution are proposed as convergence parameters. At the end of each coupling cycle, the parameters chosen are checked against the following convergence criteria

(i) $k_{eff} = 1.0 \pm 0.001$ in two successive iterations

(ii) $Q < 0.1\%$

(iii) $T_f, T_c, T_{Na} < 0.1\%$

for the local power and temperature distribution, the criterion should be intended for the residual of the parameter discussed. In particular, for the criterion (iii) representative temperatures must be selected, i.e. central point for pin and cladding, outlet for the sodium. When all the criteria are satisfied, the calculation is considered complete. In Tables 4 and 5, data necessary for the setup of the coupled simulations are provided.

5. Expected results

In this work, specifications for a coupled neutronics/thermal-hydraulics test case based on a generation IV sodium-cooled fast reactor were presented. The design chosen to serve as a model

Table 4. Neutronics simulation initial conditions.

	T[K]	ρ [kg m$^-$3]
MOX fuel	1500	10120.04
UO_2 fuel	900	10219.43
Cladding	750	7746.00
Helium	750	1.87
Sodium coolant	750	837.95

Table 5. Thermal-hydraulics simulation initial conditions [13][14].

Γ_{core}	7900	kg s$^-$1	Na inlet properties		
I/Outer FA power	6/3.84	MW	ρ	855	kg m$^-$3
Na inlet T	673	K	c_p	1289	J kg K$^-$1
Na outlet T	823	K	λ_t	73.11	W m$^-$1 K$^-$1
Re	$5\,10^4$		μ	2.76 10−4	Pa s

for the test case was the ASTRID, a technological demonstrator actually in development by the CEA. Two test cases were proposed: in the first one the inner zone fuel assembly conditions is studied, whereas in the second the outer zone assembly is considered. A loose coupling scheme was analysed with Serpent for the neutronic part and ANSYS-CFX for the thermal-hydraulic one. Since the computational resources needed to model the whole assembly might be too high, it is possible to restrict the thermal-hydraulics analysis to the single central fuel pin of the sub-assembly through the setup of appropriate boundary conditions. However, whenever possible, the authors advise to model the complete assembly.

The expected results for this benchmark can be summarized as the following

- The coupled neutronics/thermal-hydraulics steady-state simulation of the test cases proposed should be performed until the convergence of the parameters of interest.
- Axial distribution of volumetric power generation and fission rates in the fuel
- k_{eff}
- Axial distribution of temperature and density for the fuel, the cladding and the coolant
- Pressure drop across the channel for the sodium

The validation is conducted through code-by-code comparisons. However, results for similar configurations can be found in the literature, i.e. [7] and [14].

Acknowledgments

This work was realized during an internship of 2 months, hosted by the National Research Nuclear University "MEPhI" in Moscow, in the framework of the ENEN-RU II Project "Strengthening of Cooperation and Exchange for Nuclear Education and Training between the European Union and the Russian Federation". The realization of this work was funded from the EURATOM Research and Training programme under grant agreement number 605149. The authors would like to acknowledge the financial support of the ENEN and the hospitality of the MEPhI institution. This work was supported by Competitiveness Program of National Research Nuclear University MEPhI

References

[1] Chen, Z., et al. (2015). *Coupling a CFD code with neutron kinetics and pin thermal models for nuclear reactor safety analyses.* Annals of Nuclear Energy, 83, 4149.

[2] Fridman, E., et al. (2013). *Modeling of SFR cores with Serpent-DYN3D codes sequence.* Annals of Nuclear Energy, 53, 354363.

[3] Grahn, A., et al. (2015). *Coupling of the 3D neutron kinetic core model DYN3D with the CFD software ANSYS-CFX.* Annals of Nuclear Energy, 84, 197203.

[4] Ivanov, A., et al. (2011). *Development of a Coupling Scheme Between MCNP5 and Subchanflow for the Pin- and Fuel Assembbly-Wise Simulation of LWR and Innovative Reactors.* M&C, Rio de Janeiro, RJ, Brazil, May 8-12, 2011, pages 118.

[5] Nikitin, E.,et al. (2014). *Solution of the OECD/NEA neutronic SFR benchmark with Serpent-DYN3D and Serpent-PARCS code systems.* Annals of Nuclear Energy, 75, 492497.

[6] Rachamin, R., et al. (2013). *Neutronic analysis of SFR core with HELIOS-2, Serpent, and DYN3D codes.* Annals of Nuclear Energy, 55, 194204.

[7] Vazquez, M., et al. (2012). *Coupled neutronics thermal-hydraulics analysis using Monte Carlo and sub-channel codes.* Nuclear Engineering and Design, 250, 403-411.

[8] OECD Nuclear Energy Agency. (2014). *Technology Roadmap Update for Generation IV Nuclear Energy Systems*, 166.

[9] CEA. (2012). *4th Generation sodium-cooled fast reactors: the ASTRID technological demonstrator.*

[10] IAEA. (2006). *Fast Reactor Database 2006 Update.* Iaea-Tecdoc-1531, (December).

[11] Bertrand, F.,et al. (2016). *Comparison of the behaviour of two core designs for ASTRID in case of severe accidents.* Nuclear Engineering and Design, 297, 327342.

[12] Bortot, S., et al. (2015). *European Benchmark on the ASTRID-like Low-void-effect Core Characterization: Neutronic Parameters and Safety Coefficients.* Icapp 2015, (May), 668676.

[13] Chenaud, M. S.,et al. (2013). *Status of the ASTRID core at the end of the preconceptual design phase 1.* Nuclear Engineering and Technology, 45(6), 721730.

[14] Saxena, A. (2014). *Thermal-hydraulic numerical simulation of fuel sub-assembly for Sodium-cooled Fast Reactor*, Ph.D. thesis.

[15] Garcia-Cervantes, E.-Y. et al. (2016). *A comparison between oxide and metallic fueled ASTRID-like reactors*, 94, 350358.

[16] Gavoille, P., et al. (2013). *Mechanical Properties of Cladding and Wrapper Materials for the ASTRID Fast-Reactor Project.* International Conference on Fast Reactors and Related Fuel Cycles: Safe Technologies and Sustainable Scenarios (FR13) Presentations, (p. v). International Atomic Energy Agency (IAEA): IAEA.

[17] Popov, S. G., et al. (2000). *Thermophysical properties of MOX and Uranium Dioxide Fuels including the effects of irradiation.* CEUR Workshop Proceedings.

[18] Sobolev, V. (2010). *Database of thermophysical properties of liquid metal coolants for GEN-IV.* SCK CEN Technical Report, SCKCEN-BLG-1069, 16(12), 34963502.

[19] IAEA. (2008). *Thermophysical Properties of Materials for Nuclear Engineering: A Tutorial and Collection of Data.*

[20] Leppnen, J. (2013). *Serpent - a continuous-energy Monte Carlo reactor physics burnup calculation code.*

[21] ANSYS, Inc. (2013). Documentation for release ANSYS 15.0

IOP Conf. Series: Journal of Physics: Conf. Series **781** (2017) 012048 doi:10.1088/1742-6596/781/1/012048

The optimal approach for the processes of verification and validation of NPP software and hardware complexes

S Korolev[1], A Tolokonsky[2] and V Rogov[3]

[1] Deputy head of department «Automatics»
[2] Head teacher of department «Automatics»
[3] Postgraduate of department «Automatics»
National Research Nuclear University MEPhI (Moscow engineering Physics Institute), Moscow, Russia

E-mail: litos_mephi@mail.ru, toloconne@yandex.ru, rogovvictor@mail.ru

Abstract. Nowadays the task of increasing the quality of software and hardware complex development has not lost its urgency. The problem mentioned has still been solving by the search of an optimal structure, stage content and life cycle processes. The current system of international standards has been developed on the basis of advanced software complexes for aviation, space and defense industries being designed during the second half of the 20th century. However, it lacks a holistic approach for verification and validation. The paper presents the analysis of Russian and international regulatory framework and current verification and validation methods in the context of the life cycle of NPP software and hardware complexes. Basing on the data obtained, a new approach for verification and validation methods has been introduced.

1. Introduction

According to IEC 61513, the life cycle of a software and hardware complex (SHC) is an essential activity that is carried out from the initial stage of the development of the concept and requirements, to the final stage when the system can no longer be maintained [1].

The international ISO / IEC 15288 standard defines the framework for the agreement, organizational, design and technical processes. The ISO / IEC 15288 standard emphasizes the necessity of its customization to the tasks of a particular industry, specifically to the tasks of NPP software and hardware complexes [2].

2. Software and hardware NPP complexes lifecycle features

According to the ISO / IEC 15288 standard, technical processes should include the following:

- definition of stakeholders requirements;
- requirements analysis;
- architecture design;
- implementation of system components;
- integration;
- verification;
- implementation of the system;

- validation;
- operation;
- support (maintenance);
- decommissioning.

Figure 1 presents a generic V-model of a typical NPP software and hardware complex with the consideration of the possibility of modernization. The given flowchart does not show the technical processes №1 and 11 from the standard mentioned, however some of the processes are merged into one block. In fact, all processes shown in Figure 1 are carried out and described in the relevant contracts, terms of reference, project, engineering and operational documentation [3].

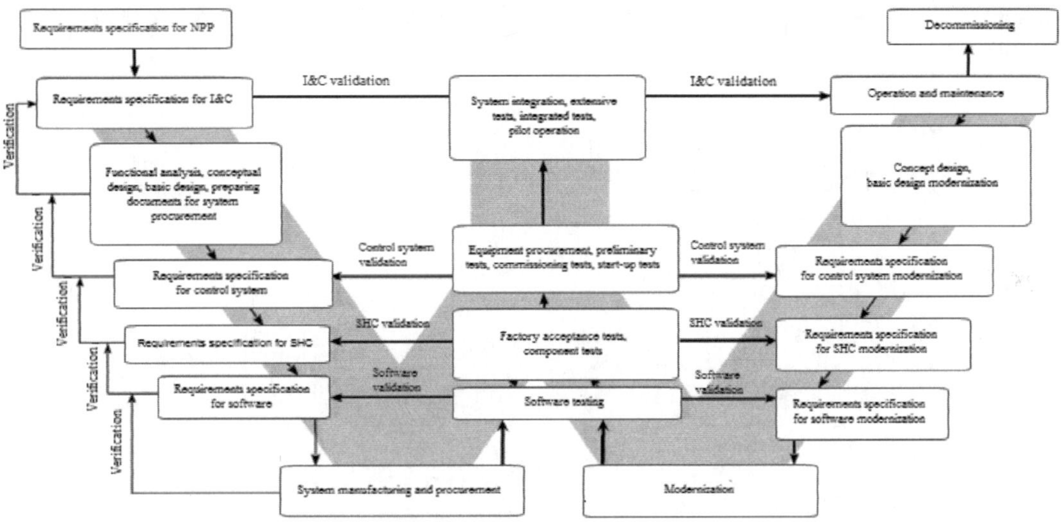

Figure 1. The life cycle of a modern nondomestic NPP software and hardware complex.

Verification of the compliance of the NPP software and hardware complex with the requirements for functional and performance characteristics is one of the important development stages of the NPP software and hardware complex life cycle.

Such verification includes the processes of verification and validation of the NPP software and hardware complex.

Different international and domestic life cycles of NPP software and hardware complexes having been analyzed, the V-model of the life cycle is the best applicable one at present.

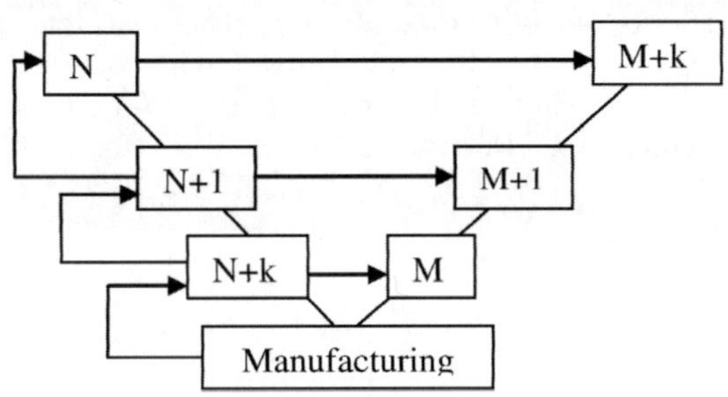

Figure 2. V-model diagram.

IOP Conf. Series: Journal of Physics: Conf. Series **781** (2017) 012048 doi:10.1088/1742-6596/781/1/012048

A systematic approach for the processes of verification and validation of the V-model of the NPP software and hardware complex envisages that the following conditions should be met:

- the verification of the N – N+k stages is carried out consistently
- the number of N stages coincide with the number of M stages
- optimal strategy is used to validate the M – M+l stages (The notes being discovered during verification and validation should be identified and implemented)

The next chapter will examine the methods of systems verification and validation that are implemented in domestic (Russian) and nondomestic reference documentation, in order to assess their applicability to the life cycle of the NPP software and hardware complex.

3. Optimal method choice for verification and validation of NPP software and hardware complexes.

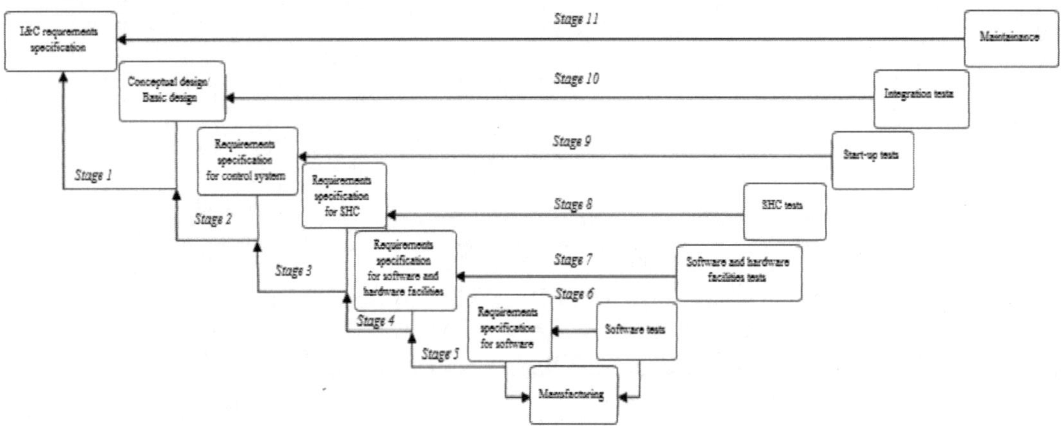

Figure 3. Verification and validation at lifecycle stages.

The following applicability criteria for verification of the NPP software and hardware complex were defined:

- Time to prepare for the usage of the method;
- Time to apply the method;
- Qualification level of the specialists;
- Possibility of using early achievements for the method;
- Efficiency of the application of the method.

Positive and negative aspects of all methods having been analyzed and the assessment of the methods according to the criteria listed in the table below having been performed, the most relevant methods were selected, in order to achieve a systematic approach for the verification and validation of NPP software and hardware complexes [4].

Table 1. Methods used for the verification of NPP software and hardware complexes.

Common method	Method	Verification				
		Stage 1	Stage 2	Stage 3	Stage 4	Stage 5
General expertise	Technical expertise	+	+	+	+	+
	Walkthroughs	+	+	+	+	+
	Traceability analysis	+	+	+	+	+
	Audit	+	+	+	+	+
	Requirements adequacy analysis	+	+	+	+	+
Specialized inspection	Usability expertise	+	+	+	+	+
Static analysis	Rules checking	-	-	-	-	-
	Model checking	-	-	-	-	-
Formal methods	Functional and timing simulation within the framework of automated design environment	-	-	-	-	-
Dynamic methods	Monitoring	-	-	-	-	-
	Unit testing	-	-	-	-	-
	Integration testing	-	-	-	-	-
	System testing	-	-	-	-	-
	Black-box testing	-	-	-	-	-
	Load testing	-	-	-	-	-
	Stress testing	-	-	-	-	-

Optimal method choice for verification and validation of NPP software and hardware complexes presented in Table 2.

Table 2. Methods used for the validation of NPP software and hardware complexes.

Common method	Method	Validation					
		Stage 6	Stage 7	Stage 8	Stage 9	Stage 10	Stage 11
General expertise	Technical expertise	+	+	+	+	+	+
	Walkthroughs	-	-	-	-	-	-
	Traceability analysis	-	-	-	-	-	-
	Audit	+	+	+	+	+	+
	Requirements adequacy analysis	-	-	-	-	-	-
Specialized inspection	Usability expertise	+	+	+	+	+	+
Static analysis	Rules checking	+	+	+	+	+	+
	Model checking	+	+	+	+	+	+
Formal methods	Functional and timing simulation within the framework of automated design environment	+	+	+	+	+	+
Dynamic methods	Monitoring	-	-	-	-	-	+
	Unit testing	+	+	-	-	-	-
	Integration testing	-	-	-	-	+	-
	System testing	-	-	+	+	-	-
	Black-box testing	+	+	+	+	+	+
	Load testing	+	+	+	+	+	+
	Stress testing	+	+	+	+	+	+

4. Conclusions

Optimal distribution of work related to verification and validation for the stages of the NPP software and hardware complex life cycle and improved quality of the implementation of these stages can be reached by the application of the approach discussed and the algorithms developed. Consequently, it raises the efficiency of the execution of the NPP software and hardware complex.

The proposed methods of a systematic approach for verification and validation can be applied both to the NPP software and hardware complexes and systems of a wide range of applications, allowing to mark a high practical usefulness of the work presented.

References

[1] IEC 61513-2011 2012 *Nuclear power plants: Instrumentation and control important to safety. General requirements for systems*

[2] ISO/IEC 15288 2015 Systems and software engineering : System life cycle processes

[3] Bozhenkov O, and Kabachnikov A 2009 System engineering in NPP I&C. Nuclear measurement and information technologies 2 27-32

[4] Kuliamin V 2008 Methods for software verification (Moscow: Institute for System Programming of the Russian Academy of Sciences) 111 pp

[5] IEC/ISO 12207-2010 Information technology. System and software engineering. Software life

cycle processes
[6] IEEE STANDARD 1012-2012 2012 IEEE Standard for System and Software Verification and Validation
[7] IEEE STANDARD 829-2008 IEEE Standard for Software and System Test Documentation
[8] ANSI/IEEE 1008 Standard for Software Unit Testing (Note: Approved 2003-04-18)

Computational simulation of thermal hydraulic processes in the model LMFBR fuel assembly

M V Bayaskhalanov, I G Merinov, A S Korsun and M N Vlasov

National Research Nuclear University MEPhI (Moscow Engineering Physics Institute), Kashirskoe shosse 31, Moscow, Russia

E-mail: mr.bayashalanov@mail.ru

Abstract. The aim of this study was to verify a developed software module on the experimental fuel assembly with partial blockage of the flow section. The developed software module for simulation of thermal hydraulic processes in liquid metal coolant is based on theory of anisotropic porous media with specially developed integral turbulence model for coefficients determination. The finite element method is used for numerical solution. Experimental data for hexahedral assembly with electrically heated smooth cylindrical rods cooled by liquid sodium are considered. The results of calculation obtained with developed software module for a case of corner blockade are presented. The calculated distribution of coolant velocities showed the presence of the vortex flow behind the blockade. Features vortex region are in a good quantitative and qualitative agreement with experimental data. This demonstrates the efficiency of the hydrodynamic unit for developed software module. But obtained radial coolant temperature profiles differ significantly from the experimental in the vortex flow region. The possible reasons for this discrepancy were analyzed.

1. Introduction

One of the possible approach to description of heat and mass transfer processes in the core and heat-exchange equipment is using of porous body model. Equations of this model are obtained as a result of a rigorous mathematical procedure of averaging initial three-dimensional equations of heat and mass transfer processes [1-2].

$$\frac{\partial}{\partial x_j}\varphi u_j = 0,\ i = 1,2,3, \tag{1}$$

$$\frac{\partial}{\partial \tau}\rho\varphi u_i + \frac{\partial}{\partial x_j}\rho\varphi u_j u_i = \rho\varphi g_i - k_{ij}u_j - \frac{\partial \varphi P}{\partial x_j} + \frac{\partial}{\partial x_i}\mu_{eff}\left(\frac{\partial u_i}{\partial x_j} + \frac{\partial u_j}{\partial x_i}\right), \tag{2}$$

$$\rho c_p\varphi\frac{\partial t}{\partial \tau} + \frac{\partial}{\partial x_j}\rho c_p\varphi u_j t = \varphi q_V - \frac{\partial}{\partial x_j}\lambda_{ij}^{eff}\frac{\partial t}{\partial x_i} - k_{Vl}(t - t_F) \tag{3}$$

where φ – porosity; u_i, u_j – components of the velocity vector; g_i – components of the gravitational acceleration; $k_{ij}=k_{\xi\xi}\delta_{ij} + (k_{\xi\xi} - k_{\eta\eta})n_i n_j$ – tensor components of resistance, where $k_{\eta\eta}(\beta)$ и $k_{\xi\xi}(\beta)$ – main components corresponding to the directions along and transverse to rods; $P=p + c\rho u^2$ – the effective pressure in the flow, the sum of the thermodynamic pressure and the pressure due to the turbulent fluctuations and variations of speeds, where "c" - the pressure coefficient; q_V – power density in the liquid; $\lambda_{ij}^{\text{eff}}$ – components of the coolant effective thermal conductivity tensor; t, t_F – temperature of the coolant and fuel rods, respectively; k_{Vl} – volumetric coefficient of heat transfer from the fuel rods to the coolant; μ_{eff} – the effective viscosity of the coolant flow.

Effective transfer coefficients in the equations (1) - (3) are determined using a specially developed integral turbulence model [3].

Software module APMod is designed to solve equation system by finite element method. APMod is developed for mathematical calculations of heat and mass transfer in the core and heat-exchange equipment of advanced NPP. The aim of this study was verification of software module at computational research of sodium coolant flow in the experimental fuel assembly with partial blockage of the flow section.

2. Experimental assembly

Experimental assembly KNS-Test [4] consisted of 169 smooth cylindrical rod elements, some of which are heated by electric current (Figure 1). The blockade was located at a distance of 40 mm from the entrance and overlapped 21% of flow section. The boundary conditions were specified by the coolant rate at the inlet, outlet pressure and the condition of the sliding cover on the surface. The temperature dependence of thermal properties of sodium [5] was taken into account in calculations.

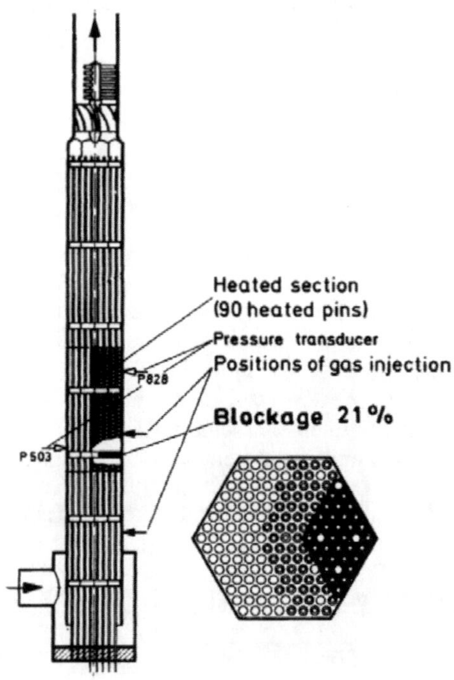

Figure 1. KNS-Test.

3. Calculation result

The resulting calculated coolant velocity distribution in a longitudinal section FA is shown on Figure 2a. It is easy to see a vortex produced behind the blockade. The existence of a vortex is also found experimentally by processing of the measured temperature fields.

The observed flow pattern can be quantitatively described by a number of characteristic parameters. The position of the top of the stagnation zone, the center of the vortex, and the reverse flow rate (Figure 2a) could be included to them. In the experiment [4], as a reverse flow velocity relative velocity U_{rR} was considered. It could be defined by the following expression:

$$U_{rR} = u_R/\overline{u} \tag{4}$$

where $\overline{u} = (u_0 + u_B)/2$; u_0 – the coolant rate at the inlet; u_B – the coolant rate in the blockade; u_R – reverse flow rate.

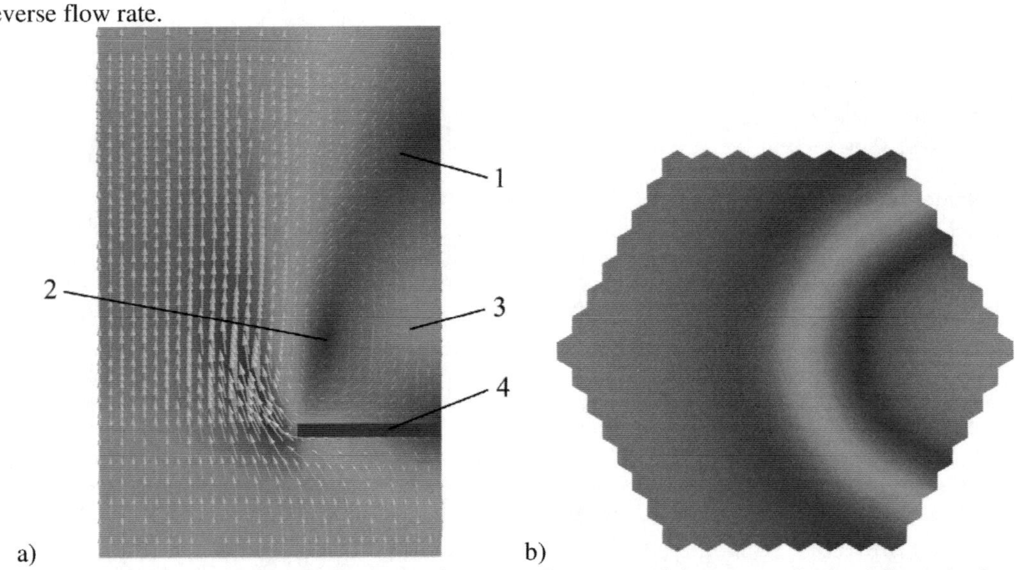

1 - the top point of the stagnation zone; 2 - center of the vortex; 3 - backflow; 4 - the blockade;

Figure 2. The coolant velocity field: FA in longitudinal section (a) is a cross-sectional area of fuel assembly in the center of the toroidal vortex (b).

The rates u_B and u_R were determined by means of the dependence of the longitudinal component of the velocity from the height of the fuel assembly in the far corner and the central cell. This dependence is also used to determine the top of the stagnation zone.

The position of the vortex center was determined by the minimum coolant velocity in its distribution in the cross section of the fuel assembly (Fig. 2b).

Comparison of the calculated and experimentally obtained parameters is given in Table 1.

Table 1. Comparison of the calculated and experimental parameters.

Parameter	Experiment	Calculation
The height of the stagnation zone, mm	100	95
The vortex center position, mm	45	35
The relative velocity of reverse flow	0.18	0.26

Quantitative and qualitative comparison of the velocity fields shows that simulation result corresponds with the experiment. This demonstrates the efficiency of the hydrodynamic unit APMod software module.

Experimentally measured coolant temperature [4] is normalized by the value of the axial temperature gradient, that is determined in the undisturbed portion of the central equivalent cell:

$$\theta = \Delta t/(dt/dz), \tag{5}$$

where $\Delta t = t - t_{\text{in}}$ – heating of the coolant at the point of measurement, dt/dz – axial temperature gradient defined by the following formula:

$$dt/dz = 4q/(\rho c_p u_0 d_G) \quad (6)$$

In the expression (6) q – heat flux into the liquid on equivalent cell to a boundary, and d_G – its hydraulic diameter.

For comparison with the experimental data, calculated temperature field was normalized in accordance with the formulas (5) and (6). Figure 2 shows a comparison of the radial profile of calculated and experimental normalized temperature at different distances from the blockade.

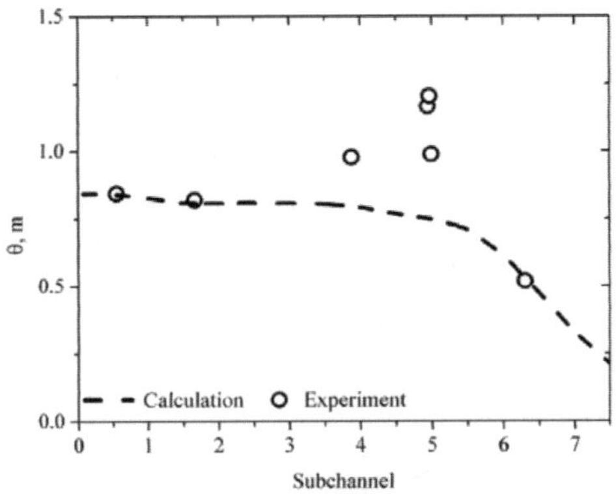

a)

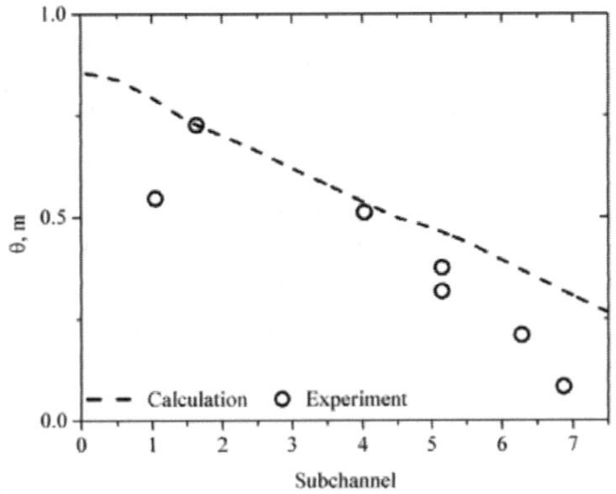

b)

Figure 3. The radial profile of the normalized temperature at different distances from the blockade: 10 mm (a), 120 mm (b).

There is a noticeable discrepancy between the obtained results and the experiment in the vortex area. One of the possible reasons for this discrepancy could be that the energy in the computational model was described as voluminous source of energy in the liquid. In experimental setup electrically heating rods was used and thermocouple was placed on the surface of the rods. So the temperature in the vortex axis could vary significantly from fluid temperature. In addition, the porous body model operates by averaged values so calculated temperature profiles will flatten.

4. Conclusion

As a result of the research a good quantitative and qualitative agreement between the calculated and experimental characteristics of the sodium coolant flow was received. Calculated radial coolant temperature profiles differ significantly from the experimental in the vortex flow region. The possible reasons for this discrepancy were analyzed.

References

[1] A.S. Korsun, V.B. Kruglov, I.G. Merinov et al., Problems of Atomic Science and Technology. Series: Nuclear and Reactor Constants, No 2, 87 (2014). (In Russian).
[2] A.S. Korsun, Yu.A. Maslov, I.G. Merinov et al., Nuclear physics and engineering, 4(7), 619 (2013). DOI: 10.1134/S2079562913070063. (In Russian).
[3] M.N. Vlasov, A.S. Korsun, Yu.A. Maslov et al., MEPhI Newsletter, 2(3), 314 (2013). DOI: 10.1134/S2304487X13030206. (In Russian).
[4] F. Huber, W. Peppler, Report No 3927 (Karlsruhe, 1985).
[5] A.A. Kazantsev, A.S. Kondratyev, Izvestia Vysshikh Uchebnykh Zawedeniy. Yadernaya Energetika, № 4, 86 (2008). (In Russian).

Test case for VVER-1000 complex modeling using MCU and ATHLET

R B Bahdanovich[1], E V Bogdanova[1], I D Gamtsemlidze[2], S P Nikonov[1] and G V Tikhomirov[1]

[1] Department of theoretical and experimental physics of nuclear reactors, National Research Nuclear University MEPhI (Moscow Engineering Physics Institute), Kashirskoe shosse 31, Moscow, Russia
[2] All-Russian Research Institute for Nuclear Power Plants Operation (VNIIAES)

RBBogdanovich@mephi.ru

Abstract. The correct modeling of processes occurring in the fuel core of the reactor is very important. In the design and operation of nuclear reactors it is necessary to cover the entire range of reactor physics. Very often the calculations are carried out within the framework of only one domain, for example, in the framework of structural analysis, neutronics (NT) or thermal hydraulics (TH). However, this is not always correct, as the impact of related physical processes occurring simultaneously, could be significant. Therefore it is recommended to spend the coupled calculations. The paper provides test case for the coupled neutronics-thermal hydraulics calculation of VVER-1000 using the precise neutron code MCU and system engineering code ATHLET. The model is based on the fuel assembly (type 2M). Test case for calculation of power distribution, fuel and coolant temperature, coolant density, etc. has been developed. It is assumed that the test case will be used for simulation of VVER-1000 reactor and in the calculation using other programs, for example, for codes cross-verification. The detailed description of the codes (MCU, ATHLET), geometry and material composition of the model and an iterative calculation scheme is given in the paper. Script in PERL language was written to couple the codes.

1. Introduction

For a long period of time simulation of nuclear reactors was carried out by means of simple mathematical models. This was mostly due to the low level of development of computer technology. However, it was enough to create a sustainable nuclear industry. Since then, methods of calculation and handy tools have changed a lot. Due to the growth of computer performance we can obtain closer to reality results. Today we see a trend to a more detailed description of simulated systems and growing complexity of calculation schemes. Modeling is gradually coming to an absolutely new level.

Nowadays scientists understand that the feedback of related physical processes occurring simultaneously could be significant. There are a number of issues, which could be solved only in the complex modeling. Coupled calculations could be done either on the basis of one calculation program, using simple physical models (BIPR, DYN3D, NESTLE), or by coupling the codes intended to simulate the processes described in different areas of physics (MCU, MCNP, ATHLET, ANSYS, etc.). Such kind of coupling was performed before for different types of reactors [1-3]. In this paper is given description of the test case for coupled calculation of VVER-1000 using MCU and ATHLET codes.

2. Codes features

2.1. MCU

MCU (Monte Carlo Universal) - is a neutron code based on the Monte Carlo method, which allows to simulate the transport of neutrons, gamma rays and electrons of different energies. The code takes into account the effects of the continuous changes in the particle energy, as well as continuous or step-energy dependence of the cross-sections. The accuracy is determined primarily by used libraries of evaluated nuclear data. MCU allows to calculate the real geometry and material composition of the complex media at various temperatures. It contains cross-sections conversion modules in the thermal and resolved resonance region, depending on the temperature of the material (using the Breit-Wigner or Adler-Adler formalisms). It can be used for multi-processor calculations (MPI technology) [4].

2.2 ATHLET.

Thermal hydraulic system code ATHLET (Analysis of Thermal-Hydraulics of LEaks and Transients) developed in Gesellschaft für Anlagen-und Reaktorsicherheit (GRS mbH) and was originally intended for the analysis of the entire spectrum of leak cases and transients in LWR reactors (PWR and BWR). However, experience has shown that code could be fully used for the Russian VVER reactors. It should be noted that at present this code is approved in Russia for use in the analysis of the safety of reactors with water coolant.

ATHLET consists of several modules that allow to describe the various phenomena in the behavior of light water reactors: thermal hydraulic module (TFD), heat transfer and heat conduction unit (HECU), neutron-kinetic unit (NEUKIN) to describe the point and the one-dimensional kinetics, module for operation of the equipment (GCSM) and a module for the numerical integration (FEBE), implements a fully implicit scheme. Other independent modules (eg, three-dimensional neutron kinetics, describe the behavior of containment, etc.) can be connected through the main interface.

The major is TFD module that is based on the use of five equations (equations of conservation of mass and energy for the liquid and vapor phase and the general equation for the angular momentum mixture drift flow) or six equations (equations of conservation of mass, energy, and pulse for the liquid and vapor in the description of the two-fluid model) with a wide range of closure ratios. In addition, the module allows to simulate the behavior of non-condensable gases, nitrogen dissolution and describe the boron transport in system. Reactor coolant system is modeled by a compound of basic thermodynamic objects (TFO), the detailed modeling is limited only by the possibility of computing - the amount of memory (RAM) and processor frequency. In addition, in TFD module it is possible for specific description of transverse coolant flow connected in parallel system [5].

3. Description of the test case

This chapter provides a description of the VVER-1000 reactor model for coupled neutron and thermal hydraulic calculation for MCU and ATHLET codes. The model is based on the fuel assemblies type 2M. Name of the test case is C2M5W.

The codes have their own features, therefore, in order to simplify the calculation, two models of the reactor have been developed. The first model represents the neutron characteristics of the reactor, and the second - thermal hydraulic. The detailed description is given next.

3.1. Model for the neutron calculation

The neutron model is an elemental cell of the reactor, consisting of the fuel part put into water moderator (Figure 1). During the calculation, the model is divided into 12 axial layers (10 layers in the fuel part and 2 in inlet and outlet moderator). Mirror reflection of neutrons is present only on the lateral border of the model, at the ends is black border. The effective neutron multiplication factor (K_{eff}) should be equal unity and it is changed by boric acid concentration. Compatibility of NT and TH models is achieved by specifying a fixed inlet temperature to the cell (290 °C in layer №0). Neutron model does not account changes in fuel and cladding density, average energy per fission is considered to be constant, the same for both uranium isotopes. Description of the initial materials and the model parameters is given in Table 1. The fuel has a porosity of 95%, the central hole and the gap between the cladding and the fuel is not considered. Nominal power is equal to 19.6319 MW (average power of the fuel assembly) [6].

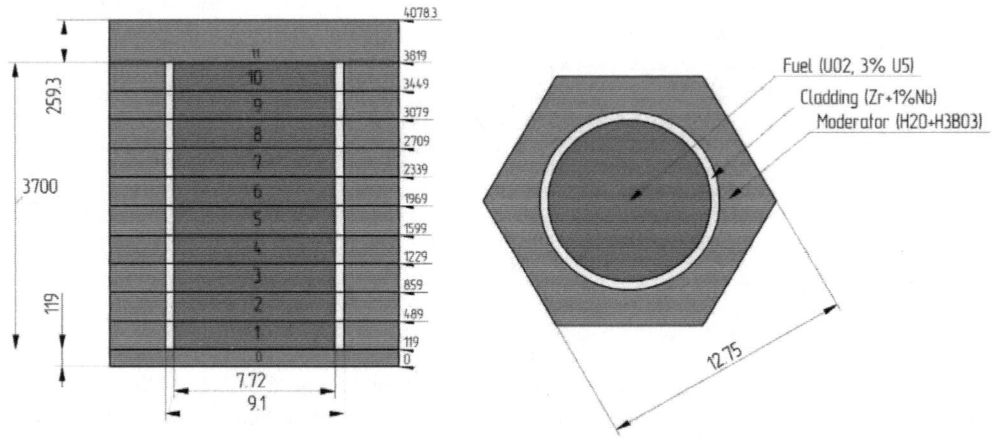

Figure 1. Neutron model of VVER-1000 (mm).

Table 1. Characteristics of materials for MCU first iteration [6-11].

Material	Composition (mass fractions)	$\rho, \frac{g}{cm^3}$
UO_2, 1.6 % ^{235}U	^{235}U – 1.4104 %, ^{238}U – 86.7411 %, O – 11.8484 %	9.445
^{90}Zr, 1% ^{93}Nb (Э-110)	Zr – 98.8925 %, Nb – 1%, Fe – 0.015%, Ni – 0.007%, Al – 0.004%, Ti – 0.003%, C – 0.02%, Si – 0.004%, O – 0.05%, N – 0.003%, H – 0.0015%	6.550
7g (H_3BO_3) /kg (H_2O)	H – 11.1577 %, O – 88.7455 %, B10 – 0. 0178482 %, B11 – 0.0789902 %	0.7136

3.2. Model for the thermal hydraulic calculation

Simplified fuel assembly is used as a TH model, which is a channel with a heat-generating structure. Real geometric dimensions are considered. Layering is the same as for the neutron calculation. Materials of rod endings are not considered. The channel simulates fuel assembly from the top of the bottom support plate to the top of the fuel rod stopper. Fuel assembly contains 312 fuel rods, 18 guide channels and a central channel (Figure 2). All the required dimensions are given in Table 2.

The heated part is located at a distance from 0.119 to 3.819 from the top of the bottom support plate. Entering flow rate is 100 kg / s. The initial temperature is 290 ° C. Inlet pressure is $162 \cdot 10^5$ Pa. Table 1 gives the main characteristics of the model materials. It should be noted that the code does not consider the impact of boron on the moderator properties.

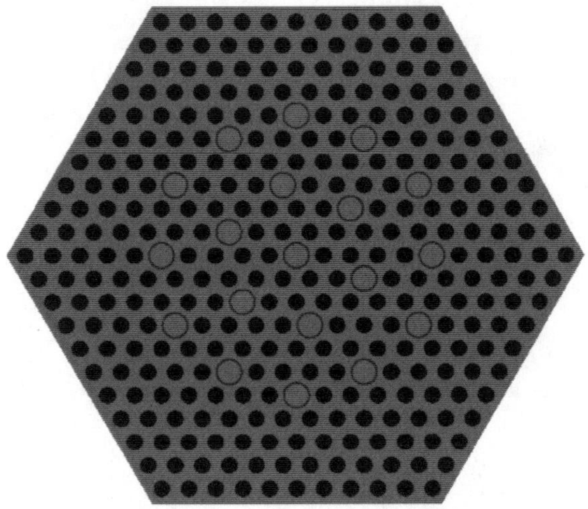

Figure 2. Layout of VVER fuel assembly

Table 2. The geometry of TVS-2M fuel assembly [6].

Element title	The distance between closer elements (mm)	The distance from the origin (mm)
Top of rods stopper	5	4078.3
Top of reflection shield	4070	4070
Top of upper spacer grid	40	4073.3
Bottom of upper spacer grid	214.3	4033.3
Top of heated rod part	3700	3819
Bottom of heated rod part	10.7	119
Top of lower spacer grid	16	108.3
Bottom of lower spacer grid	92.3	92.3
Top of lower support plate	0.0	0.0

3.2 Iterative scheme

The aim of this work is to create a test case, reflecting the physics of a real facility, which will be further used to determine the characteristics of the VVER-1000 reactor.

The iterative scheme is shown on Figure 3. The external iterative scheme is used, where transients are considered separately by each code. In fact, the transients are modeled only in TH code. Description of the calculated values and the iterative scheme is given below.

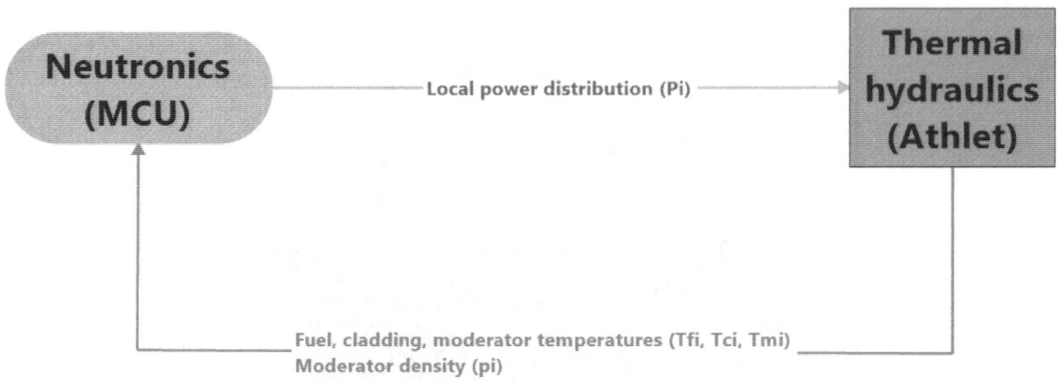

Figure 3. Iterative scheme for coupled calculation.

The input values for the MCU code are (i = $\overline{0,11}$):

- Tf_i – fuel temperature [^{0}K],
- Tc_i – cladding temperature [^{0}K],
- Tm_i – moderator temperature [^{0}K],
- ρ_i – moderator density [g/cm^3].

Calculated functionals (using MCU):

- Rf_i – sum of fission reaction rates of ^{235}U and ^{238}U normalized per 1 fission neutron [%],
- P_i – local power (i-layer) [KWt],
- Xb_i – boric acid fraction [%].

It should be noted, that the local power distribution is calculated by the formula:

$$P_i = P\frac{Rf_i}{\sum_i Rf_i} \quad (1)$$

where P - power of the entire system. This formula is valid, as the volume of fuel layers are the same, and ^{238}U fission rate is more than ten times smaller than ^{235}U fission rate. Local power calculation error obtained as the indirect calculation error:

$$\Delta P_i = \pm\left(\left(\frac{P\cdot\Delta Rf_i}{Rf}\right)^2 + \left(\frac{P\cdot Rf_i\cdot\Delta Rf}{Rf^2}\right)^2\right)^{\frac{1}{2}} \quad (2)$$

where ΔRf_i – error of total fission rate calculated in i-layer (equivalent to three standard deviations), $Rf = \sum_i Rf_i$ – cumulative total fission rate, ΔRf – error of cumulative fission rate.

The distribution of local power P_i [KWt] is transferred into the ATHLET input file, after that the averaged of all nodes parameters are calculated: Tf_i [^{0}C], Tc_i [^{0}C], Tm_i [^{0}C], ρ_i [kg/m^3].

The convergence of the scheme is tested primarily on local power distribution and fuel temperatures, in addition on the effective neutron multiplication factor, coolant temperatures and densities. Iterative scheme ending parameter (dX_i):

$$dX_i = \frac{X_{i-1} - X_i}{X_{i-1}} \quad (3)$$

where X_i - local characteristic in i-iteration, X_{i-1} - local characteristic in (i-1) iteration. Deviation is considered relative to the previous iteration (i-1), because iterative process is aimed at specifying values. Values dX_i leading to the exit from the iteration scheme are given in Table 3. Worth mentioning that all ending parameters depend on the power error (ΔP_i) caused by Monte Carlo method error. Preliminary calculations show that the power error causes a variation of other parameters and the impact is higher for the higher power level.

Table 3. Values of the output parameters of the iterative scheme.

Functional	*dX*
K_{eff}	<0.1%
P_i	<1%
Tf_i	<0.5%
Tc_i	<0.5%
Tm_i	<0.5%
ρ_i	<0.5%

Coupled calculation scheme:

1. ATHLET: Calculation on a small power level (for example, 3% of nominal) => TF_i, TC_i, TM_i, ρ_i
2. MCU: Calculation for K_{eff}= 1 by changing the boric acid fraction => Xb_i
3. MCU: fission reaction rates calculation => Rf_i
3. Calculation of local power => P_i
4. Transfer of P_i in the ATHLET input file
5. ATHLET: Calculation of the temperature and density of the moderator => TF_i, TC_i, TM_i, ρ_i
6. Transfer of TF_i, TC_i, TM_i, ρ_i in the MCU input file
7. Return to step 1 (with the increase of power).

4. Conclusions

In this paper is given description of the test case for coupled calculation of VVER-1000 using MCU and ATHLET codes. Detailed iterative scheme is provided, the necessary accuracy of calculations by MCU is given. Iterative scheme ending parameter is discussed. The test case will be used for cross-verification of the coupled NT and TH codes, study of the convergence acceleration technics and calculation of VVER-1000.

Expected results of calculation, such as power distribution, fuel temperature, cladding temperature, coolant temperature and density distribution, must meet the criteria set by the iterative scheme ending parameter and meet the real characteristics of the existing power plant with VVER-1000 reactor. Data for the comparison of the results can be found in sources [8, 12].

References

[1] Sanchez V and Al-Hamry A 2009 Development of a coupling scheme between MCNP and COBRA-TF for the predictor of the pin power of a PWR fuel assembly *Intern. Conf. on Math., Comp. Meth. & React. Phys. (M&C 2009) American Nuclear Society* Saratoga Springs New York

[2] Ivanov A, Sanchez V and Imke U 2011 Development of a coupling scheme between MCNP5 and SUBCHANFLOW for the pin- and fuel assembly with simulation of LWR and Innovative Reactors *Intern. Conf. on Math. and Comp. Meth. Applied to Nucl. Scienc. and Eng. (M&C 2011) American Nuclear Society* Rio de Janeiro Brazil ISBN 978-85-63688-00-2

[3] Waata C L Shulenberg T Cheng X Starflinger J Bernnat W 2005 Results of a coupled neutronics

and thermal-hydraulics analysis of a HPLWR fuel assembly *Proceedings of the American Nuclear Society - International Congress on Advances in Nuclear Power Plants ICAPP'05* v 1 p 300 - 306

[4] Gurevich M I and Shkarovsky D A 2012 Neutron transport calculation by the Monte-Carlo method on the MCU code: Study guide *NRNU MEPHI* p 154

[5] Lerchl G and Austregesilo H 2011 ATHLET Mod2.2 Cycle B, User's Manual *GRS* Gorokhov A K, Dragunov U G, Lunin G L, Novikov A N, Tsofin V I and Ananiev U A 2004 Rationale of neutron-physical and radiation parts of VVER projects p 496

[6] Thermophysical properties of materials for nuclear engineering: a tutorial and collection of data 2008 *IAEA VIENNA ISBN 978–92–0–106508–7*

[7] Passage G, Mandev I, Georgieva M, Stefanova S, Kamenov K and Ivanova A Computational analysis of the VVER-1000 fuel rod thermomechanical behaviour under LOCA Conditions

[8] Vygovskiy S B, Kalish M A and Strashnysh V P Thermal engeneering reliability study of the reactor core VVER-1000 (1200) in the implementation of advanced fuel cycles using full simulation complexes

[9] Booklet of the project NPP-2006 2008 *Leningrad NPP-2 Atomenergoproekt* p 81

[10] Vygovskiy S B, Ryabov N O, Semenov A A, Chernov E V and Bogachek L N 2011 The physical and constractual especially nuclear power plants with VVER: study guide *NRNU MEPHI* p 376

[11] Savander V I, Shumskiy B E and Pinegin A A Shaping of the axial power distribution in the core to minimize the vapor volume fraction at the outlet of the VVER-1200 fuel assemblies *Probl. of At. Sci. and Techn.* ISSN: 0205-4671

Acknowledgments
This work was performed at the Department of Theoretical and Experimental Physics of Nuclear Reactors in the framework of 5-100 project. BASOV and CHERENKOV clusters of MEPhI were used for the preliminary calculations.

IOP Conf. Series: Journal of Physics: Conf. Series **781** (2017) 012051 doi:10.1088/1742-6596/781/1/012051

Non-destructive methods of control of thermo-physical properties of fuel rods

A B Kruglov[1], V B Kruglov[1], V S Kharitonov[1], P G Struchalin[2] and A G Galkin[2]

[1]Docent, National Research Nuclear University "Moscow Engeneering Physics Institute, Moscow, Russia

[2]Engeneer, National Research Nuclear University "Moscow Engeneering Physics Institute, Moscow, Russia

E-mail: pstruchalin@mail.ru

Abstract. Information about the change of thermal properties of the fuel elements needed for a successful and safe operation of the nuclear power plant. At present, the existing amount of information on the fuel thermal conductivity change and "fuel-shell" thermal resistance is insufficient. Also, there is no technique that would allow for the measurement of these properties on the non-destructive way of irradiated fuel elements. We propose a method of measuring the thermal conductivity of the fuel in the fuel element and the contact thermal resistance between the fuel and the shell without damaging the integrity of the fuel element, which is based on laser flash method. The description of the experimental setup, implementing methodology, experiments scheme. The results of test experiments on mock-ups of the fuel elements and their comparison with reference data, as well as the results of numerical modeling of thermal processes that occur during the measurement. Displaying harmonization of numerical calculation with the experimental thermograms layout shell portions of the fuel cell, confirming the correctness of the calculation model.

1. Introduction

Justification of safety and efficiency of the reactor`s cores require reliable data on the thermo-physical properties of the fuel elements and change them during the campaign. At present, such data are available for the fuel rods with the oxide nuclear fuel, but in the range of averages burnup to about 5% of heavy atoms. For new fuel types (modified uranium dioxide, dispersion, dense uranium-plutonium) data on changes in the thermal conductivity of the fuel (λ_f) and the contact thermal resistance (r_c) between the fuel and cladding during the campaign are practically absent.

Earlier studies have been carried out changes of thermal properties of nuclear fuel (λ_t and r_c) from burning out. The measurements were performed on samples obtained after opening the fuel cladding. On the thermal conductivity of the fuel pellets affect accumulated fission products produced gas pores, radiation damage, changes in the morphology of the initial process of porosity and cracking of tablets. The thermal conductivity of the contact between the fuel and cladding depends on the gap between the fuel and cladding due to swelling of the fuel cladding geometry changes under the shell and release of fission gases. These changes, occurring in a fuel rod during operation, shows that authentic data only non-destructive measurement techniques are available for thermal conductivity and thermal contact resistance.

2. Current development

Heat-conducting properties of the problem of measuring rods and solved earlier. In particular, in [1, 2] is a description of non-destructive methods of measurement of thermal properties. These procedures are based on heating the cladding current pulse duration of 0.01 seconds and registration of changes in the surface temperature of the fuel element after the pulse. Typical process times reduce the temperature naturally associated with the contact thermal resistance and thermal conductivity of the fuel, can determine λ_t and r_c in a single experiment. However, the technique of heating current pulse envelope applicable to fuel elements containing the electrically conductive nitride, carbide, metallic fuel. Non-destructive methods of investigation of fuel rods with a conductive fuel can be accomplished using a shell fragment heating fuel rod laser pulses.

3. Method local heating by laser radiation

At the Department of Thermal Physics of MEPhI the technique of implementing non-destructive measurement of thermal properties of the fuel rods, which can be applied in a protective box. Schematic diagram of the experimental setup is shown in Figure 1.

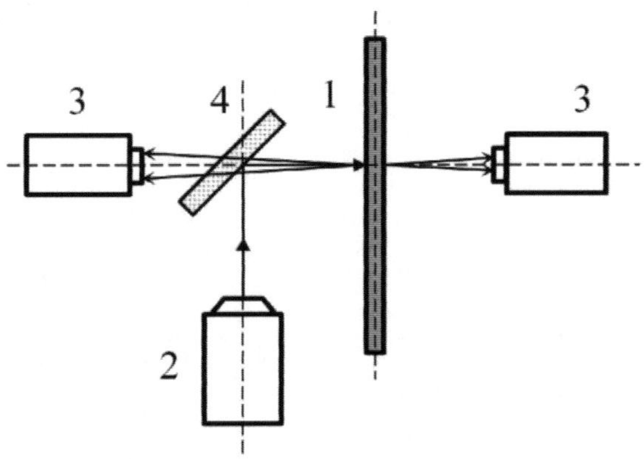

1 – fuel rod; 2 – laser; 3 – pyrometer; 4 – selective mirror

Figure 1. Schematic of the Installation.

On the surface of the fuel rod cladding given pulse of laser radiation, which heats a shell fragment. With pyrometer temperature heating is carried out during the registration point of the laser pulse duration of action 0.5 - 1 s and after its completion. You can determine the value of the contact thermal resistance r_c, defining characteristic decay time temperature heating point after the laser pulse. Fuel thermal conductivity measurement is carried out by the characteristic temperature rise time at the point opposite the point of heating. In this experiment, using heat pulses 4 - 6 seconds.

Determination r_c and λ_t by the experimental thermograms carried out using the results of numerical simulation of the thermal process in a fuel rod.

4. Test experiments

Experiments were carried out on cylindrical samples and mock-ups of the fuel elements for the verification procedure efficiency. The test samples and layouts cylindrical rods were placed in a special temperature controlled oven and inspection openings for introducing laser radiation pyrometer and recording temperature.

In the first stage, experiments were conducted on cylinders №1 and №2 of steel 321 to demonstrate the correctness of the pyrometer and the computational model. Cylinder №1 had ∅8,1 mm diameter and 40 mm of height and cylinder №2 had ∅7,4 mm diameter and 15 mm of height.

Figure 2 shows a comparison of the calculated and experimental thermograms in terms of heating and diametrically opposite point.

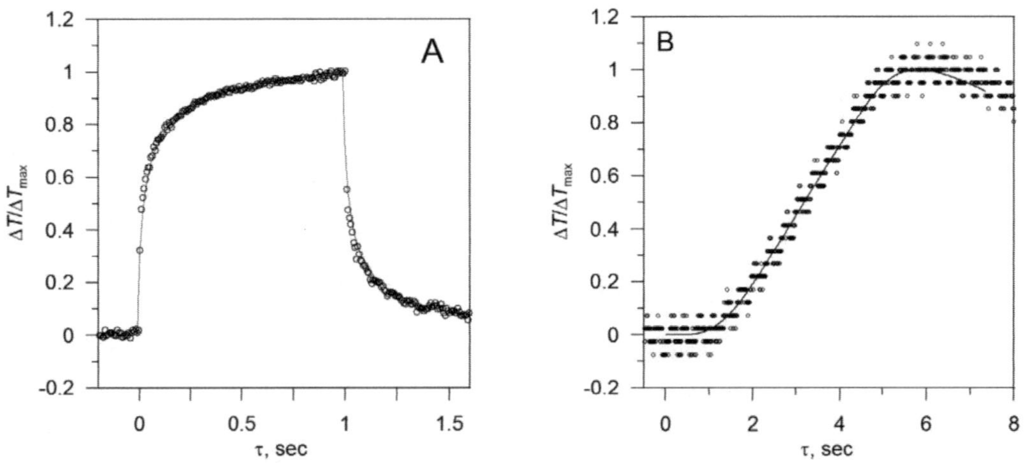

□ - experimental thermogram; ——— calculated thermogram

Figure 2. Thermograms of heating (A) and the diametrically opposite point (B) (cylinder №1).

Figure 2 shows a satisfactory agreement between the calculated and experimental thermal images, which indicates the accuracy of modelling of thermal processes in the program of a numerical calculation.

Figure 3 shows the results of measurement of the thermal conductivity by the described method on the cylinders of steel 321.

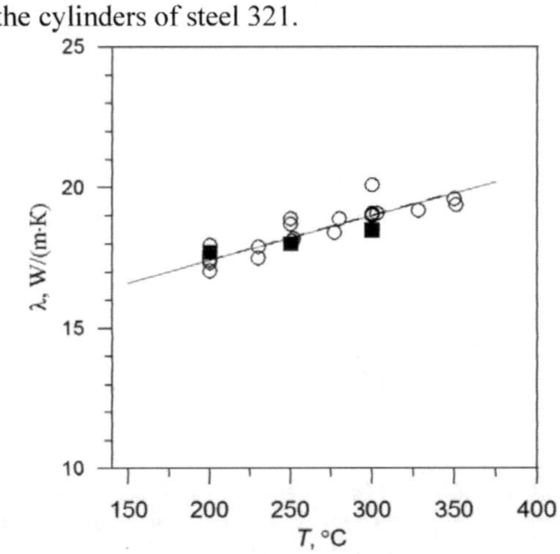

○ - cylinder ∅8,1×40 mm;
■ - cylinder ∅7,4×15 mm;
——— - reference data [3]

Figure 3. Thermal conductivity of steel 321.

The obtained data according with the reference data [3] within the 5% error.

In the second stage were measured on mock-ups of the fuel elements from the shells of steel 321, EK-164, filled cylinders of steel 321 that simulated nuclear fuel. Internal volumes of fuel rods filled layouts helium or argon pressurized 2 atmospheres. The diameters and shell thicknesses were ∅8.1×0,35 mm (st. 321) and ∅9.38×0,49 mm (EK-164). The outer diameter of the cylinder inside the sheath was made of 0.05 - 0.1 mm less than the inner diameter of the shell.

Figure 4 shows the results of measuring the thermal conductivity of the cylinder of steel 321, Figure 5 - contact thermal resistance between the cylinder and the shell.

Figure 4 shows a satisfactory match of data of thermal conductivity for steel 321 obtained in the experiment, with reference data.

Measurement error magnitude r_c does not exceed 30%, λ_t - 10%.

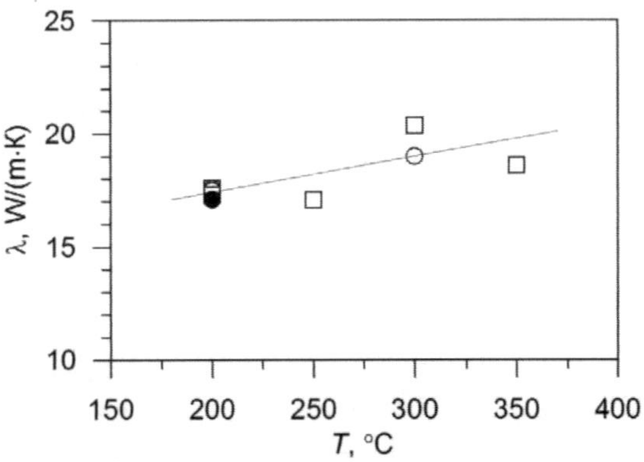

○, ● - model ∅9,38 mm, filling by He;
□ - model ∅8,1 mm, filling by He;
■ - model ∅8,1 mm, filling by Ar.
——— - reference data [3].
Figure 4 – Thermal conductivity of steel 321. Measurments on model of fuel rods.

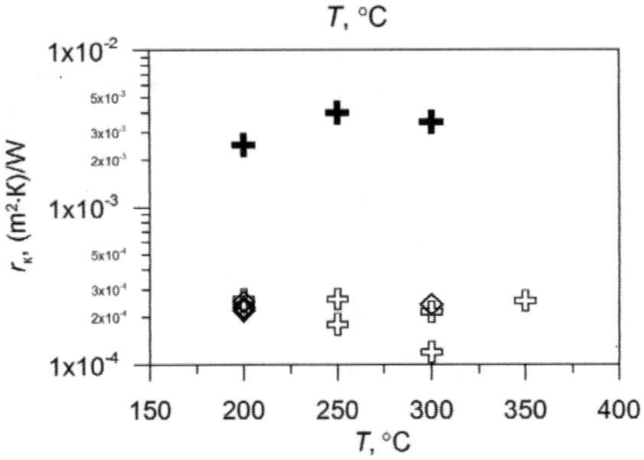

◇- model ∅9,38 mm, filling by He;
✥ - model ∅8,1 mm, filling by He;
✚ - model ∅8,1 mm, filling by Ar.
Figure 5 – Contact thermal resistance.

5. Conclusion

We propose a method of non-destructive testing of thermal properties of the fuel elements from nuclear power conductive fuel. The measurement results obtained in the model fuel rods, show quite satisfactory performance techniques.

References

[1] Korsun A., Kiselev, NP, Korsun VA, et al. Study of unsteady heat conduction process in two-layer systems under pulsed heat dissipation. Actual problems of physics and nuclear fusion reactors. M .: Energoatomizdat, 1991, p. 53 74.(In russian)

[2] A. Yu Leshchenko, Kuzmin IV Determination of thermal characteristics of SNF of VVER - 1000 by radial heating. Coll. works NIIAR, Dimitrovgrad, 2006, vol. 1, p. 33 44 (In russian)

[3] *Table of reference data. Stainless steel brand 12X18H10T. Thermal conductivity at temperatures of 340 K 1100 M .: Publishing Standards. 1994 (In russian)*

IOP Conf. Series: Journal of Physics: Conf. Series **781** (2017) 012052 doi:10.1088/1742-6596/781/1/012052

Adaptive Neural Network Algorithm for Power Control in Nuclear Power Plants

Al Masri Husam Fayiz

The Department of Automation, National Research Nuclear University MEPhI, Russia, Moscow. PhD student from the Hashemite Kingdom of Jordan

E-mail: Husam_almasri@hotmail.com

Abstract. The aim of this paper is to design, test and evaluate a prototype of an adaptive neural network algorithm for the power controlling system of a nuclear power plant. The task of power control in nuclear reactors is one of the fundamental tasks in this field. Therefore, researches are constantly conducted to ameliorate the power reactor control process. Currently, in the Department of Automation in the National Research Nuclear University (NRNU) MEPhI, numerous studies are utilizing various methodologies of artificial intelligence (expert systems, neural networks, fuzzy systems and genetic algorithms) to enhance the performance, safety, efficiency and reliability of nuclear power plants. In particular, a study of an adaptive artificial intelligent power regulator in the control systems of nuclear power reactors is being undertaken to enhance performance and to minimize the output error of the Automatic Power Controller (APC) on the grounds of a multifunctional computer analyzer (simulator) of the Water-Water Energetic Reactor known as Vodo-Vodyanoi Energetichesky Reaktor (VVER) in Russian. In this paper, a block diagram of an adaptive reactor power controller was built on the basis of an intelligent control algorithm. When implementing intelligent neural network principles, it is possible to improve the quality and dynamic of any control system in accordance with the principles of adaptive control. It is common knowledge that an adaptive control system permits adjusting the controller's parameters according to the transitions in the characteristics of the control object or external disturbances. In this project, it is demonstrated that the propitious options for an automatic power controller in nuclear power plants is a control system constructed on intelligent neural network algorithms.

1. Overview

Adaptive control, being one of the primary tasks of intelligent control principles, was created to ensure the quality of the control process in conditions of non-stationary parameters of the control object. Therefore, modern researchers use adaptive controls which are characterized by the ability to correct the parameters depending on the level of external disturbances and the current state of the control object. The process of adaptation of any technical system can be divided into two phases; firstly, to compile information about the state of the control object and secondly, to determine the control parameters of the system. In the former stage, the collection of the system information and the operational analysis regarding the status of the inputs and outputs of the control object, as well as the levels of environmental disturbances are determined. The latter stage is to define the controlling parameters by minimizing the selected quality criteria. For each adaptive control system a designer should ascertain the objectives of the system according to an adaptation strategy as a whole. However, the purpose of an adaptive control system for some difficult and dangerous objects which are under the influence of non-stationary disturbances of the environment, is extremely difficult to be determined in quantitative terms. In this case human factors in the control process of the dynamic object play a crucial role by identifying the correct solution as qualitative assessments and to retain and achieve the goal of the control algorithm expeditiously. This corroborates the effectiveness of these algorithms and the necessity of building control systems and mechanisms based on artificial intelligent principles.

The main features of intelligent control systems are the ability to learn, self-organize and change behavior but more specifically, according to the change in circumstances, by using mechanisms similar to those of the human mind which is based on how to learn and how to use the experience attained in the future [1].

Data for design, test and evaluation of the adaptive system for the APC reactor will be obtained from the multifunctional simulated computer analyzer of the VVER. The analyzer of operating nuclear power is a full scope, high fidelity and certified simulator for those types of reactors [2]. MATLAB and Neural Network Toolbox (software provided by MathWorks Inc) will be used for designing and modeling the system.

2. Introduction

A traditional approach to design any adaptive control system is to obtain the initial characteristics of the control system and subsequently perform an analysis to determine their nature. Secondly, based on the results of the analysis, a suitable adaptive strategy should be suggested to improve the quality of the regulation process.

2.1. Methodology and tools used in the project

The analysis of the adaptive control system was conducted under varied settings of the APC in the multifunctional computer analyzer of the VVER, which allows changing the complete controller's parameters for the purpose of researching and analyzing the functioning principle of the entire system. Figure 1 illustrates the interface of the automatic power controller's parameters, such as the amplification gain ***K***, the dead band ***Z*** and the external perturbations (increment value of the reactor reactivity) or disturbances ***A***.

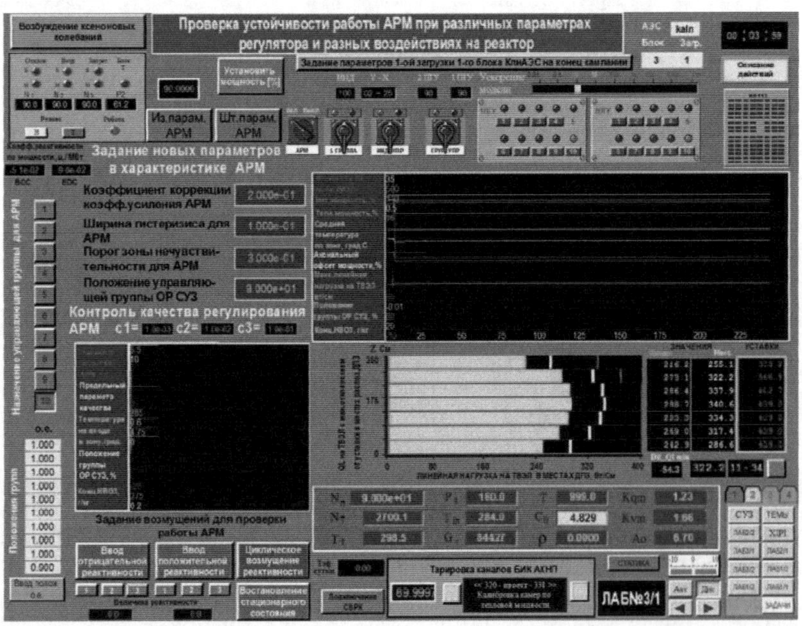

Figure 1. The interface of the automatic power controller in the multifunctional computer analyzer.

Values of steady state error ***Epsilon*** of the reactor power ***P*** were obtained at various reactor power levels with diversity of the controller's settings. This data is considered as a database for design, test and evaluation of the entire project. Table 1 illustrates part of the experimental results of the multifunctional simulated computer analyzer.

Table 1. Experimental results for the multifunctional computer analyzer of reactor VVER 1000.

P_d	*K*	*Z*	*A*	*Epsilon*	P_r
90	0.1	1	3	2.45	92.45
90	0.1	1	2	1.58	91.58
90	0.1	1	-3	-2.08	87.92
90	0.1	1	-2	-1.27	88.73
90	0.1	2	3	1.86	91.86
90	0.1	2	2	1.56	91.56
90	0.1	2	-3	-2.59	87.41
90	0.1	2	-2	-1.75	88.25
90	0.1	3	3	2.44	92.44
90	0.1	3	2	1.66	91.66
90	0.1	3	-3	-2.53	87.47
90	0.1	3	-2	-1.82	88.18
90	1	0.1	-3	-2.07	87.93
90	1	0.1	-2	-1.25	88.75
90	1	0.5	3	0.2	90.2
90	1	0.5	2	0.38	90.38
90	1	0.5	-3	-2.11	87.89
90	1	0.5	-2	-1.29	88.71
90	1	1	3	0.23	90.23
90	1	1	2	0.4	90.4
90	1	1	-3	-2.06	87.94
90	1	1	-2	-1.27	88.73
90	1	2	3	1.12	91.12
90	1	2	2	1.07	91.07
90	1	2	-3	-2.09	87.91
90	1	2	-2	-1.5	88.5

The table shows the desired power level as $\boldsymbol{P_d}$ and actual power level as $\boldsymbol{P_r}$., ***A*** is the type and nature of the disturbance which acts on the reactor (increment value of the reactor reactivity), ***K*** and ***Z*** are the automatic power controller's parameters, ***Epsilon*** shows the difference between the desired and actual or real power values. The table demonstrates part of the complete experimental results with numerous diversities of reactor power levels and controller's parameters

Analysis of the results of experiments shows that there are a sufficiently large number of cases with finite control errors (Epsilon). Figure 2 shows histograms for ***Epsilon*** under different types of disturbances, where A^{+} is a positive perturbation and A^{-} is a negative perturbation.

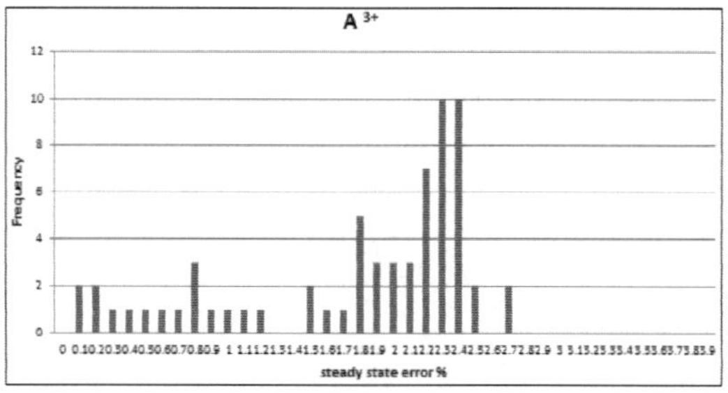

a) Histogram under disturbance A^{3+}

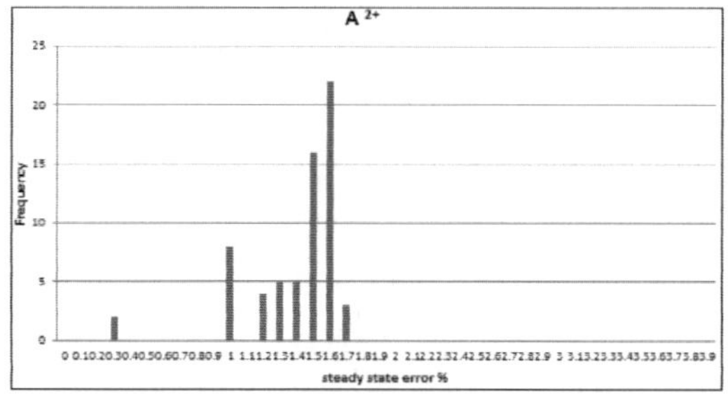

b) Histogram under disturbance A^{2+}

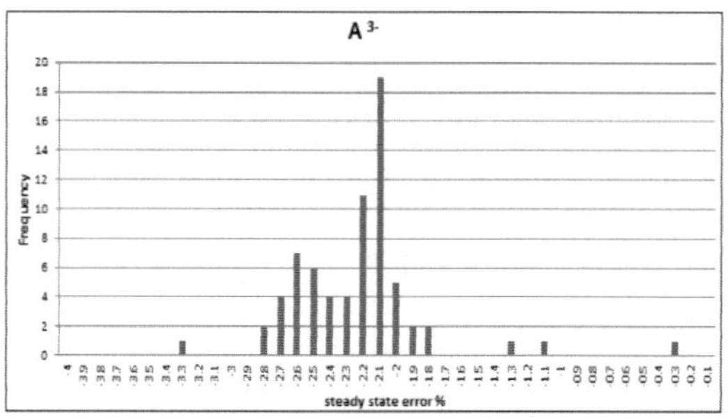

c) Histogram under disturbance A^{3-}

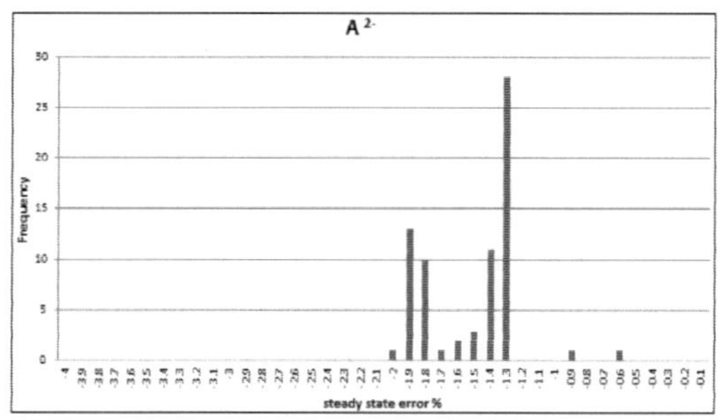

d) Histogram under disturbance A^{2-}

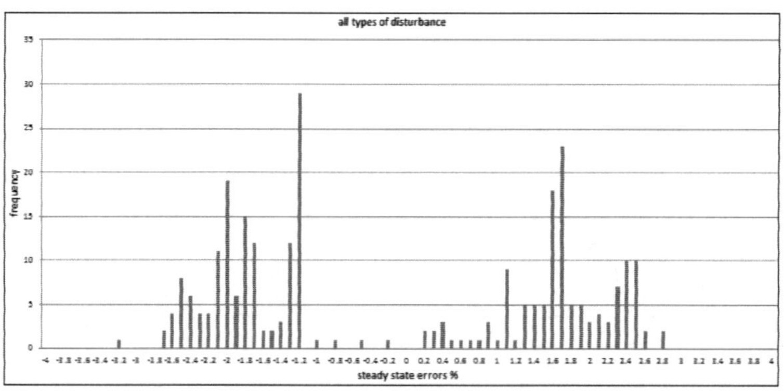

e) Histogram under all types of disturbances

Figure 2. Histogram of the experimental results for different kinds of disturbances obtained in the multifunctional computer analyzer.

2.2. Analysis of experimental results

The analysis of the experimental results shows that there are a sufficiently large number of cases with ***Epsilon***, approximately 2 – 3% of the reactor power value. To improve the quality of regulation or the control process it is paramount to decrease the steady state error .As a solution, implementation of the algorithm based on neural network techniques was offered. In order to use the implementation of neural networks as an intelligence approach, it is necessary to select the structure of the neural network (topology), form a training set and to choose the learning algorithm. The process of building, training and testing the designed model was undertaken using Neural Network Toolbox in MATLAB. Neural Network Toolbox provides functions and applications for modeling a variety of complex and non-linear systems which are problematic to be described by equations [3].

3. Design of the adaptive neural network controller

3.1. Neural network model

The aim of designing an adaptive neural network for the APC system is to improve the control quality by decreasing ***Epsilon***. Therefore, it is suggested to design an adaptive model that can give a suitable power correction value (suggested correction power value C_v) based on the history and the working principle of the APC to compensate and reduce the control steady state error.

Based on the design, a neural network model was created to study the APC behavior and to construct a database of all possible control cases which can help to make a decision about the suitable power correction value. This model will be the core to design the adaptive controller which can act to correct the power value and minimize ***Epsilon.***

The experimental data which was obtained in the APC simulator will be the base to train the neural network model. To construct such a neural network model, it is necessary to select the input and the output and to choose the network topology and structure. Figure 3 shows the general scheme for the suggested model, which illustrates the input-output relation.

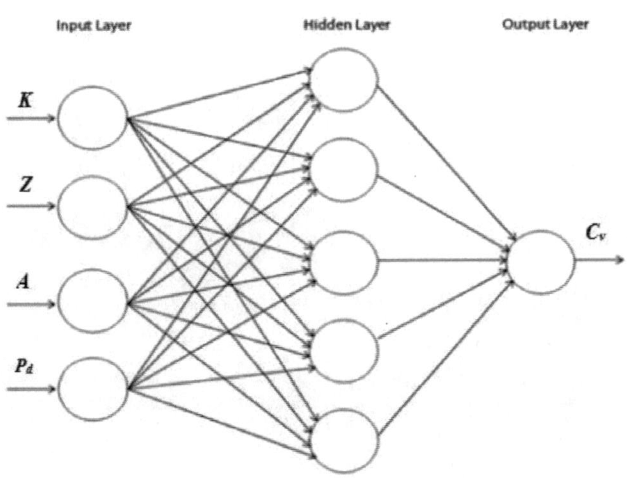

Figure 3. Structural diagram of the neural network model.

The task of this model is to determine the suggested correction power value C_v, based on the controller's current parameters K and Z, the external perturbations or disturbances A and the current power value P_d. The primary concept of the learning process of this neural network model is to provide input data which contains all possible situations and incidences including a variety of the controller's configurations, divergent types and amplitude of the disturbances. The input parameters of this model consist of the controller's parameters K and Z, the external perturbations or disturbances A and the desired power value P_d. which was determined to be the real power value P_r in the learning

IOP Conf. Series: Journal of Physics: Conf. Series **781** (2017) 012052 doi:10.1088/1742-6596/781/1/012052

process of the neural network model as illustrated in figure 3. The output of this model will be the correction power (pre-adjusted) value C_v, which is the desired power value P_d as illustrated in table 1.

A two-layer four inputs and one output feed-forward network was chosen with 24 sigmoid hidden neurons and linear output neurons (newfit), which can fit multi-dimensional mapping problems arbitrarily, given consistence data and sufficient neurons in its hidden layer. The network will be trained with Levenberg-Marquardt backpropagation algorithm (trainlm). Figure 4 shows the block diagram of the suggested model, which was designed using Neural Network Toolbox in Matlab [3].

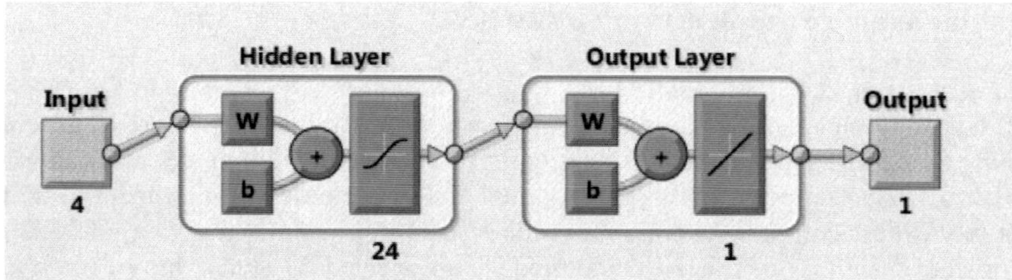

Figure 4. Block diagram of the designed model in neural network toolbox in Matlab.

Figure 5 illustrates the performance evaluation of the neural network model. In the evaluation, available experimental data obtained from the APC simulator was divided into three subsets, the first being training which was utilized for computing the gradient and updating the network weights and biases.

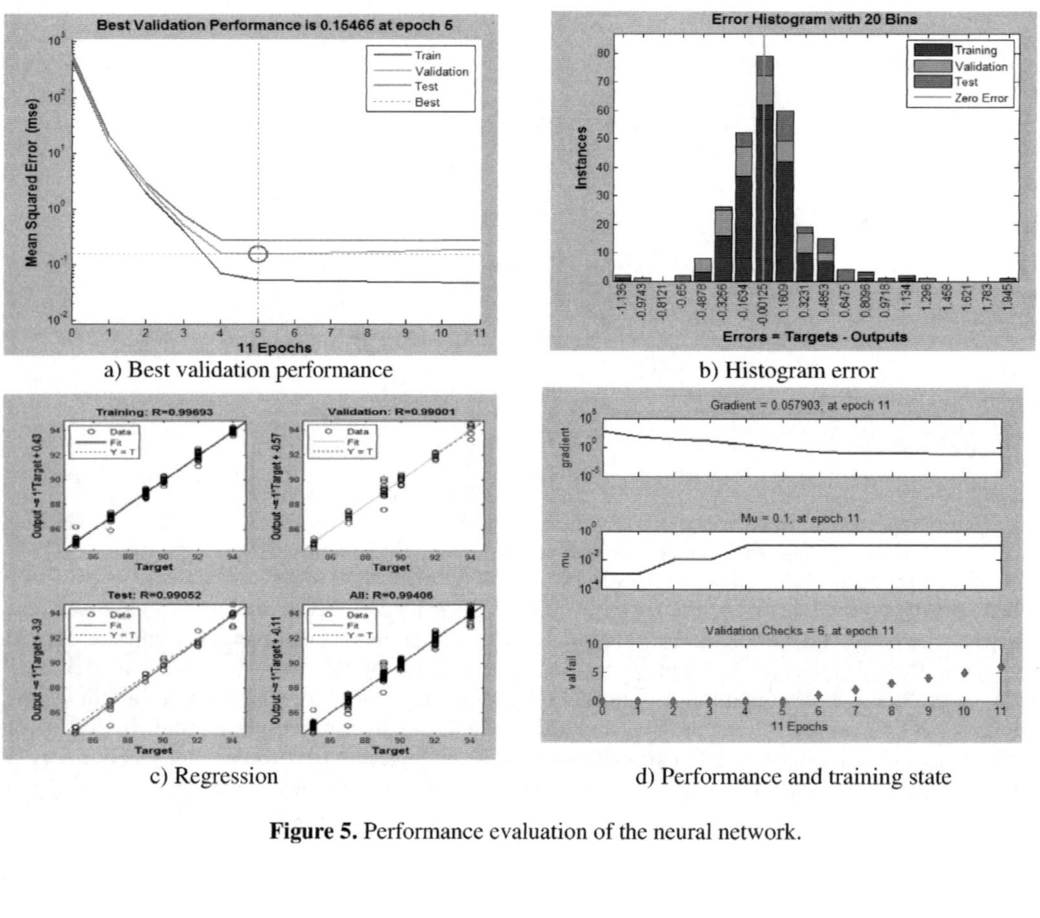

a) Best validation performance

b) Histogram error

c) Regression

d) Performance and training state

Figure 5. Performance evaluation of the neural network.

The second subset, validation, ascertained that when validation error increased for a determined number of iterations training was terminated, and the weights and biases at the minimum of the validation error were returned. The third subset, test, was utilized to substantiate the network design. Figure 5(a) illustrates the best validation performance demonstrating mean squared error of the network beginning at a significant value and decreasing to an insignificant value and plots subset errors. The distribution of subset errors in ranges of instances in the modeling process are illustrated in Figure 5(b). Figure. 5(c) shows regression plots between the network outputs and the subsets. The performance and training state of the network is demonstrated in Figure 5(d) [3].

3.2. The adaptive algorithm for the automatic power controller APC

The task of the designed adaptive power controller is to measure the current power level P_r of the reactor based on the controller's current parameters K and Z and the external perturbations or disturbances A and suggest a correction power value (pre-adjusted value) C_v based on the knowledge and experience gained by the neural network model (the adaptive algorithm), in order to compensate the error signal before entering the controller. In this way ***Epsilon*** can be minimized. Figure 6 shows a general block diagram of the system structure.

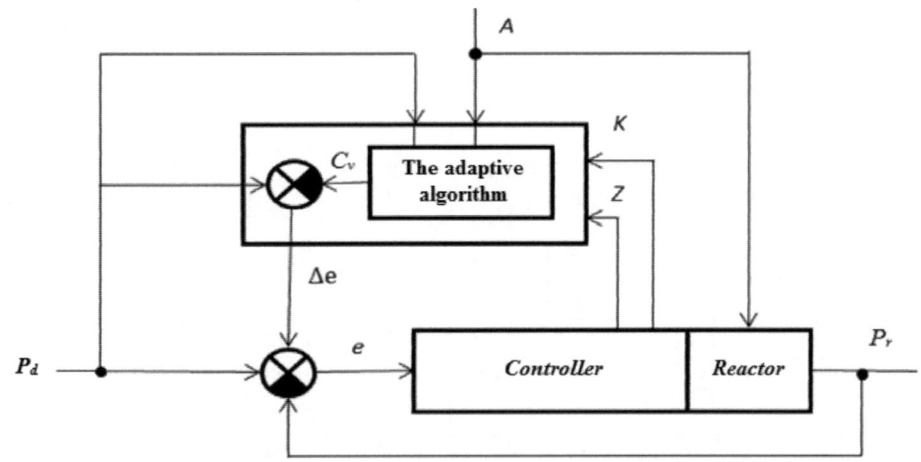

Figure 6. Block diagram of the adaptive power controller.

Using the pre-adjusted value C_v obtained from the adaptive neural network algorithm to correct the controller's input value by compensating a correction value Δe which can add or subtract a certain value from the error signal e, ***Epsilon*** can be minimized.

4. Project results

As a result of the adaptive power controller system simulation after using the adaptive values recommended by the neural network algorithm, it is evident that the final ***Epsilon*** decreased significantly compared to the classical controller without utilizing adaptive principals. A more decisive comparison of the simulation results of the adaptive neural network algorithm is shown in table 2 and figure 7. Table 2 illustrates an arbitrary chosen part of the complete results of the adaptive control algorithm based on neural networks, where P_d is the desired power level and P_r is the actual or real output power level without using an adaptive control principle. In other words the ordinary output power level using the classical APC, C_v is the recommended power level by the adaptive algorithm, P_a is the actual output power level of the control process after using the recommended power level by the adaptive algorithm, and ***Epsilon 1*** and ***Epsilon 2*** are the steady state errors before and after using the principle of adaptive algorithm.

Table 2. Evaluation results comparing the adaptive neural network power controller with the classical power controller of reactor VVER 1000.

Case study	P_d	K	Z	A	P_r	*Epsilon 1*	C_v	P_a	*Epsilon 2*
1	85	0.1	0.1	3	87.06	**2.06**	82.67	84.71	**-0.29**
2	85	1	1	3	85.8	**0.8**	84.08	85.57	**0.57**
3	85	1	2	3	87.08	**2.08**	83.69	85.76	**0.76**
4	87	0.1	1	3	87.3	**0.3**	85.48	87.32	**0.32**
5	87	1	0.5	-3	84.97	**-2.03**	89.13	87.19	**0.19**
6	87	1	2	-3	84.96	**-2.04**	89.28	87.19	**0.19**
7	87	1	10	2	88.58	**1.58**	85.31	86.98	**-0.02**
8	89	0.1	1	-2	87.73	**-1.27**	90.32	88.82	**-0.18**
9	89	1	1	3	89.83	**0.83**	88.19	89.03	**0.03**
10	89	1	2	-2	87.75	**-1.25**	90.33	89.07	**0.07**
11	89	1	10	-3	86.8	**-2.2**	91.58	89.05	**0.05**
12	90	0.1	10	3	92.45	**2.45**	87.68	89.99	**-0.01**
13	90	0.1	10	-3	87.4	**-2.6**	92.57	90.23	**0.23**
14	90	1	1	3	90.23	**0.23**	89.36	90.02	**0.02**
15	90	1	1	-3	87.94	**-2.06**	92.37	90.1	**0.1**
16	90	1	3	3	91.87	**1.87**	87.73	89.71	**-0.29**
17	90	1	3	-3	87.65	**-2.35**	92.41	90.24	**0.24**
18	90	1	10	3	92.73	**2.73**	87.21	89.62	**-0.38**
19	90	1	10	-3	87.37	**-2.63**	92.64	90.11	**0.11**
20	92	0.1	0.5	-2	90.7	**-1.3**	93.43	92.11	**0.11**
21	92	0.1	3	2	93.67	**1.67**	90.42	91.88	**-0.12**
22	92	1	0.5	-3	88.8	**-3.2**	94.38	92.2	**0.2**
23	92	1	3	3	93.86	**1.86**	89.99	91.82	**-0.18**
24	94	0.1	0.5	-2	92.67	**-1.33**	94.43	93.02	**-0.98**
25	94	1	0.5	2	95.66	**1.66**	92.76	94.18	**0.18**
26	94	1	1	3	96.25	**2.25**	92.96	94.45	**0.45**
27	94	1	3	-3	91.89	**-2.11**	96.17	93.95	**-0.05**

It is evident from table 2 that comparing ***Epsilon 1*** and ***Epsilon 2*** of the APC for both cases before and after using the adaptive algorithm, shows an obvious enhancement to the quality of the control process by decreasing ***Epsilon.*** Figure 7 illustrates a clearer picture of the comparison between the two cases for APC of the reactor.

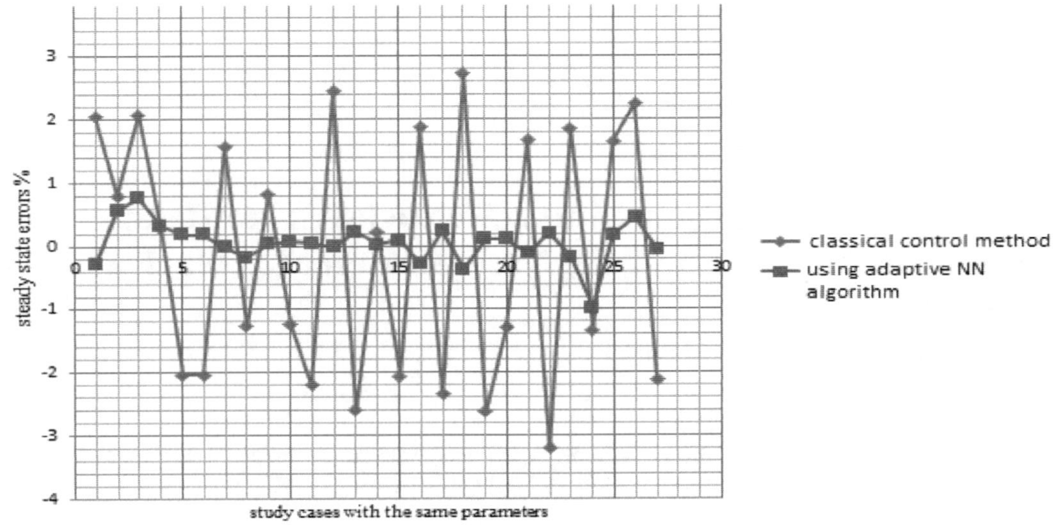

Figure.7. Comparative graph of the steady state error of the reactor power with and without using neural network algorithm.

5. Conclusions

The analysis of the project results shows that using adaptive control algorithms for the automatic power controller based on neural network principles can offer a significant improvement to the quality of the control system in comparison with conventional or classical control methods.

References

[1] Gavrilova T.A, Khoroshivsky V.F. Bazy znanij intellektual'nyh sistem - SPb.: Piter, 2000 (in Russian).
[2] Vaykovcke S.A., Karalov S.A., Chernov E.V. ychobnaya labratornaya na baze mnogophunkcanalna analzatora reaktorne uctanovak AES VVER [EDUCATIONAL LABORATORIES BASED ON MULTIFUNCTION ANALYZER THE REACTOR VVER] [VESTNIK NIYAU MIFI], 2012. V. 1, № 1, p. 104-110. (in Russian)
[3] Neural Network Toolbox User's Guide © COPYRIGHT 1992 - 2009 by The MathWorks, Inc

Experience in implementation of «Nuclear Knowledge Management» course at the National Research Nuclear University MEPhI

N I Geraskin and A N Kosilov

Department of theoretical and experimental physics of nuclear reactors, National Research Nuclear University MEPhI (Moscow Engineering Physics Institute), Kashirskoe highway, 31, Moscow, Russia

E-mail: NIGeraskin@mephi.ru

Abstract. This paper describes the experience of teaching «Nuclear Knowledge Management» course at the National Research Nuclear University MEPhI (NRNU MEPhI). Currently, the course is implemented both in engineer and master degree programs and is attended by over 50 students. Goal, objectives and syllabus of the course are discussed in detail. A special attention is paid to practical exercises and final examination options in the case of small and large student groups. The course is supported by the Cyber Learning Platform for Nuclear Education and Training (CLP4NET), developed by the IAEA. The experience of NRNU MEPhI lecturers assisting in conducting the International School of Nuclear Knowledge Management, held annually in Trieste (Italy), is described with a special attention to the fact, that the course has passed the certification process at Academic Council of NRNU MEPhI. In 2014 and 2015 the course has been recognized as one of the best ones in NRNU MEPhI. Finally, perspectives of «Nuclear Knowledge Management» course are considered. They include increase of the course duration, introduction of the course into the learning process of other departments and institutions of the university, and transferring the course to other members of the Association «Consortium of ROSATOM supporting universities».

1. Introduction

International Atomic Energy Agency (IAEA) in collaboration with leading nuclear universities aims at establishing International Nuclear Management Academy (INMA). The purpose of INMA is to create and implement, in collaboration with the members of the academy, the master degree program in the field of nuclear technology management in compliance with IAEA standards, as well as to develop uniform requirements for master degree programs with an emphasis on modern aspects of management in the field of nuclear science and technology.

At IAEA General Conference held in September 2015 an agreement has been signed, under which National Research Nuclear University MEPhI (NRNU MEPhI) becomes an official member of INMA. It is assumed that a new master degree program on nuclear technology management will be implemented in NRNU MEPhI in September 2016. This new program is highly anticipated as currently there is no uniform international program that takes into account the peculiarities of highly qualified managers training in technical universities, and such professionals are required for the successful development of nuclear energy.

An important part of the new master degree program on nuclear technology management is the course «Nuclear Knowledge Management» (NKM), which is already being implemented at the department of theoretical and experimental physics of nuclear reactors since 2012. The experience of teaching this course should be used in the development of a new master degree program in NRNU MEPhI. The present paper summarizes this experience.

2. General information about the course «Nuclear Knowledge Management», implemented in NRNU MEPhI

The goal of the course «Nuclear Knowledge Management» is studying main approaches, methods and tools of knowledge management (KM) as a key resource in the enterprises of the nuclear industry.

Currently, the course is implemented both in engineer and master degree programs: ninth semester of engineer degree program (educational direction 14.05.01 «Nuclear reactors and materials»); first semester of master degree program (educational direction 14.04.01 «Nuclear power and thermal physics» and 14.04.02 «Nuclear physics and technology»).

In 2015 the course was attended by 58 students. The geographical scope is wide enough: there were students not only from Russia, but also from other CIS countries and far abroad.

Objectives of the course include studying issues of effective KM, aimed at ensuring and improving safety, reliability and efficiency of nuclear facilities operation with a view of their long-term competitiveness and sustainability by means of using knowledge resource.

According to IAEA recommendations the course consists of eight lectures and practical exercises [1]. The course plan is presented in table 1.

Table 1. «Nuclear Knowledge Management» course plan, implemented at the NRNU MEPhI.

Title	Main topics
Knowledge Concepts and Definitions	What is Knowledge? Knowledge as a Resource Nuclear Organization and Nuclear Knowledge
KM History and Development	KM History, Definition and Models KM Elements, Importance, Benefits, Tools and Techniques KM Approaches
KM in Nuclear Science & Technology	Nuclear Knowledge & Stakeholders Nuclear Knowledge Management Knowledge Loss Risk Assessment
Managing Tacit Knowledge	Tacit Knowledge Techniques to identify, capture and transfer tacit knowledge
Managing Nuclear Information	History of Nuclear Information Methods and Tools for Information Management Knowledge Organization Systems
NKM Organizational Challenges	Integration of NKM in Organizational Management System Knowledge-sharing Culture Set Up, Implementation and Assessment of NKM Projects
Implementing KM in Nuclear Organizations	KM in Nuclear Power Plant KM in R&D Organization KM in University
NKM Maturity Assessment	KM Assessment: Elements and Steps IAEA Self-Assessment Model

During practical exercises the students acquire skills how to preserve critical knowledge and build concept knowledge maps.

IOP Conf. Series: Journal of Physics: Conf. Series **781** (2017) 012053 doi:10.1088/1742-6596/781/1/012053

The goal of the first practical exercise is to assess risk of knowledge loss on the basis of interview data and to propose measures for critical knowledge preservation [2, 3]. Students work in small groups of 3-5 people and answer the following questions using interview data: what knowledge is critical, what are the consequences of losing this knowledge, and how to mitigate the negative effects of losing this knowledge. After completing the task one of the groups presents their results, while other groups may discuss the presentation.

The goal of the second practical exercise is to build a concept map for some area of knowledge, for example, nuclear reactors designing. For this exercise students use free specialized program CmapTools (available at http://cmap.ihmc.us).

The final examination is an essay. The following topics are possible:

- Methods of explicit and tacit knowledge management.
- Tacit knowledge preservation using concept map technique.
- NKM in research institutes / NPP.

Authors of the best essays present their work as an oral report at the International School-seminar «Nuclear Energy and Non-proliferation: answer for today challenges», traditionally held by the department «Theoretical and Experimental Physics of Nuclear Reactors» in January-February.

However, essay is not an appropriate solution for large student groups. That's why we proposed to use projects in autumn 2015. Each group of 4-6 students has developed their own project. Project's topics are as follows:

- Implementation of knowledge management program in NPP / nuclear regulatory body / nuclear research and development organization.
- Knowledge map development for the leading staff of department of theoretical and experimental physics of nuclear reactors (NRNU MEPhI).
- Knowledge portal implementation for department of theoretical and experimental physics of nuclear reactors (NRNU MEPhI).
- Taxonomy development for course project on nuclear reactors designing at department of theoretical and experimental physics of nuclear reactors (NRNU MEPhI).

Practical significance and good understanding of the projects (many students had course project on nuclear reactors designing in one of the previous semesters) ensured high interest of students and great quality of presentations.

On the basis of experience of using e-learning/distance training systems the course «Nuclear Knowledge Management» is realized on Cyber Learning Platform for Nuclear Education and Training (CLP4NET), developed by IAEA [4]. Advantages of the CLP4NET platform are support by IAEA, no fee for IAEA members, as well as multi-language interface. NRNU MEPhI is the point of Russian localization of the platform supporting national and international educational activities and trainings.

The CLP4NET platform can be used for development, management and dissemination of the educational on-line materials in the operation mode of joint access. The educational materials may be created in visual environment with definition of the learning sequence. Various forms of individual tasks, projects for small teams, any educational elements (including both cognitive and communicative components) can be placed on the CLP4NET platform for all trainees. The CLP4NET platform is characterized by high level of interactivity and, thus, the CLP4NET platform allows the trainees from other countries to participate in the educational process through the Internet system.

It should be noted that some lectures of the course are conducted through Internet by means of video conference technology. For the convenience of students user manual to CLP4NET platform has been developed. User manual explains the main issues (access to the platform, access to the courses, passing tests), is detailed (74 pages) and contains many illustrations.

On the basis of many years of experience in nuclear knowledge management issues in year of 2015 the first-ever in Russia textbook «Knowledge Management in Nuclear Organizations» has been published [5]. The textbook is intended for students and PhD students specializing in the field of nuclear technology, as well as for employees of nuclear industry enterprises. It is expected that the textbook will be published also in English.

3. International School of Nuclear Knowledge Management

Teachers of the course «Nuclear Knowledge Management» serve as lecturers and consultants at the International School of Nuclear Knowledge Management, held annually in Trieste (Italy). The School is organized jointly by The Abdus Salam International Centre for Theoretical Physics and IAEA. The School is very popular: in 2015 only one fifth from more than 200 submitted applications was accepted. The participants represented 28 countries.

The School consists of distance training using CLP4NET platform (during which students learn the basics of knowledge management), as well as full-time tuition in Trieste. The School program, covering 12 modules including implementation and defence of projects, has passed the certification process as an advanced training course for nuclear industry specialists at the Academic Council of the NRNU MEPhI. In this regard, the participants of the International School are awarded official advanced training certificates of the NRNU MEPhI.

The IAEA Director General Yukiya Amano, in his address to the participants of the School, noted the importance of effective knowledge management, aimed at increasing long-term competitiveness and sustainability by means of using a knowledge resource for high performance and safe operation of nuclear facilities, and praised the initiative of NRNU MEPhI on certification of the course «Nuclear Knowledge Management» as an advanced training course for nuclear industry specialists.

4. Perspectives and recognition of «Nuclear Knowledge Management» course

Experience of teaching the course «Nuclear Knowledge Management» at the department of theoretical and experimental physics of nuclear reactors allowed us to organize International Workshop on nuclear knowledge management «Formation of competences in the field of nuclear knowledge management within the frames of university educational programs» in NRNU MEPhI. The aim of the workshop was acquaintance of the experts of ROSATOM supporting universities with the modern techniques of intellectual capital management, possibilities of using various information resources in the process of nuclear specialists training, as well as participation in practical exercises on nuclear knowledge management.

Based on the experience of teaching the course «Nuclear Knowledge Management» in 2012-2015 one can identify the main directions of its development, partially already implemented:

- Increase in course duration from the initial 16 classroom hours to 20 classroom hours at present.
- Introduction of the course into the curricula of engineer degree programs in the autumn of 2015 that allows us to cover a wide audience.
- Possible introduction of the course into the learning process of other departments and institutions of NRNU MEPhI (department of automation and Obninsk Institute for Nuclear Power Engineering are considered foremost).
- Experience transfer to other members of the Association «Consortium of ROSATOM supporting universities».

As recognition, it should be noted that the course «Nuclear Knowledge Management» was a winner of the contest «Young Lecturer of NRNU MEPhI 2014». The course was one of 78 best courses submitted to the contest by young lecturers (under 35 years of age) from all NRNU MEPhI branches. As a result the course materials should be available on NRNU MEPhI educational portal http://online.mephi.ru/.

The course «Nuclear Knowledge Management» was also a winner of the contest «Young Lecturer of NRNU MEPhI 2015». This time the selection process was more rigorous (only 57 best courses were selected). The aim of this contest is to create video-lectures in English, which will be available as a massive open online course on leading world educational and training platforms such as EdX and Coursera.

References

[1] International Atomic Energy Agency 2016 *Knowledge Management and Its Implementation in*

Nuclear Organizations (IAEA Publications STI/PUB/1724) p 52
[2] International Atomic Energy Agency 2006 *Risk Management of Knowledge Loss in Nuclear Industry Organizations* (IAEA Publications STI/PUB/1248) p 41
[3] International Atomic Energy Agency 2011 *Comparative Analysis of Methods and Tools for Nuclear Knowledge Preservation* (IAEA Publications STI/PUB/1494) p 102
[4] NKM e-Bulletin September 2014. URL: http://www.iaea.org/nuclearenergy/nuclearknowledge/NKM_NEWS/e-bulletin/2014-03-e-Bulletin-Septl.pdf
[5] Geraskin N I, Kosilov A N, Kulikov E G and Tolstenkov A N 2015 *Knowledge Management in Nuclear Organizations: Textbook* (Moscow: NRNU MEPhI) p 408

Experience and advantages in implementation of educational program in network form at Department «Closed nuclear fuel cycle Technologies» of National Research Nuclear University «MEPhI»

A G Beygel'[1], K V Kutsenko[2], A A Lavrukhin[2], E P Magomedbekov[3], V A Pershukov[2,4], V L Sofronov[2,5], E A Tyupina[2,3] and A N Zhiganov[2,5]

[1] Stock Company «Siberian Group of Chemical Enterprises», 636039, Russia, Siberian Federal District, Tomsk region, Seversk, st. Kurchatov, 1
[2] National Research Nuclear University «MEPhI», 115409, Russia, Moscow, Kashirskoe highway, 31
[3] D.Mendeleev University of Chemical Technology of Russia, 125047, Russia, Moscow, Miusskaya sq., 9
[4] ROSATOM State Atomic Energy Corporation, 119017, Russia, Moscow, st. Bolshaya Ordynka, 24
[5] Seversk Technological Institute of National Research Nuclear University «MEPhI», 636036, Russia, Siberian Federal District, Tomsk region, Seversk, pr. Kommunistichesky, 65

E-mail: aalavrukhin@mephi.ru

Abstract. The experience of implementation of the basic educational program of magistracy on direction «Nuclear Physics and Technologies» in a network form is presented. Examples of joint implementation of the educational process with employers organizations, other universities and intranet mobility of students are given.

1. Introduction

The interaction with key employers, requiring a high and multidisciplinary level of training of graduating students, is an important component of the educational process. For students training usually it takes a form of attracting the external specialists and concluding the agreements on student research work, practice and final qualifying works at the enterprises of industry.

The evolution of educational standards of higher education has been steadily moving in the direction of an increasing account of requirements of professional industry standards, and of employers' requirements. The Federal Law «On Education in the Russian Federation» (№ 273 - FZ of December 29, 2012) [1] has introduced the possibility of implementing the educational programs in a network form. The requirements for the educational program, which is realized in a network form, are provided with a set of resources granted by the organizations involved in its implementation. Such organizations may be geographically distributed units of the University (the intranet mobility), other educational organizations, as well as other (non-educational) organizations.

The ability to consolidate resources of various organizations in implementation of the educational program allows us to expand the range of educational services and to train personnel in accordance with the programs of development of enterprises and industries, as well as in accordance with international educational cooperation programs.

2. Educational Master Program «Nuclear physics and technologies», program name «Nuclear technologies of new generation»

The department «Closed nuclear fuel cycle technologies» of National Research Nuclear University «MEPhI» (MEPhi) was founded in 2014. The main purpose of foundation of the department and its educational program is a target training for the organizations of the «Proryv project» in the framework of the Federal target program «Nuclear power technologies of a new generation for 2010-2015 and for the future till 2020» [2] and the «Innovation Development Program of Rosatom» [3]. The objects of professional activity include nuclear power technologies of new generation based on fast neutron reactors (BN, BREST) with closed nuclear fuel cycle for nuclear power plants to ensure the country's needs for energy and more efficient use of natural uranium and spent nuclear fuel.

In accordance with the objective of training the following system of courses was developed:

- the basic courses of educational standard and some classic courses of MEPhI;
- the special general courses on closed nuclear fuel cycle technologies (CNFC), that are further being developed in special courses with the deepening on directions of CNFC;
- teaching in the sites of practical training in accordance with the themes of master's works and with specialization of responsibility centers of the «Proryv project».

To read the special courses the Department actively engages external specialists of such organizations as the A.A. Bochvar All-Russia Scientific Research Institute of Inorganic Materials (VNIINM), Leading Research Institute of Chemical Technology (VNIIKhT), State Research Center of Russian Federation - Institute of Physics and Power Engineering (IPPE), Innovation and Technology Center for Proryv Project, Nuclear Safety Institute of the Russian Academy of Sciences, National Research Center «Kurchatov Institute», State Atomic Energy Corporation ROSATOM.

However, for effective implementation of the educational program it was necessary to attract not only the staff, but also the material resources of partner organizations. For that the relevant agreements on the network implementation of the educational program have been concluded. As an example, consider a block of disciplines that for convenience can be called «Radiochemistry and nuclear fuel cycle» (Fig. 1).

Implementation of the block of disciplines lasts for three semesters with the increase of amount. The course «Radiochemistry» is given in MEPhI by lecturers of D. Mendeleev University of Chemical

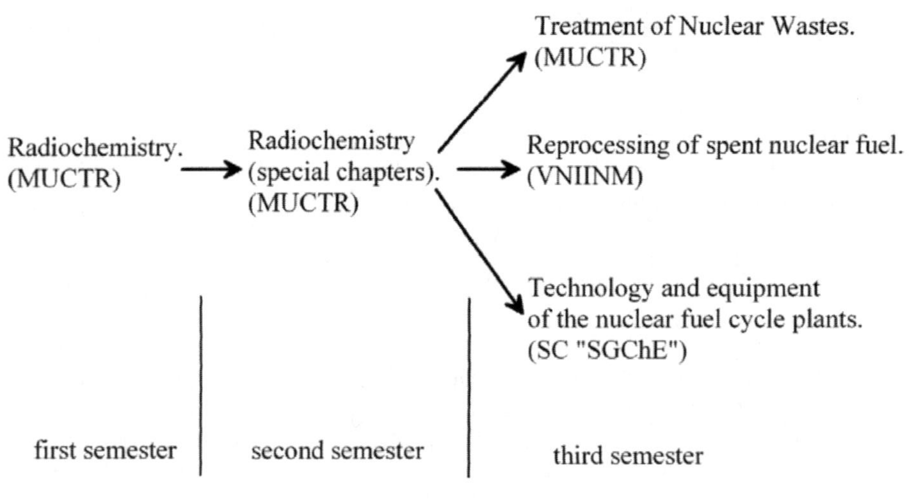

Figure 1. The structure of the block of disciplines «Radiochemistry and nuclear fuel cycle».

Technology of Russia (MUCTR). The discipline «Radiochemistry» is dedicated to acquaint students with the peculiarities of behaviour of radioactive isotopes in ultra small concentrations in solution, gas and solid phase, their distribution between the phases in the co-precipitation process, adsorption, ion and isotope exchange, electrochemistry, which is associated with the processing technology of spent nuclear fuel, radioactive waste and analysis of the quality of coolant in nuclear power plants. The discipline is levelling the incoming students, the account of the course is a two hour lecture per week.

The knowledge gained by students in the study of the course «Radiochemistry» is in demand later for the course «Radiochemistry (special chapters)», which is delivered on the basis of MUCTR in the amount of five classroom hours per week, including laboratory work. Obtained in the performance labs the ability and skills of radioactive drugs, the safety basis for work with open sources of radiation are the basic and necessary in the future to prevent a variety of emergency situations involving possible radioactive contamination of the atmosphere laboratory, clothing, skin, equipment at work at nuclear facilities. Complex laboratory works included in the curriculum will allow students to practice to get acquainted with the specifics of work in the radiochemical laboratory, learn radioactive drugs and radiometric equipment to determine the entire spectrum of radiation, and experimental reinforce the knowledge gained in lectures, in qualitative and quantitative radiochemical analysis, and use of radioactive tracers to solve various analytical problems [4]. The purpose of discipline «Radiochemistry (special chapters)» is the study of behaviour radioactive isotopes in the technologies of the nuclear fuel cycle, the methods of analysis, concentration and separation [5].

Three disciplines are already given in the block «Radiochemistry and nuclear fuel cycle» during the third semester. The course «Treatment of Nuclear Wastes» [6] is lectured on the basis of MUCTR in the amount of five classroom hours per week. The purpose of the discipline is to introduce students to technologies of management with radioactive waste, formed as a result of working with radionuclides. At present the possibility of practical training in the framework of the course on the basis of VNIIKhT is under consideration.

The course «Reprocessing of spent nuclear fuel» is given by leading experts of VNIINM on the basis of this organization in the amount of four classroom hours per week. The purpose of the discipline is to prepare students for the engineering and chemical-technological problems related to development of reprocessing technologies of spent nuclear fuel and equipment for their implementation [7, 8].

The skills, obtained in the first two semesters, are developed during mastering the course «Technology and equipment of enterprises of nuclear fuel cycle» [9, 10]. At present the course is developed as follows. Students leave for Seversk (Seversk Technological Institute of MEPhI) where they undergo theoretical training for the first week six hours per day. The second week is devoted to a familiarization practice at Stock Company «Siberian Group of Chemical Enterprises» (SC «SGChE») with a visit to the radiochemical plant, sublimation plant, isotope separation plant, industrial reactor and chemical-metallurgical plant. Construction of a pilot demonstrational power complex is being carried out just at the site of SC «SGChE». The complex consists of a fast reactor installation «BREST-OD-300» with in-situ closed nuclear fuel cycle, including the production of uranium-plutonium nitride fuel (fuel fabrication), as well as the processing of irradiated nuclear fuel to produce a secondary uranium-plutonium nitride fuel (re-fabrication of fuel). Students can get acquainted with the construction process and, at the moment, with the technology of fuel production and the fuel assemblies from nitride fuel. It is planned to move the course to a remote platform CLP4NET in the amount of four hours per week with a weekly full-time practice at the SC «SGChE».

3. Conclusion

Various forms of network implementation of the educational program are applied for the realization of block of disciplines «Radiochemistry and nuclear fuel cycle» of the basic educational program of Master's in the direction «Nuclear Physics and Technology» on the profile «Nuclear power technologies of new generation». National Research Nuclear University «MEPhI» is a basic educational organization, D. Mendeleev University of Chemical Technology of Russia is an educational organization and a partner, A.A. Bochvar All-Russia Scientific Research Institute of Inorganic Materials and Stock Company «Siberian Group of Chemical Enterprises» are not educational organizations, but they are partners, Seversk Technological Institute of MEPhI is a division of MEPhI (intranet mobility). These joint actions have allowed these organizations to form a unique block of disciplines, aimed at the comprehensive training of graduating students, that it would be impossible to implement on the basis of a single organization.

References

[1] 2012 Federal Law *On Education in the Russian Federation (with changes and additions) № 273-FZ* (System GARANT: http://base.garant.ru/70291362/#ixzz3mSbal7bQ)

[2] 2010 Federal target program *Nuclear power technologies of a new generation for 2010-2015 and for the future till 2020* (Federal target programs of Russia: http://fcp.economy.gov.ru/cgi-bin/cis/fcp.cgi/Fcp/ViewFcp/View/2010/298)

[3] 2016 *Innovative Development Program Passport and technological modernization of the State Corporation "Rosatom" for the period till 2030* (ROSATOM State Atomic Energy Corporation: http://www.rosatom.ru/about/innovatsionnaya-deyatelnost/)

[4] Ochkin A V, Tyupina E A 2009 *Laboratory work on radiochemistry: study guide* (Moscow: D. Mendeleyev University of Chemical Technology of Russia)

[5] Gelis V M, Magomedbekov E P, Ochkin A V 2014 *Chemistry of radionuclides: study guide* (Moscow: D. Mendeleyev University of Chemical Technology of Russia)

[6] Obruchikov A V, Tyupina E A 2014 *Radioactive waste handling: study guide* (Moscow: D. Mendeleyev University of Chemical Technology of Russia)

[7] Zemlyanukhin V N, Ilienko E N, Kondratyev A N 1989 *Reprocessing of nuclear fuel nuclear power plants* (Moscow: Energoatomizdat)

[8] Kopyrin A A, Karelin A I, Karelin V A 2006 *Production technology and radiochemistry reprocessing of nuclear fuel: study guide* (Moscow: ZAO "Publishing Atomenergoizdat")

[9] Turaev N S, Zherin I I 2005 *Chemistry and technology of uranium: study guide* (Moscow: TSNIIATOMINFORM)

[10] Rakov E G, Khaustov S V 1993 *Processes and devices of production of radioactive and rare metals: study guide* (Moscow: Metallurgy)

Acknowledgement

Study was partially supported by the project of Russian Scientific Foundation (RSF №16-17-10270).

Computer modeling and simulators as part of university training for NPP operating personnel

M. Volman

Ivanovo State Power Engineering University, Ivanovo, Russia

E-mail: maria_volman@mail.ru

Abstract. This paper considers aspects of a program for training future nuclear power plant personnel developed by the NPP Department of Ivanovo State Power Engineering University. Computer modeling is used for numerical experiments on the kinetics of nuclear reactors in Mathcad. Simulation modeling is carried out on the computer and full-scale simulator of water-cooled power reactor for the simulation of neutron-physical reactor measurements and the start-up - shutdown process.

1. Introduction

Russian nuclear power industry is developing actively, on a very large scale: by 2030 it is intended to more than double electric power generation from nuclear power plants [1]. Among the preconditions for fulfilling these plans anticipative training of personnel engaged in operating nuclear power plants' power units. Furthermore, highly trained personnel servicing nuclear facilities are a major factor guaranteeing their reliable and safe operation. It has been variously estimated that 15–40% of all accidents and 20–80% of all case of malfunction at nuclear power plants are attributable to the operators concerned [2]. The effective work of higher educational institutions should be directed towards solving the above problems. Computer technologies are commonly known to be highly effective in university training of future nuclear power plant personnel [3 – 6]. We offer a unified set of training programs intended to intensify university training of future senior reactor control engineers. The subject of our study comprised water-cooled power reactors as the most widespread worldwide.

2. Set of training programs

The set of training programs is based on a competency approach, which presupposes that the results of education should be evaluated not by the sum total of acquired knowledge, but according to the graduate's competence, defining his or her ability to apply knowledge and skills in successful professional activity as a result of profound understanding of the physics of the process under examination. This is achieved by integrating the use of computer and imitative simulation within a single conceptual framework.

The set of training programs consists of three parts. The first includes a laboratory-based computer practical for water-cooled reactor kinetics that we developed on the basis of numeric experiments [7, 8]. The second part of the training programs set is devoted to simulation of neutron reactor measurements [9]. The third part provides procedural guidelines for simulation of generating unit startup and shutdown on computer and full-scale simulators for nuclear power plants using water-cooled reactors [10].

2.1. Numerical experiments on the kinetics of nuclear reactors

An in-depth study into neutron processes is based on numerical experiments conducted as part of an in-house computer software package. Mathematical models encompass the cold and hot reactor kinetics. They make it possible to analyze the role of delayed neutrons, the fuel and coolant temperature effects on the nature of transients, fuel burnup, the reactor xenon and samarium poisoning, as well as xenon oscillations and the reactor xenon stability. Mathematical models are based on a system of stiff nonlinear differential equations integrated using the respective Mathcad algorithms. The use of this environment makes the process of modeling highly graphic and convenient for numerical experimentation while allowing students to manipulate with different input parameters, assess their role and undertake a research of their own. When transients are modeled, attention is given to the problem of closing the system of differential equations, which is not just a mathematical problem but is also of a practical importance allowing conditions to be formulated for analyzing the operation of interacting components of the unit's process circuit. Besides, students learn to be able to consider interlinked transients in components not only based on numerical experiments but also using a qualitative analysis of differential equations without solving them.

For illustration, we shall consider a mathematical model for excitation and suppression of axial xenon oscillations in a nuclear reactor allowing students to explore this challenging and operationally important problem. Numerical experiments to model xenon oscillations are based on a system of two similar coupled reactors with an equal energy release and contacting ends. This system of reactors is a model of one reactor divided vertically into two identical halves. The exchange of neutron fluxes between these reactors is through the neutron leakage via the contacting ends. Xenon oscillations between the reactor's upper and lower halves are excited through the excitation of one reactor half by an oscillation introduced into it in the form of a reactivity leap, while the suppression of the introduced oscillation is modeled by boron regulation acting on both reactor halves. The consideration is based on a "pointwise" model of the reactor in a two-temperature approximation for fuel and coolant with the xenon birth and death equations added to it. Since the transient is slow, no delayed neutrons affect it in any way, and these may be neglected whatsoever, that is, all neutrons may be treated as instantaneous or a single-group approximation may be used. In the investigated model, the axial xenon oscillations were suppressed through a temperature reactivity effect and by reducing the time of the neutron flux exchange between the coupled reactors. Based on the presented numerical experiments, curves are plotted by students to define the boundary between the xenon stability and the reactor instability. Oscillations are growing in the xenon instability region and fading in the stability region. The transition from one region to the other is characterized by sustained oscillations.

2.2. Simulation of reactor measurements

The second component deals with the reactor neutron measurements based on a full-scale simulator and a computerized functional analytical simulator.

The simulator complexes are based on mathematical simulation of neutronic, thermal-mechanical, thermal and other processes and the resulting creation of a generating unit functioning model for various real-time operating modes. The analytical simulator is developed on a PC using 3KeyMaster™ software from Western Services LLC [11]. The extensive capabilities of the analytical simulator permit experiments above the reactive core and simulation of normal and emergency situations, while the mobility and convenience of using personal computers facilitates effective use of analytical simulator during training. 3KeyMaster™ is a modeling platform, multi-purpose environment for developing, debugging, execution, testing, integration and final operation of models. Tools of this environment are a set of object-oriented graphical modeling tools with built-in algorithms for numerical solution. 3KeyMaster™ is widely used in a variety of simulators worked worldwide during the training of operators. Simulator interface includes operators and instructors' rooms. The main working window in both modes can be used for:

– the output of layout for panels and consoles with the ability to transition to their graphic imitations, repeating the relative position of the control and management, technological lines, alarm display, panels and consoles of real block control panel;
– the output of concept displaying the main technological parameters and operational condition of the equipment with the ability to transition to the unwrapped schemes of most systems.

Management of simulated processes can be made both by panels and consoles impacting on the image of control keys, mode switches, setting devices and controls; and by the schemes of technological systems by changing the options. Full-scale simulator can be used for simulating of reactor measurements. Full-scale simulator has identical software in addition to a full-scale model of a real control room.

It should be noted that reactor measurements are not quite adequately covered by the existing university programs and respective guides. However, a nuclear reactor can be safely started up and further operated only when its neutron performance is known as exactly as required by nuclear safety and core thermal reliability regulations. Familiarization with and mastering of the physical experimentation technology during education is one of the most important tasks in training of personnel who will be in charge of nuclear reactor control.

Each laboratory task concerning reactor measurements includes a theoretical section; an experimental section that includes existing methods of neutron measuring on existing equipment adapted to computer imitators; a practical section comprising programs for and methods of processing experimental data. In the course of fulfilling each task, students perform appropriate actions on simulated control units, controlling current processes both with the help of control units and the available possibility of graphically recording the dependence of various parameters on time. Results are processed on the basis of saved graphic or digital experimental data. Computing required neutron characteristics by means of the simulated reactive core, creating graphic data, determining measurement errors are carried out.

The practical includes the following experiments:
– simulating reactivity meter work;
– determining temperature, barometric and density coefficients of reactivity at various reactor power levels;
– determining differential and integral efficiencies of reactor control and safety system control units and reactivity coefficient regarding reactor boric acid concentrations at minimal controlled power level;
– determining efficiency of reactor emergency safety and of the most efficient reactor control and safety system unit;
– determining the power coefficient and power effect of reactivity with reactor power increased from minimal controlled level to 1% of nominal value.

2.3. Simulation of unit startup and shutdown

The closing component is simulation of the NPP unit processes, specifically the unit startup and shutdown. The guidance for this component has been developed based on the plant's model stepwise startup/shutdown program with regard for the standard list of startup/shutdown operations, as well as the procedures for and the sequence of these. This training component makes it possible for students not only to study the action of the unit components but also to explore the interlinking among them, solidify and systematize theoretical knowledge, and acquire primary skills in operating complex facilities.

Simulator training promotes the formation of occupational mentality and is an effective tool for the personal development of future staff and is an aid for the improvement of occupationally important traits. Therefore, simulation of startup/shutdown operations includes investigations to identify the individual and personal qualities that define the degree of success achieved in performing simulator tasks. Personal qualities are a component of competences that has an effect on the rate and efficiency of their formation. Students with different simulator work efficiencies exhibit a greatly differing extent

IOP Conf. Series: Journal of Physics: Conf. Series **781** (2017) 012055 doi:10.1088/1742-6596/781/1/012055

to which their personal qualities are pronounced. Analyzing these qualities makes it possible to evolve the guidance aspect of training based both on integrated solutions and on individual recommendations to trainees.

3. Conclusion

The computer technologies developed and described above have been introduced into the whole range of subjects and are used as the basis for students' research training and graduation papers. The share of job-related subjects which involve mathematical modeling and simulation amounts to more than 35%. As reported by the leaders of respective plant departments, this leads to the time for the university graduates to adapt themselves to the real plant environment cut by a third or even by half.

References

[1] Information on http://www.rosenergoatom.ru/wps/wcm/connect/rosenergoatom_copy/site_en.

[2] Ostreykovskiy, V.A. Teoriya nadezhnosti [Reliability Theory]. Moscow, 2003.408 p. (in Russian)

[3] Korovin Yu.A., Tikhonenko A.V. Sovremennye kompyuternye tekhnologii v podgotovke spetsialistov yaderno-energeticheskogo profilya i nauchnykh issledovaniyakh [Modern computer technologies in research and training for nuclear power personnel]. Izvestiya vuzov. Yadernaya energetika. 2011, No. 2, pp. 188–193.

[4] Naumov V.I., Smirnov V.Ye. Modelirovaniye nestatsionarnykh i avariynykh protsessov v yadernykh energeticheskikh ustanovkakh: laboratorny praktikum [Simulation of transients and emergency processes in nuclear reactors: computer training]. Moscow, MIFI Publ., 2007. 104 p. (in Russian).

[5] Kryakvin L.V. Sbornik programm dlya rasshchyota kinetiki i dinamiki yadernogo reaktora [Collection of programs for calculation of the kinetics and dynamics of a nuclear reactor]. Available at: http://www.exponenta.ru/educat/systemat/kryakvin/index.asp. (Access date: 1 June 2016).

[6] Tashlykov O.L., Shcheklein S.E., Titov G.P., Nosov D.A., Tuchkov A.M. Metodika ispol'zovaniya programmno-trenazhernykh sredstv pri izuchenii spetsial'nykh distsiplin [Methods for using computer training facilities in studies of special disciplines]/ Izvestiya vuzov. Yadernaya energetika. 2016, No. 3, pp. 63–72.

[7] Semenov V.K. Kinetika yadernykh reaktorov: metodicheskiye materialy dlya provedeniya kompyuternogo laboratornogo praktikuma [Kinetics of nuclear reactors: guidance materials for computer training]. Ivanovo, IGEU Publ., 2013. 52 p. (in Russian).

[8] Semenov V.K., Volman M.A. Programmny kompleks simulyatsii perekhodnykh protsessov v yadernykh reaktorakh [A software package for simulation of transients in nuclear reactors]. Svidetelstvo No. 2014618789, 2014 (in Russian).

[9] Semenov V.K., Volman M.A. Reaktornye izmereniya: metodicheskie materialy dlya provedeniya kompyuternogo laboratornogo praktikuma [Reactor measurements: guidance materials for computer training]. Ivanovo, IGEU Publ., 2014. 80 p. (in Russian).

[10] Volman M.A. Pusk i ostanov energobloka VVER-1000: metodicheskiye materialy dlya provedeniya laboratornogo praktikuma [Startup and shutdown of the VVER-1000 reactor: guidance materials for laboratory training]. Ivanovo, IGEU Publ., 2014. 60 p. (in Russian).

[11] Information on http://www.ws-corp.com.

Open web system of Virtual labs for nuclear and applied physics

I S Saldikov, V V Afanasyev, V I Petrov and M Yu Ternovykh

National Research Nuclear University "Moscow Engineering Physics Institute" (NRNU MEPhI), 31 Kashirskoe highway, Moscow 115409, Russian Federation

Corresponding author's e-mail address: ISSaldikov@mephi.ru

Abstract. An example of virtual lab work on unique experimental equipment is presented. The virtual lab work is software based on a model of real equipment. Virtual labs can be used for educational process in nuclear safety and analysis field. As an example it includes the virtual lab called "Experimental determination of the material parameter depending on the pitch of a uranium-water lattice". This paper included general description of this lab. A description of a database on the support of laboratory work on unique experimental equipment which is included this work, its concept development are also presented.

1. Introduction

One of the important tasks of modern higher education is to provide equal access to knowledge, the formation of a unified educational space for kindred institutions. At the same time universities are letting specialists for a number of high-tech industries such as nuclear, should provide a high level of teaching fundamental and applied physics, supported by modern teaching laboratories. In the National Research Nuclear University "Moscow Engineering Physics Institute" (NRNU MEPhI) stuff are working on the implementation of the above objectives. As an example, the publication of papers on teaching methods in some areas of applied physics can be considered [1, 2].

The university has a number of unique experimental facilities and the accumulated experience of their use in the educational process. Such plants often have a high metal and power consumption and cost for their operation is necessary to ensure safety, obtaining the appropriate licenses and the availability of highly qualified scientific and technical personnel. This makes it difficult to implement the learning process of these plants.

As a rule, incorporate it in the educational process is preceded by many years of scientific and engineering research. This includes installation of equipment, experiments, development labs and its description. Often leading educational institutions have their own unique assemblies. It reflects the history of the specificity and the history of this institution. It is obvious that the effectiveness of training would be improved if information about the unique experimental facilities, and conduct them on labs will be available to other institutions. In this complex devices like approach to the student, the effect of presence, involvement in a major development, facilitated by simulation and activity-oriented vocational training in an obstructed area. To this end, the Faculty of Physical Engineering developed a database of Support system for laboratory work on unique experimental equipment in nuclear power plants - is unique educational project in the field of nuclear energy and training system. It is created on the basis of specialties and laboratory work at the National Research Nuclear University MEPhI and combines everything a student need for both locally and the Internet access to laboratory work in the study of nuclear power plants. With photo and video materials of real facilities it allows for conducting virtual work to enhance the effect of presence. The project is available online at http://vlr.mephi.ru/, in English and Russian versions. Figure 1 shows the main page of the project.

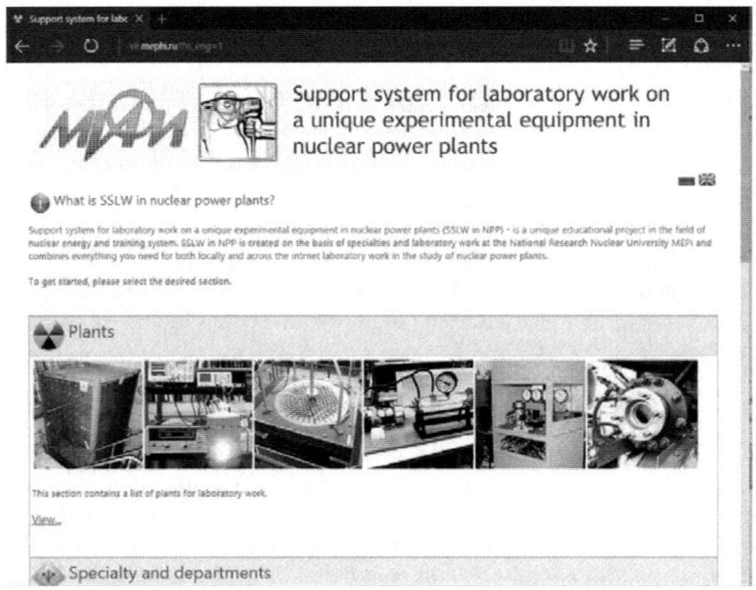

Figure 1. Main web page of support system for laboratory work on unique experimental equipment in nuclear power plant.

The database has a modern interface (the system of nested menus) and provides information in a variety of file formats. The files can be viewed, printed and / or downloaded for offline use.

In preparing students for laboratory work the database can be used as an alternative to the standard "paper" tutorial, because information in it fully reflects the content. The theoretical material which is applied to each lab, as well as photo and video materials allow you to see the process of work and can be used as teaching aids in the process of lectures and seminars. Each lab is applied checklist of questions that the student should know the answers before doing the lab.

The database is an open system and can be updated with information as the acquisition and development of new facilities and laboratory work.

2. Concept development of virtual labs

It is now possible to develop mathematical models that adequately describe the physical processes taking place in many experimental systems. A scientist can simulate a set of experimental results - the results of virtual experiments for different values of "input" parameters. The simulation is sometimes carried out in real time, but it is not always possible. In the case of complex mathematical models of time getting "virtual" results may be much greater than the time of the actual experiment. In this case, the scientist can use the approximation of the 'virtual' results, with a preliminary mass calculations based on a mathematical model of the assembly. It is also possible to use an approximation of the actual experimental results. An adequate mathematical model allows obtain the 'virtual' results that are not achievable in the existing experimental assembly, which can greatly enhance the ability of a 'virtual' laboratory.

In any case, the mathematical model is needed to create a virtual laboratory facility and should precede its development.

The development of the virtual laboratory work can be carried out using a variety of means, but the structure of this work should include the following elements:

- Initial data (installation parameters of studied materials, detector parameters, etc.).

- Set of different modes for the experiment (presence or absence of a source, the heater power, etc.).
- Simulation of the measurement (animation, photo or video materials).
- Saving (exporting) the results of virtual dimensions for further processing.

Virtual lab simulates the real laboratory work. Thus the design of the experimental setup is often described schematically, making it difficult to understand the physical processes occurring in the measurement process. To demonstrate the "insides" of the experimental setup and physical processes can use the additional multimedia educational materials made on the technology of computer graphics. The development of such materials is a separate problem. Thus, the concept of the development can be represented in figure 2.

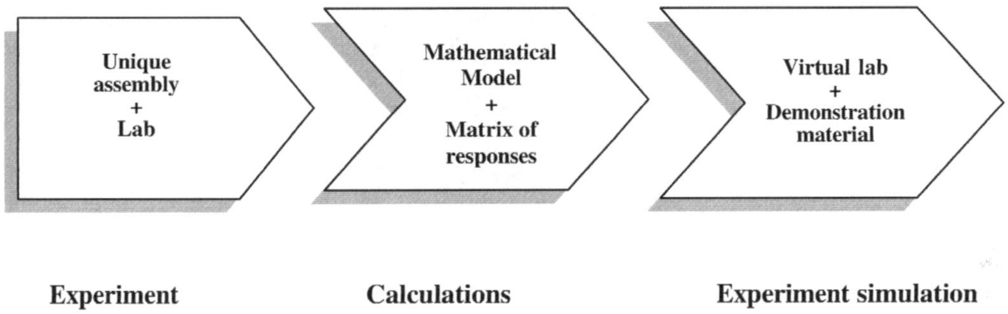

Figure 2. The development concept of virtual lab works.

3. Mathematical modelling of physical plants

The establishment and improvement of the workshop learning laboratory of the Department "Theoretical and experimental physics of nuclear reactors" NRNU MEPhI occurred simultaneously with the development of most of the department and was an integral part and a reflection of its scientific and educational work.

Lab "Experimental determination of the material parameter depending on the pitch of a uranium-water lattice" was created on the basis of uranium-water subcritical assembly that was developed in MEPhI.

To simulate a number of experimental settings, one can use a general-purpose professional programs. For example, for the simulation of systems with fissionable materials and/or sources of neutrons or gamma rays can be used Russian code MCU[3]. This program is based on the Monte Carlo method and is used for modelling the transport of neutrons and gamma rays in different systems. A scientist can adequately describe the geometry of the experimental facilities in MCU and conduct "virtual" experiment. However, the calculations are time-consuming, making it difficult to use the program MCU during laboratory work. So the following scheme in the development of "virtual" laboratory work was applied:

- Development of the model of the experimental assembly for the program MCU.
- Conduct a series of calculations for all possible "states" of the experimental plant, resulting in laboratory work.
- Filling of the "matrix of responses" of experimental values.
- Develop a program-simulator lab, in which "matrix of responses" will be used when performing a "virtual" laboratory work.

It should be noted that the "response matrix" can be completed without the calculations, and using only the real experimental data. However, the using of computer codes extends the "possibility" of experimental facilities. It allows enter a new "variable". For example, for subcritical assemblies containing natural uranium, it will be possible to perform the calculations for the enriched uranium and model the experimental indications of units for this event. Similarly, it will be possible to carry out computational studies on the effect of impurities in water at the experimental values. Thus, it can be argued that the use of mathematical modeling will increase the capacity of the experimental assemblies and develop "virtual" lab work with the new features.

4. Mathematical models of subcritical stands

Consider the process of mathematical modelling on the example of the uranium-water subcritical assembly diagram is shown in figure 3.

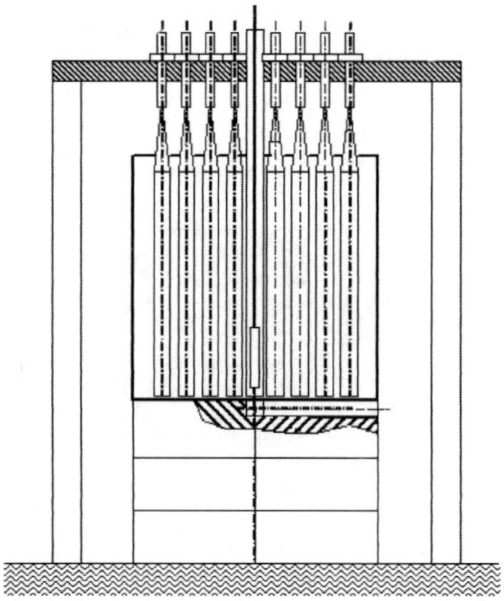

Figure 3. Uranium-water subcritical assembly scheme.

To create a virtual laboratory works on the assembly of mathematical models were developed for the program MCU. The visualization of the model is showed in figure 4. The "matrix of responses" was received for assemblies, based on calculations by MCU code. The calculations were performed for different states determined by the initial parameters.

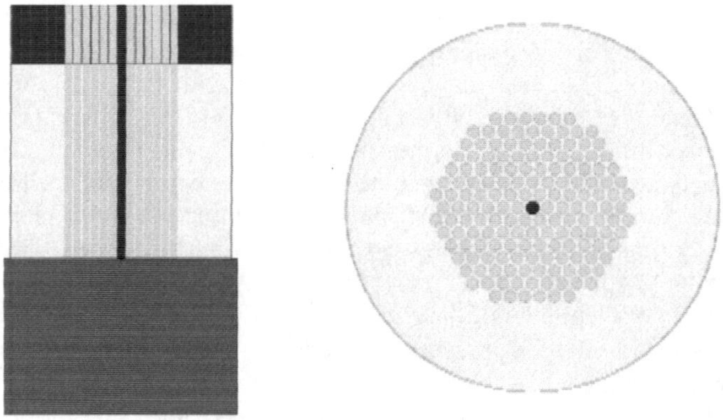

Figure 4. Visualization of the model of Uranium-water subcritical assembly (a) – left view, (b) – top view.

The number of structural elements in all the models corresponded to the number of elements in the corresponding experimental assemblies. However, the properties of model elements can differ from reality. Calculations of model assemblies with non- real parameters have been carried out to extend the range of the input data in a virtual lab, compared to the real one. Table 1 shows the values of the properties considered structural components for assembly.

IOP Conf. Series: Journal of Physics: Conf. Series **781** (2017) 012056 doi:10.1088/1742-6596/781/1/012056

Table 1. Fuel type and enrichment (%) of uranium rods in subcritical assembly for real and virtual assemblies.

	Fuel Type	Fuel Enrichment, %
Real	U_{me}	0.71
Virtual	UO_2, U_{me}	0.71, 0.9, 1.1, 1.3

5. Creating a Virtual Lab

For the development of virtual labs was created software environment Modeling of the Experiments based on the Monte-Carlo Method (MEMCM) to simulate real-world performance laboratory at subcritical assemblies. MEMCM allows you to receive experimental data as if it were actually performed by the lab, "Determination of the material parameters of the uranium-graphite assemblies" and "Experimental determination of the material parameter depending on the pitch of a uranium-water lattice" in a real experimental setup.

MEMCM opportunities for modeling subcritical assemblies:

- Calculation of the count rate (flux density in a given period of time) of neutrons in the assembly.
- Displays the number caught in the neutron detector in real time (figure 4).
- Measurement of the neutron count rate with cadmium filter on the detector.
- Measurement of background.
- View information about the performed laboratory work locally (without the Internet connection).

MEMCM opportunities for simulation of virtual assembly for virtual lab "Experimental determination of the material parameter depending on the pitch of a uranium-water lattice":

- Changing the fuel enrichment in the assembly of 0.72% to 1.3%;
- Change fuel type of U metal and UO_2;
- Changing the measurement time of 15 to 45 seconds.

The main working area of the program is presented in figure 5. Here a user can set the channel number for the detector. Channel selection is displayed on the diagram. The neutron source can be set or removed (for background measurements). The cadmium filter can be put on source for removing the thermal flux. When the user starts the experiment the number caught in the neutron detector is showed in the area. At the completion of the virtual measuring its results can be recorded and viewed in the table of the results.

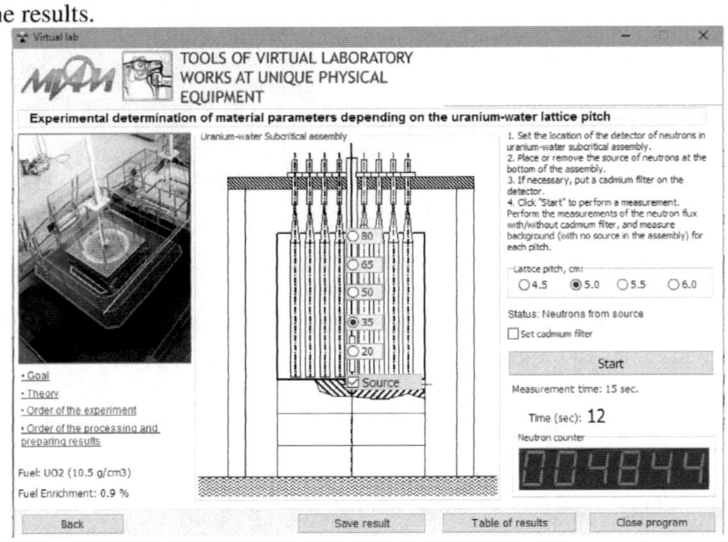

Figure 5. Main area of virtual experiment.

A general view of real experimental assembly is showed in figure 6.

Figure 6. Uranium-water subcritical assembly.

6. Conclusions

A similar scheme was applied for a database of laboratory works of engineering and physical training on the unique experimental equipment, the Faculty of Physical Engineering: 16 units, 24 labs. Mathematical models in English are built for six units and three prototypes of virtual labs:

- Uranium-graphite subcritical assembly [4].
- Spherical combustion research chamber.
- Thermophysical stand.
- Uranium-water subcritical assembly (described above).
- Spherical combustion chamber.
- Atmospheric shock tube.

Developments have been successfully used in the educational process NRNU MEPhI. Students not only learn good theoretical material, but also gain valuable engineering competencies neat course and prepare the final projects. Anyone can try to do the virtual lab in his/her University.

This work was carried out within the framework of 5-100.

References

[1] Petrov V I, Samossadny V T and Shalnov A V 1997 Training nuclear physics engineers in Moscow. Nuclear Europe Worldscan. No. 9-10, p 44.

[2] Sysoev A A, Sysoev A A, Petrov V I, Poteshin S S (2012). Engineering education technique based on professional activity imitation. Procedia. Social and Behavioral Sciences. London, vol 55. pp 707-709.

[3] Kalugin M A, Oleynik D S, Shkarovsky D A 2015 Overview of the MCU Monte Carlo Software Package. Annals of Nuclear Energy. vol 82 pp 54-62.

[4] Afanasyev V V, Kryuchkov E F, Petrov V I, Saldikov I S, Ternovykh M Yu, Tikhomirov G V (2014) Virtual labs on unique experimental equipment. Procedia. Social and Behavioral Sciences. vol 128 pp 482-488.

Effectiveness evaluation of the R&D projects in organizations financed by the budget expenses

D Yakovlev, E Yushkov, A Pryakhin and M Bogatyreova

National Research Nuclear University MEPhI (Moscow Engineering Physics Institute), Kashirskoe shosse 31, Moscow, Russia

E-mail: DYYakovlev@mephi.ru

Abstract. The issues of R&D project performance and their prospects are closely concerned with knowledge management. In the initial stages of the project development, it is the quality of the project evaluation that is crucial for the result and generation of future knowledge. Currently there does not exist any common methodology for the evaluation of new R&D financed by the budget. Suffice it to say, the assessment of scientific and technical projects (ST projects) varies greatly depending on the type of customer – government or business structures. An extensive methodological groundwork was formed with respect to orders placed by business structures. It included "an internal administrative order" by the company management for the results of STA intended for its own ST divisions. Regretfully this is not the case with state orders in the field of STA although the issue requires state regulation and official methodological support. The article is devoted to methodological assessment of scientific and technical effectiveness of studies performed at the expense of budget funds, and suggests a new concept based on the definition of the cost-effectiveness index. Thus, the study reveals it necessary to extend the previous approach to projects of different levels – micro-, meso-, macro projects. The preliminary results of the research show that there must be a common methodological approach to underpin the financing of projects under government contracts within the framework of budget financing and stock financing. This should be developed as general guidelines as well as recommendations that reflect specific sectors of the public sector, various project levels and forms of financing, as well as different stages of project life cycle.

1. Introduction

Knowledge management deals with the task of defining the most valuable knowledge generated in the course of R&D projects implementation. The quality of generated knowledge in the future as well as the competitiveness of will-be specialists depend on how efficiently new projects are assessed. Previous discussions embraced competitive education using advanced network technologies at the NRNU MEPhI [1]. The current study analyses in more detail methodological aspects of evaluating highly promising R&D projects.

Quite recently, **this** country with its huge and, as it seemed, unlimited reserves of raw material resources, was heavily oriented towards the development of industries associated with the extraction and processing of natural raw materials. However, at present, an increasing implication is the idea of shaping a new economic development model, in which intellectual property is both one of the main catalysts for the processes occurring in the economy and part of the enabling environment for enterprise development, including small innovative business.

It should be borne in mind that the mere existence of intellectual property is a necessary but not sufficient condition for the achievement of goals. The most essential step is the establishment of a sound and workable system of innovation activity support at all key stages, and most importantly – at the stage of testing its effective functioning in practice.

The practice of public R&D funding brought about a demand for the guidance that have been developed and reflected in the activities of state authorities, state financing funds, Russia's Fund of Technological Development and others. Overall, the findings reveal that no common methodological base has been made available so far – with respect to decision-making procedures on the effectiveness of public funding of scientific-research, scientific-technical and innovative activities (STA).

It turns out that currently there is not any clear notion of R & D rating methodology excepting for separate methods. Methodology assessing the effectiveness of R & D dates back to the Soviet period of the 60-ies - 80-ies. By the end of 80-ies a formal approach had rooted in "the Methodical Recommendations for the Integrated Assessment of the Efficiency of Measures Aimed at Accelerating Scientific and Technical Progress". It was approved by the resolution of the State Committee for Science and Technology and the Presidium of the Academy of Sciences of the USSR of March 3, 1988, N 60/52 [2]. The early 90-ies were marked by a sharp decrease in scientific research on this issue. The only guidelines "Methodological Recommendations on the Evaluation of Investment Projects Efficiency" (1999) [3], as the name suggests, were not focused on STA distinctive features.

Suffice it to say, the assessment of scientific and technical projects (ST projects) varies greatly depending on the type of customer – government or business structures. An extensive methodological groundwork was formed with respect to orders placed by business structures. It included "an internal administrative order" by the company management for the results of STA intended for its own ST divisions. Regretfully this is not the case with state orders in the field of STA although the issue requires state regulation and official methodological support.

2. Evaluation of projects in accordance with the methodology of the State Committee on Higher Education in Russia

The methodology developed in 1996 was adapted to assess the quality of scientific and technical projects according to the materials of business plans [4]. The indicators of the project are divided into four groups:

- scientific and technical;
- organizational and those of production;
- market-based;
- economic.

A number of scientific and technical indicators included:

- direction of development;
- scientific and technical level of development;
- scientific and technical potential of the team;
- legal protection.

Each indicator involved two levels and scoring of these levels (5 and 1 point).

With respect to the indicator "direction of development", we can see that following from the description is nothing else but the scale of the scientific and technical problem solved. In fact, two levels of the indicator are proposed:

- Scientific and technical production (STP) is a technical tool at the system level. Therefore, it is not just the improvement of STP properties that counts, but also the change in the basic structure, functional principle (5 points).
- STP is a technical tool at the system item level. The existing properties are retained (1 point).

In our opinion, it is essential to replace the name of the indicator by "scale development" and to expand the number of levels, at least up to three, and even better – five. Listing two levels is evidently insufficient.

IOP Conf. Series: Journal of Physics: Conf. Series **781** (2017) 012057 doi:10.1088/1742-6596/781/1/012057

As for the indicator "scientific and technical level of development", here the authors simplified the real state of things by only taking into account two levels:

- the STP being created is substantially different from products of similar purpose in Russia;
- the STP being created is substantially different from products of similar purpose abroad.

In general, this methodology does not fully meet necessary requirements due to the lack of novelty, non-obviousness, a differentiated approach to the properties of the projects concerning their scale and technical level.

3. Methodology developed within the framework of Russia's Program of Economic Research

The methodology originates from economic research and suggests the following set of features underlying the choice of projects:

- the importance of the project for solving urgent problems of national economy;
- the clarity of problem statement;
- the use of modern scientific methodology, the detail and originality of the method of study;
- the availability of the necessary data, the degree of their objectivity;
- applicants' experience and their acquaintance with scientific work in the relevant field;
- realism and effectiveness of the project design;
- a certain level of the project presentation;
- performers' professional growth promotion [5].

Thus, the list has two new characteristic features:

- The first sign is "the clarity of the problem statement", which describes the application in terms of the quality of expressing its content.
- The second sign is "sustaining performers' professional growth"; it points to the existence of a special kind of scientific studies effect – an educational one, which has been conventionally neglected in Russia's practice of research funding, as compared to foreign practice.

4. Methodology developed within the framework of Russia's Foundation for Basic Research (RFBR)

This methodology involves a multi-stage examination and the rating of projects [6]. The first level is a preliminary review of the project, where the following tasks are accomplished:

- Selecting projects for the second level of expertise.
- Preparing motivated conclusions for rejected projects.

However, it is not quite clear, what criteria serve the basis for the rejection of projects on the first level. On the second level the expertise determines the overall rating of the project (OR):

$$OR = R1 + R2 + R3 \quad (1)$$

where R1 is the rating of the scientific value of the project, R2 is the rating of the feasibility of project execution on time, R3 is the rating, correcting the total value of the two previous ratings.

The first rating assesses whether the implementation of the project is to:

- lead to new fundamental results;
- ensure significant progress in this area;
- exercise impact on progress in this or related scientific field (for example, grade "2" means "sufficient usefulness of the project, grade "5" is a request for an outstanding result).

The second rating takes into account the following:

- scientific level of the chief of the staff;
- potential of the team headed by him;
- scientific contribution and publications on the topic;
- information, laboratory and material security of the project;
- correctness of task distribution over stages, results and terms of work.

The overall ranking value OR can take values from points 2 to 13.

In fact, when creating this method, the authors relied significantly on the practical experience of functioning funds for financing research rather than scientific theory. It appears that the major drawback of the methodology is that it does not explicit such a property as the scope of the problem. In addition, it is difficult to speak about the universal character of the given system. Universality could have been tackled upon in case of an expert assessment that allows correlating the rating of the project and the cost for its funding in the form of a function: "cost – scientific effect." However, the range of funding for projects in the RFBR system is not great.

Doubtful as it might seem is the criterion of "a number of publications on the research topic" even with the prestige edition. In fact, the inclusion of the researcher in a new direction is in parallel with generating new ideas that have not yet manifested in published works. That is why the approach adopted in the RFBR system creates benefits to researchers continuously operating in the appropriate direction, as the trend of "diminishing returns" comes into being. Thus, the criterion of publications on the topic prevents researchers from switching over to new directions, because it appears to be no easy matter to get a state support for such projects.

The literature on the issue is extensive. A good idea of research projects assessment is given in an overview by M. Gibbons & L. Georghiou in Evaluation of research. P.: OECD. A selection of current practices (1987) [7], as well as in the studies from IMEMO RAN A.A. Dynkin and others.

In 1967, the United States approved the criteria for research project evaluation in relation to the needs of the National Science Foundation (NSF), which were specified further. In the first place, there are such indicators as:

• the scientific value of the proposed research;
• the possibility of its implementation by a separate scientist or a team of researchers.

This takes into account the following:

• the likelihood that the study will lead to important discoveries or significant conceptual generalizations in the field of science and their dissemination to other areas;
• the ability of a scientist to carry out an investigation, given his academic achievements, experience, education, capabilities, alternative approaches to the problem;
• the ability of the organization where the scientist is affiliated to provide the required conditions for project implementation;
• the practical value of the research results to national goals (the likelihood that the study could become the basis for new inventions or improved technologies);
• the role of the project in solving social problems;
• the role of the project in strengthening the scientific potential of the country, educational and scientific level of the personnel.

Of interest is the criteria system of research projects used in the US Ministry of Defense (Naval Research Department), with a ten-point scale evaluating a number of factors:

• the quality of the study – the importance of the problem, its uniqueness, the degree of fundamentality as compared to the achieved level of knowledge in this area;
• the quality of the scientific staff (a number and quality of principal investigators' publications, rewards, acquired experience, etc.);
• the degree of compliance of the resources with the objectives, the degree of innovation and operability of technical facilities, etc.;
• the need for resources necessary to accomplish the project;
• a potential impact of the project on satisfying the needs of the Department in new technology, research and operations.

The study, conducted by the Organization for Economic Cooperation and Development (OECD), examined the evaluation criteria for the selection of research projects in the member of countries of the above mentioned international organization. The classification of evaluation criteria of the projects makes up three groups.

The first group comprises the criteria for the scientific value of the project:

• compliance of the scientist's competence level with the project requirements;
• value of the work for the relevant field of science (originality);

• impact of results on other fields of science.

The second group is valid for the compliance of required resources with the assigned scientific task.

The third group is concerned with the impact of the project on reaching economic and social goals.

The analysis of domestic and foreign data on this issue shows that the methodology for project evaluation in R & D (contract and budget-based) must satisfy the following requirements:

- be scientifically sound;
- ensure "absolute" assessment to determine the feasibility of the project;
- be universal (applicable to projects of different scale, different social orientation, funded by using a variety of institutional forms (state contract, estimate and budget, stock grant, venture capital) at different stages of the life cycle of the project;
- have a quantitative nature;
- the complexity and intensity of the system must be justified (experts' participation in the evaluation process is costly);
- be informationally efficient (is to ensure the complete utilization of available information both directly and indirectly characterizing the project, to mobilize information, which is owned by the initiator of the project);
- possess full characteristics of the projects;
- take into account the differences in value of different effects and costs;
- take into account the specifics of public-oriented projects (the difficulty of comparing the types of effect, cost and inevitability of the expert-scoring approach).

No doubt, the existing methodological approaches bring us to the conclusion that they do not satisfy the given requirements as they:

- do not have satisfactory scientific justification;
- are not absolute, but are of locally-comparative nature;
- are far from universality;
- their simplicity is not justified;
- do not use all available information about the project (in particular the assessment of a potential project implementer);
- fail to give a full description of the projects (the scope of the project is omitted).

5. The concept of the proposed methodology

A new method of evaluating project implementation by determining the index of cost-effectiveness (IEC), which is the ratio of budgeted cost-effectiveness of the project (BEC) for regulatory cost-effectiveness (NEC) is proposed.

$$IEC = BEC/NEC \quad (2)$$

where budget cost-effectiveness (BEC) is the ratio of the budget effect of the project (BEP), expressed in points to the cost of the project (the CP) (in millions of rubles):

$$BEC = BEP/CP \quad (3)$$

Regulatory efficiency project (NEC) is a minimum acceptable ratio of the budgetary effect of BEP (min) expected from the implementation of the project, expressed in points, to the cost of the project (one million rubles):

$$NEC = BEP\ (min)/CP \quad (4)$$

The condition of acceptance of the project financing is:

$$1 < IEC \quad (5)$$

The higher the IEC score, the more the project is perspective for the funding.

Therefore, the valuation is assumed to be based on the following criteria:

- novelty;
- the scale of the problem being solved;
- scientific and technical level;
- scientific and technical potential;
- executor's potential (intellectual);
- executor's potential (level of experience);
- technical base;
- information base;
- target criteria (relevance of the objectives targeted by the project and the project's contribution to their achievement);
- competitiveness of results;
- planned period of use of the results (takes into account the period of their obsolescence);
- positive / negative side effects of the project implementation (environmental, social, connected with the state security, etc.);
- quality of feasibility studies;
- the assessment made by the executor of the project and the level of trust in them;
- risks associated with the project.

Normally the sum of scores corresponds to the cost of the project. This means that points are getting value. For example, if you set the standard of 100 points on one million rubles, it means that 100 points is equivalent to one million rubles, and 1 point – to 10 thousand in ruble terms. The project cost of two million rubles should ensure the effect, which is not less than 200 points. Accordingly, the project cost in 500 thousand ruble needs to ensure the effect of 50 points.

Thus, the study reveals it necessary to extend the previous approach to projects of different levels – micro-, meso-, macro projects. The preliminary results of the research show that there must be a common methodological approach to underpin the financing of projects under government contracts within the framework of budget financing and stock financing. This should be developed as general guidelines as well as recommendations that reflect specific sectors of the public sector, various project levels and forms of financing, as well as different stages of project life cycle.

References

[1] D. Yakovlev, A. Pryakhin S. Korolev, Y. Shaltaeva, N. Samotaev, E. Yushkov, Engineering competitive education using modern network technologies in the NRNU MEPhI, in Proceedings of the 2015 IEEE Workshop on Environmental, Energy, and Structural Monitoring Systems, EESMS 2015

[2] Guidelines for integrated assessment of the efficiency of measures aimed at accelerating scientific and technical progress, in SCST Decree of the USSR Presidium (March 3, 1988), No. 60/52. Available at https://www.lawmix.ru/sssr/56905

[3] Methodical recommendations for estimation of effectiveness of investment projects (Ministry of Economy, Ministry of Finance, Gosstroy of the Russian Federation (N VK 477, 1999). Available at: http://www.zakonprost.ru/content/base/77305

[4] Procedure for forming, financing and implementing innovative ST programs and projects of the RF State Committee for Higher Education, Normative-methodical materials (Moscow, 1996)

[5] Conditions for competition and application requirements. Economic research contest "Spring 2015", see the website of the consortium for economic research and education at http://www.eercnetwork.com/default/download/creater/how_to_apply/file/8f5cc0b81dcd59a7f8076845326f75b07846994d.pdf (date of access: 17.03.2016)

[6] Regulations on examination procedure of RFBR competitions in Decision of the RFBR Bureau of the Board (May 17, 2012). Available at: rfbr.ru

[7] M. Gibbons, L. Georghiou, Evaluation of research. P.: OECD, in a Selection of Current Practices, 1987

IOP Conf. Series: Journal of Physics: Conf. Series **781** (2017) 012058 doi:10.1088/1742-6596/781/1/012058

Specifics of MS training in the area of nuclear materials safe management for new-comers in nuclear power

N I Geraskin and V B Glebov

National Research Nuclear University MEPhI (Moscow Engineering Physics Institute), Kashirskoe shosse, 31, Moscow, 115409, Russia

E-mail: VBGlebov@mephi.ru

Abstract. The issues of specialists training in the field of nuclear materials safe management for the countries, who have taken a way of nuclear power development are analyzed. Arguments in justification of a need of these specialists training for the new-comers are adduced. The general characteristic of the reference MS program "Nuclear materials safe management" is considered. The peculiar features of the program, which is important for graduates from the new-comers have been analyzed. The best practices got as a result of implementation of the program in recent years for the students from Kazakhstan, Belarus, Azerbaijan, Tajikistan, Iran, Turkey and other countries are presented. Finally, the directions of international cooperation in further improvement and development of the program are considered.

1. Introduction

Nuclear materials are the integral component of any nuclear power system. One of very important aspects for development of these systems is providing the conditions of nuclear materials safe management. It relates especially to the countries, who have taken a way of nuclear power development (new-comers) and where the staff deficit in nuclear area is often observed. For these countries it is important to have their own staff, which capable to make professional solutions on safety assessment of the existing conditions and steps taken in development of nuclear power. Therefore the new-comers are strongly interested in training of national personnel in the field of nuclear materials safe management.

The notion of the nuclear materials (NM) safe management is many-sided and includes nuclear, radiation, technical, fire safety, etc. Among the sides of NM safe management the safe keeping and also full and reliable knowledge about NM play an important role.

All these sides of NM safe management include one common component – personnel. The extensive operational experience shows that a permanent attention and purposeful efforts are required to maintain a high level of reliability of personnel at nuclear sites. This activity is reflected in the notion of nuclear safety culture. At that, the nuclear safety culture is considered as an active mean for increase of the safety level of NM and nuclear installations.

The MS program "Nuclear Materials Safe Management" is devoted to training of masters in all noted above areas of the NM safe management. The program is especially important for the students from new-comers, because it is focused on complex specialists training on NM safe management.

IOP Conf. Series: Journal of Physics: Conf. Series **781** (2017) 012058 doi:10.1088/1742-6596/781/1/012058

This program is already implemented in NRNU MEPhI (the Moscow site) for a number of years. During this time the students of Kazakhstan, Belarus, Ukraine, Azerbaijan, Tajikistan, Iran, Nigeria and other countries have successfully graduated, and also are training according to this program now.

The program "Nuclear Materials Safe Management" is developed as a reference educational program of a magistracy in the field of NM safe management. The program corresponds to the educational standard of the higher education of NRNU MEPhI in the direction 14.04.01 "Nuclear Power and Thermophysics" [1]. In October, 2015 the program has successfully passed the state accreditation. Thus, it has been recognized that the program conforms to the federal educational standards for a master's degree and it is suitable for the academic training at NRNU MEPhI.

2. Professional activity and competences of the graduates of the program

As a rule, an educational program is developed by universities depending on the required kinds of activity of graduates and a spectrum of the competences allowing the graduates to solve effectively professional tasks.

Taking into account noted specific conditions and needs of the labor market in the new-comers the program "Nuclear materials safe management" is focused on the following kinds of graduates activity: management; design; production and technology, expert, pedagogical work.

The condition of a successful solution of the professional tasks within noted kinds of activity is expressed as a spectrum of the competences of program graduates. Among them there are special competences reflecting specific conditions and needs of labor market in the new-comers. Note some of them here:

- The ability to formulate criteria for an assessment of NM safe management conditions;
- The ability to act in non-standard situations, to be responsible for adopted decisions;
- The ability to creation of theoretical and mathematical models describing the processes of NM management;
- The ability to carrying out physical experiments for the determination of NM characteristics;
- The readiness to analyze and estimate the state of nuclear materials physical protection, control and accountability systems (MPC&A systems);
- The ability to analysis and management of the data, characterizing of NM state, on the basis of modern information technologies;
- The readiness to carry out the organization and technical equipment of workplaces, to develop operational plans of the work of production divisions;
- To have knowledge of the basics of the power installation equipment design;
- To possess test methods for control of the capital equipment of nuclear power plants and other power installations;
- The readiness to carry out technical and economic calculations by production of thermal and electric energy with use of nuclear fuel.

3. Characteristic features of the program "Nuclear materials safe management"

3.1. General structure and training conditions in the program

The training of students of a nuclear profile for the new-comers has to account the existing circumstances in these countries, including: weak traditions (or their absence) of the NM safe management; existence only of some elements of nuclear infrastructure; existence of small experience in nuclear area; lack of the large-scale contingent of experts and specialists in nuclear area, etc. The program "Nuclear materials safe management" takes into account the specifics of nuclear activity in these countries.

The entrants are accepted for training in the program, if they have education at the bachelor level (or above) in the field of technical science and they successfully passed an introductory interview. The term of training in the program is 2 years. The volume of the MS program is 120 credits. At full-time courses the program volume, realized in one academic year, is 60 credits.

Structurally the MS program is constructed according to modular structure recommended by the educational standard 14.04.01 and includes obligatory and variable parts. The disciplines relating to variable part of the program (including the scientific-research work and practice), define its profile. This profile is reflected in the name "Nuclear Materials Safe Management".

Specific filling of structure of the program is presented in the form of working curriculum. The curriculum defines all sequence of two years' training of undergraduates and takes into account the restrictions and recommendations of the educational standard. The curriculum of the program "Nuclear Materials Safe Management" includes more than 30 disciplines. Among them 22 disciplines are included in the variable part of the program. The central place in variable part of the program is taken by the courses having key value for the students from the new-comers: the block of the disciplines devoted to NM safe management, "Nuclear technologies", "Legal and international aspects of NM management", "Materials of nuclear reactors", "Methods and devices for NM measurement" and others.

The students from different countries are trained within the program. And as the educational program is realized in Russian, the students, who are not knowing Russian, learn it previously. It should be noted, nevertheless, that teaching of the courses in English is planned in the nearest future.

Often it turns out that students from the different countries and universities have the different level of initial knowledge, but they study together. The practice of adaptive training is provided in these circumstances. It is expressed in a form of the education according to individual working curricula. The similar flexible system of training allows to master successfully the program to the students with different initial professional knowledge.

3.2. Complex consideration of the issues of NM safe management

Specifics of the issues of NM safe management are that all sides of the NM safe management should be considered equally: nuclear, radiation, information, technical, fire safety, NM safe keeping, full and reliable knowledge about NM, the issues of non-proliferation and reliability of personnel. It means that the program "Nuclear materials safe management" has to give to students a complex of knowledge, abilities and skills on all these components of the NM safe management.

In fact, the curriculum of first year of training included the following disciplines:

- basics of nuclear and radiation safety;
- Regulation of nuclear and radiation safety;
- Basics of information security of critical technologies;
- Legal and international aspects of NM management (nuclear non-proliferation);
- Influence of radiations;
- Introduction to the systems of NM physical protection, control and accountability for;
- Methods and procedures for NM control and accountability for;
- Computerized systems for NM control and accountability for;
- Methods of the vulnerability analysis and optimization of physical protection subsystem;
- Safety criteria and risk assessment in the tasks NM physical protection, control and accountability for.

On the second year of training within the cycle "NM safe management" students study:

- Technical means of physical protection of nuclear objects;
- Basics of radioactive waste management;
- Active methods of NM control.

The subject "Nuclear Safety Culture " is common for the courses of the cycle "NM safe management". Much attention is paid to this subject in the program. Different versions of this subject are included in several courses of the cycle of NM safe management.

3.3. Practical trend in teaching

As it is noted in the standard 14.04.01 "The nuclear power and thermophysics", the area of professional activity of the program graduates includes development, creation and operation of the devices and installations, where thermal and nuclear energy are used. In this regard the program "NM Safe Management" is focused, first of all, on practical tasks in nuclear power, including its safe development.

Therefore the considerable part of courses of the program is followed by laboratory and practical training. The last allow students to acquire, in the mode of active work, the knowledge and skills required for further successful professional work. We will note here that a special attention to the practical side of education plays a crucial role for formation of an self-sufficiency in the work of students. This quality is of great importance, first of all, for the students from the new-comers, owing to deficiency of skilled experts in nuclear area there.

The exit classes at the enterprises of nuclear power branch are given for students of the program. There, they get acquainted with the modern technologies of a nuclear fuel cycle, talk to experts, can gain deeper impression about specifics of NM safe management at the sites of various profiles. Besides studying of purely technical issues, the students can feel attention of the management to staff of nuclear enterprise and estimate the significance of a human factor in its safety.

3.4. Using of advanced technologies for distance training of foreign students

The educational standard 14.04.01 "The nuclear power and thermophysics" provides a possibility of the network form of education. Within this educational form, the technology of remote training is allowed. Since fall of 2015 NRNU MEPhI, within the program "NM safe management", realizes the network form of education for the staff of Kazatomprom (the Republic of Kazakhstan).

Besides NRNU MEPhI, the Universities of Kazakhstan participate in training in the program: Al-Farabi Kazakh National University (Almaty); Euroasian national university of name L.N. Gumilev (Astana) and East Kazakhstan state university of S. Amanzholov.

Training courses of the program: "Basics of nuclear technologies"; "Basics of nuclear non-proliferation and NM safe management"; "Basics of information security of critical technologies; "Special chapters of higher mathematics"; "Methodology of scientific cognition"; "English language" are assigned to NRNU MEPhI. These courses are delivered to listeners in Kazakhstan by teachers of NRNU MEPhI with application of remote technology of training.

4. International cooperation for further development and implementation of the program

Training of specialists according to the program "NM safe management" is carried out with active support from the Department of Energy (DOE) of the USA. Collaboration is carried out within the Basic Agreement between Memorial Institute Battelle and NRNU MEPhI. The cooperation between NRNU MEPhI and DOE is expressed in a definition and further solution of the specific tasks, which arise during development and implementation of this program. These tasks correspond to certain directions of cooperation, allowing to realize effectively the possibilities of both organizations in achievement of common goals.

Below some of these directions are briefly noted.

- Strengthening of educational potential of the program by development and modernization of laboratory base;
- Formation and improvement of methodical provision of the program. Development of working curricula;
- Further improvement and development of the program by contacts, survey, collection of opinions, estimates of disciplines of the program from students, its graduates and employers;
- Transfer of experience of the Moscow site in NRNU MEPhI branches and other nuclear universities.

Being reference, the program includes extensive international experience in the field of nuclear power. A number of courses of the program "NM safe management" has been created and is realized now thanks to joint efforts of the teachers of NRNU MEPhI and the specialists of IAEA.

One of such courses is "Nuclear knowledge management". The course consists not only of lectures, but contains also practical tasks, and also research projects with their subsequent defense.

The course "International Nuclear Cooperation" considers the international system of nuclear cooperation and activity of the IAEA in this area. The course shows the ways, how the results of international cooperation can be used in the practical activities of organizations and experts, especially young specialists.

The courses "Nuclear Knowledge Management" and "International Nuclear Cooperation" are given by the experts of IAEA in the Russian and English languages. These courses are delivered by means of videoconference technology on the Internet. For lecturing support the computer CLP4NET platform developed by IAEA for nuclear education is used.

References

[1] Facer R I, Phillips J, and Pieroni N 2007 Milestones in the Development of a National Infrastructure for Nuclear Power pp 5-19 (Vienna, IAEA Nuclear Energy Series, No.NG-G-3.1)

[2] Management of Nuclear Materials 2009 (US Department of Energy, Order DOE O 410.2, Washington) pp 1-15 file:///C:/Users/User/Downloads/o4102.pdf accessed: 08.11.2016

[3] General provisions for ensuring NPP Safety 2015 pp 2-15 (Rostekhnadzor RF, NP-001-15, Moscow, in Russian) file:///C:/Users/User/Downloads/0001201602050003.pdf accessed: 08.11.2016

[4] Geraskin N I, Krasnoborodko A A, and Glebov V B 2015 Nuclear security culture enhancement: the role of culture coordinators at Russian nuclear sites (10 years of work) *Defense and Security Analysis* **31** 4 pp 330-345

[5] The educational standard of the higher education of NRNU MEPhI, educational level – Master of Science, the direction 14.04.01 "Nuclear Power and Thermophysics", the protocol No. 13/07 from 12/27/2013 with the changes and additions, approved by the Scientific Council of the University, the protocol No. 14/10 from 12/30/2014 (in Russian).

[6] DOE Handbook, 2014 Training Program Handbook: A Systematic Approach to Training pp 5-62 (Washington, US Department of Energy, DOE-HDBK-1078-94) pp 1-62 http://energy.gov/sites/prod/files/2016/03/f30/DOE-HDBK-1078-94_Reaffirm_June_2014.pdf accessed: 07.11.2016

[7] Knowledge Management for Nuclear Industry Operating Organizations 2006 pp 1-20 http://www-pub.iaea.org/MTCD/Publications/PDF/te_1510_web.pdf accessed: 08.11.2016

The quantum-mechanical approach to construction of quantitative assessments of some documentary information properties (on example of nuclear knowledge)

A A Lebedev[1], N V Maksimov[2] and E V Smirnova[2]

[1] Lebedev Physical Institute of Russian Academy of Sciences, 119991, Russia, Moscow Leninskiy pr., 53

[2] National Research Nuclear University MEPhI, 115409, Russia, Moscow, Kashirskoe highway, 31

E-mail: lebedevalex@live.ru

Abstract. The paper presents a model of information interactions, based on a probabilistic concept of meanings. The proposed hypothesis about the wave nature of information and use of quantum mechanics mathematical apparatus allow to consider the phenomena of interference and diffraction with respect to the linguistic variables, and to quantify dynamics of terms in subject areas. Retrospective database INIS IAEA was used as an experimental base.

1. Introduction

To present day there is no generally accepted constructive definition of a phenomenon of information and issue of its physical nature is still open that does not allow to quantify its properties. Among the works devoted to this problem [1-4] the common is following: information exists objectively and itself is an objective physical quantity among others – such as mass, energy, momentum, etc. In article [5] the analogy between information and light was considered. The concept of information quantum was proposed as a kind of signal having the meaning of a simple piece of information, which arises due to the transition of a system from one stable state to another.

In article [6] a physical nature of information was considered from the perspective of physical fields variety, their superposition and nonlinear interactions, the concept of information as a physical field. It is shown that all information generated and perceived is a physical field or superposition and interaction of the fields.

2. Description of the model

The objects of this study are terms of natural-scientific language of documentary database INIS IAEA. Each term initially has single meaning, single interpretation. But using of the original term in combination with others in various circumstances alters the perceptual meaning of the term.

Changing of term context perception can be attributed to the presence of non-point (probabilistic) nature of its existence in space of meanings. [7] Additionally, the context change takes place in the space of patterns (meanings) as a result of information objects activity (objects that have the ability to reflect and change state of another physical object or it relationships), but this change observing (registration) takes place by means of fixed (in the form of documents) knowledge. Thus, the projection of term space on the space of fixed knowledge will be determined by a distribution function, which is represented by description of all meanings contained in the term. By determining of

IOP Conf. Series: Journal of Physics: Conf. Series **781** (2017) 012059 doi:10.1088/1742-6596/781/1/012059

term meaning through a description of the circumstances in which it is used, we come to the different subject areas where the term is have a sense.

This approach is similar to probabilistic description of particle characteristics, which is used in quantum mechanics [8]. It considers particles wave functions instead of material particles considering. Although the wave function has no physical meaning (meaning has only square of the wave function, defining the probability density of quantum system observing in different states), the wave properties of quantum systems help to explain some of phenomena.

2.1. Basic concepts and definitions

As subject area (SubjAr) we will assume a set of objects (and / or state) of natural or artificial origin, existing either in form of sustainable natural formation exists in result of evolution or allocated to some entity in accordance with its activity goals.

In informational practice the SubjAr is traditionally identified by area headings. The rubrics in documentary systems denote attachment of a document to one or more classes defined by classification system. That is each SubjAr acts as a fully autonomous entity and a person may hold a mental border between two non-identical SubjAr. Meanings that terms express associated with a single SubjAr, but rubrics only indicates the SubjAr where the meaning likely takes place.

As information objects (InfObj) will named those objects of SubjAr that have property (ability) to reflect and, subject to certain conditions, change the state of another physical object feature or relationships, and that will have the role of objects-originals patterns.

Informational interaction (InfInt) will be called processes in which the pattern objects are used as operating. Unlike physical interactions occurring between individual bodies (originals) at a particular time, informational interactions will involve their patterns – objects that reflect properties of some of the originals, but do not possess them. InfInt differs from the physical by the fact that we are not interested in the specific state of an information object, but in probability state of SubjAr, changing as a result of interaction [1].

Information - a superposition of possible states of InfObj in n SubjAr, characterized by function:

$$\Lambda = c_1\psi_1 + c_2\psi_2 + \cdots + c_n\psi_n \qquad (1)$$

where ψ_i – InfObj distribution function for the *i-th* subject area, in particular, those may be language terms, representing a linguistic variables; c_i – probabilistic measure of SubjAr reffering to *i-th* subject area. It should be noted that the SubjAr can be represented as InfObj, consisting of a set of information objects forming some integrity given by rubric. I.e. in general case rubric can act both as artificially given set of terms, and as a separate SubjAr.

Then it is possible to characterize the function of informational interaction:

$$\Omega = c_1\psi_1\Theta_i^T + c_2\psi_2\Theta_i^T + \cdots + c_n\psi_n\Theta_i^T \qquad (2)$$

where Θ_i is the distribution function for InfObj, which interacts with the original InfObj, for the *i-th* SubjAr. I.e. the information by interacting with specifically SubjAr, in fact, takes the only one state of many possible. Such information state resulting of the selection (InfInt) *recorded* in the form *of context caused information object* should be called *knowledge* [1].

Further, for term "information" using one needs to understand that we are talking about InfObj, which has a characteristic duality state: before interaction – it is some integral object, and during the interaction, it`s macroobject "quanta". Moreover, these related by certain way microobjects, at the same time can be used as elements of other patterns (image), present or future, the original objects. For the text form the information quanta essence are concepts, to be exact - denoting them language terms.

The term wave function (by analogy with the particles in quantum mechanics) determines its state at any time, to within a phase factor. By analogy with [9,10] we assume that the state of the term is defined by the wave function:

$$\psi(x,t) = A(x,t) \cdot exp(-i \cdot \varphi \cdot t) \qquad (3)$$

so that:

$$|\psi(x,t)|^2 dx = P(x,t) \quad (4)$$

where: x – coordinate of term in the semantic space, which is defined by rubricator (named, sometimes, as Subject index headings) of SubjAr,
t – time; $A(x,t)$ – probability amplitude; φ – term phase; $P(x,t)$ –probability that the term has a meaning corresponding to a neighborhood dx, given by rubric x (at the present moment we propose to evaluate this value as identification part of documents with the term which belong to particular rubric x - Subject index field of INIS record).

The wave function is complex and represents the probability wave that describes the objectively existing capabilities of a particular result.

Coordinate of term is equal to x if document to which it belongs, is related to the rubric h_x from set $\{h_1, h_2,..., h_m\}$ x=1,...,m (m – the number of rubrics).

Term phase $\varphi \in [0; \frac{\pi}{2}]$. The phase is associated with the presence of the term opportunity to join other terms. In general, it is a cyclic quantity that is characteristic of the wave process of term perception. In this paper, cyclic recurrence is not taken into account, as the value of the period is great in comparison with considered time intervals.

The magnitude of phase in particular time depends on the condition of its life cycle where term is, whether it is a structural unit which can be coupled with other terms to form a new meaning. The value depends on the phase relation between the real and imaginary part of the term, that is, how it is specific. The more real is part of the term, the more specific it is. Since the used database does not allow to identify roles and relationships between terms, the phase of the wave function was evaluated as quotient of L_1 and L_{12}, where L_1 is a number of documents that contain the term and related to rubric h_x at time t; L_{12} is a number of documents that contain the term and at the same time belong to the rubric h_x and to any of rubrics set $\{h_1,...,h_{x-1}, h_{x+1},...,h_m\}$ at time t.

Thus, the probability distribution of term S_i gets in rubric h_x at time t:

$$P_i\,(x,t)=|A(x,t)|^2=(Re\ \psi(x,\,t))^2+(Im\ \psi(x,t))^2 \quad (5)$$

Based on foregoing, InfInt process can be represented as a movement of some "information wave" from source to receiver, consisting of oscillating particles, which are terms of incoming message, constituting the wave front.

Each wave is characterized by a spectrum – descriptor distribution by subject areas (rubrics).

The set of terms representing descriptors environment (associated with them) are secondary wave front, etc.

Primary wave $W_1=\{S_1,S_2,...,S_n\}$, where $S_1,S_2,...,S_n$ - the terms included in the message.

Secondary wave $W2=\{S_{11},...,S_{1m},S_{21},...,S_{2k},...,S_{n1},...S_{nl}\}$, where S_{ij} – terms that constitute the S_i term environment, i.e. they are with him in the same document in the same rubric. For the term S_i $j=1,...,m$.

2.2 Interference and diffraction of information

The foregoing idea of "information wave" in form of classical waves of mechanical nature is a good tool for macrophenomena illustrations, for example: meanings overlay, rounding, outrunning, etc.

By "interference of information" we assume this type of InfInt, in which there is mutual increase or decrease of resultant amplitude of two or more information waves due application to each other.

By "diffraction of information" we mean a type of InfInt at which a penetration of the waves in an area of "information shadows" takes place, that is, the information distribution in the source affects the information distribution to the receiver.

In the quantum-mechanical description of "information wave" it is a wave of probability, describing the objective existing possibilities of one or another result.

For the primary wave W_1 (Fig. 1), the resulting probability distribution is defined as the total distribution for all terms $S_1, S_2, \dots, S_n$: $P_{W1} = \sum_{i=1}^{n} P_i(x,t)$.

For the secondary wave W_2 (see Fig. 2): $P_{W2} = \sum_{i,j=1}^{n,m} P_{ij}(x,t)$.

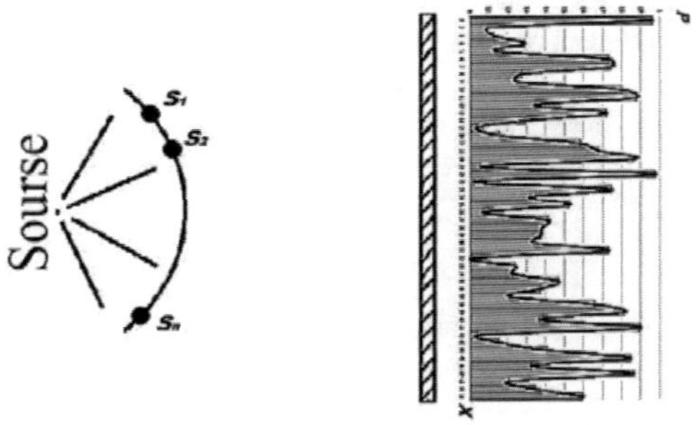

Figure. 1. Distribution of meanings on the rubrics for the primary wave.

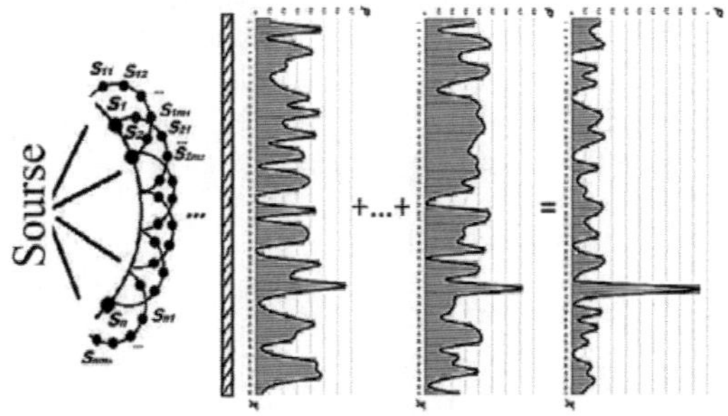

Figure 2. Distribution of meanings on the rubrics for the secondary wave.

Thus, information interference is InfInt in which there is a relative increase or decrease resulting probability of meaning finding in a given subject area.

Considering the "wave of information" as the flow of "quantum information" (a pair of terms), we obtain:

$$P_1 = |A_1|^2, \quad P_2 = |A_2|^2, \quad P_{12} = |A_1 + A_2|^2 \tag{6}$$

$$P_{12} = P_1 + P_2 + 2\sqrt{P_1 P_2}\cos(\delta - \tfrac{\pi}{2}) \tag{7}$$

where $\delta = \varphi_1 - \varphi_2$ – the difference of terms phases; P_{12} – the probability of finding the meaning which get into a neighborhood dx near x coordinate of semantic space at joint use of these two terms.

From (7) it follows that the greater the difference between the terms phases, the interference will be observed stronger. This is generally consistent with the practice of terms-phrases constructing when

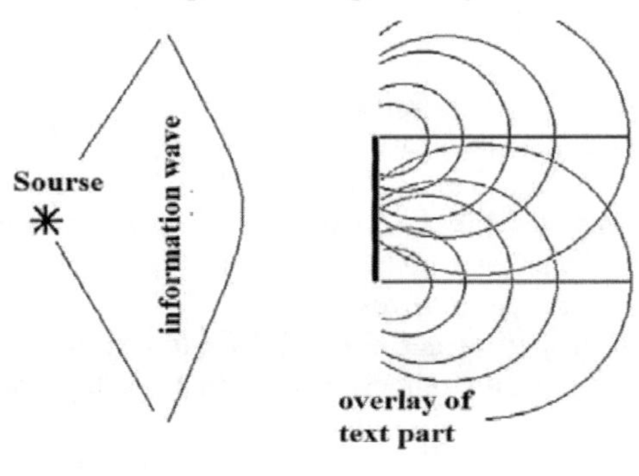

Figure 3. The interference of "information wave".

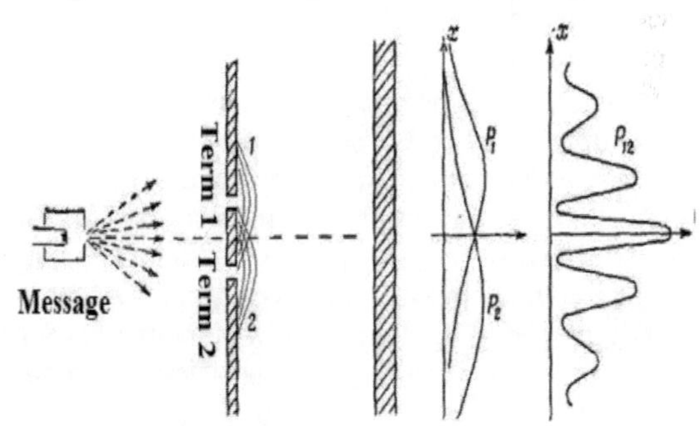

Figure 4. The rounding of obstacle by "information wave".

one is a metaconcept term and another term denotes a particular object or property.

The phenomenon of diffraction (fig.4) can be treated as a special case of interference, namely, as the interference of secondary waves.

At the "overlay" of text, carrying information it is possible to restore original meaning of the message on the remaining part. I.e. "wave", which carries the information, like goes around an obstacle and gets in area behind him.

IOP Conf. Series: Journal of Physics: Conf. Series **781** (2017) 012059 doi:10.1088/1742-6596/781/1/012059

Consider an example that illustrates the outlined phenomenon.

For example, in message "NUCLEAR REACTION YIELDS" term "REACTION" will be dropped out. The terms included in the environment of terms "NUCLEAR" and "YIELDS", i.e. co-occur with them in a same document in the same category (secondary waves), interfere with each other. Let us construct the distribution of probability that the meanings get in specific SubjAr, resulting in the interference of secondary waves, and on the same graph shows the actual distribution for term "NUCLEAR REACTION YIELDS" (see Fig. 5). Based on how close these distributions are one can judge on the possibility of message with omission of original meaning restoring.

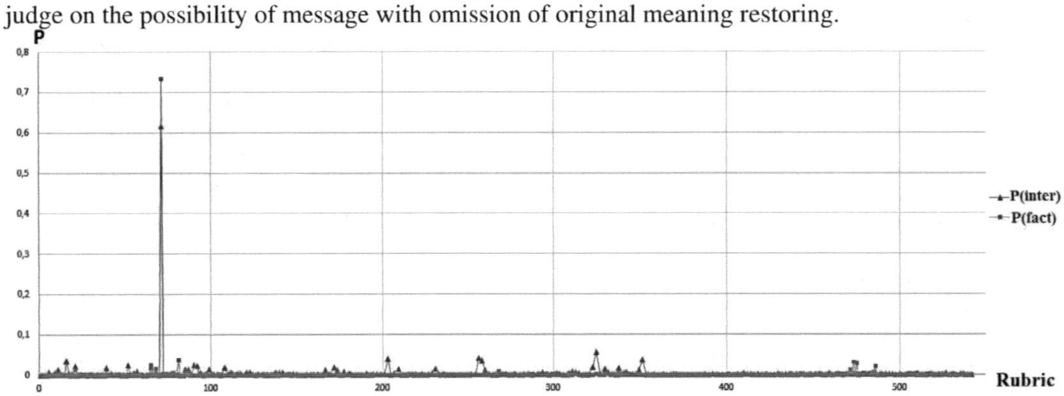

Figure 5. Comparative schedule of the probability distribution for entering in semantic area for messages without omission (P(fact)) and for case of waves interference for message with omission (P(inter)).

Resulting distributions are close to each other, get into the same SubjAr, which indicates that the meaning of the message "NUCLEAR REACTION YIELDS" can be restored with omission of term "REACTION" due to the interference of secondary waves (generated by terms associated with the terms "NUCLEAR" and "YIELDS").

3. Conclusion

A model of a documentary information properties (in particular, for terminology systems), based on the assumption that term of language may be defined as a physical particle, to which basic laws of physics can be applied was proposed. With using of wave and quantum concepts the phenomena of documentary information interference and diffraction was described. On the INIS IAEA database array a probability distribution of meaning in the subject areas as a result of terms interference and diffraction that experimentally confirm adequacy of the model was constructed.

The proposed approach for documentary information properties definition allows to improve the quality of information request, in particular through a more effective request expansion procedures and dynamically definable issuance thresholds.

References

[1] Maksimov N V 2010 *Automatic Documentation and Mathematical Linguistics* **44**(4) pp 177-186

[2] Brilloin L 1956 *Science and Information Theory* (New York: Academic Press)

[3] Colin K K 2010 *Alma mater (High Scholl Herald)* **1** pp 1-8

[4] Ursul A D 2010 *Nature of information: a philosophical essay* (Chelyabinsk: Chelyab. State Academia of Culture and Arts)

[5] Romanenko V N, Nikitina G V 2011 *Modern high technologies* **2** pp 1- 4

[6] Vildyaykin G F 2010 *Scholarly Notes of Komsomolsk-na-Amure State Technical University* III-**1**(3) pp 46-56

[7] Nalimov V V 1989 S*pontaneity of consciousness. Probabilistic Theory of meaning and the Meaning architectonics of personality* (Moscow: Prometheus)
[8] Landau L D, Lifshitz E M 1989 *Theoretical physics, quantum mechanics (non-relativistic theory)* (Moscow: Nauka)
[9] Lebedev A A, Maksimov N V and Smirnova E V 2016 *Automatic Documentation and Mathematical Linguistics* **1** pp 26-34
[10] Gabora L. and Aerts D 2002 *Journal of Experimental and Theoretical Artificial Intelligence* **14** pp 327-358

Cognitive competence of graduates, oriented to work in the knowledge management system in the state corporation "Rosatom"

V Kireev, A Silenko and A Guseva

[1]National Research Nuclear University MEPhI (Moscow Engineering Physics Institute), Moscow, Russian Federation

E-mail: VSKireev@mephi.ru

Abstract. This article describes an approach to the determination of the level of formation of competences of university graduates, oriented to work in the state corporation "Rosatom" in a knowledge management system. With the use of cluster analysis graduate classes were identified, focused on knowledge transfer, analysis and the search for new knowledge, creative transformation of knowledge. In addition, the class innovators were identified, which were fully formed the necessary cognitive competences.

1. Introduction

Knowledge management is a fundamental factor in the development of innovative high-tech enterprise of any industry and solves two main tasks. The first problem is related to the increase of efficiency, use of knowledge for productivity growth to increase at the expense of performance or reduce costs. The second objective is focused on the creation of innovations, new products and services [1].

Especially true knowledge management for the state corporation "Rosatom", the world leader in the level of scientific and technical developments in the field of reactor design, nuclear fuel fabrication, construction and operation of nuclear power plants and nuclear power plant personnel qualifications, creation of radiation technologies [2].

The question of what cognitive competencies should have staff in the field of treatment and the capitalization of corporate knowledge, is still under discussion. This is vividly demonstrated by the III International Forum of Rosatom Knowledge Management, held in April 2016.

The works devoted to the professional development of cognitive competencies is understood beneath the mastery of staff separate ways of acquiring new knowledge [3, 4].

Three types of professional cognitive competencies are considered: the professional competence of the transfer of knowledge, professional competence and knowledge discovery search, professional competence of creative transformation of knowledge [3]. Diagnosis of professional cognitive competencies can be performed on the basis of the analytical methods professiografii including a description of the professional features, professional tasks and problems of performance features. Each professional task is a generalized model of professional situation that a specialist can meet in their professional practice.

In this regard, one of the goals of higher education is to prepare graduates as the innovators - people who are able to generate, locate, implement innovations [4].

Cognitive competence-graduate innovator must lie in the integration of different knowledge, skills, belonging perhaps to different areas of activity. These include the ability to identify problems and methods for solving them; ability to work in a team, to participate in the project activities, creative independence, the ability to take a critical look at the current state of affairs, etc. In [1] the basic cognitive competencies of innovators author includes technical skills, leadership, people skills and conceptual abilities, ie skills analysis of the behavior of people, structures and dynamics of long-term relationships.

This study focuses on evaluating the level of formation of a number of cognitive competencies of graduates in physics disciplines needed innovators. These include the ability to work with people (team work, team management, conflict resolution) and conceptual abilities (problem solving, decision making, self-motivation). Assessment of leadership skills is the next stage of our research.

2. The proposed approach

Assessing the level of formation of cognitive competences was conducted as follows. Using standard test Eysenck, Thomas and Belbin following individual personality traits have been identified: extraversion-introversion, neuroticism-stability, choice of team role preferences, preferences, behavioral strategies in conflict [5, 6].

Eysenck test enabled with basic indicators (extraversion-introversion and neuroticism-stability) to evaluate the direction of the person on the inner or the outer world, as well as to identify the level of emotional stability. These characteristics affect the fulfillment of their obligations, including to carry out professional activities. Belbin test reveals preferred for human role in the success of the team: coordinator (Co), shaper (S), plant (P), monitor evaluation (ME), implementer (I), resource investigator (RI), team worker (T), completer finisher (CF).

To select the properties of human behavior in a conflict situation is used Thomas test. According to the results determined as preferable to test one of the two final strategies of conflict resolution - co-operation and compromise. In a compromise, one party wins and the other loses, both sides lose, or (strategy B). With the cooperation of both parties of the conflict wins (strategy A).

To determine the level of formation of the respective competences of the students were brought 15 jobs situation analysis. Assignments are made of two types. The first type is related to the performance of analytical tasks in the form of classification problem solving, which involves thinking, observation, comparison and analysis of many of the facts, the second type of job requires the acceptance of the test unequivocal military solution in extreme situations.

Informally, these tasks are divided into three parts, under which the subjects are determined to achieve with the management team, the ability to work in teams, manage conflicts. motivation level was determined based on the identified gaps between the inclinations of man and the results of his achievements. Thus, the results obtained allow to measure the value of the level of formation of appropriate cognitive skills.

3. Research outcomes and discussion

The study was conducted among a large number of senior students of the two faculties of MEPhI. The first sample constituted the Faculty of Management students and high-tech economy (Y), studying in the areas of training management, economics and business informatics. The second sample was made up of students of the Faculty of Physics and Technology (F), which is the parent in Russia to train specialists in the development of nuclear power plants, nuclear non-proliferation, physics separation processes, physical materials. These training areas were selected for the reason that on the one hand, students are oriented to work in the nuclear industry, on the other - have a well-formed professional competence in the field of collective management, teamwork, conflict management.

Figure 1 shows the results of a comparative analysis of the number of different personality types in the samples for students to managers and future graduates of physical specialties. For administrators the highest number of students corresponds to the normal subjects (25%), a melancholy, choleric (28%) and choleric (11%) types.

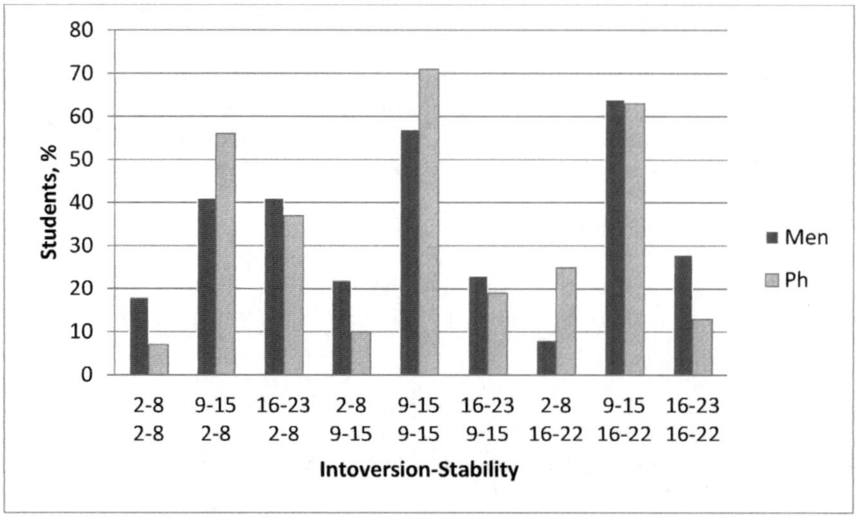

Figure 1. Personality types among students of different specialties.

For graduates of physical specialties of the highest number of students belongs to the normal (42.5%), sanguine, phlegmatic (17.2%), choleric (11.5%) and sanguine (11.5%) types.
Table 1 shows results of various tasks accomplishment. In both groups no less than 80% of students accomplished no less than 60% of tasks. Successful accomplishment of tasks both in terms of type and category differs in both sample groups significantly. Managers successfully cope with the tasks in the resolution of conflicts. At that time, both for the future physicists these tasks are difficult.

Table 1. Results of tasks accomplishment for M and Cis students.

	M	Ph
Conflict resolution tasks	73%	40%
Team working tasks	38%	39%
Team leading tasks	56%	53%
Analytical tasks	35%	47%
Decision making tasks	65%	53%

Next, cluster analysis was carried out in the field of 12 attributes: introversion, stability, conflict conduct strategy A, strategy B, roles of "Coordinator" (Co), "Shaper" (S), "Plant" (P), "Monitor Evaluation" (ME), "Implementer" (I), "Resource Investigator" (RI), "Teamworker" (T), "Completer Finisher" (CF).

Clustering was performed for different numbers of clusters, from 2 to 10. For each cluster identified the mean values of clustering variables in the original scales, as well as the average values of the variables sumA and sumS characterizing the level of response to questions on Analytics and decision making. If the number of points given to the role, less than 5 – it is considered spam when more than 10 is preferred.

All samples are clearly separated main clusters, in which the subjects coped well with analytical tasks and assignments on decision making. Have been allocated to clusters for which both tasks were impossible. In addition, were selected as clusters where the subjects coped well with only analytical tasks or with the tasks on decision making.

As a result of clustering identified five classes:

• "Successful" that can easily cope with all types of jobs;

• "Analysts", who can easily cope with the analytical tasks, but poorly on tasks with decision-making;

IOP Conf. Series: Journal of Physics: Conf. Series **781** (2017) 012060 doi:10.1088/1742-6596/781/1/012060

• "Solvers", to which job decisions were considerably easier than analytical;
• "Losers", for which any type of task was very difficult;
• "Neutral", for which all the tasks were of average difficulty.

In general, it can be noted that of all variables the largest contribution to the characteristics of the classes made just by the role values; strategy type and the personality type of the subjects (introversion and stability) is influenced not so pronounced.

The ratio of these classes for different departments is shown in Fig. 2.

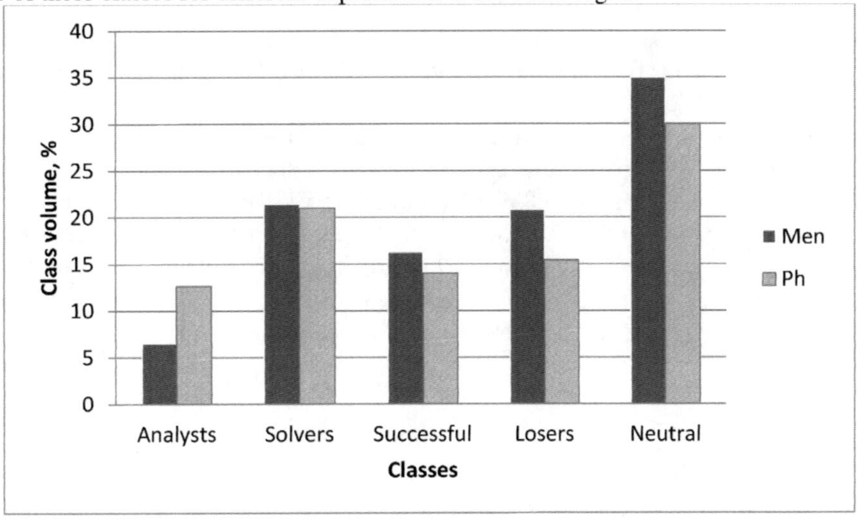

Figure 2. The ratio of allocated classes.

For each of the classes were identified rules for deciding the ownership of the test to a particular class. Overall, this study has allowed to achieve objectivity in the process of estimating the level of formation of relevant cognitive competencies.

Class "Neutral" have formed cognitive competence for knowledge transfer. Class "Solvers" (20% - 21%) more in line with efforts to transform the creative knowledge. For the class of "Analysts" (6% - 13%) is characterized by the search and knowledge discovery.

The amount of class "Successful" is 14% -16% of the total volume of graduates depending on the areas of training. This class corresponds to the fully formed cognitive competencies of innovators.

The results allowed to formulate a number of recommendations on the use of educational technologies for various areas of training.

The results of the tasks on the ability to work in team showed less than 40% - not too high a result. To form the respective competencies necessary to enter into the learning process of project learning in small groups for some subjects. This project will allow training in practice to form a teamwork skills, conflict management, etc.

The results obtained in solving these types of tasks, such as analytical and decision-making, talk about the balance of the educational process at the faculty of "F".

For the training areas "Y" of the Faculty of superiority in the decision-making tasks for natural and meet the requirements. Nevertheless, the proportion of students who have successfully completed the analytical task is small and is only 35%. This means that the natural science disciplines unit formally implemented, teachers do not take into account the peculiarities of training of graduates of economics.

4. Conclusions

Thus, the survey has made it possible to assess the level of formation of cognitive competencies of graduates, oriented to work in the state corporation Rosatom. In the aggregate, 80% to 85% of graduates of MEPhI have cognitive competence, sufficient to work in the field of nuclear knowledge management.

References

[1] Milner B 2003 The concept of knowledge management in the modern organization *Russian Journal of Management* **1** pp 57-76.

[2] Geraskin N, Kosilov A, Kulikov E and Tolstenkov A 2015 Knowledge management for nuclear organizations: a tutorial (Moscow: Mephi) p 405

[3] Pogorelova E 2015 Evaluation of formation of professional cognitive competencies of staff *Naukovedenie* **6 (31)** p 77.

[4] Glukhikh I and Pryakhina E 2015 Innovative educational environment of universities and competencies of graduates-innovators *Historical and social educational ideas* **7(4)** pp133-136.

[5] Guseva A, Silenko A and Kireev V 2014 System for assessing the socio-personal competence for certification qualifications of specialists-managers in the nuclear industry *Life Sci J* **11(9)** pp 859-864 doi:10.7537/marslsj110914.129

[6] Kireev V, Guseva A and Silenko A 2015 Social and personal competence assessment within qualification certification of nuclear industry university graduates *Procedia - Social and Behavioral Sciences* **214** 2015, pp 150-158 doi:10.1016/j.sbspro.2015.11.608

IOP Conf. Series: Journal of Physics: Conf. Series **781** (2017) 012061 doi:10.1088/1742-6596/781/1/012061

Modernization of physical protection educational laboratories in the National Research Nuclear University MEPhI

N I Geraskin and A A Krasnoborodko

Department Theoretical and Experimental Physics of Nuclear Reactors, National Research Nuclear University MEPhI (Moscow Engineering Physics Institute), 115409, Moscow, Kashirskoe highway, 31, Russia.

E-mail: aakrasnoborodko@mephi.ru

Abstract. Non-proliferation of nuclear materials includes, in addition to accounting and control, the Physical Protection (PP) of one. The paper considers the experience by MEPhI in application the practical educational in the area of PP technical systems. The following aspects are discussed in the paper: specific features graduate program in nuclear security area; overview of the practical course curricula in the special laboratory.

1. Background

MEPhI has the 20-year experience in developing and implementing the educational programs of different graduate levels in the area of Nuclear Security. There are the Master of Science Graduate Program and Engineer Degree Program. To train high level specialists is needed laboratory with modern equipment. Joint efforts undertaken by the RF State corporation Rosatom, US Department of Energy, and MEPhI have resulted in creating the MS Graduate Program (MGP) in Nuclear material (NM) Physical Protection, Control and Accountability at MEPhI Department "Theoretical and Experimental Physics of Nuclear Reactors" in 1997. The educational program takes two academic years and has been developed for training those, who have already completed technical education at the level of Engineer or Bachelor of Sciences in the associated directions. The graduates receive the MS degree in Physics. In 2001 further joint efforts was started the Engineer Degree Program (EDP) "Security and Non-Proliferation of Nuclear Materials". This educational program takes 5,5 academic years. Under the EDP the students have an opportunity to have a specialized training in the following two directions: nuclear materials (NM) control and accountability and physical protection of nuclear objects and materials.

2. The main direction of PP training

The graduates from the areas of nuclear security and NM physical protection shall:

- Know all the regulatory requirements to PP systems of nuclear objects, methodology for PPS development and upgrading, have a clear notion about a variety of PP engineering and technical tools;
- Have a capability to prepare the technical requirements on PP systems creating or upgrading, have a capability to estimate the proposed decisions, organize and conduct the works on PPS creating or upgrading;

IOP Conf. Series: Journal of Physics: Conf. Series **781** (2017) 012061 doi:10.1088/1742-6596/781/1/012061

- Have a capability to apply the methodologies for assessment of PP systems effectiveness.

Currently, these areas are covered by the following main disciplines:

- Fundamentals of nuclear materials physical protection, control and accountability.[1]
- Methods of vulnerability analysis and optimization of PP systems.[2]
- Equipment and devices of physical protection systems.[3]

The training course "Fundamentals of NM physical protection, control and accountability" is the first course of the direction that contains introduction into the specialty. The training course "Methods of vulnerability analysis and optimization of PP systems" covers the methodologies and basic principles used in PPS designing, creating and effectiveness assessment.

The course "Equipment and devices of physical protection systems" contains deep knowledge the engineering and technical tools, which are the mostly used in PP of nuclear objects, and specific features in their applications.

In present time, PP training laboratories, created in 2003, consist of separate training places, in which the following laboratory works are carried out:

- Interior sensors for physical protection systems;
- Access control and management system;
- Video-surveillance systems;
- Data acquisition and processing systems.

Also students learn the simplest methods of integration of separate PP subsystems, such as "dry contacts" (relay). For example – start of camera work by a signal from the sensor alarm. Such methods were widely used in real PP systems at nuclear facilities at the beginning of the millennium.

3. Modernization of PP training laboratories

Over the past decade, new equipment and methods have been established on nuclear facilities PP system. For instance, integration of subsystems based on Ethernet technologies is used overall, web-cameras are widely applied, biometric and video analysis technologies have been developed. These changes force us to improve the educational process. Therefore, MEPhI is able to update laboratory equipment under financial support from the US Department of Energy in 2016.

The new laboratory complex will be based on several workstations. Every workstation deals with its own PP subsystems (access control, alarm sensors, video cams). All subsystems are connected with each other by network.

The educational process in the PP laboratory training includes the following PP components:

- detection systems and tools for physical protection of rooms, buildings and perimeters, methodologies and specific features in their applications;
- access control systems and tools: identification tools, control and management systems, devices of biometrical identification, executive means, methods and technologies for creation of the distributed access control networks;
- main components of video-surveillance system: camera-system, tools for transmission of video-pictures, commuters, recorders, methods and technologies for creation of the distributed video-surveillance networks, remote monitoring system;
- methods and technologies for creation of the integrated security systems: alarm systems through radio-channels and telephone lines, the integrated PP including systems for protection and situation assessment at lengthy perimeters;
- fire alarm systems.

Fire alarm system is not formally PP subsystem, but uses similar principles of operation. Moreover, fire systems are equipped at all the nuclear and other dangerous facilities.

Figure 1 shows the flowchart of interaction access control and alarm subsystems.

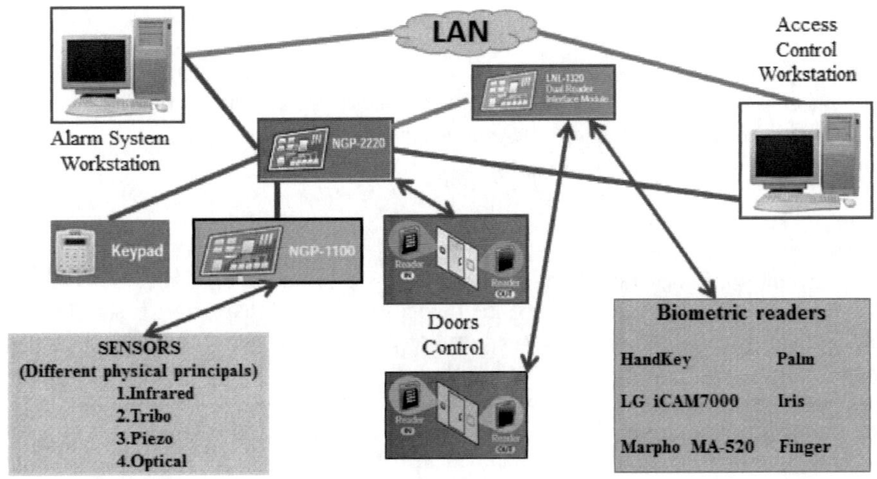

Figure 1. The Access Control and Alarm System laboratory facilities scheme. Lines show the LAN (Local Area Network) connection. Arrows show information transfer under firmware protocols, the interface is provided by intelligent NGP controllers family.

Video surveillance is one of the most complex PP subsystems. Figure 2 shows the flowchart of designing video parts for laboratories.

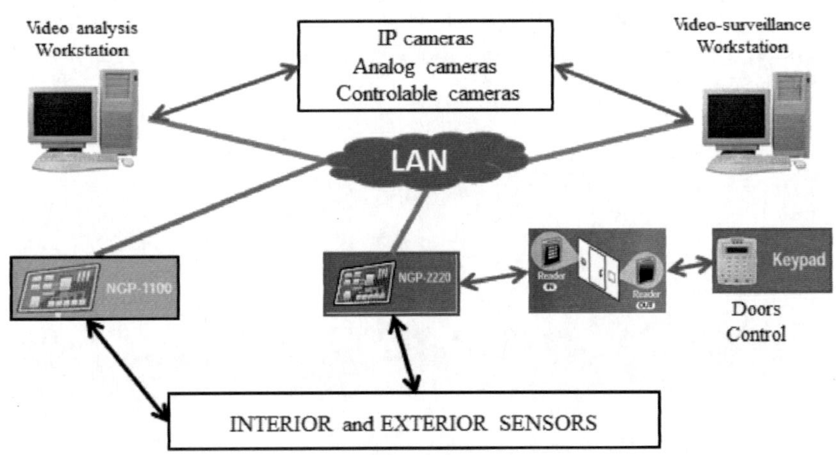

Figure 2. Laboratory video facilities scheme. Designations correspond to the previous figure. As a signal source to be used for above ten cameras.

During training, students consistently study separately all subsystems and, principles of operation of each equipment element. However, since all the parts are integrated with a common manager server, learning tasks can be more complex. It is possible to organize complicated scenarios of interaction of subsystems. Possible options for such modeling scenarios are listed below.

a) The sensor detects intrusion into the protected area. Video system receives an alarm signal, selects the appropriate camera and displays the picture on the operator workplace. Simultaneously, the picture is transmitted to the video analysis work station. After analysis, the characteristic can be given, is the event false or not.

b) The employee enters an area which is protected by sensors. Access control system checks whether the entry is authorized or not. If access is denied, the sensors are switched off. Otherwise, there is an alarm and other actions perform, such as blocking doors and locks.

The modeling of those and similar tasks allows students to successfully take on the challenges of real objects.

4. Conclusion

For the modernization of the training PP laboratories the modern equipment has been selected. Characteristics of equipment correspond to same devices used at nuclear sites. It should be noted that the PPS training and research laboratories were structured as a "sand-box", i.e. the students can create and simulate various PP system structures within the limited frames of available laboratorial equipment. These capabilities are of special importance in the course of the training research works, pre-diploma internship works, preparation of the graduate qualification works or preparation of the MS thesis. New equipment is going to be installed in the PP laboratories for the Fall-2016 semester, and currently activity on creation and development of new laboratorial practical works is underway.

References

[1] Glebov V B, Izmailov A V, Rumyantsev A N 2001 *Introduction To The System Of Accounting, Control and Physical Protection of Nuclear Materials: Textbook* (Moscow: MEPhI)

[2] Izmailov A V 2002 *Methods for the Design and Analysis of the Effectiveness of Physical Protection Systems of Nuclear Materials and Facilities: Textbook* (Moscow: MEPhI)

[3] Bondarev P V, Izmailov A V, Tolstoy A I 2008 *Physical Protection of Nuclear Facilities: a Textbook for Universities* ed. N S Pogozhin (Moscow: MEPhI)

[4] Badikov A V, Bondarev P V 2007 *Video Surveillance Systems: the Possibility of Equipment and Construction of Distributed Systems: Laboratory Workshop.* (Moscow: MEPhI)

[5] Geraskin N I, Krychkov E F, Krasnoborodko A A, Piskureva N F, Pogozhina I N 2014 *Nucl. Phys.and Engineering* **5** 29

IOP Conf. Series: Journal of Physics: Conf. Series **781** (2017) 012062 doi:10.1088/1742-6596/781/1/012062

Improvement of the course "Management of intellectual property" based on the mixed state contract in the field of scientific R&D

D Yakovlev, A Ageev, E Yushkov and M Bogatyreova

National Research Nuclear University MEPhI (Moscow Engineering Physics Institute), Kashirskoe shosse 31, Moscow, Russia

E-mail: DYYakovlev@mephi.ru

Abstract. Intellectual property (IP) is one of the forms of storing knowledge – intangible assets of knowledge economy. The translation of IP knowledge to the young generation is one of the challenges of nuclear knowledge management. At the NRNU MEPhI (National Research Nuclear University MEPhI), the subject is studied within the framework of the academic course "Management of intellectual property. The aim of the course is to train qualified specialists, ready to use modern methods of strategic management of IP in commercial firms. The article is devoted to the strengthening and transfer of IP rights for scientific output application in industry and commerce. The state remains the main source that finances all the significant developments in the field of science and technology. Therefore, the primary task is to effectively utilize the R&D output created at the expense of the Federal budget, both within the state order and the estimated budget financing. Currently, there exist contradictions in the area of strengthening and transfer of rights for R&D in the field of science. The newly borne concept dealing with the strengthening of IP rights for the scientific R&D output, centers on the theory of a mixed state contract and is conveniently integrated with the academic course mentioned. Knowledge itself takes the form of IP, as soon as it becomes formalized. The academic course when supplemented with various approaches to strengthening and transfer of IP rights for the scientific R&D output certainly gives a better understanding of the commercialization process of the intellectual capital and structures relevance to intellectual property. The research material is integrated with the educational process, and the academic course "Management of intellectual property" is designed both for economic and engineering specialties.

1. Introduction

Intellectual property (IP) is one of the forms of storing knowledge – intangible assets of knowledge economy. The translation of IP knowledge to the young generation is one of the challenges of nuclear knowledge management. At the NRNU MEPhI (National Research Nuclear University MEPhI), the

IOP Conf. Series: Journal of Physics: Conf. Series **781** (2017) 012062 doi:10.1088/1742-6596/781/1/012062

subject is studied within the framework of the academic course "Management of intellectual property", which is delivered to will-be economists, physicists and engineers. The aim of the course is to train qualified specialists, ready to use modern methods of strategic management of IP in commercial firms.

In the result of acquiring the fundamentals of the discipline, undergraduates are supposed to:

- be aware of the legal, economic, organizational maintenance of IP management;
- understand the characteristics of strategic management of IP as a developing business resource;
- get an idea of the valuation, insurance, accounting and tax accounting of intangible assets, aimed for commercial use [1].
- be able to use methods of civil law regulation of relations arising in connection with the creation and use of IP objects.

The state remains the main source that finances all the significant developments in the field of science and technology. Therefore, the primary task is to effectively utilize the R&D output created at the expense of the Federal budget, both within the state order and the estimated budget financing [2]. Currently, there exist contradictions in the area of strengthening and transfer of rights for R&D in the field of science. In particular, these include the incompleteness of the Russian legislative base regulating the aspects of commercialization of the intellectual activity output.

The newly borne concept dealing with the strengthening of IP rights for the scientific R&D output, centers on the theory of a mixed state contract and is conveniently integrated with the academic course mentioned.

Knowledge itself takes the form of IP, as soon as it becomes formalized. The academic course when supplemented with various approaches to strengthening and transfer of IP rights for the scientific R&D output certainly gives a better understanding of the commercialization process of the intellectual capital and structures relevance to intellectual property.

There are a few major provisions in our new approach to be considered below.

2. Practical approaches to strengthening rights for using R&D outputs in the field of science

In the XXI century intellectual capital is associated with the source of gaining competitive advantage – a value to be turned into productive force, contributing to the export of state-of-the-art technologies and products of innovative performance. Nevertheless, there are some contradictions in the way of Russia's economic advancement that stem from both nationwide and sectoral challenges affecting innovative activity of enterprises:

- underdevelopment of the Russian market of high technology products, as well as the lack of effective demand for high-tech products;
- complicated financial and economic condition of industrial enterprises, lack of internal working capital, funds for the implementation of innovative projects;
- incompleteness of the Russian legislative base regulating all the aspects of intellectual activity results commercialization [3,4];
- insufficient number of companies and specialized units in enterprises for technology commercialization, as well as highly qualified specialists in the IP management.

In principle, a vigorous involvement of the R&D output in the economic turnover requires a clear regulation of the issues of exclusive rights for the intellectual product by law. For businesses to succeed in the dynamically changing economic environment it is important to use various forms of public-private partnership in the sphere of science and innovation. There are substantial benefits to be gained from different approaches in order to endorse IP rights for scientific outputs [5]:

- concluding a state contract;
- providing the condition of an agreement, according to which all the rights belong to the state, if the parties do not envisage otherwise;
- providing the condition of an agreement, according to which all the rights belong to the contractor, unless the parties envisage otherwise;
- the rights for commercial use belong to the state;
- the state retains the rights to support consumers receiving new products at a reduced price:

IOP Conf. Series: Journal of Physics: Conf. Series **781** (2017) 012062 doi:10.1088/1742-6596/781/1/012062

• the rights belong to the state, which may transfer them to economic management by the state unitary enterprise;
• the rights should belong to the contractor: it can stimulate them to identify the scientific R&D output, allow them to exercise these rights better than the state does, since it has adequate capacity; increase tax revenues and expand employment.

Note, that the latter option is reflected in the Government Resolution No. 685 "On the procedure of using rights for R&D output obtained at the expense of the Federal budget". It implies that the state customer unilaterally exercises the option of acquiring rights, provided that:
• the R&D output is essential to perform public functions related to ensuring defense and security of the country;
• the introduction of the R&D output into economic turnover with the state participation is beneficial to the state;
• the given output is withdrawn by virtue of the law or may be limited in circulation. In other cases, rights can be assigned to the organization or to the Russian Federation and the performing organization in conjunction [4].

Central to the theoretical concepts of endorsing rights for R&D outputs is the concept of a mixed state contract.

In accordance with article 421 (item 3) of the RF Civil Code, "Parties may enter into a contract containing elements of various contracts provided for by law or other legal acts (mixed contract)" [6]. In fact, the state customer, ordering R&D works, is primarily interested in the public use of the obtained results (for example, in the field of defense, healthcare, environmental protection, social assistance, etc.). Simultaneously, the output can have a commercial application that attracts the contractor – a commercial organization.

Thus, in case of a mixed agreement including elements of the agreement on joint activities, its participants acquire rights to use common result in exchange for their deposits [7].

3. The use of a new type of a contract for the transfer of Federal technology from state scientific organizations to the sphere of business

The evidence points to the importance of regulating by law the use of a new type of a contract. Necessarily, the contract must be well documented and signed by the head of the scientific institution with the commercial organization on behalf of the Russian Federation. Such a contract is of a mixed character and includes a number of elements:
• contracting out, the first phase of which is the transfer of a pre-commercialized unfinished technology;
• contract on the RF joint activities, including both the institution and the commercial firm acting on behalf of the state;
• license agreement (a contract that provides a given commercial firm with the full state commercial license for the use of Federal technology).

Still there is a certain inaccuracy, which needs clarification with regard to the widespread treatment of this agreement abroad – "agreement on the joint R&D". The interpretation seems to reflect only one element of the given mixed agreement, along with the Agency and license elements.

Upon the whole, Figures 1-5 present options for transferring public technologies to the two segments: state contract and estimate-budgetary (administrative) segments. We will examine three possible ways of technology transfer in the state contract segment, depending on the distribution of functions between the parties.

Under the first option, the R&D contractor is the only participant in the commercial refinement of the technology (CRT). This option was reflected in the Federal Draft Law "About technology transfer". The evidence points to a deficiency that speaks for itself: the factor of considerable transaction and transfer costs (costs of developing the technology) is practically ignored in this scheme (See Fig 1).

IOP Conf. Series: Journal of Physics: Conf. Series **781** (2017) 012062 doi:10.1088/1742-6596/781/1/012062

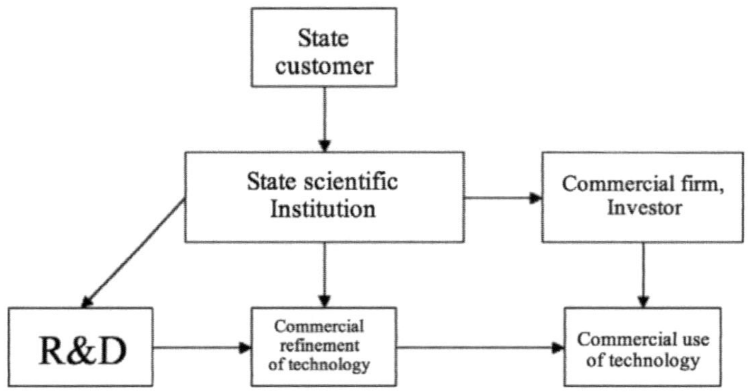

Figure 1. The first option of technology transfer in the state control segment.

Under the second option, the commercial firm is a further user of technology. Even though the contractor is involved in its commercial refinement, this procedure is unlikely to fully solve the problem (See Fig. 2).

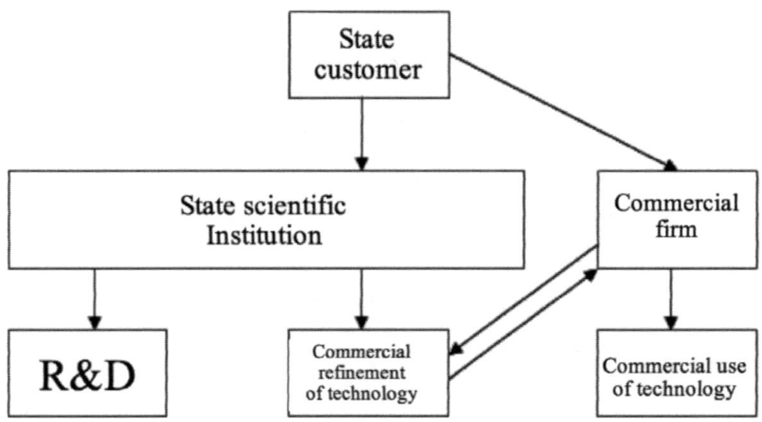

Figure 2. The second option of technology transfer in the state contract segment.

Eventually, the third option may be recognized as the most adequate one, because it is based on the diversification of the participants' activities and in addition to performing R&D assumes commercial refinement of the technology as well as its commercial use. Regretfully, this option was not reflected in the bill (see Fig. 3).

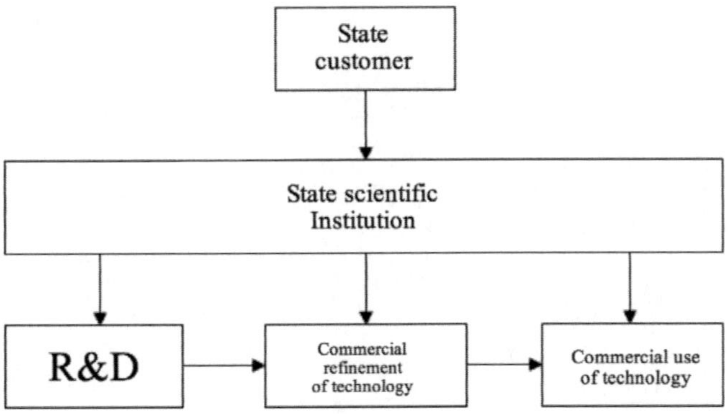

Figure 3. The third option of technology transfer in the state contract segment.

As for the estimate-budgetary segment, it would be beneficial to make use of the US experience, based on the partnership between public research institutions and commercial firms for commercial refinement of technology. The advantages also suggest a subsequent transfer of modified technologies to commercial firms to be used in the framework of a licensing agreement (see Fig.4 and 5).

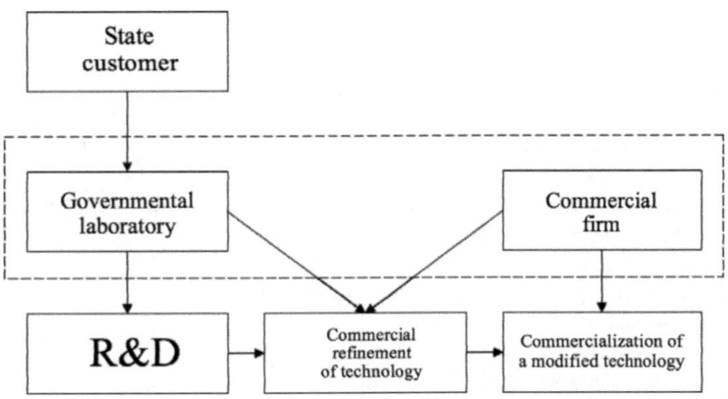

Figure 4. Commercialization of technologies developed in the US government research laboratories.

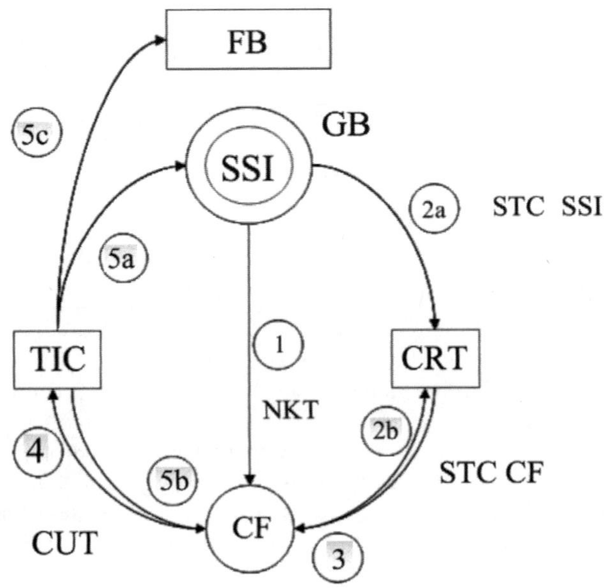

Designations:
GB – government body;
SSI – state scientific institution;
STC SSI – scientific and technical contribution of the SSI to the commercial development of technology;
STC CF – scientific and technical contribution of a firm to the commercial refining of technology;
CRT – commercial refinement of technology;
CF – commercial firm;
TIC – the total income from the commercialization of technology;
NKT – commercially non-modified technology;
CUT – commercial use of the technology;
FB – Federal budget. 1 2a 2b 3 4 5a 5b 5c
The sequence of actions:
1 – provision of commercial firms with NMT;
2A, 2B – scientific and technical contributions of CF and SSI to the commercial refinement of technology;
3 – transfer of NMT to a commercial firm;
4 – commercial use of the technology;
5a, 5b, 5c – distribution of the earned income between CF, GB, SSI and the Federal budget.
Figure 5. The proposed model of the commercialization of technologies developed in public research institutions.

Summarizing the importance of theoretical concepts of strengthening IP rights for the R&D output, the study investigated the features of a mixed state contract, where scientific and commercial interests are met.

We chose to support these theories with a sound practical base as applied to technological innovations. As was shown, there are different ways to transfer a technology from a public research institution to the sphere of business. The most appropriate option, in our opinion, is based on the diversification of activities: transfer of commercially non-modified technologies to a firm, commercial refinement of the technology and its commercial use. Thereby, it is logical to suggest that this procedure needs a further regulation by law.

The development of these ideas in an academic module aims at teaching graduates to use operational interdisciplinary knowledge in science-intensive branches of industry and business.

References

[1] E.S. Yushkov, A.E. Yushkov, A.E., Medvedeva, D.U. Yakovlev, Status and trends of IP insurance development in Russia: Technology of mechanical engineering, Vol. 4, 2014, pp. 68-71

[2] E.S. Yushkov, G.A. Eremenko, IP rights in the field of scientific research and technology development (Poligrafics RPK, 2009), 127 p.

[3] RF Government Resolution No. 685 "On the procedure of utilizing rights for scientific R&D output (November 17, 2005)

[4] Provision on consolidation and transfer to business entities of rights for scientific R&D output obtained at the expense of the Federal budget" No. 685, Approved by the RF Government Decree (November 17, 2005)

[5] A.E. Yushkov, E.S. Yushkov, V.A. Tupchienko, Creation of the effective system of utilizing rights for scientific R&D output in the development of innovation management in Russia: Scientific Bulletin of Moscow State Technical University of Civil Aviation, No 131(7), 2008, pp. 113-119

[6] Civil code of the Russian Federation, part.1, N 51-FZ (Nov.30, 1994, amended on May, 23, 2016)

[7] A. E. Yushkov, E.S. Yushkov, Approaches to the practical solution of consolidating rights for R&D output: Scientific Bulletin of Moscow State Technical University of Civil Aviation, No 131(7), 2008, pp. 113-119

International academic program in technologies of light-water nuclear reactors. Phases of development and implementation

N I Geraskin and V B Glebov

National Research Nuclear University MEPhI (Moscow Engineering Physics Institute), Kashirskoe shosse, 31, Moscow, 115409, Russia

E-mail: VBGlebov@mephi.ru

Abstract. The results of implementation of European educational projects CORONA and CORONA II dedicated to preserving and further developing nuclear knowledge and competencies in the area of technologies of light-water nuclear reactors are analyzed. Present article addresses issues of design and implementation of the program for specialized training in the branch of technologies of light-water nuclear reactors. The systematic approach has been used to construct the program for students of nuclear specialties, which corresponding to IAEA standards and commonly accepted nuclear principles recognized in the European Union. Possibilities of further development of the international cooperation between countries and educational institutions are analyzed. Special attention is paid to e-learning/distance training, nuclear knowledge preservation and interaction with European Nuclear Education Network.

1. Introduction

One of the basic strategic initiatives advanced by National Research Nuclear University Moscow Engineering Physics Institute (NRNU MEPhI) consists in forming a system of the academic programs and teaching appliances to train high-quality, internationally recognized specialists in nuclear area. Within the frames of this strategic initiative, NRNU MEPhI has participated during the past four years in the European projects CORONA and CORONA-II related with creation and development of the European Centers of Competence on technologies of VVER-type nuclear power reactors.

Technology of VVER-type light-water power reactors has been worked out in the 1950-60s by the USSR specialists. Now, the VVER technology continues upgrading and serves as a basis for many nuclear power plants currently in operation in several European countries [1]. Some innovative projects of VVER-type reactors are currently under implementation in Finland, Hungary, Czech Republic, Turkey, Vietnam, China and India [2]. Multi-year reliable and safe operation of VVER-type reactors has given a foundation to regard the VVER-technology as one of main directions towards further development of nuclear energy systems in European states and in some other countries. Therefore, creation of the European Centers of Competence and their activity in training and qualification upgrading of the specialists involved in operation of VVER-type reactors became an actual task for many countries.

Besides the Russian Federation, the following European countries have participated early and are participating now in the CORONA and CORONA-II Projects because they already apply or interested in future application of the VVER-technology:

- Bulgaria: Kozloduy NPP, Institute for Nuclear Research and Nuclear Energy (INRNE), Risk Engineering Ltd.

- Germany: ESIS GmbH.
- Hungary: Budapest University of Technology and Economics (BME).
- France: European Nuclear Education Network (ENEN).
- Belgium: Joint Research Centre JRC-IE.
- Czech Republic: Research Center CV Rez.
- Finland: Fortum Oyj.
- Austria: PM Dimensions GmbH.
- Spain: Consortium Tecnatom.

2. The CORONA project (2012-2014)

The CORONA and CORONA-II Projects foresees a wide envelopment of many educational and scientific problems related with the VVER-technologies. The works on creating the European Center of Competency were carried out in several large directions. These directions were presented in the form of the Work Packages (WP). Experts from several countries, members of the CORONA Project, participated in performing the Work Packages. For example, within the WP1 frames the works were carried out to define the demands for training of various target groups and the requirements to professional competencies of the specialists graduated from the educational programs. The work packages WP2 (the leading organization - ESIS GmbH, Germany) and WP3 (the leading organization - INRNE, Bulgaria) were related with developing the academic programs and teaching appliances for two target groups, namely for nuclear specialists directly involved into NPP operation and for non-nuclear specialists (sub-contractors, for instance). The work package WP5 was devoted to studying the human factor effects on safe operation of VVER-type reactors and the problems of nuclear security culture. The Consortium Tecnatom (Spain) is the leading organization of this direction, responsible for creation of appropriate training module and further realization of the works under the WP5.

3. The academic program "VVER-technologies for the students of nuclear specialties"

NRNU MEPhI is the leading organization in realization of the work package 4 (WP4) which is related with development and implementation of the academic program in VVER-technology for the students of nuclear specialties ("nuclear" students).

From the very beginning, creation of the academic program in VVER-technology for the "nuclear" students has been based upon the widely used systematic approach to the educational process.

That is why the first phase in developing the academic program in VVER-technology for the "nuclear" students was devoted to analysis of the demands for training (including identification of potential target groups of the "nuclear" students) and exact definition of the educational aims. The educational aims were briefly expressed in the form of competencies acquired by the specialists graduated from the program. Under the CORONA Project the academic program was created for two main target groups, namely for the students of nuclear and non-nuclear specialties. The present paper is limited by consideration of the first target group, i.e. of the "nuclear" students only.

The "nuclear" student group included two target sub-groups: the energy sub-group ("VVER-technology") and non-energy sub-group ("Technical aspects of NPP with VVER-type reactors"). For both sub-groups the requirements were prepared to the program entrants. In general, preliminary qualification of the program entrants was defined at the level of the Bachelor degree (or above the level of the Bachelor degree) in nuclear specialties. Correspondingly, the sets of the competencies acquired by the specialists graduated from the academic program were formulated for both target sub-groups.

This formulation of the competencies for the graduates from the academic program for the "nuclear" students was essentially a focal point in implementing the systematic educational approach. Just a list of the competencies defines, in the end, full spectrum of knowledge acquired by the graduates, namely what they must be able to do, what methods and equipment units for research, development and designing they must be able to apply. In this connection, it was very important to define the contributions from each training course and from each educational module of the academic program to forming full spectrum of the graduate's competencies. These contributions to the competencies formed during the course or the module delivery can demonstrate significance of the courses and modules for accomplishing a final aim of the educational process under the academic program.

The competency models of the graduates, the educational programs and associated methodological materials were created with application of the experiences gained by NRNU MEPhI in implementing the BS Graduate Programs and the MS Graduate Programs in nuclear physics and technologies. These experiences allowed us, in a lot of aspects, to use professional competencies of the specialists graduated from the BS and MS Graduate Programs which were worked out in tight cooperation and concurred with Russian nuclear employers for many years of joint activity. Of course, the works on formulation of professional competencies for the specialists graduated from the academic program "VVER-technologies for the students of nuclear specialties" were carried out in tight collaboration with other partners of the CORONA Project.

In accordance with the systematic approach to the educational process, the next phase consisted in forming the educational modules, the program as a whole and the curricula for both target sub-groups, namely for the sub-group "VVER-technology" and for the sub-group "Technical aspects of NPP with VVER-type reactors".

At this phase a modular principle was adopted as a basic methodology for creating the program. Each educational module consisted of a series of the training courses intended to form a certain type of professional competencies including production and technological competencies; research, development and designing competencies; organizational and managerial competencies; expert competencies, for each target sub-group. The modular structure of the academic program has allowed us without any serious difficulties to embody the systematic approach mentioned above because the flexible modular structure is able to satisfy completely all the industrial demands for the graduates with various nuclear specialties and with various professional competencies.

An important final phase in creating the academic program was the establishment of the compliance between the educational results and the problems to be solved by the graduates in their future professional activity. A fund of the evaluating means could be used to control and assess the professional competencies acquired by the program's trainees. Development of such a fund was the final phase in creating the academic program "VVER-technologies for the students of nuclear specialties". Assessment of the professional competencies acquired by the students of both target sub-groups ("VVER-technology" and "Technical aspects of NPP with VVER-type reactors") has demonstrated that the competencies in VVER-technology were in full compliance with the 3-6 levels of the European Qualification Framework [3] depending on the graduate's specialization.

4. Pilot implementation of the program "VVER-technologies for the students of nuclear specialties". Analysis of its effectiveness

This phase in realization of the academic program includes the pilot implementation of the educational module devoted to nuclear and radiation safety of VVER-type reactors.

Topical actuality of the educational module "Safety-related aspects of VVER-technologies" aroused a keen interest from the side of potential trainees and instructors of the training course. The number of applications from different countries on participation in the training courses exceeded our capabilities. The entrants who passed selection successfully included the students from Sevastopol National Technical University, Odessa University, Belarus State University, Al-Farabi Kazakh National University, Nizhny Novgorod State Technical University, National Research Nuclear University MEPhI and so on.

The module's curriculum has been compiled by such a way that, in addition to brief overview of the fundamental principles related with physics and safety of light-water reactors, significant attention was given to some practical aspects of safety ensuring at NPP with power reactors of VVER-1000 type. The lessons on the multi-functional simulator of VVER-1000 [4] at NRNU MEPhI have allowed

IOP Conf. Series: Journal of Physics: Conf. Series **781** (2017) 012063 doi:10.1088/1742-6596/781/1/012063

the students to feel the real difficulties in safe operation of VVER-1000 and potential consequences from their erroneous actions and deviations from nominal regime of the reactor operation.

The next phase consisted in assessing the results obtained in the pilot education and included the following three significant components:

- Quality assessment of learning the educational materials by the students with application of tests and control tasks from the fund of the evaluating means that was previously worked out.
- Expert assessment of the educational module by the invited specialists from the Open Joint-Stock Company "Concern ROSENERGOATOM", the National Research Center Kurchatov Institute, Institute of Physics and Power Engineering, Rostechnadzor, Bulgarian NPP "Kozloduy" and Bulgarian Institute for Nuclear Research and Nuclear Energy (Sofia).
- General evaluation and comments on quality of the submitted methodological materials by the students.

When the certificates on successful completion of the educational module, official representative of the Open Joint-Stock Company "Concern ROSENERGOATOM" has marked large quantity of correct answers to the tests and, as a consequence, high level of learning the educational materials. This opinion was supported by other experts.

As it was noted in the student comments, the lessons and practical exercises conducted in the pilot delivery of the educational module have given additional pulse to them and called interest from the standpoint of their future participation in the studies related with safe operation of NPP with VVER-type reactors.

5. The CORONA-II project (2015-2018)

Further development of the European Centers of Competency in VVER-technology foresees creating the International Association of the educational institutions in the area of VVER technologies, so called VVER Academy. The VVER Academy must become not only an organizational basis for rendering the educational services but also the VVER Academy can make it possible to involve many other countries (outside of the European Union, for instance) which are interested in the use of power VVER-type reactors and associated nuclear technologies. Tight collaboration with and further connection of the CORONA Consortium up to the European Nuclear Education Network (ENEN) can play here a significant role. Such an association could form a common approach to the European certificate for Master of Science degree in nuclear engineering. The common approach should be based on the following four main requirements:

- Development of common qualification criteria in the area of nuclear engineering.
- Creation of common system for mutual recognition of the European MS certificates.
- Promotion for more mobility of the professorial staff and trainees throughout the European Union.
- Promotion of more tight collaboration with and feedbacks from nuclear employers.

The European Nuclear Education Network, within the frames of the EUROATOM – ROSATOM agreement, implements already several projects with Russian organizations for developing the joint educational programs under the conditions on mutual recognition of the acquired qualifications and credits.

The integration of the CORONA Consortium with the European Nuclear Education Network could promote the wider implementation of the educational programs in VVER technology to the EU countries and outside. The CORONA-II Project foresees developing a structural scheme of the VVER Academy and mechanisms for the CORONA-ENEN cooperation. It is presumed that intense activity of the VVER Academy and ENEN work teams will form a foundation for further well-planned and fruitful functioning of the interacting bodies.

One of main directions towards further perfection of the academic program "VVER-technologies for the students of nuclear specialties" consists in performing the approbation (pilot implementation) of the program for different target groups. Application of the remote education techniques can offer the larger prospects for the wider dissemination and use of the program in practical training activities.

IOP Conf. Series: Journal of Physics: Conf. Series **781** (2017) 012063 doi:10.1088/1742-6596/781/1/012063

As is planned, the nearest task of the VVER Academy included the works on final development and pilot implementation of the academic program with application of electronic and remote educational forms under the work package WP5 of the CORONA-II Project. This task was entrusted to National Research Nuclear University MEPhI, as the leading organization of the work package WP5, and to its partners in the Project.

The pilot implementation of the academic program with application of electronic or remote educational forms is complicated by needs to resolve some methodological, organizational and technical problems. Amongst them, the demands for technical educational appliances should be mentioned in the first turn.

In the CORONA-II Project the Cyber Learning Platform for Nuclear Education and Training (CLP4NET) [5] was selected as a main instrument for conduction of remote (distant) lessons. The CLP4NET platform is currently and successfully used to solve some problems of the University.

In addition to the implementation of technical appliances for remote education, the work package WP5 foresees forming a library of the educational modules and their pilot application for electronic or remote education of foreign students. Here, like previously, National Research Nuclear University MEPhI is the leading institution responsible for creating the educational module in tight collaboration with foreign partners. The educational module for "nuclear" professionals is under joint development now with the Engineering Support Company (Germany), for "non-nuclear" professionals – with experts from Institute for Nuclear Research and Nuclear Energy (Bulgaria).

It should be noted that, under the previous CORONA Project, each educational module has passed a detailed analysis performed by all participants of the Project. Such a multilateral analysis gave an objective estimate to the modules in the term of their efficiency or helped to elaborate some concrete corrective measures. It is planned to use a corporative approach for analysis and estimation of the education quality for the modules with remote educational forms.

One else significant factor for successful mission of the academic program "VVER-technologies for the students of nuclear specialties" consists in wide disseminating the best practices developed under the Project. The specialized informative portal of the CORONA and CORONA-II Projects can serve as a quite suitable instrument for the best practice dissemination. In this connection the work package WP5 stands before National Research Nuclear University MEPhI the task on rendering support and future advancement of the Project's informative portal http://vverportal.com [6]. Performance of this task includes the following steps:

- Integration of the CLP4NET platform and the informative portal http://vverportal.com.
- Informational support to the portal.
- Extension of the portal's functional capabilities and provision of its reliable operation.

Currently, partners of National Research Nuclear University MEPhI from Czech Republic, Hungary, Germany and Bulgaria are rendering assistance in further development of the Project's informative portal.

References

[1] Nuclear power in the European Union, section: EU nuclear generation capacity 2016,mhttp://www.world-nuclear.org/information-library/country-profiles/others/european-union.aspx accessed: 03.11.2016

[2] Ershov G A The Project NPP-2006 – Technology and Philosophy of Safety 2015, http://nuklearis.hu/sites/default/files/docs/Jersov_MNT_2015szeptember10.pdf accessed: 02.11.2016.

[3] Descriptors defining levels in the European qualifications Framework, https://ec.europa.eu/ploteus/content/descriptors-page 2016 accessed: 02.11.2016.

[4] Chernov E Utilization of WWER-1000 Reactor Department Multi-Functional Analyzer (MFA-RD) as Advanced Educational Tool 2014, https://www.iaea.org/NuclearPower/Downloadable/Meetings/2014/2014-05-19-05-22-TM-NPTD/Session1/Chernov_MEPhI_Simulator_Part_2.pdf accessed: 03.11.2016.

IOP Conf. Series: Journal of Physics: Conf. Series **781** (2017) 012063 doi:10.1088/1742-6596/781/1/012063

[5] Sbaffoni M CLP4NET Introduction to a cyber-learning platform for nuclear education and training, https://www.iaea.org/NuclearPower/Downloadable/Meetings/2012/2012-01-TM-WS-Vienna/Day-2/11.IntroductiontoCyberLearningPlatform-MSbaffoni.pdf accessed: 03.11.2016.

[6] Dedicated training on VVER technology, http://vverportal.com accessed: 03.11.2016.

Application of the Institution of Exclusive Rights in the Field of Science

D Yakovlev[1], E Yushkov[1], A Zanardo[2] and M Bogatyreova [1]

[1] National Research Nuclear University MEPhI (Moscow Engineering Physics Institute), Moscow, Russian Federation
[2] Ca' Foscari University of Venice

E-mail: DYYakovlev@mephi.ru

Abstract. The problem of legal protection of scientific research results is of growing interest nowadays. However, none of the three hitherto existing rights (the right for trade secrets, patent and copyright) is able to fully take into account the characteristics of scientific activities. In Russia, the problem of legal protection of scientific research results has been developed actively since the 50-ies of the last century, in connection with the introduction of the system of state registration of scientific discoveries. A further concept allowed for not only the registration of discoveries, but also the entire array of scientific results. However, theoretical applicability of exclusive rights institutions in the sphere of science remained unstudied. The article describes a new system, which is not fixed in legislation and remains unnoticed by the vast majority of researchers. That is the institution of scientific and positional rights, focused on the recognition procedure of authorship, priority, and other characteristics of intellectual scientific results value. In case of complex intellectual results, comprising scientific results, the recognition of result-oriented exclusive rights proves to be unsustainable. This circumstance urges us to foreground the institution of scientific and positional exclusive rights. Its scope is budget science where non-fee published scientific results are generated. Any exclusive right to use open scientific results is out of the question. The sphere of open (budget) science is dominated by scientific and positional exclusive rights, sanctioned both by the state (S-sanctioned), the bodies of the scientific community (BSC-sanctioned) and scientific community (SC-sanctioned) rights.

1. Introduction

In Russia, the problem of legal protection of scientific research results has been developed actively since the 50-ies of the last century, in connection with the introduction of the system of state registration of scientific discoveries. A further concept allowed for not only the registration of discoveries, but also the entire array of scientific results [1].

However, theoretical applicability of exclusive rights institutions in the sphere of science remained unstudied.

As early as the 20-ies of the last century, there arose a discussion in foreign literature on the issue of the so-called "scientific intellectual property". It implied the right of the authors of scientific results, used in creating inventions, to share the income generated from the commercial use of these inventions [2].

However, those discussions ultimately came to naught, mostly because of the poor theoretical elaboration of the problem.

About half a century ago, there was a significant turn that took place rather in sociology of science, than in the field of law. In 1957, the American sociologist Robert Merton put forward the concept of symbolic intellectual property in the field of science [3] and summarized, that "... the social component of scientists' moral income takes the form of crumbs of recognition of colleagues. By accumulating gradually, they shape the wealth of its reputation."

According to R. Merton, intellectual property right is based on the recognition on behalf of peers and is protected by a set of unwritten conventional laws. These create a system of social incentives, which, along with internal interest in research, encourage scientists to achieve feasible results and transfer the fruits of their labor to the common fund of scientific knowledge in the form of an open publication, available to anyone who would like to take advantage of them. "These unwritten laws impose symmetric obligations on anyone, who uses the fruits of other labor; oblige them to specify the source and thereby reward the authors by recognizing their contribution to science [4]. The only way to establish the author's ownership of the discovery traditionally involves the recognition of this right by the scientific community and the consequent colleagues' respect".

Notwithstanding R. Merton's highly stimulating concept, it appears necessary to take an alternative approach and focus on some of its weaknesses, for example, the statement, according to which the object of intellectual property is related to the result of intellectual activities.

We are reasonably sure that the recognition of contribution to science can be hardly regarded as such a result. In fact, recognition is caused by the evaluation activities of the members of the scientific community, with the right to use the results belonging to their owners. It is interesting to note that the recognition of a contribution to science gives evidence of a special undeniable character of the result. Its emergence signals a turning point in the researchers' position in the "smart market", i.e. promotes professional advancement in a competitive environment and provides benefits, freedoms and opportunities associated with this promotion.

The researcher's new professional position is determined by the recognition of obtained scientific results (SR) due to certain characteristics – novelty, level of creativity, scientific validity, etc. Just these features serve the ground for recognizing the fact of owning a certain scientific qualification by the author of these results. Following from the recognition of the mentioned facts is the researcher's professional position – the object of rights that are not intellectual and therefore were wrongfully attributed to the results of intellectual activity.

2. The proposed theoretical concept of exclusive rights in the field of science

In the current study, we will follow the logic of theoretical research, including such stages, as identifying significant factors; defining the most important influencing factors; hypothetical extension of the concept and assessment of its explanatory possibilities. Given the fundamental differences between commercial (business) and budget science, it might be worthwhile considering these areas of science separately.

2.1. The entrepreneurial science sector

The sector should include research funded by commercial enterprises and organizations. Regretfully, there is not any statistics on the extent of research results protection in the commercial secret regime. In the Russian Federation, no surveys of commercial enterprises are conducted on the issue. One can only assume that the major part of obtained results is protected by means of the commercial secret regime.

Obviously, the copyright plays a somewhat limited role in the sphere of entrepreneurial science. Income from commercial editions of scholarly monographs is not sufficient to cover the costs for relevant research to be conducted.

2.2. The budget science sector

One of the main concerns is the fact that despite the viability of the copyright, it does not protect the contents of published scientific works.

Thereby, quite justified is a widespread tendency for official categorization of the subjects of scientific activity. Underlying is the system of scientific degrees and titles, particularly those conferred on members of reputable associations of scientists, primarily state academies.

The system also includes the categorization of SR, which is concerned with awarding granted scientific degrees and titles, various state prizes.

2.3. Factors for nominating hypothetical concepts

Among the factors to be referred to as a starting ground in nominating hypothetical concepts are a number of features of scientific research. The scientific characteristics include:

- a worldwide cooperation of researchers;
- coincidence of a community of creators and users of scientific results;
- a higher complexity of new results evaluation than in the field of technology development;
- a higher complexity of new results isolation from previously created ones, than in the field of technology development,
- information-based use of research results against innovative ones, typical of the technology development field;
- the ease of unauthorized borrowing of published scientific results;
- a wider range of result users than in the field of technology development;
- a possibility of obtaining independent SR, missing in the field of literature and art.

2.4. The hypothetical concept of intellectual property rights in the entrepreneurial science sector

The foregoing factors suggest that the institution of commercial scientific secrets and that of copyright can operate in this sector, though to a limited extent.

The factors limiting the use of commercial scientific secrets are:

- a worldwide cooperation of researchers and a limited inter-company exchange of SR (such an exchange is practiced abroad);
- a wider range of result users against the sphere of technology development; the need to attract the scientific community in evaluating obtained results and qualifying researchers working in this sector.

The main factor limiting the use of copyright is an adverse cost ratio for the creation of scientific works and the amount of money proceeds, depending on the number of users. Resulting is a significant share of commercial editions of textbooks, popular scientific and review papers.

2.5. The hypothetical concept of scientific and positional exclusive rights in the science budget sector

The hypothetical concept is to show that the action of the preceding factors leads to effects that manifest in observable facts.

The factor of complexity in isolating new SR from previously created not only excludes SR patenting, but also complicates the evidence-based assessment and, consequently, the assessment of scientific workers' qualification.

The use of SR in new scientific works is a silent, inconclusive and not objectified assessment of such SR, which is critical to normal science budget functioning.

The protection of valuable results is accomplished in an indirect way – by ignoring non-admissible novelty and creative level of works. The first author of a new creative SR gets a monopoly position in the scientific community as the owner of scientific and positional exclusive rights.

The exclusion of non-original works from the area of those used as well as of their authors from the list of contenders for the professional promotion occurs in a natural way, because of the lack of interest in such works by the majority of researchers who become familiar with them.

IOP Conf. Series: Journal of Physics: Conf. Series **781** (2017) 012064 doi:10.1088/1742-6596/781/1/012064

Owing to the fact that the object of these rights is related to the researcher's position in the scientific community (in a certain "intellectual market") rather than SR themselves, scientific and positional rights cannot be attributed to intellectual property rights. This allows us to refer scientific and positional rights to the category of exclusive rights of a special kind.

So, in case of complex intellectual results, comprising SR, the recognition of result-oriented exclusive rights proves to be unsustainable. This circumstance urges us to foreground the institution of scientific and positional exclusive rights. Its scope is budget science where non-fee published SR are generated. Any exclusive right to use open SRs is out of the question. The sphere of open (budget) science is dominated by scientific and positional exclusive rights, sanctioned both by the state (S-sanctioned), the bodies of the scientific community (BSC-sanctioned) and scientific community (SC-sanctioned) rights.

The positional nature of exclusive rights means that the creator of SR enjoys a monopoly position, enabling him/her to act in the intellectual market as the author of the result that has certain value characteristics. The author of the SR becomes the owner of the primary scientific position of exclusive rights. The author of the totality of research SR acquires a certain scientific reputation and becomes the owner of the secondary scientific and positional exclusive rights.

The complexity of SR as positional objects of exclusive rights (is it correct or do you mean: as objects of positional exclusive rights?) accounts for the fact that the authors of the results only submit applications for obtaining SR to the scientific community. After a certain time confidence increases, in so far as the stated results possess specific characteristics that testify to their value. These features include the following: innovation (priority), the scale of the scientific tasks solved, level of creativity, evidence, etc.

It appears that none of the three existing rights (the right for trade secrets, patent and copyright) has been able to fully take into account the characteristics of scientific activity. In practice, the sphere of science admits a certain likely combination of institutions: S-sanctioned, BSC-sanctioned and SC-sanctioned scientific and positional exclusive rights. In their turn, BSC-sanctioned exclusive rights are possible in a number of variants, differing in the role of individual bodies of the scientific community. The choice of the combination is defined **by the right-acquiring performance** (the ratio of costs and specification results of exclusive rights). These are the factors that predetermined the fate of the institution of state registration of scientific discoveries in this country.

Thus, throughout history various means were developed to prevent unauthorized use of the intellectual results, based on uncontrolled access to the intellectual output – commercial trade secrets, patents, the institution of non-patent monopoly, copyright. However, none of them solves the problem of countermeasures in case of scientific results generated in the field of science budget. Essential for the budget science can be a system, which is not fixed in legislation and has remained unnoticed by the vast majority of researchers. That is the institution of scientific and positional exclusive rights, focused on the recognition of authorship, priority, and other characteristics such as the intellectual value of SR. This system of exclusive rights is suggestive of a monopoly opportunity that the first SR creator enjoys: to use the statement reflecting the recognition by the scientific community. The procedure undoubtedly turns to a recognition of the priority and of other characteristics of the SR, which allows the author to reach a scientific qualification level.

References

[1] E.L. Belilovsky, State registration of scientific results: experience and prospects (VNIIPO, Moscow, 1990).

[2] V.A. Rozenberg, Scientific property. Russian Philosophy of Property (Ganza, Saint Petersburg, 1993), pp. 401-426

[3] I.I. Lapshin, Philosophy of invention and invention in philosophy: Introduction to the history of philosophy (Republic, Moscow, 1999), 399 p.

[4] R. Merton, The Matthew effect in science, II: Accumulation of advantages and symbolism of intellectual property, The World of Man. Vol. 1, Issue 3 (THESIS, 1993), pp. 256-276

IOP Conf. Series: Journal of Physics: Conf. Series **781** (2017) 012065 doi:10.1088/1742-6596/781/1/012065

Model and technology of lingware support in tasks of the nuclear knowledge management system

A S Gavrilkina, O L Golitsyna and V A Fedorova

National Research Nuclear University MEPhI, 115409, Russia, Moscow, Kashirskoe shosse, 31

E-mail: asgavrilkina@yandex.ru

Abstract. Model and technology of lingware application in the tasks of the knowledge management are presented. The used model of thesauri paradigmatic relationships allows measuring semantic similarity of documents. An algorithm for compiling a semantic intersection of conceptual patterns based on the coinciding maximum principle is described. The semantic neighborhood construction in accordance with the value of semantic similarity measure and descriptors ranking allows using in the expansion of the search query the descriptors, whose semantics are closer due to the links, which reflect the common semantics of genus-species and associative relationships. An algorithm for thesauri used in automatic classification tasks is considered. The technology of conceptual hierarchical dynamic formation of structures is proposed. The proposed method of constructing a terminological network structure with hierarchical relationships for the collection of texts of a particular subject area is based on the sequential formation of lexicographical neighborhoods of "nuclear" terms, which are used as high-frequency nouns. To build up a conceptual structure a documental databases are used.

1. Introduction

Considering the knowledge management as a set of mechanisms for a selective and varying knowledge elements used in different times and in different circumstances [1], it is necessary to note that for the solution of this problem not only the knowledge but also the means (primarily linguistic) used for knowledge representation and transmission must be preserved. Accordingly, the technology and the means to monitor changes and adaptation means of knowledge representation should be done as a part of knowledge management systems (KMS). It will eventually provide:

- necessary and sufficient specification to a potential user so he/she can accurately find and use an object in his/her practical activity (in other circumstances different to it was created and described);
- continuous update, i.e. it must allow modifications that reflect the refinement and expansion of knowledge;
- identifiability within commonly accepted international, national, and sectoral classification systems, i.e. the attributes that describe an object of knowledge must conform to systematization practices of the professional community (for example, in the case of knowledge in the nuclear field, these are the recommendations that are issued by the IAEA unit for the management of nuclear knowledge).

Knowledge preservation and transfer are inevitably associated with their physical media representation as well as with the identification at the level of the whole sample and at the level of a separate knowledge fragment. For this purpose the following components are used in the information practice:

- conceptual and terminological system (glossaries, thesauri, ontologies) — the instruments of learning and fixing knowledge on the media;
- classification or categorization scheme providing atomic uniformity of the subject area based on goal, organizational or methodological concepts.

These components traditionally are a part of the information systems lingware (and therefore the KMS) and are used in such tasks as automatic indexing, automatic classification, query expansion, etc.

In practice of information activities thesauri are the most widely used. The thesauri's main purpose is to reflect the immanent relationships (paradigmatic - at the linguistic level), i.e. permanent, stable for the subject area links between objects.

However, the objects in a subject area also have the so-called situational relationships, which are perceived as "temporary" and are connected with the certain situation. If the immanent relationships are inherent in the subject area in general, the situational relationships link the objects in a subject area within its fragment (corresponding to a certain process), which in addition is time-varying.

Any object exists at least in two states – as "integrated" into the current conceptual view of the subject area in general and as participant represented in a real situation. This makes it necessary to support not only the strict conceptual structures (which include thesauri), but also to have generating means of operative (situational) mobile terminological networks, for example, which enable to identify the property of novelty on the basis of the analysis of commonalities and differences between the current terminology situation and the "established" terminology of the subject area.

Let us consider some of the models and the technology of knowledge management tasks using both thesauri and rubricators (subject headings indexes) and mobile terminological structures, being built "on the fly" on base of information files.

2. Model of conceptual images comparison based on semantic similarity

2.1. Semantic similarity measure

Semantic similarity measure is a numerical value, which represents the measure of similarity between two objects. It is often used in fuzzy matching algorithm both in information retrieval and analytic problem. Semantic similarity is mentioned when objects are presented as descriptions, built by using denoting subjects (classes of subjects) terms, which are the stable concepts signs in the subject area [2].

The formalization and comparative assessment of semantic similarity make the subject of many studies. Note can be given to papers that contain quite extensive overviews on semantic similarity [3, 4].

The measure analytical expression depends on the used features of semantic similarity (for specialized conceptual and symbolic resources such features might be related to the types of relationships between concepts).

The conceptual network in information retrieval thesaurus is formed mainly by hierarchical and associative relationships.

The additive measure is proposed to use in [5] to calculate the semantic similarity of thesaurus descriptors:

$$S(d_1, d_2) = \alpha \cdot S_H(d_1, d_2) + \beta \cdot S_A(d_1, d_2) \tag{1}$$

where $S_H(d_1, d_2) = \frac{|UC(d_1, H^{d_1}) \cap UC(d_2, H^{d_2})|}{|UC(d_1, H^{d_1}) \cup UC(d_2, H^{d_2})|}$ is the contribution of hierarchical relationships; $S_A(d_1, d_2) = \frac{|UC(d_1, A^{d_1}) \cap UC(d_2, A^{d_2})|}{|UC(d_1, A^{d_1}) \cup UC(d_2, A^{d_2})|}$ is the contribution of associative relationships.

Multiplicity $UC(d_i,H^{d_i})$ contains descriptor d_i, and all its parents (based on transitive hierarchical relationships) in a hierarchical chain – H^{d_i}:

$$UC(d_i,H^{d_i}) = \{d_j \in H^{d_i} |\ \exists m = (d_i, x_{i_1}, x_{i_2} \dots, x_{i_k}, d_j) \cup (d_i = d_j)\} \tag{2}$$

where m is path, connecting d_i and d_j through broader terms descriptors $x_{i_1}, x_{i_2}, \dots, x_{i_k}$.

The association has the property of symmetry, however, the specific semantics of symmetry cannot be determined since the actual type of communication is not specified. For the construction of the multiplicity $UC(d_i,A^{d_i})$ [3] proposes this concept and rules of "semantic double" construction

$$UC(d_i,A^{d_i}) = \{d_j \in A^{d_i} |\ \exists a = (d_i, \tilde{d}_j) \cup d_j \cup (d_i = d_j)\} \tag{3}$$

where a is associative relationship, connecting d_i and $\tilde{d}_j$ ($\tilde{d}_j$ – «semantic double» of descriptor d_j.

To determine the values of the coefficients α and β there are «vertical» and «horizontal» development of specific information retrieval thesaurus:

$$\alpha = \frac{S_H}{S_H + S_A} \tag{4}$$

$$\beta = \frac{S_A}{S_H + S_A} \tag{5}$$

where $S_H = \max_{d_1,d_2 \in T} S_H(d_1,d_2)$, $S_A = \max_{d_1,d_2 \in T} S_A(d_1,d_2)$, T – set of thesaurus descriptors.

2.2. Algorithm for constructing the semantic intersection of conceptual images

In [5] one also uses an algorithm to measure the semantic similarity of fuzzy matching descriptions of two objects, based on the coinciding maximum principle.

Descriptions of objects represented by their linear conceptual images - a thesaurus descriptors sets - $D_1 = (d_1^1, d_2^1, \dots, d_n^1)$ и $D_2 = (d_1^2, d_2^2, \dots, d_m^2)$. For all descriptors pairs in the thesaurus a semantic similarity matrix may be calculated measuring $n \times m$: $W = (w_{ij})$, where $w_{ij} = S(a_i, b_j), i = \overline{1..m}, j = \overline{1..n}$.

In contrast to the different approaches, the sufficiency of the value of semantic similarity to include a pair of descriptors in the semantic intersection can be determined not by defined fixed threshold, and based on the local context, formed descriptors of each image. A set of semantic intersection is formed from the descriptor for which the matching condition highs:

$$\max_{j=\overline{1,n}}(w_{ij}) = \max_{i=\overline{1,m}}(w_{ij}) \tag{6}$$

This algorithm can solve the problem of establishing a semantic similarity measure threshold to include a pair of descriptors in the fuzzy intersection.

2.3. Semantic concepts neighborhood formation

The concept, which is included as a descriptor in the corresponding thesaurus, is generally accompanied by a descriptor section showing there is a direct connection (hierarchical and associative) with other descriptors. However, using this connection it is necessary to consider that the thesaurus's hierarchical relations link concepts and do not conform to the objects or subjects, so disturbed the correct structure of the tree - the same term can have several "parents" - the broader terms at the previous level. Meanwhile associative relationships do not have the property of transitivity. Therefore, an automatic query expansion built on an automatic thesaurus using direct links, in practice do not lead to constructive results.

The semantic neighborhood construction in accordance with the value of semantic similarity measure and descriptors ranking allows using in the expansion of the search query not just the

IOP Conf. Series: Journal of Physics: Conf. Series **781** (2017) 012065 doi:10.1088/1742-6596/781/1/012065

descriptors, combined descriptor entry, and those whose semantics are closer due to the links which reflect the common semantics of genus-species and associative relationships (for example more than one of the generic descriptor, a different set of subordinate descriptors for ones concepts). Figure 1 shows a descriptor entry NEUTRONS from the INIS thesaurus [6] and semantic fragment of the neighborhood (the terms are arranged in descending order of measure), showing that the descriptors in a descriptor entry of the same level (narrow terms) have different values of semantic similarity measure.

0,56-THERMAL NEUTRONS	**NT (narrow terms):**
0,54-COLD NEUTRONS	*ANTINEUTRON*
0,54-RESONANCE NEUTRONS	BETA-DELAYED NEUTRONS
0,54-PILE NEUTRONS	FAST NEUTRONS
0,54-INTERMEDIATE NEUTRONS	COSMIC NEUTRONS
0,54-POLYNEUTRONS	SLOW NEUTRONS
0,54-FISSION NEUTRONS	EPITHERMAL NEUTRONS
0,54-EPITHERMAL NEUTRONS	FISSION NEUTRONS
0,54-SLOW NEUTRONS	POLYNEUTRONS
0,54-FAST NEUTRONS	INTERMEDIATE NEUTRONS
0,54-BETA-DELAYED NEUTRONS	RESONANCE NEUTRONS
0,53-NUCLEONS	PILE NEUTRONS
0,48-PHOTONEUTRON	SOLAR NEUTRONS
0,48-*ULTRACOLD NEUTRONS*	THERMAL NEUTRONS
0,48-TRINEUTRONS	COLD NEUTRONS
0,48-TETRANEUTRONS	PHOTONEUTRON
0,48-PROMPT NEUTRONS	**BT (broader terms):**
0,48-DELAYED NEUTRONS	NUCLEONS
0,45-PHOTONUCLEONS	**RT (related terms):**
0,45-PROTONS	*NEUTRON FLUX*
0,42-DINEUTRONS	*NEUTRON DENSITY*
0,42-BARYONS	*NEUTRON BEAMS*
0,4-PROMPT PROTONS	*CINDA*
0,4-TRAPPED PROTONS	*NEUTRON SPECTRA*
0,4-RETARDED PROTONS	*NEUTRON TEMPERATURE*
0,34-SOLAR NEUTRONS	*NEUTRON SEPARATION ENERGY*
0,34-COSMIC NEUTRONS	

Figure 1. Descriptor entry NEUTRONS and fragment of semantic descriptor neighborhood NEUTRONS (denoted as mismatched descriptions in italics).

3. Technology of lingware support

3.1. The technology of joint use of thesauri and rubricator in a task of automatic classification
Automating the process of the input information flow distribution by a predetermined classes (for example, by subject classifier categories) is based on the formation of a preliminary descriptive specification of classes (categories) on the one hand, and the input documents on the other hand [7].

Method of classification using thesaurus involves the construction of descriptive specification headings on the basis of the thesaurus descriptors. In this case the intersection of a row and column matrix of "descriptor-category" contains the value which characterizes the semantic proximity of the concrete descriptor to the category.

To construct this matrix it is necessary to solve two problems:

- to distribute preliminary thesaurus descriptions according to categories;
- to calculate semantic similarity measure of "descriptor-category".

To solve the first problem an abstracts database can be used, in which each abstract is provided with a search pattern formed on the basis of expert estimation. Such a search pattern is usually

IOP Conf. Series: Journal of Physics: Conf. Series **781** (2017) 012065 doi:10.1088/1742-6596/781/1/012065

represented as a set of descriptors and subject indexes. In this case technology of comparison of the descriptor with the category includes the following steps:

1. Building a list of descriptors belonging to the category.
2. Identification of descriptors relating to more than one category.
3. Calculation of value of descriptor-category correspondence for descriptors identified at step 2 (for example, as such measures coefficient of linear correlation can be used).
4. Descriptors distribution by categories based on the principle of correlation coefficient maximum.

However, this approach does not guarantee all thesaurus descriptors distribution by category. Algorithm application for classification descriptors of INIS thesaurus in accordance with the INIS categories [8] using the abstracts database SARI showed that near 50% of thesaurus descriptors could not be distributed by rubrics (since these descriptors do not exist in the database).

To ensure classification completeness this primary distribution can be further supplemented by the thesaurus relations and semantic similarity measures. A descriptor, to which a category is not assigned yet, is regarded as the node of a permissible conceptual path. Then previous and subsequent nodes are determined which correspond to descriptors that are already matched to the subject headings and the semantic similarity measure of these descriptors with the specified descriptor is calculated. At the last step the category is assigned to that descriptor for which the value of the measure is maximum.

3.2. Technology of the terminology systems actualization

The above-mentioned strictness of conceptual terminological structures such as thesauri does not allow to effectively track information flow changes and the formation of certain language using directions. To construct and promptly update a subject area concepts and terminology systems, an interactive and iterative technology based on the methods and means of information retrieval in the documentary resources of scientific, technological, and educational information can be used. The technology involves the following steps [9]:

- development of an information concept of a subject area at the level of documents that are selected from the document sources, which adequately and objectively reflect the state of the art in the subject area;
- the design of a terminology base in the form of glossaries, i.e., the lists of terms that express the basic meanings of the selected relevant documents;
- construction of concepts and terminology networks (thesauri and ontologies) for the subject area.

The proposed method of constructing a terminological network structure with hierarchical relationships for the collection of texts of a particular subject area is based on the sequential formation of lexicographical neighborhoods of "nuclear" terms, which are used as high-frequency nouns. Lexicographical neighborhood includes the phrase with a noun as the main word. Construction of phrases is made according to patterns [10, 11] with the following basic rules:

- adjective precedes the noun;
- in phrases with two or more nouns without prepositions, second and subsequent nouns should be in the genitive, dative or instrumental case (place names are not considered);
- nouns related by prepositions may be in any case.

Table 1 shows examples of word combinations of patterns corresponding the above rules by which combinations of operating documentation were extracted and assessed the correctness of their selection. The constructed combinations were sorted in descending order of frequency of occurrence and reviewed the first 300 for each pattern. The right column shows the number of incorrect phrases per considered 300.

Table 1. Number of incorrect phrases according to patterns.

Pattern	Number of incorrect phrases
adjective + noun	0 (0%)
noun + noun in the genitive, dative or instrumental case	30 (10%)
noun + prepositions + noun	85 (28.3%)
noun + adjective + noun in the genitive, dative or instrumental case	18 (6%)
adjective + noun + noun in the genitive, dative or instrumental case	33 (11%)

Figure 2 shows the lexicographical neighborhood of term NEUTRONS built for glossary.

NEUTRONS	NEUTRON ABSORPTION
FAST NEUTRONS	NEUTRON FLUX
NEUTRON INTERACTIONS	NEUTRON FLUX DENSITY
NEUTRON DIFFUSION	NEUTRON FLUX DENSITY MEASUREMENT
NEUTRON MODERATION	NEUTRON FLUX IN ACTIVE ZONE
NEUTRON CAPTURE	NEUTRON FLUX IN REACTOR
NEUTRON SOURCE	NEUTRON MULTIPLICATION
PROMPT NEUTRONS	NEUTRON REACTION
SLOW NEUTRONS	RESULT OF NEUTRON REACTION
EPITHERMAL NEUTRONS	REGISTRATION OF NEUTRONS
NEUTRON IRRADIATION	NEUTRON VELOCITY
NEUTRON REFLECTOR	THERMAL NEUTRONS
NEUTRON REFLECTION	THERMAL NEUTRON CAPTURE
NEUTRON TRANSFER	THERMAL-NEUTRON ABSORPTION
NEUTRON ABSORBER	REGISTRATION OF THERMAL NEUTRONS

Figure 2. Fragment of lexicographical neighborhood for descriptor NEUTRONS.

There is a fairly insignificant intersection with descriptors thesaurus entry. This lexicographical neighborhood allows associating terms in different aspects. Relationships exist not only genus-species, but also situational ones which characterize the use of the object in the present context.

4. Conclusion

The proposed models and technologies are focused on the independent lingware objects contour formation in the KMS framework. These objects support requires the development of models and technologies from the point of view of their subject area adequacy. It should be done in two directions: updating stable conceptual terminology structures for the subject area as a whole (with the aim of embedding knowledge in the current vision), and dynamic situational terminological structures formation (searching for and fixing similarities and differences related to the concrete knowledge objects).

The addressed models and algorithms have been tested on an array of full text dissertations conceptual search images.

References

[1] Golitsyna O L, Kupriyanov V M and Maksimov N V 2015 *Autom. Doc. Math. Linguist.* **42** (3) 150-161

[2] SINTOL 1968 *Sbornik perevodov po voprosam informacionnoj teorii i praktiki* (Moscow: VINITI) no 10 p 177

[3] Kryukov K V, Pankova L A, Pronina V A, Sukhoverov V S and Shipilina L V 2010 Measures of semantic similarity in ontology *Probl. Upr.* **5** 2-14

[4] Ngok N B and Tuzovskii A F 2013 The model of information retrieval based on semantic meta descriptions *Upr. Bol'shimi Sist.: Sb.Tr.* **41** 51-92

[5] Golitsyna O L, Maksimov N V and Fedorova V A 2016 *Autom. Doc. Math. Linguist.* **50** (4) 139-153
[6] INIS Thesaurus 2016 *IAEA-INIS-01 (2016/10)* (Vienna)
[7] Gulyaev O V and Loukachevitch N V 2013 *Novye informacionnye tehnologii v avtomatizirovannyh sistemah* **16** 238-244
[8] Subject Categories and Scope Description 2010 *INIS/ETDE Joint Reference Series* no 2 (Rev 1)
[9] Golitsyna O L, Maksimov N V, Strogonov V I and Tikhomirov G V 2011 *Sistemy Upr. Inf. Tekhnol.* **44** (1.1) 126-134
[10] Khokhlova M V 2012 *Proc. of the Annual Int. Conf. "Dialogue"* (Moscow: RSUH)
[11] Nayhanova L V 2005 *Internet Portals: Content and Tehnologies* **4** 452-479